Libellula auripennis

Erythemis simplicicollis

A MANUAL OF THE *Dragonflies of North America* (ANISOPTERA)

Including the Greater Antilles and the Provinces of the Mexican Border

JAMES G. NEEDHAM · MINTER J. WESTFALL, JR.

UNIVERSITY OF CALIFORNIA PRESS
BERKELEY, LOS ANGELES, LONDON

UNIVERSITY OF CALIFORNIA PRESS
BERKELEY AND LOS ANGELES

UNIVERSITY OF CALIFORNIA PRESS, LTD.
LONDON, ENGLAND

CALIFORNIA LIBRARY REPRINT SERIES EDITION, 1975
ISBN: 0-520-02913-5
LIBRARY OF CONGRESS CATALOG NUMBER: 54-6674
PRINTED IN THE UNITED STATES OF AMERICA

PREFACE

It is the purpose of this book to provide for all who are interested in the animal life of North America a means of cultivating acquaintance with the order of insects called Odonata, or dragonflies. All over our continent they are known by sight to every observant person. They enliven every inland landscape at the waterside. Their beauty of coloration, their mastery of flight, and their singularly interesting life history should be a part of the common knowledge of educated persons.

This book is first of all a means of finding the names of dragonflies, but it is our hope that those who use it will not be content merely to learn their names. Names are necessary, and lead us to the sources of information. But getting the names of species is only a first step toward acquaintance with them.

Knowing dragonflies merely as dried specimens in a collection is missing the great pleasure of seeing them alive and in action. Our readers will enjoy being out of doors with the dragonflies where they haunt natural parks, spring-fed brooks, grass-bordered ponds, placid lake shores, and winding riverbanks. Dragonflies, like ourselves, revel in clean places by the waterside.

Indoors we study them as species and varieties; we see how their likenesses and differences place them in natural groups. And if we go far enough in such study, we are rewarded by glimpses of the evolutionary processes which make endless varieties of living things. So their study, if diligently pursued, will yield wholesome pleasure and intellectual satisfaction.

The study of the immature stages (nymphs) of our dragonflies has been too much neglected in the past—perhaps for lack of an introduction such as is hereinafter offered. More knowledge of the aquatic nymphs is urgently needed by conservation workers engaged in stream and lake surveys, and by anglers generally. The characters by which nymphs may be recognized are stated in keys and tables. A good photograph is placed near the key to nymphs in each genus; and for the average user these fine photographs will be worth more than many pages of technical description.

In the preparation of the keys and tables much time and thought have been given to the convenience of the user and to the limitations of the beginner. We have tried to use language as simple as is consistent with clearness, and to give meaning to necessary technical words by the use of diagrams and abundant photographic illustrations. Except when otherwise stated, the photographs are the work of the junior author. We have worked together on the manuscript, and have spent many hours gathering, condensing, and verifying the data for the tables of species under each large genus—a unique feature of this book.

For practical reasons we have stated the range of species in terms of states, which the user can find on his maps, rather than in terms of the more significant but less clearly mapped zones and provinces of climatology and zoögeography.

This book has been a long time growing. Many graduate students in the Department of Entomology of Cornell University have had a share in the spadework for it. Those who have chosen to do graduate work on odonatological problems under the guidance of the senior author and who have published the results of their studies on dragonflies will find their names and their papers cited on some of these pages. Those who did not publish are remembered by him for the pleasant personal relations their helpful interest engendered.

We have had much help from many sources. There are those who have labored with us on the making of the book itself. Chief of those is Mrs. Margaret Shepherd Westfall, wife of the junior author, who did most of the careful typing of the numerous tables, and much of the more delicate retouching of the photographic prints. Mr. A. L. Smith, then of the Cornell University Photo-Science Service, devoted much time and great care and skill to the making of our negatives and prints. Dr. May Gyger Eltringham, of Perris, California, at the beginning of the writing of the text was a co-worker with the senior author, and made the best of our pen drawings, as their legends will show. The last typing of the completed manuscript was made by Mrs. Bertha Beasley, who while typing made many desirable textual improvements. At the end Miss Edith Beasley gave invaluable help in assembling text and illustrations and correlating headings and cross references. To all these helpers we extend our hearty thanks.

Those who have contributed the most valuable specimens needed to cover our field were our ever helpful friends and correspondents, Dr. P. P. Calvert, of the University of Pennsylvania; Dr. E. M. Walker, of the University of Toronto; Mrs. Leonora K. Gloyd, of the Illinois Natural History Survey; Dr. Septima Smith and Dr. Robert S. Hodges, of the University of Alabama, Mr. Paul N. Albright, of the Public Health Department of San Antonio, Texas; Miss Alice Ferguson, of the State College for

Teachers, at Commerce, Texas; Mr. Carl Cook, of Crailhope, Kentucky; and Dr. Paul R. Needham (son of the senior author), of the Department of Zoölogy of the University of California, Berkeley, California.

A number of institutions have been of assistance in the production of this volume. Cornell University has, almost from the beginning, supplied laboratory, library, and museum facilities, and the aid of the office staff of its Department of Entomology. Our thanks are due to Dr. C. E. Palm, head of the department, for continuing that coöperation since the retirement of the senior author from teaching; also to Dr. Henry Dietrich, curator of the department's collections, for continuing care of our materials and other helpful services.

The senior author has been provided with field headquarters (and much besides) at several distant parts of the field covered by this volume, namely, seven seasons at the Archbold Biological Station near Lake Placid, Florida, where Mr. Richard Archbold, director, and Mr. Leonard J. Brass, botanist of the Station, gave him much personal assistance.

Two seasons were spent on the west coast of Florida: the first at the Bass Biological Station, Englewood, and the second at the Hegener Research Supply, Sarasota, Florida. In both places Mr. and Mrs. William Hegener gave much practical help.

Two seasons were spent at the University of Puerto Rico, where Dr. and Mrs. Julio Garcia Diaz were constant guides and co-laborers. One short but noteworthy month (June, 1940) was spent in Santo Domingo, where Dr. and Mrs. Garcia accompanied and aided, and where Sr. Manuel Tavares, of Santiago City, lent us the use and sustenance of his mountain house at San José de las Matas for a week of superb dragonfly collecting.

Two seasons were spent at the Atkins Institute (Tropical Gardens of Harvard University), Soledad, Cuba, by courtesy of the late Dr. Thomas Barbour.

Financial support toward the completion of the manuscript came in two grants of funds made by the American Philosophical Society, three made by the Research Committee of the Cornell University Graduate School, and a final grant through Dean William I. Myers from the New York State College of Agriculture. All these grants are hereby gratefully acknowledged.

James G. Needham
Minter J. Westfall, Jr.

Ithaca, New York
Gainesville, Florida
July 27, 1954

CONTENTS

Part One
Dragonflies in General

Part Two
Systematic Classification

TABLES

Part One

Dragonflies in General

I

INTRODUCTION

Dragonflies are rather large predacious insects, of ancient lineage and of unique form and habits. They are beautiful in coloration and agile in action almost beyond comparison.

They were probably first called dragonflies because of superstitious fear of them; yet they are harmless to man and can frighten only the timorous. The very large ones may seem to threaten sometimes when they sweep through the air near one's head on sharply rustling wings. And it is an ugly visage that some of them present when hovering directly before one's face, close enough to show their big eyes, their huge mouths, their be-whiskered lips, and their ogre-like expression. Perhaps near-contacts like these led credulous folk to invent such names for them as "devil's darning needles" and "horse stingers." They certainly have no sting, no long needle with which to "sew up venturesome small boys' ears." Probably the name "mosquito hawks" arose among people who knew something of the beneficial work of dragonflies as destroyers of mosquitoes. They are in fact real dragons in the world of little insects, flying dragons, predators supreme, that disdain to take anything but living prey.

The scientific name Odonata (Greek *odon,* a tooth) refers to the strong and sharply toothed jaws that are the outward sign of their predatory habits. An earlier ordinal name that for a long time included the Odonata was Neuroptera (Greek *neuron,* a nerve, and *pteron,* wing). This also is meaningful: it refers to the rich network of veins or nervures in their wings.

More than any other creatures, dragonflies are dependent on their wings for meeting the needs of life. They not only hunt on the wing for food and glide through the air for pleasure, but they also capture their prey on the wing, and many of them lay their eggs while flying by dropping them on the surface of the water in ponds and streams. Most of their time is spent in pursuit of food or of mates.

Some of the larger dragonflies are among the fleetest of living creatures. However, there are insects that can fly faster; and it is not without mechanical significance that the best flyers have developed as monoplanes.

The true flies (Diptera) have no hind wings. In bees and their kin (Hymenoptera) the hind wings are reduced to a very small size and are hooked on the rear of the large fore wings so that the two act as one. Dragonfly wings keep to the more primitive biplane plan: two pairs of free wings. And in them this plan appears to have been carried to the known limits of perfection. Their wings have acquired a marvelous richness of venational characters far beyond that of any other insect.

The coloration of living dragonflies is infinitely varied and often surprisingly beautiful. But alas! Some of it fades badly in preserved specimens. The surface colors (diffraction colors) keep very well, as does the iridescence of the clear wing membrane, but the deeper pigment colors suffer from drying. Greens fade to yellow, purples to black, and bright blues darken with internal post-mortem changes. Especially disappointing to anyone who has seen the glow of the eyes in the living dragonfly is the dulling of their fine opalescent tints.

The wings of many dragonflies are attractively patterned in brown, which is often streaked and tinged with a bright flavescence. The most beautiful and permanent bits of color are found on the top of the head and face of mature males: flaming red, deep dark purple, metallic as of copper and bronze that change with the angle of our vision or shine like burnished gold in the sunlight. Much of the color and all the color pattern can be saved by quickly and thoroughly drying the specimens after they have been killed.

Most of the beauty of butterflies is in their wings; in dragonflies, there is quite as much beauty in the color and decorations of their shapely bodies.

All dragonflies are aquatic in their immature stages. Their young are found in all kinds of unpolluted shoal waters: streams, lakes, ponds, and ditches; and the adults fly and forage about the shores. Thus, in the course of their lives, dragonflies play two successive roles, so to speak, in the world: as young (nymphs) they live in the water for the greater part of their lives; when grown they leave it to become free-roving adults in the air. So different are adult and nymphal stages that the nymphs will be treated in a separate chapter.

The eggs are laid in the water. After two weeks or more of incubation, they hatch into pale little nymphs of a form that will not change very much until they are grown. They are generally thinly hairy at first, and the hairs gather silt for a covering that hides them from their many enemies which live in the water with them.

Most nymphs do not chase their prey. They lie still and wait for it to come to them; and they eat little or much according to the bestowal of Providence.

Fig. 1. A Florida Didymops (*Didymops floridensis*), newly transformed; recently come out of nymphal skin, seen below it clinging to sedge stems. Now at end of its youthful career in the water, at beginning of adult life in the air. Taken at daybreak, its wings still soft, its abandoned nymphal skin still wet. It stood on middle and hind legs, its fore legs held aloft, a first resting attitude; its pale wings fully expanded but not yet outspread, nearly ready for its first flight. (Photographed from life by Mr. Richard Archbold.)

When fully grown they creep out of the water onto some solid support, fix their claws to it firmly, and shed their loosened nymphal skins. Then, with a marvelous bit of reshaping, they become adult dragonflies. How different the sprawling stiff-legged nymph is from the fleet adult that in half an hour's time or less after leaving the water rises on glistening wings and flies away!

Dragonflies are like other insects in these well-known hexapod characters: they wear an armor-like skeleton on the outside of the body; there are three main divisions in the body—head, thorax, and abdomen—and

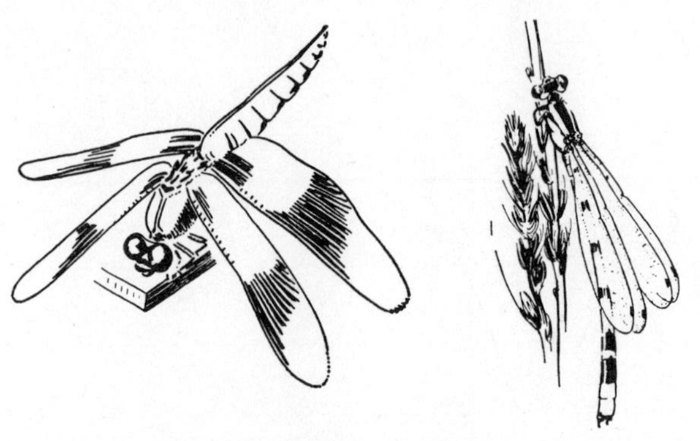

Fig. 2. Dragonfly and damselfly.

each division is made up of rings or segments; the head bears eyes, antennae, and mouth parts; the thorax bears three pairs of legs (a pair on each segment) and two pairs of wings, a pair each on the second and third segments; the abdomen bears the caudal appendages.

THE ORDER ODONATA

Dragonflies and damselflies together make up the order Odonata. It is a well-marked order, characterized structurally by the following: biting mouth parts, three pairs of spiny legs, and two pairs of large free wings with copious venation; great development of the eyes along with little development of the antennae; the side plates of the body askew, with the legs shoved far forward below and the wings drawn far to rearward on the back of the thorax; and a long abdomen with the accessory copulatory apparatus of the male lodged in a pocket on the under side of the second segment, far removed from the opening of the sperm ducts on the ninth segment.

The nymphs are distinguished by an enormous labium, or lower lip, that can be thrust far forward to grasp living prey; also, by the form and location of the tracheal gills, by means of which they breathe under water.

Transformation from the nymphal form to the adult is direct, with no intervening pupal stage.

All these external features sharply distinguish dragonflies and damselflies from other insects. There are distinctive internal features as well, but with these we are not concerned in this volume.

Most like adult Odonata in stature, in form of wing, and in general superficial appearance are the equally free-winged Ascalaphidae of the order Neuroptera; but these lack all the structural peculiarities mentioned above.

The order Odonata, thus briefly characterized, is divided into two suborders of recent forms:*

Suborder ANISOPTERA, dragonflies (this common name is often used in a broad sense to cover all Odonata), the subject of this volume.

Suborder ZYGOPTERA, damselflies, the subject of another volume now in preparation.

The common features of these two suborders seem to indicate a common origin in past geologic ages, since the two groups are alike in their development and composed of like parts throughout; but their divergence in recent times is very great.

The dragonflies all have large compound eyes that tend to overspread the top of the head until in extreme development they meet in a long eye-seam on the middle line. The damselflies all have the eyes far apart and tending farther and farther apart until in some cases they become hemispherical. The part of the head between the eyes becomes reduced and hollowed out behind until the head capsule and also the brain within it may become almost stalked at each side.

In the dragonflies the hind wings are larger than the fore wings, and the increase in breadth has come as an expansion of the basal rearward part (cubito-anal area) of the wing. In the damselflies the fore and hind wings are about of equal size (the difference, if any, being in smaller hind wings), the reduction being mainly in the cubito-anal area in both pairs. In most damselflies the wings have become stalked.

A difference in the position of the wings when at rest makes the suborders distinguishable at a glance. The sitting dragonfly holds its wings horizontally outspread; the damselfly sits with wings lifted, and, in all but a few species, folded together above its back.

In the nymphs, also, there are striking differences. Dragonfly nymphs are stout. They breathe by tracheal gills that line the inner walls of a rectal gill chamber, a terminal portion of the alimentary canal. They

* We are leaving fossil forms altogether out of account; also, one very peculiar living species not found in the New World, *Epiophlebia superstes*, found in Japan and in northern India, for which a separate suborder, Anisozygoptera, is generally recognized.

swim by jet propulsion of water from this gill chamber. Damselfly nymphs are slender in form, widest across the head. They breathe by means of three more or less platelike external tracheal gills attached to the end of the abdomen. They swim by sculling with these gill plates.

With these differences noted we now leave the suborder Zygoptera from further consideration in this volume.

Within the suborder Anisoptera of the order Odonata the features mentioned above have been developed in very diverse ways that will now be considered.*

THE DRAGONFLY ADULT

Head.—The head has for its framework a strong chitinous capsule that is hollowed in the rear for the insertion of the neck, bulged at the front by the benchlike prominence of the face, and expanded at the sides where covered by the huge eyes. It bears three ocelli and a pair of slender bristle-like antennae in front, and the usual mouth parts beneath. Of the latter, only the upper and lower lips (*labrum* and *labium,* respectively) are exposed to view, and only these will be noted in our descriptions. The two pairs of included jaws (*mandibles* and *maxillae*), which show their toothed tips between the lips, are of a strictly flesh-eating type. The one-piece mandibles, which may be exposed by lifting the labrum, are heavy and hard-biting, with cutting teeth and ridges on their opposed surfaces. The maxillae may be seen by turning the specimen over and pulling down the labium. Each maxilla is in two parts: a five-pronged inner piece, perfectly shaped for a meat fork, used for holding a captured insect and for turning it as the mandibles cut it up; and a bladelike outer covering piece, concave on its inner surface, that seems to serve mainly to prevent leakage of the chips.

The labrum is in one piece, but the labium is plainly divided in its free portion into three parts: median and lateral lobes. The presence or absence of a cleft in the front margin of the median lobe is a small character of great phylogenetic interest, the labium, in ages long past, having been developed (evolved) out of a second pair of free-swinging maxillae. This little cleft in the front margin is the last remaining sign of their original separateness, and is a sort of historic landmark, so to speak.

The face above the labrum should be more carefully mapped, for details of its coloration will be told in many of the descriptions of species which follow. The most prominent feature of the face is the bulging *frons.* Its broad vertical surface is more or less flat and smooth and shining; its upper horizontal surface often bears a median longitudinal

* The external anatomy of a dragonfly may be quickly and easily learned by taking a specimen in hand and comparing it part by part with the descriptions and figures hereinafter presented.

frontal furrow, and the angle of meeting of vertical and horizontal surfaces is sometimes raised in the *cross ridge* of the frons. Between labrum and frons is the wrinkled area of the *clypeus*, divided by a flexible suture into *anteclypeus* and *postclypeus*, the latter (next to the frons) widened at each end and rounded off below to form two prominent *facial lobes*.

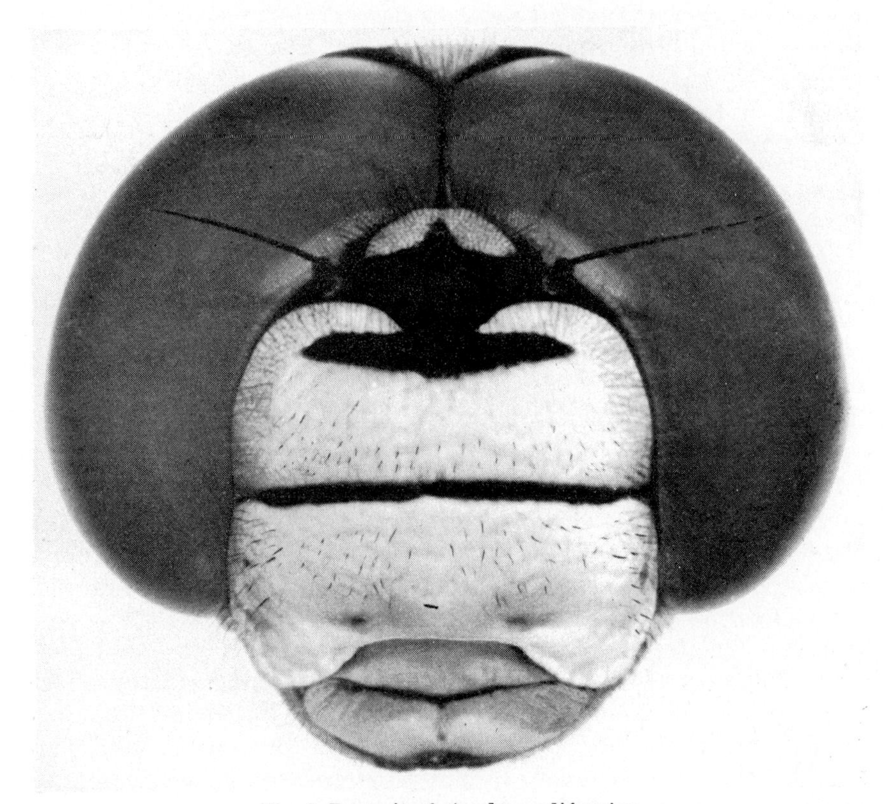

Fig. 3. Portrait of *Aeschna californica*.

These parts must be learned by the user of this book, for the color pattern of the face is described in terms of them and is important—often the most important character for matching males with females when the sexes differ greatly in coloration.

The top of the head between the compound eyes is covered by the *vertex*, which rises often in more or less pyramidal form and bears on its base three simple eyes, the *ocelli*, one at each side and one on the median line in front. The *middle ocellus* is at the rear end of the frontal furrow. That furrow widens the field of its vision.

Behind the vertex, the *occiput* completes the upper surface of the head and also the rear. No line of demarcation between vertex and occiput remains because of their complete fusion. As in the frons, the occiput has horizontal and vertical surfaces, with the angle between them often raised in a transverse ridge, the *crest of the occiput.*

All the above-mentioned parts are well shown in the photograph of *Aeschna californica* (fig. 3). The face is clear cerulean blue in life, though it looks white in the photograph. Its parts, read from below upward, are:

labrum, narrow, white, streaked with black along both front and rear margins.
clypeus, in two parts hinged together:
anteclypeus, small, transverse, trapezoidal in form.
postclypeus, large, white, with a pair of rounded *facial lobes* in front and a heavy line of black along its rear margin.
frons, white, with an inverted T-shaped black spot on its receding upper surface.
vertex, black, moundlike, with white top, with *middle ocellus* at its front dimly showing in midst of black. (Other two ocelli, turned laterally, do not show in photo.) The vertex narrows to rearward, ending in a long black eye-seam that leads to the occiput. The topmost prominence is called the vesicle of the vertex.
occiput, triangular, white in middle and black at sides, which meet at end of eye-seam and form with it a large Y mark.

On the black field of the vertex at each side next to the compound eye, a slender black antenna rises on its narrowly white-ringed pedicel.

Thorax.—The thorax is composed of three body segments, as in other insects: prothorax, mesothorax, and metathorax. The small prothorax is narrowed downward to the base of the first pair of legs, and flattened on the back into a shieldlike disc, which is divided by two transverse depressions into *front, middle,* and *rear* lobes. At the rear it is often hair-fringed or sculptured.

The synthorax (composed of mesothorax and metathorax) is remarkable for the consolidation of its two segments into a single unit that is a veritable power plant for flying operations. This strong box is filled with muscles, mostly wing muscles. Its thin walls are stiffened and braced in many ways to withstand the pull of these muscles.

The mesothorax and the metathorax are distinct in the nymph, but in the adult they become so completely fused that the suture between them on the sides is lost. The side wall of each of these segments is formed, as in other insects, of two plates (*sclerites*), the *episternum* and the *epimeron.* In each segment the infolded edge has been extended and thickened to form a stiff ridge that stands edgewise inside the thorax. The inner edge of that thickening is further stiffened to rodlike or ladder-like form (*scalariform*), which provides an inflexible support to withstand the pull of the powerful muscles that move the wing. In scantily pigmented darners like *Anax,* the rungs of a ladder, so to speak, may be seen from the outside through the transparent skin.

Furthermore, these internal braces in mesothorax and metathorax, standing at the four corners of the synthorax, are set at such an angle with its middle plane that they oppose each other and balance the stresses with those opposed on the other side of the body. No muscles attach to the scalariform part of these braces; they serve only for support.*

The side view of *Aphylla protracta* (see fig. 4) shows the principal external features of the thorax in relation to head and abdomen. The head

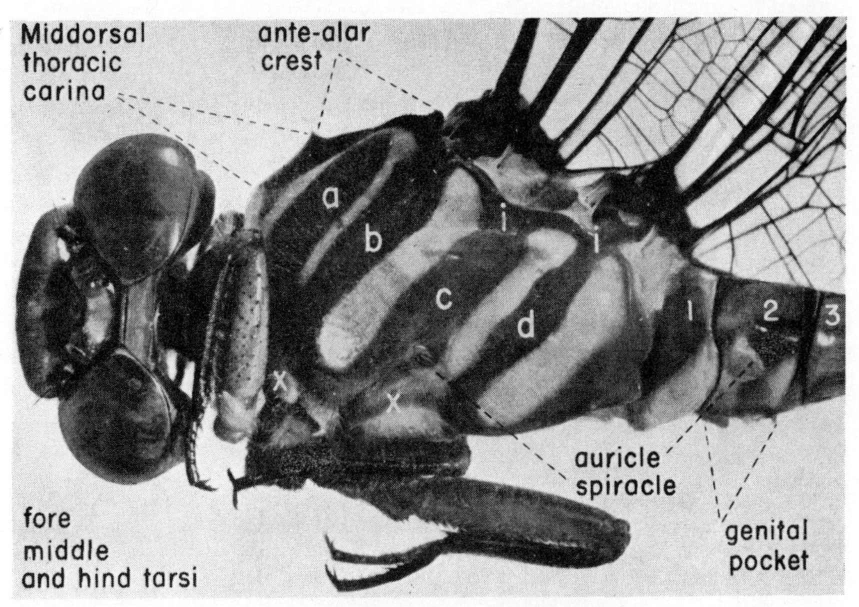

Fig. 4. *Aphylla protracta,* photograph showing especially make-up of synthorax and structural basis of its color pattern.

is turned to show its dorsal surface. The abdomen is cut off near the base of segment 3. The prothorax is hidden by the closely folded fore legs.

The *carina* lies lengthwise upon the blackish *middorsal* stripe of the synthorax. The near end of the *crest* rises in a recurving prominence before the left fore-wing root. From it a *subalar carina* (*i,i*) extends far to rearward, carrying a black band that connects the stripes on the lateral sutures. At *a* is the *antehumeral stripe;* it is not on a suture. At *b* is the *first lateral,* or *humeral* (mesopleural); at *c,* the *second lateral,* or *midlateral* (interpleural); at *d,* the *third lateral,* or *femoral* (metapleural). These three stripes overspread sutures that bear the same names. Suture *c* is obsolete, as explained on page 10. The *supracoxal plates* (*x,x*) overlie

* A fuller account of these skeletal parts and their development may be found in William D. Sargent's paper, "The Internal Thoracic Skeleton of the Dragonflies," *Ann. Ent. Soc. Amer.,* 30:81–95, 1937.

the leg bases. *Spiracle* and *auricle* are named on the photograph. The location of the *genital pocket* is faintly indicated; it lies on the midventral line of abdominal segment 2.

The strong backward slant of the synthorax to the wing bases is brought about in this way. The side plates (*episterna*) of the mesothorax grow forward and upward to meet on the back. They crowd the wings backward far from the prothorax and push in between to meet on the middorsal line in a sharp elevated ridge, the *middorsal thoracic carina* (often in descriptions called simply the *carina*). At the front end of this carina, next to the prothorax, a cross ridge is formed, called the *collar*. At its upper or rearward end the carina merges right and left into a very remarkable structure, the *crest,* a narrow brace-shaped area bounded front and rear by a serrulated rim. Front and rear edges of the rim swing out in a graceful curve to meet at their outer ends and form there on each side a protecting shield in front of the wing roots. An exposed corner is thus strengthened to meet the shocks of the dragonfly's hard-driving flight.

Correspondingly, the side plates at the rear of the metathorax (metathoracic *epimera*) are expanded downward to meet behind the leg bases and below the level of the abdomen. These meet and completely fuse in a flat median area, the *intersternum,* at each side of which is formed a low and smooth longitudinal ridge at the postero-lateral boundary of the thorax.*

* The backward slant of the side pieces of the thorax was measured by Needham and Anthony (*J. N. Y. Ent. Soc.,* 11:117–126, 1903) and was found to increase progressively throughout the order, and to parallel other trends in specialization. A summary of their measurements of this slant was published by Needham and Heywood in their *Handbook of the Dragonflies of North America* (p. 13) as follows:

Families	*Angle of humeral suture*[1]			*Angle of tilt of wing bases*[2]		
	Min.	*Max.*	*Av.*	*Min.*	*Max.*	*Av.*
AESCHNIDAE	21	50	38	22	35	26
LIBELLULIDAE	29	52	40	18	38	26
AGRIONIDAE	43	64	56	35	61	39
COENAGRIONIDAE	59	72	64	38	62	51

[1] This is the angle that the humeral suture (the foremost of the three lateral sutures), viewed from the side, makes with the perpendicular.
[2] This is the angle that a line drawn through the wing bases makes with the axis of the body.

When we consider the perching habits of dragonflies, their momentary pauses and sudden flights, we appreciate the advantage of this skewness. The legs are put forward where they readily reach and grasp vertical stems; and the wings are moved backward and tilted, with their cutting edges directed obliquely upward. In this position a simple sculling action lifts the body instantly from its support and launches it in the air.

The coloration of the synthorax follows the structural plan. There are three *lateral sutures:* one between mesothorax and metathorax, and one within each of these segments. As the dark color spreads from the sutures it narrows the intervening areas and causes them to become pale stripes, or, by fusion at the ends, to become spots of the pale disappearing ground color. In some of the more specialized dragonflies, this skeletal basis of the color pattern becomes wholly obscured.

Legs.—The legs of adult dragonflies are used for standing, for climbing, for perching, and for capturing flying insect prey—very little for walking, if at all. The wings are used for traveling from place to place.

The legs are composed, as in other insects, of two short basal joints (*coxa* and *trochanter*) that make the upward turn from beneath the body; two long joints (*femur* and *tibia*) that meet at the knees; and three short joints that form the foot (*tarsus*). The last joint ends in a pair of claws.

In dragonfly legs there is great variation in size, length, and development of spines. Each femur and tibia may have a double row of spines, one on the anterior and one on the posterior aspect, beneath. The six legs are so bunched as to form a leg basket for capturing prey.

The three pairs are much alike, their length increasing from front to rear. Their spines are generally longer in the female than in the male. The inner row on the front tibiae of males is composed in part of spines that differ from all the others in being flattened, decurved, and close-set to form a sort of eye-brush that is used to sweep the surface of the compound eyes and keep them clean.

Though all the legs conform rather closely to a common hexapod pattern, there is evolutionary progress within the order: from roughness of surface and crudeness of form to more neatly turned joinings of the long segments, more smoothly tapering margins, and more beautifully graded ranks of slender spines.

Wings.—Two pairs of long free wings are present in all Odonata. As in insects generally, each wing is rigid at its front margin, where strong supporting veins are crowded closely together, and thin to rearward, where pliancy is required and where the veins are weak and wide apart. At the front stands a triad of strong veins well braced against each other by crossveins, and still more firmly bound together by consolidation at both ends. These three veins act as one. They support the wing much as a main mast supports a sail. At their base is a hinge on which the wing swings up and down.

Behind this triad, veins diverge and weaken and form wide interspaces filled with crossveins. The membrane becomes thinner and more pliant to the rounded hind margin. The action of the wing is primarily sculling, but the broadening to rearward of the base of the hind wing adds a large surface for planing.

Wing venation.—Nothing will help more toward knowing dragonflies than a careful study of the venation of their wings. Vein characters are very definite. They are plain as the printed page. Nearly all the genera may be recognized by the venation of their wings alone. If at first glance the rich network of veins and crossveins appears complicated, a brief study of our diagrams and comparison with the real wings should make it easy to understand. The veins in our figures bear the following names and designations:

C	Costa	M	Media
Sc	Subcosta	Cu	Cubitus
R	Radius	A	Anal vein

The *costa* (C) is marginal, and coincides with the front border of the wing. The notch in the middle of that border is called the *nodus* (*n*).

The *subcosta* (Sc) ends at the nodus.

The *radius* (R) is a strong vein that parallels the two preceding, and appears simple; but at its base it is fused with the media as far as the arculus (*ar*), and at the nodus it gives off a strong branch to rearward, the radial sector (Rs). This branch descends by way of the subnodus (*sn*) and the oblique vein (*o*) to its definitive position behind the first two branches of the median vein. At the point where Rs bends outward it is connected proximally by a brace, the bridge (*br*), which appears like its true base and which joins it to the median vein farther back.

The *media* (M) is a four-branched vein. At its base it is fused with the radius, but the fusion is not quite complete; it has been aptly compared to the union of the barrels of a double-barreled gun. At the arculus (*ar*) the media descends to meet a crossvein and then bends sharply outward again toward the wing tip. It gives off to rearward a strong branch (M4) at its departure from the arculus, another (M3) at the middle fork (Mf), and another (M2) at the subnodus (*sn*). From this terminal fork, one branch (M1) runs parallel to the main radial stem (R1), and the other (M2), diverging, parallels the radial sector. The two branches formed at its first forking (technically M1–3 and M4) are known as the *sectors of the arculus* (*k*).

The *cubitus* (Cu) is a two-branched vein. Like the median vein, it is strongly bent. Its base is stout and straight to the arculus or beyond.

Then it turns sharply to rearward, forming the inner (proximal) side of the triangle (T). At the hind angle of the triangle it forks, and its branches arch outward, running more or less parallel to reach the hind margin.

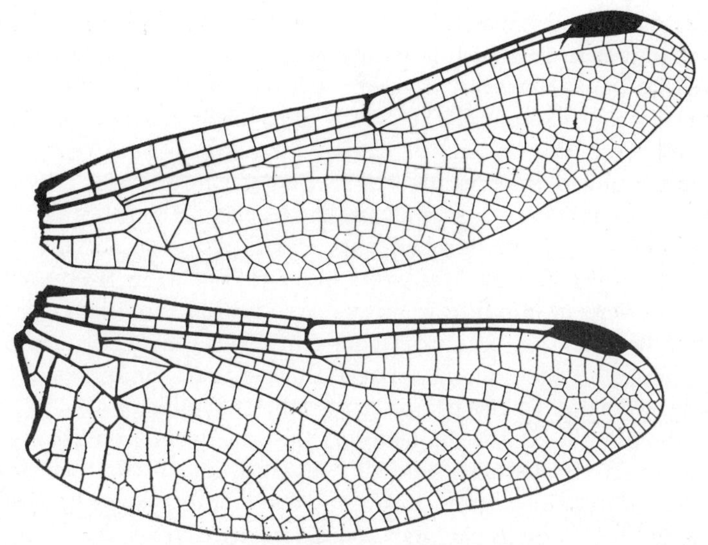

Fig. 5. *Erpetogomphus coluber*, showing relatively primitive wing structure.

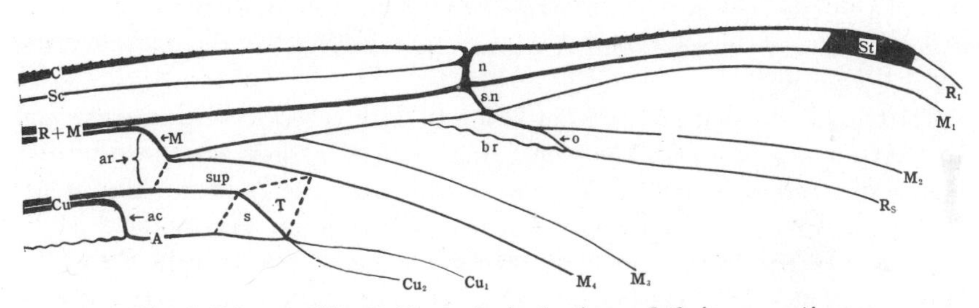

Fig. 6. Diagram illustrating principal veins and their connections.

The *anal* vein (A) is here treated as a single vein, though its branches are perhaps the equivalent of the separate anal veins of other orders of insects. It is convenient to designate them as A1, A2, and A3, from front to rear. The course of these branches and their degree of development vary greatly in the different groups.

The long veins of the wing are connected crosswise at three levels. At the stigma three long veins are conjoined; at the nodus, five; at the arculus, all six. At the stigma the brace vein (*b*) is obviously a modified crossvein.

At the arculus a short crossvein (its lower part) supports the basal bend of M4.

Two crossveins that no longer look like crossveins, between M and Cu, have converged to meet on M4 and form the triangle (T). They now form the front and outer sides of the triangle. The front one is so far aslant that it is almost in line with the base of Cu and appears to be a part of that vein. However, its swing out of line is only a little greater than that of the cubito-anal crossvein that forms the front side of the subtriangle (*s*).

The inner side of the triangle is a portion of vein Cu bent to rearward. The front side might be mistaken for the basal portion of M4 if one did not know the story of its development.

The nodus (*n*) is a grooved cross brace at the wing front. A double row of crossveins occupying two interspaces behind the costa is in two divisions, antenodal (*an*) and postnodal (*pn*). In figure 7* it may be seen that the first and fifth antenodals are thicker than the others, and only these two are matched in position across the subcosta.

The triangle (T) is the mainstay of the broad basal cubito-anal area of the wing. It lies between veins M4 and Cu, and binds them together in a strong yet flexible union. Bordering the triangle along its front side is a longer, narrower triangular space, the *supertriangle,* or *supratriangular space* (not labeled in the figures), and on the inner (proximal) side of the triangle is the *subtriangle* (*s*). The stout perpendicular crossvein connecting veins Cu and A nearer the wing base is the *anal crossing* (Ac).

Thus far in our progress from front to rear across the wing, fore and hind wings are alike. In the remaining anal area they are quite unlike. In the fore wing the anal vein runs out almost directly to the hind angle of the triangle and ends there. There is little or no development of its branches. In the hind wing it is three-branched. It joins vein Cu at the hind angle of the triangle (T), as in the fore wing, but does not stop

* For quick finding of both names of parts and abbreviations for them, they are here arranged in alphabetic order:

Ac anal crossing
an antenodal crossveins
ar arculus
b brace vein of stigma
br bridge
g gaff
Mf middle fork
mq intermedian crossveins
n nodus
ob oblique vein
pn postnodal crossveins
pr paranal cells of fore wing
n, o, p, and l paranal cells of fore wing
1, 2, 3 postanal cells of hind wing
s subtriangle
sn subnodus
st stigma
T triangle
tr trigonal interspace
x, y, z 1st, 2d, 3d anal interspaces, respectively

there. Its terminal branch (A1) fuses with the base of vein Cu2 for a distance and then frees itself and runs out to the hind-wing margin. The fused portion is known as the gaff (*g*). In figure 7 the cells labeled *n, o, p,* and *l,* which border the anal vein on its posterior side, are called *paranal* cells; the cells labeled 1, 2, and 3 on the proximal side of A1 are the postanal cells.

The other two branches of the anal vein (A2 and A3) divide the anal area into three interspaces marked *x, y,* and *z.*

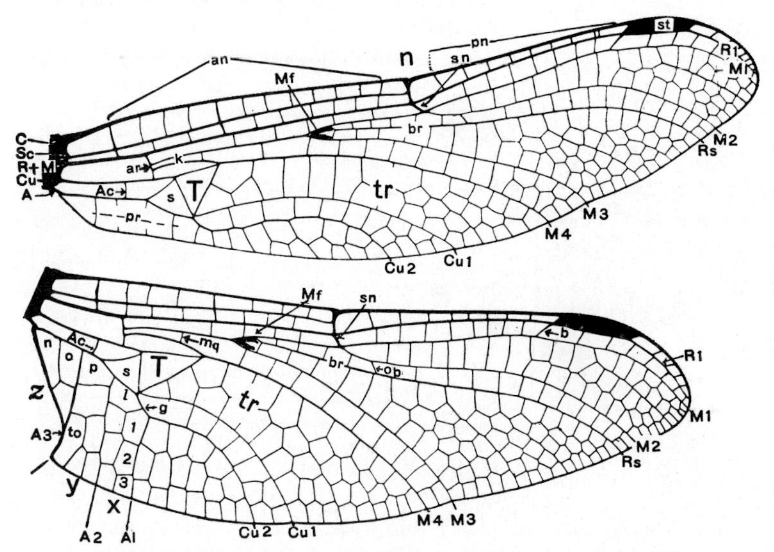

Fig. 7. Wings of *Gomphus cavillaris,* illustrating venational characters for Anisoptera in general. Middle fork (Mf) blackened for easier recognition.

In all the venational features thus far considered, males and females are very much alike; but in the third anal space of the hind wing (*z*) the males have a special venational feature that females lack. It is the *basal triangle.* In the species shown in figures 5 and 7, it is three-celled. Its boundaries are greatly strengthened. Its area is contracted in such a way that a portion of area *y* reaches the inner border of the wing and forms there a prominent angle. That angle is occupied by an enlarged cell (sometimes, a cell group), called the *tornal cell* (*to*), or merely the *tornus.*

If the main veins seem to be lost in a superabundance of crossveins, figure 6 (illustrating their course and conjunctions) will aid in recognition. The veins bear the same labels as in the following figure, and crossveins are omitted except for a few (shown in dotted lines) that have become special braces.

This may seem a bit puzzling at first glance. It long was so to entomologists. The puzzle was solved by studying vein origin. Wing veins develop about the air tubes (*tracheae*) that traverse the wing buds of the dragonfly nymph. Chitin is deposited about these tracheae, forming the stiff, rodlike, but hollow supporting veins; the areas between expand and become thin and transparent membrane. Crossveins develop late; if one examine with a microscope the wing pad of a well-grown dragonfly nymph, he may see the course of the antecedent tracheae very clearly. Figure 8 shows the tracheae of the six principal veins. It shows also how the three cross connections at stigma, nodus, and triangle are formed. The stigma is

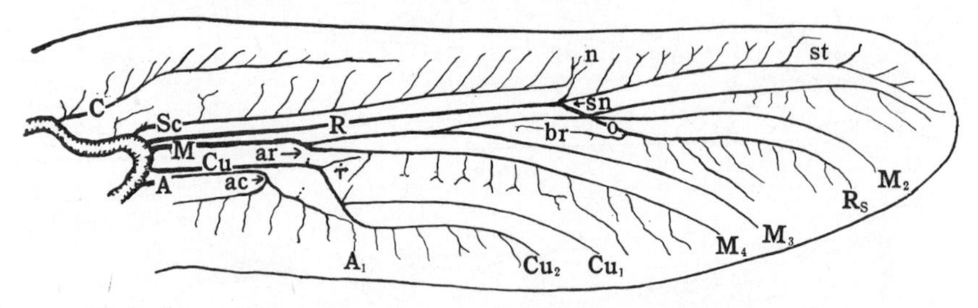

Fig. 8. Tracheation of nymphal wing of *Gomphus*, showing lines on which principal veins of adult wing are all laid down.

merely a thickening—an area of heavy chitin deposition. The subnodus (*sn*), oblique vein (*o*), and bridge (*br*) are all formed about the base of the radial sector. Arculus (*ar*) and triangle (T) are initiated by basal bends in the media and the cubitus, respectively. These bends are very gentle at first, and become sharply angulated only in the adult wing.

Figure 9 of nymphal wings of *Anax* shows the tracheation. The air-filled tracheae appear as black lines; the veins are translucent. The correspondence between veins and tracheae is obvious. It may be seen in the wings of any near-grown dragonfly nymph, with no preparation whatever save cleanly clipping off whole wings, mounting them on a glass slide in water, adding a cover glass, and slipping them under a microscope.

It is not alone veins and crossveins that tell relationships. The interspaces between the long veins add their testimony. Their shaping is important and is generally far more constant than the number of crossveins included in them.

Figure 10 shows diagrammatically the named principal interspaces that are common to all Anisoptera. In this figure the paired veins are numbered 1, 2, and 3. The fusions are lettered *a, b, c,* and *d* (*d* is the gaff). The extent of each fusion is indicated by an extra line drawn alongside it. The interspaces are fully labeled: *antenodal* and *postnodal, midbasal, inter-*

median, cubito-anal, trigonal, anals (1st, 2d, and 3d, labeled *x, y,* and *z,* respectively), and *bridge,* which ends at the *oblique vein* (*o*). Large T is the triangle, small *t,* the subtriangle; *n, o,* and *p* are the first three paranal cells; two additional paranals lie within the anal loop (L). The wide middle space (*tr*) beyond the triangle (sometimes called "discal space") is the *trigonal interspace.*

Figure 10 is the figure of a female wing. It lacks the basal triangle of the male which is the principal difference between the sexes. The hind

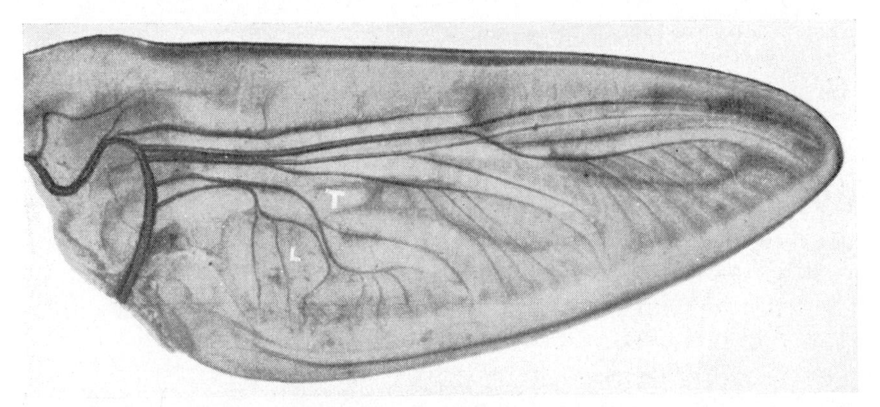

Fig. 9. Hind wing of nymph of *Anax junius.* Tracheal system well developed; veins just beginning to appear; triangle foreshadowed at T and anal loop at L; also nodus, stigma, and planates.

angle of the wing is rounded and not angulate. In the female the third anal interspace (Z) is composed of more than three cells and is less sharply defined.

Venational characters are used far more constantly than any others by students of dragonflies generally. Veins have a definiteness that is gratifying. They are always outspread and easy to see. They are hardly ever lost by breakage. The heads of dried specimens seem to drop off with a touch and sometimes get lost; the legs and abdomens break and fall away far too frequently; but wings stay on to the last, with their venation unchanged. And they come in quadruplicate, the four, originally alike, never completely differentiated.

Our pleasure and satisfaction in the study of Odonata may be greatly increased by careful attention to their marvelous wings, for in the veins of the wing we may read the story of the development of the group. It is a long story. The earliest chapters are missing; but some very interesting later chapters, written in characters of veins and crossveins, tell of progress along different lines in the several families. Thus we get sometimes glimpses and sometimes clear views of the evolutionary processes that have wrought form and fitness in all living things.

Abdomen.—The long abdomen of a dragonfly is composed of ten distinct segments and a portion of an eleventh that carries terminal appendages. Segments 1 and 10 are shorter than the others. Segments 2 and 3 are swollen to form a more or less spindle-shaped enlargement of the base of the abdomen. In the male dragonfly they bear on the ventral side the unique copulatory organs that are a distinguishing characteristic of this

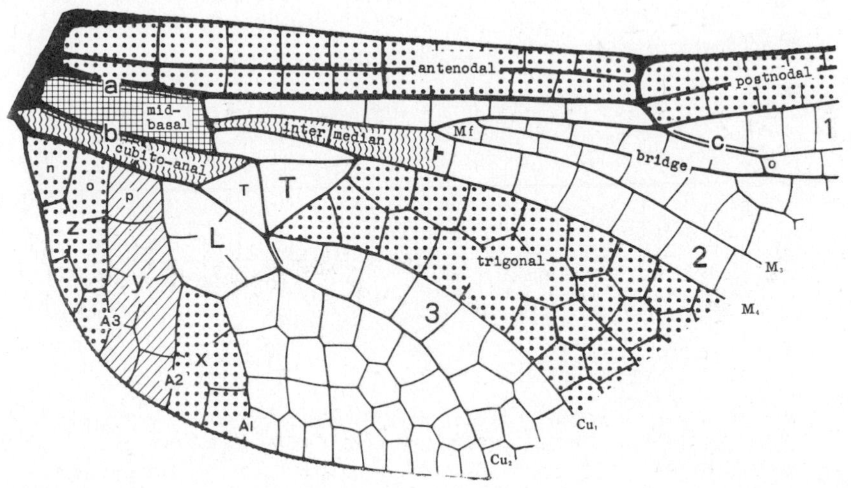

Fig. 10. Diagram showing paired veins, fused veins, and principal interspaces in venation of dragonfly wing. (Based on wing print of female *Ophiogomphus mainensis.*)

order of insects. Since the ultimate criteria of species are often to be found in the form of the accessory genital apparatus, it is important to know something of these parts.

The males have three appendages at the end of the abdomen that are forceps-like in action, and are used for seizing and holding the female. The males have also, in a cleft on the ventral side of the swollen second abdominal segment, paired hamules for grasping and a penis for use in copulation. Previous to copulation the sperm must be transferred from the spermaries (whose ducts open on the ventral side of segment 9) to a cavity in the tip of the penis. This is done by bending the abdomen downward and forward, bringing the two orifices together.

A generalized form of the genitalia of abdominal segment 2 is found in the genus *Ophiogomphus*. It will serve for introduction. (See fig. 12.) The genital pocket lies in the sternum of segment 2. A penis springs from the front end of the sternum of segment 3 and extends forward, folded double upon itself when at rest, on the middle of the pocket floor. The front half of the sternum of segment 2 develops as an anterior lamina that completes the rim of the pocket in front. A rigid, more or less spoon-shaped, upcurving guard rises from the middle line of the pocket floor; it covers and protects the U-shaped bend of the penis much as the guard of a bicycle wheel protects the rear of the tire. Converging over the tip of the guard are two pairs of hamules. These parts differ greatly according to species.

The female is not very different from females of other orders. The genital opening is at the apex of abdominal segment 8 on the ventral side.

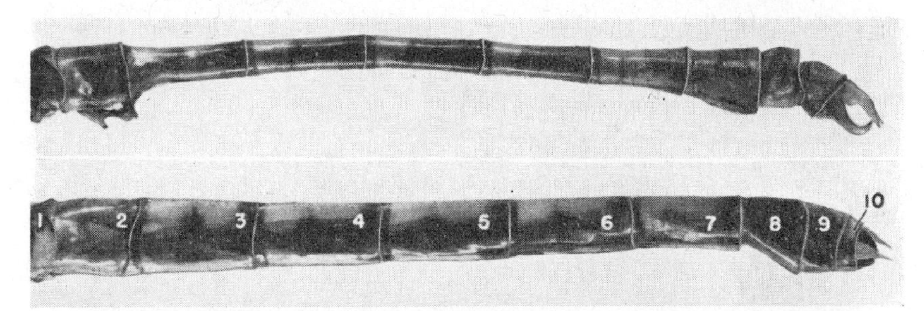

Fig. 11. Abdomen of *Erpetogomphus designatus*. Above, male; below, female.

The opening has a specialized covering. The coverings are of two kinds: a relatively simple subgenital plate (*vulvar lamina*) in Gomphidae and Libellulidae, and a more or less complicated *ovipositor* in the other families.

The subgenital plate (vulvar lamina) is a flat or scoop-shaped prolongation of the sternum of abdominal segment 8. Above its base lies the opening of the oviduct (fig. 236, p. 392). Many variants of its form accompany descriptions of species in Part Two. From it the eggs are shed, mostly in flight and as successive batches.

The ovipositor is an instrument for puncturing soft tissues of plants and placing the eggs singly in the punctures, mostly under water. It consists, at its best development, of two pairs of sharp-pointed stylets and a pair of covering genital valves.

In figure 13, note the black caudal appendages (*cerci*) and the scooplike prolongation of segment 10, its under side covered with stiff spinules. The strongly divergent black palps are on the blunt tips of the transparent genital valves through which upper and lower stylets of the ovipositor show plainly, the result of boiling the specimen in KOH. (See also fig. 177, p. 295.)

Fig. 12. Genitalia of *Ophiogomphus bison*, male.

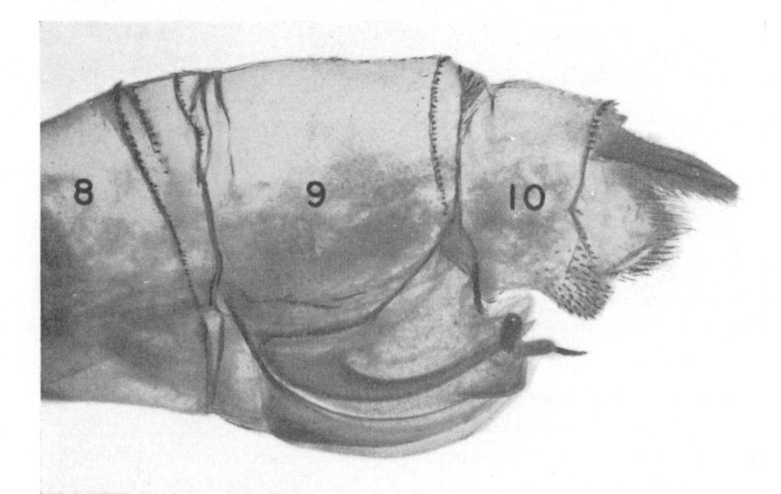

Fig. 13. End segments (8, 9, and 10) of abdomen of *Boyeria vinosa*, female, showing ovipositor as viewed from side.

The stylets are so grooved on their opposed edges that together they form a channel for the passage of the eggs in single file. The upper pair is pointed and awl-like; the lower pair, tapered and grooved like a rasp to make holes for the eggs.

In copulation the female is swung, suspended beneath the body of the male, in an inverted position, the reverse of his own, his caudal appendages still holding her in front. Her genital segments (8 and 9) are grasped and held by the hamules of the male during the transfer of the sperm. The form of hamules and caudal appendages in the male and of the plate

Fig. 14. Diagram of copulatory position, *Aeschna constricta.* (After Calvert.)

that covers the genital opening of the female is shown in many figures on succeeding pages.

The primitive form of the odonate abdomen was probably more or less cylindric, perhaps flattened little on the ventral side, and without notable enlargements, constrictions, or sculpturing. Each of the several rings of which it is composed may have first acquired the transverse ridges (carinae) at front and rear ends. These stiffen the walls where strength is most needed for lateral movements. The abdomen of *Tanypteryx* has not greatly progressed beyond that condition. Greater or lesser enlargement of the basal segments is general in the order. Dilatation of segments 8 and 9 reaches its maximum in the male of *Gomphus ventricosus.* A waistlike slenderness of segment 3 is characteristic of *Aeschna* and its allies, and a long slenderness of the middle segments, with greater widening at the rear than at the front, has produced an abdomen shaped like an Indian club (its handle being at the base) in the more specialized Cordulines.

In general the middle segments are most typical. Each segment is like a Quonset hut in cross section, its roof (*tergum*) all-over arching, the flat floor (*sternum*) not quite reaching the sides, and the interval between roof and floor filled by a soft and flexible *pleural* membrane variously infolded. A spiracle is found near each anterior corner of the tergum in segments 1 to 8.

Longitudinal ridges (*carinae*) are variously developed along the middle segments, and are named *middorsal, lateral,* and *submedian ventral* according to their location; these find their best development in the Libellulidae.

Such in brief are the principal external features of the structure of adult dragonflies. The lesser details will be taken up in the discussion of the groups to which they pertain.

THE DRAGONFLY NYMPH

When we take up the study of the nymphs of dragonflies, we seem to be entering another world of life, with nearly everything changed. The nymphs are very different in appearance and in habits. They are not less interesting than the adults, though far less beautiful in coloration. Theirs is another kind of beauty: the beauty of fitness to life in the water. They are like the adults in only one particular habit: they are equally voracious carnivores. As the adults eat mosquitoes in the air, so the nymphs eat mosquito "wrigglers" (larvae) in the water, and for a much longer time.

There are three stages in the life history of a dragonfly: egg, nymph, and adult. The nymphal stage is far the longest. The intervals between successive sheddings of the nymphal skin are called *instars.*

The nymphs* of Odonata may be distinguished from all other aquatic nymphs by two very striking characters: the huge grasping labium, and the form of the tracheal gills that serve for respiration.

In Part Two the form of the nymph for most genera of Anisoptera is shown in a photograph or drawing at the head of the genus.

The plan of the body is, of course, the same as in the adult, and its parts bear the same names. The three segments of the thorax are more nearly equal in size than in the adult. There is hardly any consolidation of mesothorax and metathorax into a synthorax. The head is less freely movable than in the adult. Its freedom of movement is limited by the enormous labium, which is well shown in figure 16. The two lateral lobes of the labium cover the face up to the eyes, and its middle hinge extends backward between the bases of fore and middle legs; the broad triangular area between is called its *mentum.*

* Called also *larvae* (singular, *larva*) when one wishes a more general term, applicable to the young of all animals which undergo a transformation (*metamorphosis*). "Nymph" is the special name for that type of insect larva that has the wings developing externally upon its back. Aquatic nymphs are called "naiads" by some authors.

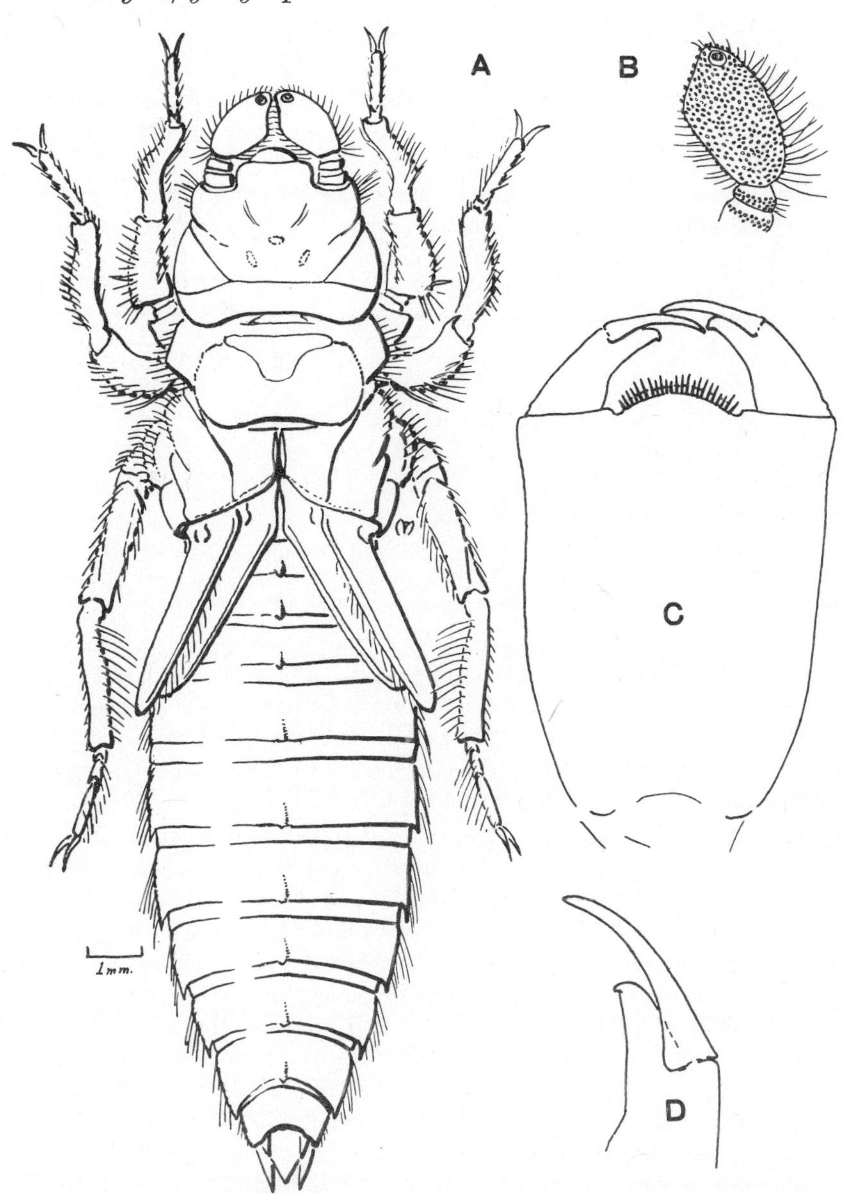

Fig. 15. Nymph of *Progomphus serenus*. A, whole nymph; B, right antenna, showing flattened segment 3 (segment 4 vestigial); C, labium; D, right lateral lobe, showing strong movable hook and minute end hook at tip of incurved blade. (Drawings by Dr. May Gyger Eltringham.)

The compound eyes are very small at hatching, and grow with every casting of the nymphal skin. They are built up fastest on the upper inner side, and the successive additions are visible upon the surface at the front of the head, where instars may be counted like growth rings in the heartwood of a tree. The layers are quite evident in such pale nymphs as those of *Celithemis, Erythemis,* and *Coryphaeschna.* The new layers of eye-stuff are crowded close together under the tough cuticle, and their expansion and release come with the final shedding of the skin at time of transformation.

Perhaps the most remarkable feature of nymphal anatomy is the labium, or lower lip. It is folded upon itself like a hinge at the mid-length, and then turned backward beneath the front legs. At its front end is a pair of strong grasping lateral lobes that are variously armed with teeth, hooks, and spines, according to families, genera, and species, as will be told and further illustrated in detail in Part Two. The most useful systematic characters often are labial characters.

The labium is a grasping organ of unique design. When extended, it is almost as long as the fore leg. It is thrown out and drawn back again with such swiftness that the eye cannot follow it. With it the nymph reaches for a victim, clutches it between the armed lateral lobes, and draws it back right to the jaws. The hooks and spines of the labium hold; the jaws devour; and if any fragments fall, they are retained on the mentum as on a tray. The labium is finally thrust forward a little, and even these fragments are gathered up. The strong incurving bristles are called *raptorial setae,* or merely *setae.* They are lined up in four ranks and set in miniature sockets. The species shown in figure 17 has four *lateral setae* on the upper rim of each lateral lobe and a broken row of eight or nine *mental setae* on each side of the mentum within. The innermost four or five mental setae are always feebly developed. At the front end of the row of lateral setae and above the teeth stands the stout black clawlike *movable hook.* It has about the same curvature as the setae but many times their strength.

This combination of hands, carving tools, and serving table is highly efficient. But it must have room to swing in. It is liable to entanglement in the threads of filamentous algae and slender sedges such as *Websteria,* and sometimes cannot free itself. Furthermore, it is quite useless for taking crevice dwellers, however toothsome, out of their retreats.

The jaws are of a distinctly flesh-eating design. The mandibles with their **Z**-shaped sharp chitinous ridges cut the victim to pieces, while the maxillae, shaped like meat forks, turn it conveniently for cutting. So it is made into pellets and swallowed.

The antennae are much larger in the nymph than in the adult. They undergo regressive development at the final moult. The most primitive form in the order is that of the Petalurines, such as *Tachopteryx,* in which the joints from base to tip are all thick, and there is little differentiation between base and apex. Generally they are seven-jointed, with two short cylindric joints forming a *pedicel;* but in Gomphidae they are four-

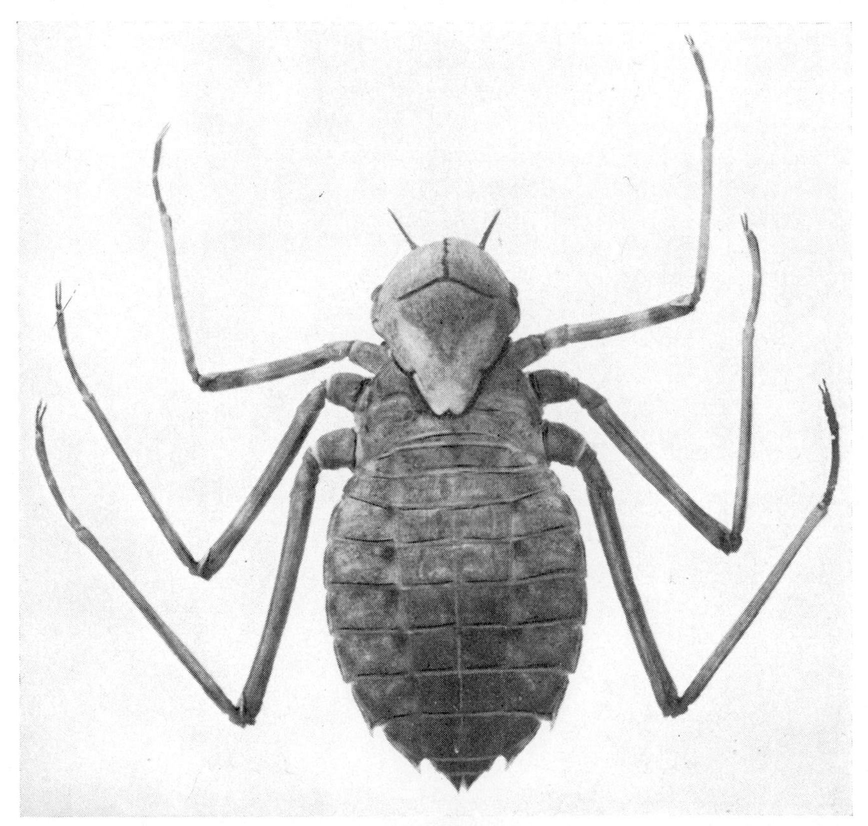

Fig. 16. Nymph of *Macromia,* inverted to show position of labium when folded beneath head.

jointed: the first and second joints short and ringlike, the third very long and often flattened, the fourth longest in *Progomphus.* In other genera of Gomphidae the fourth is very minute, sometimes so small as to be hardly visible.

In families other than Gomphidae the antennae are generally six- or seven-jointed, and the joints beyond the two basal ones (the pedicel) are slender, cylindric, and greatly elongated (*filiform*). We have already seen that the antennae in adult dragonflies are bristle-like (*setiform*) beyond the pedicel.

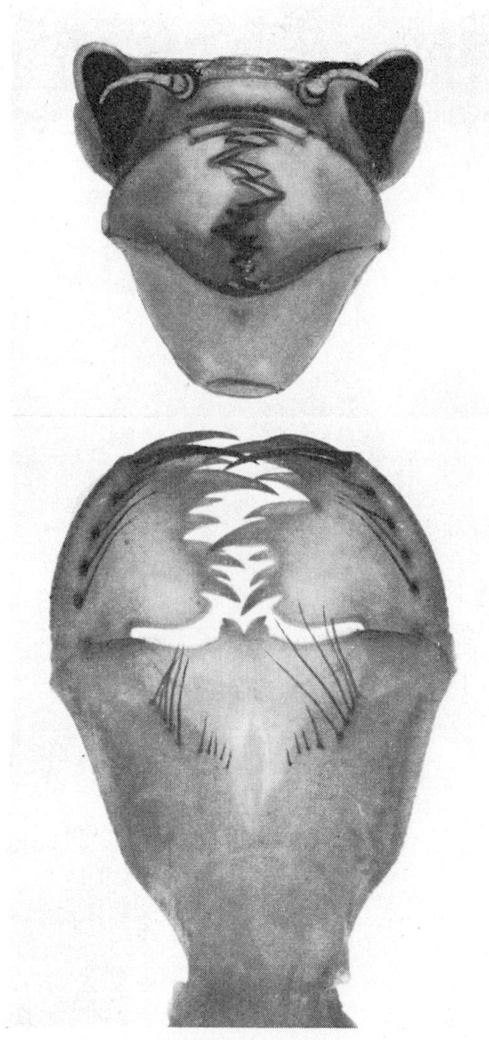

Fig. 17. Spoon-shaped labium of nymph of *Cordulegaster erroneus.* Above, face view, showing how labium covers face up to eyes; below, inside view of mentum showing huge incurving teeth on opposed edges of the two lateral lobes, and small V-shaped notch in end of prominent median lobe.

The legs have the same make-up as in the adult, but they are not spiny. They are used for standing and locomotion and not for capturing food.

The abdomen varies from nearly cylindric, as in *Cordulegaster* (fig. 40, p. 80), to very flat, as in *Hagenius* (fig. 53, p. 114). In general the abdomen is three-ridged (*triquetral*) lengthwise, with one middorsal and two lateral ridges, the under side of the abdomen being always more or less flat. The hooklike, spinelike, or stublike projections ranged along the middorsal line are called *dorsal hooks;* those at the side margins of each segment, *lateral spines* (technical terms, named for position without regard to form). Equally distinctive are the five spinous caudal appendages, but better shown in the next two figures. All these prominences offer good characters for distinguishing genera and species, and are much used in Part Two.

In the abdomen is the gill chamber, a modified portion of the hinder (*rectal*) part of the alimentary canal. It is set off from the intestine by a

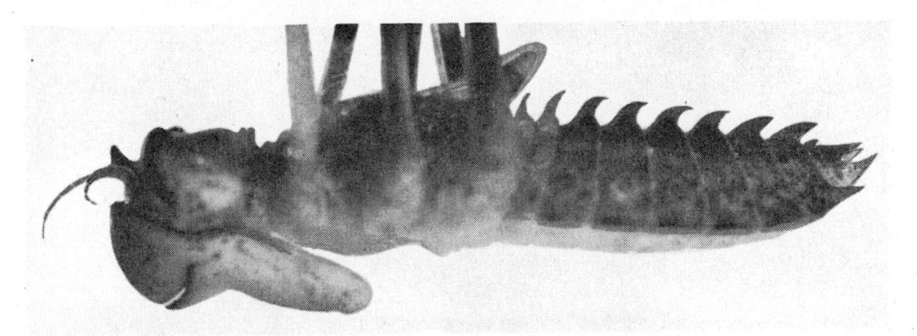

Fig. 18. Nymph of *Macromia caderita,* with wings and legs lifted to show full series of dorsal hooks on segments 2 to 10 of abdomen; showing also huge labium, middle horn on front and tubercles on rear of head.

valvelike constriction in front, and is enlarged into an oval sac that may half fill the abdomen. The gills hang from the inner walls of this chamber in longitudinal rows. They are minute, thin-walled, and very numerous. They are filled with fine air tubes (*tracheoles*) that branch off from four big air trunks running lengthwise of the body. The walls of the gill chamber are provided with muscles for changing its capacity. When it expands, water is drawn in from the rear, bringing fresh oxygen; when it contracts, the water is expelled. The posterior (*anal*) opening is guarded by a cluster of five spinous caudal appendages which terminate the abdomen: one long *superior* appendage, two smaller *laterals,* and two *inferiors.* Between these are three little scales that together guard the immediate opening. They serve as strainers; partly closing the entrance while water is flowing in, they fly open like shutters when the water is squirted out.

The regular expanding and contracting of the abdomen is easily seen in a living dragonfly nymph. The water currents may be watched if a little colored fluid be released in the water at the end of the abdomen.

If a very young and transparent living nymph be mounted in water for the microscope, its whole tracheal system, from gills to antennae, may be studied without further preparation. Tracheae and tracheoles stand out in solid lines of black. The gills may be seen in action.

Dragonfly nymphs emerge from the egg with long three-jointed antennae, small eyes, and one-jointed tarsi, each tipped with two slender claws. There are no wings or even wing buds present at hatching. The lateral abdominal appendages (*cerci*) also are wanting. The labium,

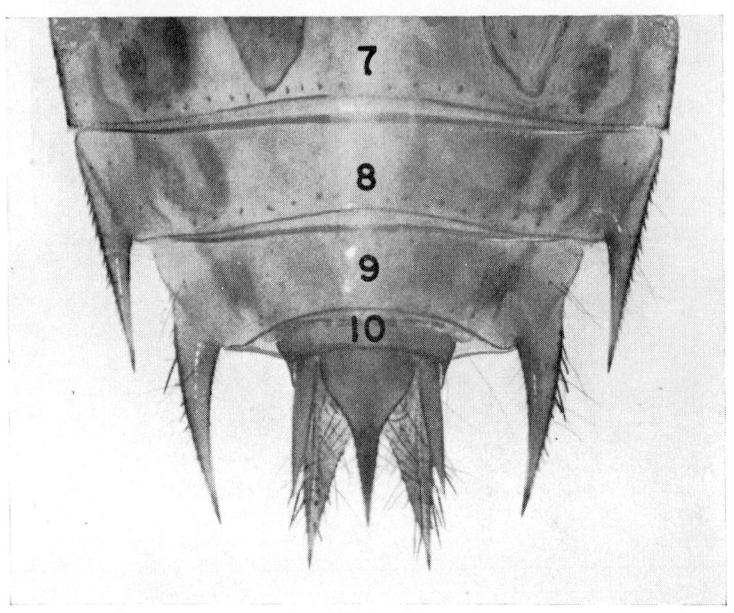

Fig. 19. End of abdomen of nymph of Saddle Bags, *Tramea*, showing lateral spines and caudal appendages. Last four abdominal segments numbered 7, 8, 9, 10. Large bristly lateral spines on 8 and 9. Superior appendage median in position. Slender laterals naked, large inferiors bristly. Wing tips overlap on 7. No dorsal hooks.

though ready for use, is often quite different in the details of its armature, as shown in figure 20.

The changes in form of the nymph that occur in the course of its growing up have been studied, moult by moult, for very few species of Odonata. There seems to be only one record of culture maintenance through successive generations, that of Evelyn George Grieve for the damselfly *Ischnura verticalis* (*Ent. Americana,* 17:123–153, 1937). It is a detailed account of culture methods and form changes (and much besides) through twelve instars and three generations.

To judge by the regular recurrence of their seasons of flight, and by the differences in size of the nymphs when taken from the water together, there seems to be one brood a year in most species; two years are required for the growth of the little Libelluline, *Nannothemis bella,* as determined by Dr. P. P. Calvert; three years or more, for the growth of some of the larger and more primitive forms.

When the dragonfly has completed its metamorphosis and has wings, it is still *teneral,* that is, it still has the paleness of all newly emerged insects. A little later in adult life its body will assume more mature tints, and its chitinous armor will harden to full strength. It must fly and forage before

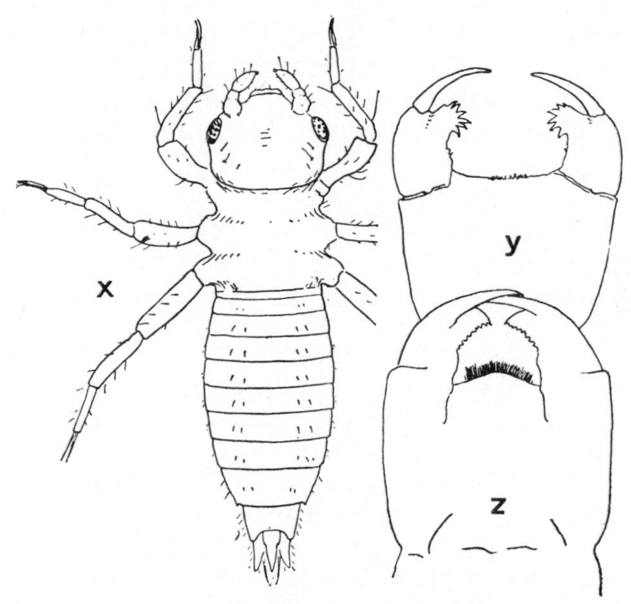

Fig. 20. Nymph of *Gomphus graslinellus. x,* newly hatched nymph; *y,* labium greatly magnified; *z,* labium of grown nymph.

its pigmentation fully develops. Then, if it escapes casualties and enemies and lives out the full measure of its days, the surface of its armor may grow *pruinose,* developing a whitish bloom; it may become hoary with age.

In haunts and habits dragonfly nymphs fall into three groups: climbers, sprawlers, and burrowers. The more active ones climb about in green vegetation in beds of waterweeds, or cling to stems of reeds or to hanging roots. Such are the darners (Aeschnidae) and many of the smaller Libellulines. Nymphs of the big Sky Pilot (*Coryphaeschna ingens*), for example, dwell among the erect stems of coarse aquatic grasses and cattails (*Typha*) or between the long leafstalks of pickerel weeds (*Pontederia*). Their bodies are striped lengthwise from head to tail by bands of pale brown and green. They habitually cling lengthwise to the stems, oftenest head downward, pattern and posture in complete accord.

Sky Pilot nymphs do not merely wait for their prey to come within range of the extended labium, but stealthily advance toward it with a movement almost as slow as the movement of the minute hand of a clock, and then catch it in a lightning flash of the terrible grasping labium.

The more sluggish sprawlers lie flat upon the bottom amid the silt, with legs outspread. Protected by their coloration and often by a coat of adherent silt that hides them perfectly, they wait in ambush until their prey wanders within reach.

The dark varied markings of *Pachydiplax* or *Tetragoneuria* fit the environment better where dead stems and fallen leafage fill the background. Other climbers, such as the darners *Boyeria* and *Nasiaeschna,* for example, cling to blackish twigs and bark or to logs and stumps in woodland streams, and they themselves are blackish, with only faint touches of paler color.

The nymphs that are sprawlers upon the bottom are not colorful. Macromias have the longest legs; the abdomen is wide, and they lie on the sand of the stream beds with legs fully outspread. Their dappled color pattern matches colored sand. Most Libellulas live in the muck that settles in the small embayments of streams or in the silted beds of ponds and pools. They have little color pattern, and that is generally hidden under a coat of ooze that clings to their hairy bodies. All these sprawlers (and there are many kinds) have more or less vertical faces, with eyes capping the fore corners of the head. They sit with antennae laid out forward upon the surface of the mud. The big flat *Hagenius* is another sprawler that is plain brown or black and quite patternless. It lies amid flakes of drifted bark and leaf fragments that settle in the eddies of woodland streams.

With the exception of *Hagenius,* all our Gomphids are burrowers. They all have flattened, wedge-shaped heads, and short thick antennae that lie flat on the face. Front and middle tibiae end in burrowing hooks. The best burrower is *Progomphus,* which digs like a mole in loose beds of drifting sand. The nymph is pale and shining on the back, well scoured by the sand, faintly tinted with green in front, and touched with brown on the edges of the abdomen toward the rear. Our other Gomphidae are less active. Their skins are scurfy, silt-covered, and patternless. Their burrowing legs are weaker, with sinuous bare scars along the sides. They burrow very shallowly, just under the top layer of silt, mostly in streams. A few are pond dwellers. One of them, *Aphylla,* is fitted better than the others for living in deep, loose muck because of its long tubular tenth abdominal segment. The upturned end lifts the respiratory opening through the silt to clear water.

All the foregoing is a rapid sketch of the main lines of development in the suborder Anisoptera, or dragonflies, with emphasis on form as related to habitat. Further details will be found in Part Two.

II

FIELD STUDIES

A good way to begin the study of dragonflies is to take an insect net and a notebook in hand and go to the nearest fresh-water pond on a sunshiny day in summer. Almost any clean and permanent pond will do. About its borders some of the big strong-flying dragonflies will be making themselves conspicuous by their superb aerial performances. Some will be flying low and others perching on weed tips along the shore. Some may be flying in couples, the male leading the female by the head. The males chase the females persistently, and males chase each other, apparently trying to drive rivals from a chosen field (thus showing a sense of proprietorship of domain); and in their fighting the males indulge in furious charges and countercharges, at their climax face to face with noses almost touching. However, these are sham battles. They never come to blows or to bites.

These activities may be seen at any pondside where there are enough dragonflies for competition. Females are generally much fewer than males, for they do most of their foraging in fields back from the water and come down to the shore when ready to lay their eggs. Often a female may be seen furtively darting here and there, low above the water, tapping its surface with the tip of her abdomen at every descent. She washes off with each tap a little cluster of a dozen or more eggs. Pairs of some species (*Celithemis eponina,* for example) continue to fly in couples after mating, the male seeming to select places for depositing the eggs. He assists in the necessary moves from place to place.

All this is after the manner of the Libellulines. Egg laying by the large Cordulegasters and Aeschnids is done quite differently. The female darner has a puncturing ovipositor by means of which she plants her eggs, one at a time, deep in the tissues of green stems or soft wood or soil, under the water. Her eggs, having greater security during development, are fewer and larger than those of the Libellulines. They contain more food and are differently shaped.

In the field, transformation will not often be seen, for it occurs mainly at night, when fewer enemies are abroad. But cast-off nymphal skins will be common enough along the shore, clinging to any solid support up which the nymphs have climbed for the final great change that makes them adult dragonflies. The empty, often mud-encrusted skins cling to their supports until beaten down by wind and rain or crushed by passing feet. Once in a while, however, a belated nymph that could not reach the shore earlier may be seen by daylight in transformation; and it is an amazing sight, well worth watching for the short time required to see it through.

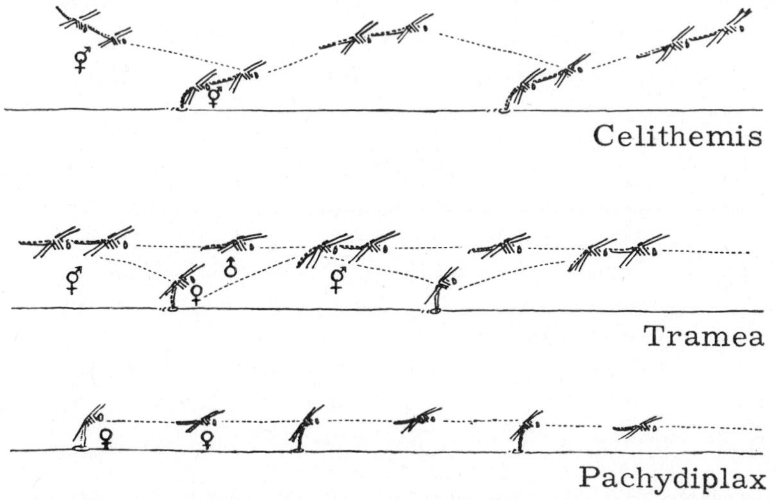

Fig. 21. Egg laying from the air. Above, *Celithemis*, pair (♂ and ♀) hitched together during oviposition; middle, *Tramea*, male unhitched between dips; below, *Pachydiplax* ovipositing alone (sometimes led by male) and keeping to one low level.

First comes a splitting of the skin down the back and across the head, and the pushing up through the split of the head and thorax. The legs are slowly withdrawn from their sheaths. Soon a mighty effort lifts the front of the body and leaves it standing on its tail, so to speak, in the loosened skin of the abdomen. After a rest of a few minutes, with legs folded, the dragonfly makes a final lurch forward and seizes a footing; then, with withdrawal of the abdomen, the entire body at length stands free.

It is a sorry-looking thing, misshapen, with tightly crumpled wings. Only the legs are ready for duty; but the abdomen will soon push out to full length; the wings will expand in length and breadth; bright colors will begin to glow through the pallid skin; the body will become streamlined and will take on an elegance of form that is not excelled by any other creature in the whole animal world.

The field is the place in which to see the dragonfly's complete mastery of aerial navigation, its manner of flight, its alertness, and its expertness in the capture of living prey. Just how jaws and leg basket coöperate in the seizure of prey is very hard to see in detail, but often a dragonfly will be taken from the net with recognizable remains of the last mosquito eaten still in its jaws.

Wonderful also are the agility and fleetness shown in escaping its enemies. Its worst natural foes are birds in the air and fishes in the water. Big dragonflies capture and eat little ones. So the perils of all of them are manifold.

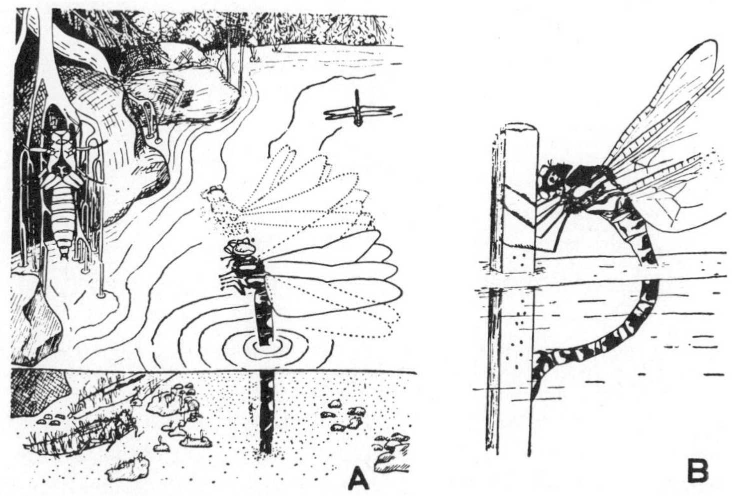

Fig. 22. Oviposition. A, *Cordulegaster dorsalis,* female ovipositing during flight on a gravelly bank (after Kennedy); B, *Aeschna constricta,* female ovipositing in a reed stem (after Calvert).

The big darners, flying in long sweeping curves, dominate the upper air, where power and speed count most. The lesser dragonflies keep close to cover, where agility, alertness, and shelter count most. It is as among birds: hawks may roam the skies at will, but sparrows must keep to the bushes.

Dragonflies are as easily recognized by their manner of flight as birds are; also by the manner of their perching and the places chosen for it; and by their postures at rest and in the air. Indeed, the full brilliance of their colors is seen only when they are sitting or flying in the sunshine.

Perching dragonflies will sometimes allow one to approach close enough to see their burnished armor that glistens in the sunshine; but the brightness of their colors and the elegance of their form and finish cannot be fully appreciated until specimens are taken in hand for closer examination.

Insect net.—The chief instrument for collecting adult dragonflies is an insect net. It must be large, light, and strong: large in diameter, to sweep a big section of the atmosphere; light in weight, that it may be swung quickly; and strong, so that it will not break easily. In pursuit of dragonflies a net is sure to get hard usage.

Collecting nets are obtainable from dealers in biological supplies, or may be made to order. Given the requisite materials, an ingenious boy could make one after the plan shown in figure 23. A collector far from home must improvise sometimes, or miss a chance to get fine specimens for his collection.

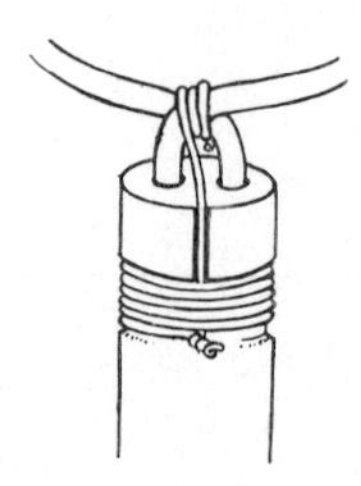

Fig. 23. A simple way of attaching rim of a light-weight collecting net to its handle.

Swatter.—Another useful instrument for collecting some species is a dragonfly collector's swatter. It is the easiest of all to make and to carry. It consists of a small square of wire cloth (window screening, old or new, will do very well) about six inches square. For a handle, a long slender bamboo cane is best. The top end of the bamboo is cut off below a node and the cane is split back down one internode. The cloth is drawn edgewise well into the split, and tied in place by a wire or thread through the mesh and around the stick.

This swatter is used, not as a fly swatter to smash with, but as a retainer to be suddenly dropped on the back of a dragonfly that is sitting flat on the ground; or it may be used to tap lightly and stun one that is hovering low over the surface of the water. The dragonfly on the ground must be held there until fingers, carefully edging their way under the swatter, can pick it up. The one knocked down on the water and stunned may be picked up on the swatter and retrieved at once before it recovers its power to fly. In striking contrast to the agility of dragonflies in the air is their helplessness when they are struck down. They are easily, though temporarily, paralyzed.

If a swatter is to be carried afield along with air nets, all raw edges should be bound with a half-inch-wide strip of waterproof adhesive. This will cover sharp wire ends and prevent entanglement with net bags in transport.

The successful use of a swatter requires great care in approach to sitting specimens, and a little knowledge of how far a dragonfly can see. The latter can be learned only in the field. But it is well to bear in mind that the mosaic-patterned retina of the dragonfly's eye is best adapted to detect *movement;* that our movements, not our pauses, startle it; especially those above, as pictured on the retina of a dragonfly sitting on the ground, may indicate the approach of a possible enemy. Therefore, it is well to make a progressively crouching approach, keeping the head from rising suddenly above the level of the dragonfly's line of vision. And,

Fig. 24. A small killing bottle.

having reached the probable limit of his tolerance of one's approach, it is time to stand perfectly still while bringing the swatter into action—time for strategy (a strategy well known to collectors of lizards). The dragonfly's attention will be so fixed on the huge enemy near at hand that he will hardly notice the little square of wire cloth slowly approaching edgewise at almost ground level, until it quickly claps down on him, holding him helpless on the ground.

This same strategy may, of course, be used with a net, but the net is more unwieldy; its bag gets entangled and its handle is shorter. Our best collecting of Gomphine dragonflies has been done with such a swatter. It has another great merit, especially for a collecting trip involving travel: only the little square of wire cloth need be taken along, for a stick that will do for a handle may be found almost anywhere.

Killing bottle.—The complement of net and swatter is the cyanide bottle that is used to kill quickly the specimens collected in the field. Various types of cyanide bottles are in use. Dealers sell them, but most collectors make their own, and in sizes suited to the specimens to be collected. At least two sizes are desirable: one for small specimens, made of a large-sized shell vial (or better, one of unbreakable celluloid), stoppered with a cork; for the large specimens, an ordinary jam or mayonnaise bottle, closed with a cap. The kind of cap that closes with a quarter turn and does not waste one's time in screwing and unscrewing is greatly to be preferred.

Whatever the size of the bottle, the placement of the cyanide may be as follows: in the bottom, a layer of sawdust or bran, which, if the bottle be rounded to the bottom, should fill it at least to the level where its sides become parallel; then a layer of pulverized cyanide of potassium (a deadly poison); then close-fitting circular discs of blotting paper, with a white disc on top for a better background against which to see the specimens. The top disc may be fastened with a drop of Duco or other transparent cement, placed at each of several points where it rests against the glass (easily done by lifting successive drops of cement on the tip of a wire, poking each down inside the bottle and depositing it at the junction of disc and glass). A POISON label should be placed on the cap of the bottle and, when the cement has dried, it is ready for use.

A ring of wire or of adhesive tape around the neck of the bottle, with a safety pin held in a loop in the ring for hanging it on the collector's belt, will free both his hands for taking specimens out of the net.

Preserving adult dragonflies.—There is no need to describe the current methods of preparing specimens of dragonflies for exhibit under glass. The pinning, spreading, and labeling process is well known to all entomologists, and is illustrated in nearly every textbook of entomology. But a simpler, speedier, and more economical way of making and keeping a study collection of dragonflies has come into use since cellophane envelopes became available. When specimens are taken from the cyanide bottle, they may be placed in envelopes made of waterproof cellophane and kept there permanently. The specimens should be dried *speedily* and *thoroughly* and kept away from all moisture (kept, of course, in pest-proof, larger containers).*

The specimens come out of the killing bottle in all sorts of postures. The wings, if flexed downward, may easily be erected. In general, the best position in which to dry them is with wings lifted and legs (in males) so placed as not to hide the genitalia of the second abdominal segment. Or the wings may be outspread, for the sake of better displaying their beauty, if

* We use envelopes of the *size shown* in figure 25. We do not seal down the flap (except when specimens are to be handed around a group of persons for immediate inspection). The flap serves as a convenient handle in removing and returning specimens to and from the files in which they are permanently stored. They are filed on edge, like cards in a filing case. Ours are kept in pasteboard trays, ranged in four columns to fit inside the standard glass-topped insect collection (Comstock) drawer.

so desired; but more trouble will have to be taken in placing wings, legs, and head in natural posture and leaving them so until dry.

The use of cellophane containers for specimens has many advantages over pinning and spreading:

1. Convenience. Both sides of the specimen may be seen without removal from the envelope, and the male hamules, so much used in species determination, are more accessible than in pinned specimens.

Fig. 25. Cellophane envelope for use as a container for dragonflies. Our standard size, suitable for all but a few of very largest species.

2. Safety. There is much less breakage in handling than with pinned specimens; and when breaks occur, the parts are kept together, and apart from fragments of other specimens.

3. Better labels. There is ample room for labeling and easier reading of data on the labels.

4. Space-saving. In envelopes, the specimens occupy only a fraction of the space that would be required if they were pinned and spread.

5. Time-saving. Here is probably the greatest saving of all: no pinning,

no spreading, no folding and unfolding of cross-kinked paper envelopes (and, incidentally, no snapping off of heads when the paper buckles).

6. Cost-saving. No spreading boards are needed; no cork bottoms are required; envelopes are cheaper than pins alone; shipping charges are greatly lessened.

Specimens may be shipped with envelopes open at the top by clipping them together in small bunches with a paper slip folded over and covering the open tops, as shown in figure 26. The clips should be thin, else they will weigh more than the specimens, and weight within the package is undesirable. The wedge-shaped bunches are placed alternately in reversed position in a box, with a little cotton filling the interspaces.

Collecting hints.—Some things will get into your nets that should not be allowed in your killing bottle if you wish clean, perfect specimens for your collection. These should be kept out:

1. Lepidoptera (moths and butterflies) of all sorts, for they shed scales that will stick to your dragonflies. The scales are difficult to remove, and may entirely hide characters of importance in classification.

2. Orthoptera (grasshoppers) of all kinds, for they spit out saliva that smears.

3. Specimens that are at all broken in capture, for even small breaks in the skin may leak badly.

4. Teneral specimens, for these are soft and more or less sticky. If these are worth keeping, they are better preserved in alcohol.

5. Large beetles and other strong-clawed insects that go rampaging around, breaking weaker bodies and tearing wings to shreds.

Moreover, if large dragonflies are to be killed in a cyanide bottle along with much smaller ones, they should first be stilled (deactivated) by turning the head once around on its pivot. This may easily be done under the light pressure of a finger. It puts the brain out of control. Though this treatment may seem a bit cruel, it probably is less so than allowing the insect to die more slowly in the bottle, unaided.

A killing bottle should be kept dry and clean. If specimens get knocked into the water while collecting, they should be dried before being put in with the others.

There are other ways and means of collecting adult dragonflies. As a last resort, some collectors use a gun loaded with dust shot to bring down the biggest ones. They obtain a fair proportion of usable specimens.

Wary males may be lured within reach of a net by tethering a female to a stick, holding her out over the water with one hand and swinging a light net with the other.

Perching species will often come to a reed tip set just offshore, if it offer a good point of outlook over the water. A better place to perch may be a sufficient lure. The collector waiting beside it must have patience to keep stock-still until time for swinging his net.

Good specimens may sometimes be taken from spider webs, but usually only the stoutest species will be found there in fit condition for preservation. Sometimes the big darners are downed in a storm and may be

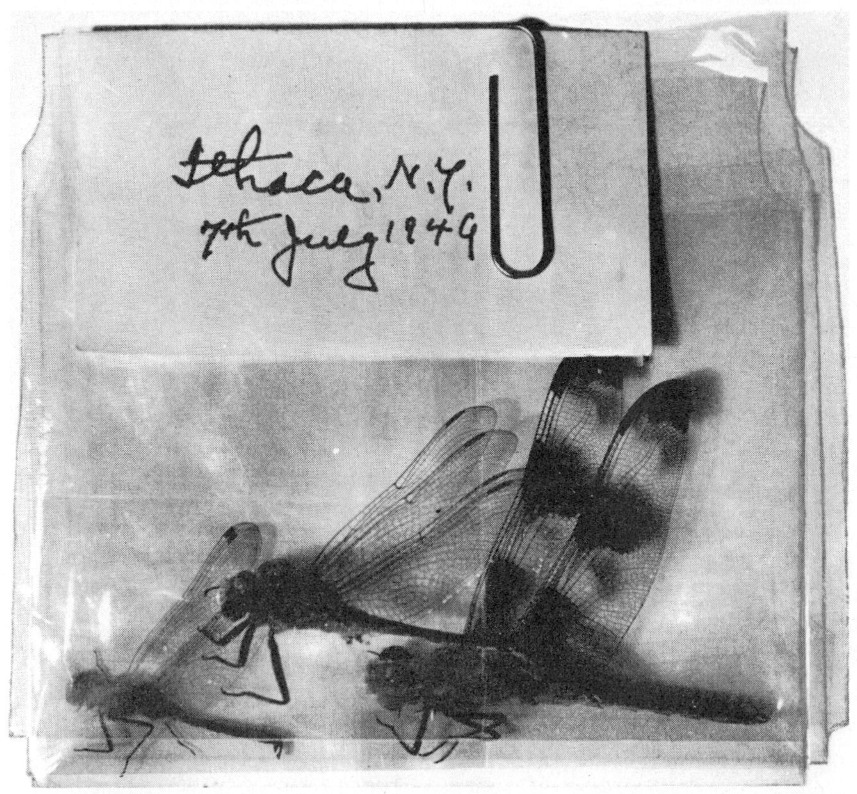

Fig. 26. Three dragonflies in separate cellophane envelopes, bunched and clipped together for safe shipment. Packed in a stout box between layers of sheet cotton, there is no breakage of specimens.

found in the drift line on a lake shore. A sun shower will sometimes send a large number of *Gomphaeschna furcillata* to alight near together on the trunk of a single tree, where they may be very approachable and whence they may be picked by hand.

There are times of abundance when dragonflies seem to be swarming, large numbers together, all or nearly all one species; these are the opportunities to seize if one is building a collection—golden opportunities to get specimens for exchange with other collectors. When a goodly number of dragonflies are flying together and have to watch each other as well as the net-swinging enemy, they are more easy to approach than when flying alone.

The large dragonflies require open areas for their aerial rounds of flight. A lane through the woods is better than little openings in the woods. On the lee side of a brushy fence row, collecting is better on a windy day, for the wind blows small insects out on that side. Freshly blossoming flowers attract flower-visitors, and the seekers after pollen and nectar attract the dragonflies.

Sometimes many dragonflies emerge at nearly the same time from a small body of water, and the shore vegetation may seem to be full of them. Then is the time to collect *with discrimination*. When tenerals flush at every step forward, the beginner is tempted to swing his net and fill his bottles (and later, his boxes) with a lot of disheveled and unattractive specimens. The weak young things are easily approached, do not fly far when flushed, and are readily taken.

Dragonflies flock to their feeding grounds. When mayflies swarm at dusk on the shore of lake or river, the big late-flying darners may be found among them, greedily foraging. And throughout the summer, at the same time of day, these big darners will gather about the bee yard, swooping about the hives and capturing the bees that are then returning heavy-laden from their last trip out.

Success in collecting dragonflies comes with knowledge of their haunts and habits,* their times and seasons and hours of flight. A few species (Gynacanthas and Neurocordulias) fly only in the twilight.

It is not difficult to make a collection of adult dragonflies. No great amount of equipment is needed. The use of cellophane envelopes for preservation and storage of specimens has greatly simplified the task and lessened its cost. The hardest task is to collect the living insects. It is perhaps also the most exhilarating, and is not lacking in basic educational value.

Dark cage.—It is hard to keep dragonflies alive and in good condition for even a short time in close confinement, for they will struggle to escape, and will soon batter their eyes and tear their wings. Often one desires to keep a nice specimen alive to await conditions favorable for taking a

* Many items of information on the habits of dragonflies (manner of flying, egg laying, and transformation), published in 1929 in Needham and Heywood's *Handbook of the Dragonflies of North America*, are not repeated in this volume. A finding list for these items was placed as a footnote to page 21 of that work. Since that material is already available in print, we have chosen to give more space here to descriptions and illustrations of species, adding only a few items concerning habits, gleaned from new observations and from more recent literature. The *Handbook* still has a place in libraries as a historic record and as a repository for such information.

photograph of it. We have used successfully the simple two-piece dark chamber shown in figure 27.

Each half is made of two parts: a small stick of wood and a rectangular sheet of heavy opaque black paper. The stick should be about five inches long, squared, and rounded across one end, and in diameter a trifle wider than the crosswise thickness of the body of the dragonfly. The sheet of

Fig. 27. A dark cage for keeping single dragonflies alive and inactive pending their use in making pictures from life.

paper should be as wide as the stick is long, and a little more than twice its length. The paper is doubled closely about the stick and glued or tacked to both sides of it. The other half of the chamber is similar. The stick is the same for both; the paper for the one that is to go outside may be a trifle longer. Each is open on two sides. In use, one is shoved over the other in reverse position, thus closing all sides and creating a dark chamber within—not absolutely dark, of course, but dark enough. Dragonflies are quiet in the dark.

The wings of the dragonfly to be kept alive are folded together above its back. It is then placed deep in the angle of one part with its feet against the stick (a long forceps will aid in placing it properly), and the other (cover) part is quickly shoved on astride.

Narrower cages will be required for smaller specimens; otherwise, they will manage to roll up in their wings and ruin them. If a dragonfly is to be held caged for more than a day, it should be taken out and fed (best taken out the way it is best put in, with a long forceps clasping the base of all its wings). If held by the wing tips and given some support to which it may cling by its feet, it will generally eat living flies when they are presented one at a time to its jaws.

Collecting nymphs.—Collecting the nymphs of dragonflies is a very different undertaking. It calls for different equipment, containers, and supplies.

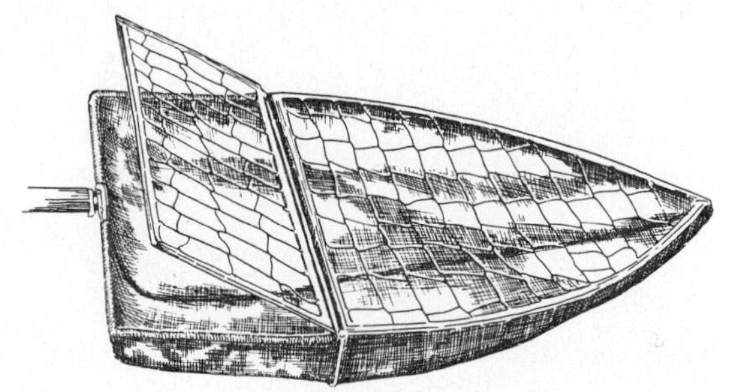

Fig. 28. Apron net for collecting nymphs. (Drawing by Esther Coogle.)

The apron net (fig. 28) is best for collecting from the water. It gathers and sifts at one operation. Its pointed front end penetrates the waterweeds and probes under loose bottom trash. Coarse trash cannot enter because of the covered top. The hinged rear section of the cover when lifted allows the contents to be examined where they lie, or to be dumped out at either rear angle. A poor substitute is the ordinary dip net, but even this will get plenty of nymphs if it is diligently used.

A bare garden rake may be effectively used to collect some of the larger dragonfly nymphs when there is plenty of trash lying loose on bottom mud. The trash is raked out on the bank, the nymphs with it, and they are picked up when they begin to show themselves by squirming. Some nymphs will be entangled in the trash, but many will be lost.

The largest nymphs are often caught by fishermen when seining for minnows among the reeds.

Dragonfly nymphs are best preserved in 70 per cent to 80 per cent alcohol; if it is stronger, it hardens them too much and makes them brittle. Stronger alcohol will be diluted by the water in the bodies of the specimens; therefore, it will not do to add so many specimens that the strength of the preservative is reduced much below 70 per cent.

Rearing nymphs.—The most generally useful form of rearing cage that we have found is the pillow cage, shown in figure 29. It is also the simplest.

It is made of a piece of wire cloth eighteen inches square. Two edges of the cloth are brought together, and folded over twice, making a "tinker's hem." This is pounded flat; then the whole is expanded to cylindric form by pressure of the hand placed inside. One end is closed and twice folded in like manner, and the fold is pounded flat. The other edge, preferably the one with a woven edge, is left open or given only a preliminary creasing until specimens are introduced at the waterside.

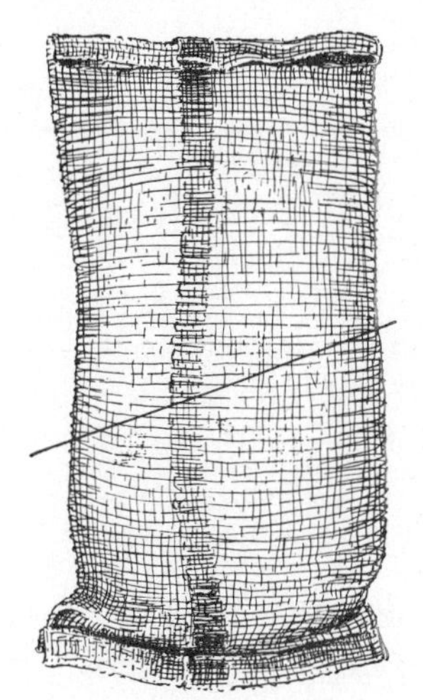

Fig. 29. Pillow cage for rearing dragonfly nymphs; to be set aslant in water, to the cross line.

The mesh of the cloth should be close enough to retain the specimens to be reared in it. Ordinary window-screen wire cloth is most used because it is everywhere available. Such a cage, one-third immersed in water, gives the nymphs a good foothold and plenty of room in which to expand their wings at transformation. It also gives protection from enemies. Moreover, it is "bug tight."

It can be used indoors; a row of pillow cages may be set in an aquarium, or in a deep pan of water in a sink, and a trickle of fresh water from a tap allowed to flow through. Grown nymphs, if introduced uninjured, will generally transform under these conditions.

Rearing cages should be inspected every morning for removal of any adults that may have appeared during the night. The teneral adult should not be lifted until the wings are expanded and dried a little, and until the first signs of color pattern appear. They should be put singly in paper bags, each with its own cast-off skin (*exuvia,* plural *exuviae*), and left there overnight until the color pattern is well developed, and the chitinous skin hardened a little; otherwise, pale and shriveled specimens will result.

If the nymphs when collected are to be carried home alive, something that will afford them a foothold should be put in the containers with them or the big ones may eat the little ones, and all will exhaust themselves clawing at the bare sides of smooth containers. If a pail of water have several half-length pillow cages set in it with tops left open, the nymphs may be sorted into them according to sizes while they are being collected in the field. The cages afford foothold and crevices into which they may creep for safety.

Nymphs may be shipped long distances alive, not in water but in a moist atmosphere such as may be provided by putting them in tight butter cartons, a few in each, with a little wet moss, and wrapping each carton in waxed paper. They will live for weeks in the summer heat of a desert, confined in a moist water bag of the sort used for cooling water by surface evaporation (Tinkham, *Ent. News,* 60:15, 1949).

Very valuable life-history material may be quickly and easily obtained at times when dragonflies are transforming. Occasionally, at the waterside, one may come upon nymphs crawling from the water or adults just quitting their old nymphal skins. Both are delicate and require careful handling.

The cast skins preserve perfectly the form of the nymphs and are equally available for specific determination. Moreover, they may be kept dry in the cellophane envelopes along with their adults. Legs will not break off too easily if they are dried outspread in the plane of the body before being placed in the envelopes.

Exuviae will usually be found head upward, their claws holding against gravity. Each should be lifted gently with a slight forward movement to release the claws. Each one that is taken in transformation should be kept isolated along with the adult that came out of it. Always keep the two together and apart from others, and so learn to associate them correctly. The dried exuviae may be carried safely in any small, strong containers, such as metal salve boxes.

If entirely dry when taken, the skins will have to be softened (as by a dip in hot water) before their legs and their posture can be arranged for permanent keeping. They should be dried under the light pressure of a sheet of paper to keep their legs down. Being thin, they dry very quickly. They must not be crushed.

A ready container in which to keep the soft adults singly while hardening is a stiff paper bag. A pocketful of grocers' small paper bags is little encumbrance in one's kit, and should be kept available.

Collecting eggs.—Eggs of Odonata for embryological study or other purpose are not difficult to obtain if females can be found in the field laying them. As there are females with ovipositor and females without, and consequently two ways of laying eggs, so also there are two ways of getting them. For example: females of the big darners such as *Anax* and *Aeschna* may sometimes be found sitting on a stem of cattail (*Typha*) at the surface of the water, busily engaged in puncturing the stem and laying an egg in each puncture. The place will be at the outer edge of the cattail patch, next to open water—a place from which she can see around and make a quick escape if danger threatens.

Females of many of the species that lack an ovipositor may be seen flying low along the edges of the pond, dropping repeatedly to strike the

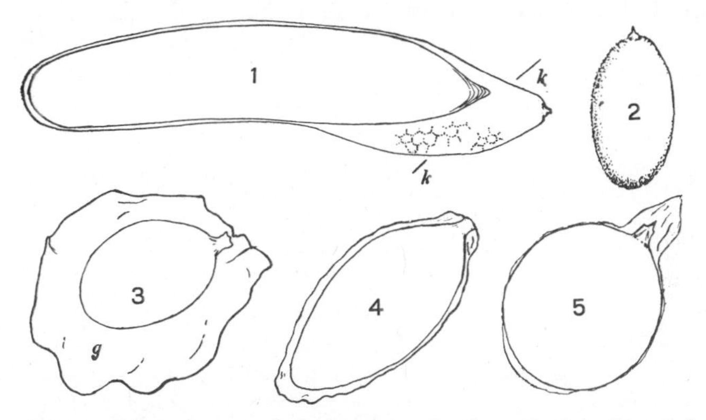

Fig. 30. Eggs of five genera of Anisoptera. 1, *Anax junius* (line *k–k* indicates depth to which egg is inserted in plant's tissues); 2, *Perithemis tenera;* 3, *Gomphus descriptus;* 4, *Hagenius brevistylus;* 5, *Tramea lacerata* (*g* indicates gelatinous envelope of egg). See also fig. 236, p. 392.

surface of the water with the tip of the abdomen. The female doing this is washing off a cluster of eggs at every descent. The eggs scatter when they enter the water and drift apart as they fall to the bottom. Silt sticks to their surface and they become at once undiscoverable.

If the female can be captured uninjured, and is held by the tips of her fore wings in one hand and dipped to touch the surface of the water in any convenient vessel held in the other hand, she will go on laying. Indeed, she will thus lay eggs in the small amount of water that the collector may hold in the hollow of his hand.

Almost any of the Libellulidae and Gomphidae will serve to show this method of laying eggs, but there are three known exceptions in our dragonflies: *Tetragoneuria, Micrathyria,* and *Cordulegaster*. *Tetragoneuria* lays eggs in long gelatinous strings (resembling those of toads but much smaller), and clusters of these strings are suspended at the surface, attached to plant stems. *Micrathyria aequalis* sits on the edge of a floating leaf, turns her abdomen down under the leaf, and lays her eggs in a broad patch on the under side of it. *Cordulegaster* has a blunt-tipped ovipositor that is used for depositing the eggs in the mixed mud, sand, and gravel at the water's edge. The weaving up and down while making punctures for the eggs is sketchily indicated in figure 22.

If "dated" eggs are wanted for the study of the nymphal moults, an *Anax junius* female will obligingly lay them in a place prepared for her convenience. It is necessary only to anchor a fresh (unused) stalk of cat-tail *aslant* (this is where her convenience is served) at the surface, and at the very forefront or a little out in front of a patch of erect emergent stems; and to provide a fresh stalk each day.

So much for the technique of collecting and handling specimens. We now turn to the procedure to be followed in Part Two.

III

PROCEDURE

The following abbreviations will be used in bibliographic references to literature of species and in the data on distribution.

AUTHORS

Burm.	Burmeister, Hermann
Calv.	Calvert, P. P.
Fabr.	Fabricius, J. C.
Garm.	Garman, Philip
Klct.	Kellicott, D. S.
Kndy.	Kennedy, C. H.
Linn.	Linnaeus, Carolus
McL.	McLachlan, Robert
Mrtn.	Martin, René
Mtgm.	Montgomery, B. E.
Mtk.	Muttkowski, R. A.
Ndm.	Needham, J. G.
N. & H.	Needham, J. G., and Heywood, Hortense Butler
Rbr.	Rambur, M. P.
Walk.	Walker, E. M.
Whts.	Whitehouse, F. C.
Wlsn.	Wilson, C. B.
Wmsn.	Williamson, E. B.
Wstf.	Westfall, M. J., Jr.
Zimm.	Zimmerman, E. C.

STATES AND PROVINCES

CANADA

Alta.	Alberta
B. C.	British Columbia
Man.	Manitoba
N. B.	New Brunswick
Nfld.	Newfoundland
NW. Terr.	Northwest Territory
N. S.	Nova Scotia
Ont.	Ontario
P. E. I.	Prince Edward Island
Que.	Quebec
Sask.	Saskatchewan

UNITED STATES*

Ala.	Alabama
Ariz.	Arizona
Ark.	Arkansas
Calif.	California
Colo.	Colorado
Conn.	Connecticut
Del.	Delaware
D. C.	District of Columbia
Fla.	Florida
Ga.	Georgia
Ill.	Illinois
Ind.	Indiana
Kans.	Kansas
Ky.	Kentucky
La.	Louisiana
Md.	Maryland
Mass.	Massachusetts
Mich.	Michigan
Minn.	Minnesota
Miss.	Mississippi
Mo.	Missouri
Mont.	Montana
Nebr.	Nebraska
Nev.	Nevada

* The abbreviations used for the states are those found in the current *U. S. Postal Guide.*

UNITED STATES—*Continued*

N. H.	New Hampshire
N. J.	New Jersey
N. Mex.	New Mexico
N. Y.	New York
N. C.	North Carolina
N. Dak.	North Dakota
Okla.	Oklahoma
Oreg.	Oregon
Pa.	Pennsylvania
R. I.	Rhode Island
S. C.	South Carolina
S. Dak.	South Dakota
Tenn.	Tennessee
Tex.	Texas
Vt.	Vermont
Va.	Virginia
Wash.	Washington
W. Va.	West Virginia
Wis.	Wisconsin
Wyo.	Wyoming

MEXICO

Baja Calif.	Baja California

ANTILLES

Dom. Rep.	Dominican Republic
P. R.	Puerto Rico

TITLES FREQUENTLY CITED

1942. Borror, Revis. Erythrodiplax = A Revision of the Libelluline Genus Erythrodiplax (Odonata), by Donald J. Borror. Ohio State Univ., Graduate School Studies, Contr. in Zool. and Ent., 4, Biol. Ser. 286 pp., 41 pls. Columbus, Ohio.

1839. Burm., Handb. = Handbuch der Entomologie: Gymnognatha, Libellulina, by Hermann Burmeister. 2:805–862. Berlin, Enslin.

1930. Byers, Odon. of Fla. = A Contribution to the Knowledge of Florida Odonata, by C. Francis Byers. Univ. of Fla. Publ., Biol. Sci. Ser., I. 327 pp., 11 pls. Gainesville, Fla.

1901–1908. Calv., B. C. A. = Biologia Centrali-Americana: Neuroptera, Odonata, by P. P. Calvert: *1901*, pp. 17–72; *1902*, pp. 73–128; *1903*, pp. 129–144; *1905*, pp. 145–212; *1906*, pp. 213–308; *1907*, pp. 309–404; *1908*, pp. 405–420. Pls. 2–10, 1 map. London, R. H. Porter and Dulau & Co.

1893. Calv., Odon. of Phila. = Catalogue of the Odonata (Dragonflies) of the Vicinity of Philadelphia, with an Introduction to the Study of This Group of Insects, by P. P. Calvert. Trans. Amer. Ent. Soc., 20: 152*a*–272, pls. 2–3.

1927. Garm., Odon. of Conn. = Guide to the Insects of Connecticut: Part V, The Odonata or Dragonflies of Connecticut, by Philip Garman. State Geol. and Nat. Hist. Surv. of Conn., Bull. 39. 331 pp., 22 pls. Hartford, Conn.

1861. Hagen, Syn. Neur. N. Amer. = Synopsis of the Neuroptera of North America, by Hermann Hagen, with a List of the South American Species. 347 pp. Washington, Smithsonian Inst.

1917–1923. Howe, Odon. of N. Eng. = Manual of the Odonata of New England, by R. Heber Howe, Jr. Mem. Thoreau Mus. Nat. Hist., II. 138 pp., figs.

1899. Klct., Odon. of Ohio = The Odonata of Ohio: A Descriptive Catalogue of the Dragonflies Known in Ohio, with Keys for Their Determination. A Posthumous Paper, by D. S. Kellicott. Ohio State Acad. Sci., Spec. Paper no. 2. 114 pp. Wooster, Ohio.

1932. Klots, Odon. of P. R. = Insects of Porto Rico and the Virgin Islands, Odonata or Dragonflies, by Elsie Broughton Klots. Scientific Survey of Porto Rico and the Virgin Islands, XIV. 107 pp., 7 pls. New York, N.Y. Acad. Sci.

1908–1910. Mrtn., Coll. Selys Aeschnines = Aeschnines, by René Martin. Collections Zoologiques du Baron Edm. de Selys Longchamps: *1908*, pp. 1–84; *1909*, pp. 85–156; *1910*, pp. 157–223. 6 col. pls. Brussels, Hayez, Impr. des Académies.

1906. Mrtn., Coll. Selys Cordulines = Cordulines, by René Martin. Collections zoologiques du Baron Edm. de Selys Longchamps. 98 pp., 3 col. pls. Brussels, Hayez, Impr. des Académies.

1908. Mtk., Odon. of Wis. = Review of the Dragon-Flies of Wisconsin, by R. A. Muttkowski. Bull. Wis. Nat. Hist. Soc., 6:57–127. 2 pls.

1929. N. & H., Handb. = A Handbook of the Dragonflies of North America, by James G. Needham and Hortense Butler Heywood. 378 pp., illus. Springfield, Ill., C. C. Thomas.

1842. Rbr., Ins. Neur. = Histoire naturelle des insectes neuroptères, by M. P. Rambur. 534 pp. Paris, Libraire Encyclopédique de Roret.

1909–1916. Ris, Coll. Selys Libell. = Libellulinen, by F. Ris. Collections zoologiques du Baron Edm. de Selys Longchamps: 1. *1909*, pp. 1–120; 2. *1909*, pp. 121–244; 3. *1910*, pp. 245–384; 4. *1911*, pp. 385–528; 5. *1911*, pp. 529–700; 6. *1912*, pp. 701–836; 7. *1913*, pp. 837–964; 8. *1913*, pp. 965–1042; 9. *1916*, pp. 1043–1245. Illus. Brussels, Hayez, Impr. des Académies.

1854–1878. Selys, Bull. Acad. Belg. = Synopsis des Gomphines and Synopsis des Cordulines, by Edm. de Selys Longchamps. Bull. de l'Académie royale des sciences de Belgique:

- *1854*. Synopsis des Gomphines, 21(2):23–112.
- *1859*. Additions au Synopsis des Gomphines, (2)7:530–552.
- *1869*. Secondes Additions au Synopsis des Gomphines, (2)28:168–208.
- *1871*. Synopsis des Cordulines, (2)31:238–316, 519–565.
- *1873*. Troisièmes Additions au Synopsis des Gomphines, (2)35:732–774.
- *1873*. Appendices aux Troisièmes Additions et liste des Gomphines, décrites dans le synopsis et ses trois additions, (2)36:492–531.
- *1874*. Additions au Synopsis des Cordulines, (2)37:16–34.
- *1878*. Secondes Additions au Synopsis des Cordulines, (2)45:183–222.
- *1878*. Quatrièmes Additions au Synopsis des Gomphines, (2)46:408–698.

1857. Selys, Mon. Gomph. = Monographie des Gomphines, by Edm. de Selys Longchamps, H. A. Hagen collaborating. 460 pp., 23 pls., fold. tables. Brussels and Leipzig, Nuquardt; Paris, Roret.

1912. Walk., N. Amer. Aeshna = The North American Dragonflies of the Genus Aeshna, by E. M. Walker. Univ. of Toronto Studies, Biol. Ser., 11. 213 pp., 28 pls.

1925. Walk., N. Amer. Somatochlora = The North American Dragonflies of the Genus Somatochlora, by E. M. Walker. Univ. of Toronto Studies, Biol. Ser., 26. 202 pp., 35 pls.

1941. Whts., Odon. of B. C. = British Columbia Dragonflies (Odonata), with Notes on Distribution and Habits, by F. C. Whitehouse. The American Midland Naturalist, 26:488–557.

1900. Wmsn., Odon. of Ind. = The Dragonflies of Indiana, by E. B. Williamson. Report of State Geologist. Pp. 231–333, 7 pls. Indianapolis.

Elsewhere we have in the main followed the standard practices of the *Zoological Record* in abbreviating titles.

OTHER PRACTICES

MAINLY SPACE-SAVING DEVICES

1. All measurements are in millimeters except when otherwise stated (25 mm. = 1 inch, approx.) The measurement for abdomen given under each species includes appendages.

2. In the descriptions of species, Arabic numerals are used to designate the segments (1 to 10) of the abdomen; often also, in keys and tables, to designate segments of tarsi and antennae. This is not only to save space, but also because Arabics catch the eye more quickly, making easier the comparison of specimens and descriptions.

3. The solidus (/) is used often as a substitute for the words "in fore and hind wings, respectively," or their equivalent. Thus, for example: "Cells in the triangle 3/1" briefly states that these cells are 3 in the fore wing and 1 in the hind wing. This device is oftenest used in the tables, where space-saving is most imperative.

4. In counting crossveins, the brace vein of the stigma is not counted among the "crossveins under the stigma"; and the anal crossing (Ac) is counted as a "cubito-anal crossvein"; it may well have been only that in the beginning.

5. In counting rows of cells in various wing areas, where rows are often obscure, there must be at least three cells in a direct line to be counted as a row.

6. Throughout Part Two the name of the genus under consideration will be found at the top of the right-hand page head, and the names of the species in each genus are placed in alphabetic order.

7. Personal names of describers of valid species and genera are all in one list beginning on page 54, where the names are first given.

8. As an aid to the pronunciation of scientific names in that list, accent is indicated by placing a sign of length over the vowel of the accented syllable; length, arbitrarily reduced to two categories, long and short; as, for example, ē, long, and ĕ, short (or at least not long).

9. Italics are used in many of the tables to indicate *prevailing numbers* in a variant series, as, for example, crossveins 3, *4*, 5; or crossveins 3, *4*, *5*, 6, when one or two numbers run far ahead of the others.

10. To save space in the tables, differences in characters compared are often arbitrarily condensed into two or three classes. Fine points in coloration are thus disregarded.

11. Distribution is broadly stated in the tables of species: N, S, E, W; NE, Northeast; N, E, North and East, without other punctuation; C, central; G, general.

12. The measuring of minute parts is time-consuming, and we have found that a rough estimate of size in comparable parts, expressed in tenths of one of the parts, taken as a standard, better answers the needs of this book. It also leaves out all complications that arise from differences in size of the specimens compared. It is used extensively in the tables for conciseness of statement.

13. When certain structures (such as the genital hamules of the male) are shown in a succession of similar figures in Part Two, there is no need to burden the legends of the figures by repetition of the names of those parts. Therefore the legends are reduced to the bare names of the species shown together. Read the names from left to right.

14. In order that the venation figures may be strictly comparable, males of clear-winged species have been selected to represent each genus. The venation as seen in females is shown only in figure 10. Aside from the basal anal triangle of the male, the sexes differ only in that the female lags a little in clearance of crossveins and has a few more of them.

15. The photographs of many nymphs were of necessity taken from alcoholic specimens, or, when none of these were available, from cast skins (*exuviae*); so of course they do not do justice to coloration, though they show well enough the structural characters cited.

SUGGESTIONS TO USERS OF THIS MANUAL

Characters stated under family and genus headings are not often repeated under the description of the species; they are only amplified there. Therefore, be sure you are in the right family and genus before going on to determine the species.

In counting cells or crossveins, examine the wings of both sides for concurrence or disagreement. In general it may be said that the shape of an enclosure is much more constant than the number of enclosed crossveins. Remember that color characters change with aging, that the pattern is more constant than the depth, and that extreme pruinosity may obscure the pattern.

Anyone who wishes to study dragonflies should equip himself with a good pocket lens, magnifying at least ten diameters, for often the characters that distinguish species cannot be seen without magnification.

If in using the keys a picture is needed to supplement the wording of a rubric, references to an illustration will generally be found under any name to which the rubric is leading the way, down the right-hand margin of the key. So, look for a figure under that name, remembering that generic names are given in the right-hand page head throughout Part

Two, and species names are in alphabetic order under the genus. When an illustration, cited by cross reference, is very distant from its citation in text, we have added the page number to make it easy to find.

A beginner may be told that in locating spots and rings on the abdomen of the adult, or hooks and spines on that of the nymph, it is easier to count abdominal segments from the rear *forward,* since the end segments are well exposed and there are always ten segments.

Do not expect always to find exact agreement of your specimens with all the characters stated in keys and tables, for characters vary more or less, and accidents cause artifacts. We could not hope to cover all the variations. Let preponderance of evidence determine your decisions, trusting photographs rather than words, and at the last trusting what several specimens show.

As to the data on distribution records: the last column in each table of species (adults) gives merely broad hints as a first aid to the user in finding the names of his specimens. At the end of the description of most species we give the record by states in the hope of encouraging the publication of state lists. (We include all the published records we have found that we consider dependable. These are manifestly very incomplete.)

LIST OF GENERA AND SPECIES TREATED IN THIS MANUAL

Genera appear under their respective families and subfamilies; species are arranged alphabetically under genera and subgenera. Names of their describers are given with them; also markings (long –, or short ◡) on the vowel of the accented syllable in each name, as an aid to pronunciation.

O-DO-NĀ-TA Fabricius
An-i-sŏp-te-ra Selys

PET-A-LŪ-RI-DAE Needham

Ta-chŏ-pter-yx Selys
1. thō-rey-i Hagen

Tan-y̆-pter-yx Kennedy
2. hă-gen-i Selys

COR-DU-LE-GAS-TĔR-I-DAE Calvert

Cor-du-le-găs-ter Leach
3. di-a-dē-ma Selys
4. di-ăs-ta-tops Selys
5. dor-sā-lis Hagen
6. er-rō-ne-us Hagen
7. fas-ci-ā-tus Rambur
8. ma-cu-lā-tus Selys
9. ob-lī-quus Say
10. sāy-i Selys

GŎM-PHI-DAE Rambur

A-phy̆l-la Selys
11. am-bĭ-gu-a Selys
12. ca-raī-ba Selys
13. pro-trăc-ta Selys
14. wĭl-liam-son-i Gloyd

Drō-mo-gŏm-phus Selys
15. ar-mā-tus Selys
16. spi-nō-sus Selys
17. spo-li-ā-tus Hagen

Er-pe-to-gŏm-phus Selys
18. cŏl-u-ber Williamson
19. com-pŏs-i-tus Hagen
20. cro-ta-lī-nus Hagen
21. des-ig-nā-tus Hagen
22. di-a-dō-phis Calvert
23. lam-pro-pĕl-tis Kennedy
24. nā-trix Williamson

Gŏm-phoi-des Selys
25. ăl-bright-i Needham
26. stig-mā-tus Say

Gŏm-phus s. lat. Leach

Ār-i-gom-phus Needham
27. cor-nū-tus Tough
28. fŭr-ci-fer Hagen
29. lĕn-tu-lus Needham
30. măx-well-i Ferguson
31. păl-li-dus Rambur
32. sub-me-di-ā-nus Williamson
33. vil-lō-si-pes Selys

Gom-phū-rus Needham
34. a-dĕl-phus Selys
35. con-săn-guis Selys
36. crăs-sus Hagen
37. dil-a-tā-tus Rambur
38. ex-tĕr-nus Hagen
39. fra-tĕr-nus Say
40. hȳ-bri-dus Williamson
41. li-ne-ā-ti-frons Calvert
42. mo-dĕs-tus Needham
43. văs-tus Walsh
44. ven-tri-cō-sus Walsh

Gŏm-phus Leach
45. aus-trā-lis Needham
46. bo-re-ā-lis Needham
47. brĭm-ley-i Muttkowski
48. ca-vil-lā-ris Needham
49. con-fra-tĕr-nus Selys
50. de-scrĭp-tus Banks
51. di-mi-nū-tus Needham
52. e-xī-lis Selys
53. flā-vo-cau-dā-tus Walker
54. grăs-li-nĕl-lus Walsh
55. hŏd-ges-i Needham
56. kū-ri-lis Hagen
57. lĭ-vi-dus Selys
58. mi-li-tā-ris Hagen
59. min-ū-tus Rambur
60. ō-kla-ho-mĕn-sis Pritchard
61. quăd-ri-co-lor Walsh
62. spi-cā-tus Hagen
63. wĭl-liam-son-i Muttkowski

Hȳ-lo-gŏm-phus Needham
64. ab-bre-vi-ā-tus Hagen
65. brĕ-vis Hagen
66. păr-vi-dens Currie
67. vi-rĭd-i-frons Hine

Sty-lū-rus Needham
68. am-nĭ-co-la Walsh
69. in-tri-cā-tus Hagen
70. ī-vae Williamson
71. laŭ-rae Williamson
72. no-tā-tus Rambur
73. ol-i-vā-ce-us Selys
74. pla-gi-ā-tus Selys
75. po-tu-lĕn-tus Needham
76. scŭd-der-i Selys
77. spī-ni-ceps Walsh
78. tŏwnes-i Gloyd

Ha-gē-ni-us Selys
79. brev-i-stȳ-lus Selys

Lăn-thus Needham
80. al-bi-stȳ-lus Hagen
81. păr-vu-lus Selys

Ŏc-to-gŏm-phus Selys
82. spe-cu-lā-ris Hagen

Ō-phi-o-gŏm-phus Selys
83. a-nŏ-ma-lus Harvey
84. a-ri-zō-ni-cus Kennedy
85. as-pĕr-sus Morse
86. bī-son Selys
87. ca-ro-lī-nus Hagen
88. că-ro-lus Needham
89. co-lu-brī-nus Selys
90. ed-mŭn-do Needham
91. hŏwe-i Bromley
92. main-ĕn-sis Packard
93. mon-tăn-us Selys
94. mŏr-ri-son-i Selys
95. ne-va-dĕn-sis Kennedy
96. oc-ci-dĕn-tis Hagen
97. ru-pin-su-lĕn-sis Walsh
98. se-vē-rus Hagen

Prō-gŏm-phus Selys
99. a-la-chu-ĕn-sis Byers
100. bo-re-ā-lis McLachlan
101. clĕn-don-i Calvert
102. ĭn-te-ger Hagen
103. ob-scū-rus Rambur
104. se-rē-nus Hagen
105. zĕ-phy-rus Needham

AĔSCH-NI-DAE Selys

Aĕsch-na Fabricius
106. ă-ri-da Kennedy
107. cal-i-fōr-ni-ca Calvert
108. ca-na-dĕn-sis Walker
109. clĕ-psy-dra Say
110. con-strĭc-ta Say
111. dū-ges-i Calvert
112. e-re-mī-ta Scudder
113. in-tĕr-na Walker
114. in-ter-rŭp-ta Walker
115. jŭn-ce-a Linnaeus
116. li-ne-ā-ta Walker
117. mănn-i Williamson
118. mŭl-ti-co-lor Hagen
119. mu-tā-ta Hagen
120. ne-va-dĕn-sis Walker
121. ŏc-ci-den-tā-lis Walker
122. pal-mā-ta Hagen
123. psī-lus Calvert
124. sep-tĕn-tri-o-nā-lis Burmeister
125. sit-chĕn-sis Hagen
126. sub-ărc-ti-ca Walker
127. tu-bĕr-cu-lĭ-fe-ra Walker
128. um-brō-sa Walker
129. ver-ti-cā-lis Hagen
130. wălk-e-ri Kennedy

Ā-nax Leach
131. a-ma-zĭ-li Burmeister
132. jū-ni-us Drury
133. lŏn-gi-pes Hagen
134. wăl-sing-ham-i McLachlan

Ba-si-aĕsch-na Selys
135. ja-nā-ta Say

Boy-ĕr-i-a McLachlan
136. graf-i-ā-na Williamson
137. vi-nō-sa Say

Cor-yph-aĕsch-na Williamson
138. ad-nĕ-xa Hagen
139. ĭn-gens Rambur
140. lū-te-i-pĕn-nis Burmeister
141. vī-rens Rambur

Ep-i-aĕsch-na Hagen
142. hē-ros Fabricius

Gomph-aĕsch-na Selys
143. an-ti-lō-pe Hagen
144. fur-cil-lā-ta Say

Gy-na-cănth-a Rambur
145. e-re-ā-gris Gundlach
146. ner-vō-sa Rambur

Nas-i-aĕsch-na Selys
147. pen-ta-căn-tha Rambur

Ŏp-lon-aĕsch-na Selys
148. ar-mā-ta Hagen

Trī-a-cănth-a-gȳ-na Selys
149. sĕp-ti-ma Selys
150. trī-fi-da Rambur

LI-BEL-LŪ-LI-DAE Selys

MA-CRO-MĪ-NAE Needham

Dĭ-dy-mops Rambur
151. flo-ri-dĕn-sis Davis
152. trans-vĕr-sa Say

Ma-crō-mi-a Rambur
153. ăl-le-ghā-ni-ĕn-sis Williamson
154. an-nu-lā-ta Hagen
155. ca-de-rī-ta Needham
156. geor-gī-na Selys
157. ĭl-li-noi-ĕn-sis Walsh
158. mag-nĭ-fi-ca McLachlan
159. mar-ga-rī-ta Westfall
160. pa-cĭ-fi-ca Hagen
161. rĭck-er-i Walker
162. taē-ni-o-lā-ta Rambur
163. wa-bash-ĕn-sis Williamson

COR-DU-LĪ-NAE Selys

Cor-dū-li-a Leach
164. shŭrt-leff-i Scudder

Dō-ro-cor-dū-li-a Needham
165. lĕ-pi-da Hagen
166. lī-be-ra Selys

Ĕp-i-cor-dū-li-a Selys
167. prĭn-ceps Hagen
168. re-gī-na Hagen

Hē-lo-cor-dū-li-a Needham

169. sē-ly-si-i Hagen
170. ūh-le-ri Selys

Neū-ro-cor-dū-li-a Selys

171. ă-la-ba-mĕn-sis Hodges
172. clā-ra Muttkowski
173. mo-lĕs-ta Walsh
174. ob-so-lē-ta Say
175. vir-gin-i-ĕn-sis Davis
176. xăn-tho-sō-ma Williamson
177. ya-măs-kan-ĕn-sis Provancher

So-măt-o-chlō-ra Selys

178. al-bi-cĭnc-ta Burmeister
179. căl-vert-i Williamson & Gloyd
180. cĭn-gu-lā-ta Selys
181. ē-lon-gā-ta Scudder
182. en-sĭ-ge-ra Martin
183. fi-lō-sa Hagen
184. for-ci-pā-ta Scudder
185. frănk-lin-i Selys
186. geor-gi-ā-na Walker
187. hin-e-ān-a Williamson
188. hud-sŏ-ni-ca Hagen
189. in-cur-vā-ta Walker
190. kĕn-ne-dy-i Walker
191. li-ne-ā-ris Hagen
192. mī-nor Calvert
193. ō-zar-kĕn-sis Bird
194. prŏ-vo-cans Calvert
195. săhl-berg-i Trybom
196. sĕ-mi-cĭr-cu-lā-ris Selys
197. sep-tĕn-tri-o-nā-lis Hagen
198. te-ne-brō-sa Say
199. wălsh-i-i Scudder
200. whīte-house-i Walker
201. wĭl-liam-son-i Walker

Tĕt-ra-go-neūr-i-a Hagen

202. cā-nis McLachlan
203. cy-no-sū-ra Say
204. mō-ri-o Muttkowski
205. pe-te-chi-ā-lis Muttkowski
206. sem-i-ā-que-a Burmeister
207. sē-pi-a Gloyd
208. spi-nĭ-ge-ra Selys
209. spi-nō-sa Hagen
210. stĕl-la Williamson
211. wĭl-liam-son-i Muttkowski

Wĭl-liam-sōn-i-a Davis

212. flĕtch-er-i Williamson
213. lĭnt-ner-i Hagen

LI-BEL-LU-LĪ-NAE Selys

Brach-y-mē-si-a Kirby

214. fur-cā-ta Hagen

Brech-mor-hō-ga Kirby

215. mĕn-dax Hagen

Can-nă-cri-a Kirby

216. gră-vi-da Calvert
217. hĕr-bi-da Gundlach

Can-nă-phi-la Kirby

218. fu-nē-re-a Carpenter
219. in-su-lā-ris Kirby

Ce-lĭth-e-mis Hagen

220. a-măn-da Hagen
221. bĕr-tha Williamson
222. e-lī-sa Hagen
223. ep-o-nī-na Drury
224. fas-ci-ā-ta Kirby
225. le-o-nō-ra Westfall
226. măr-tha Williamson
227. mon-o-me-laē-na Williamson
228. or-nā-ta Rambur
229. vĕr-na Pritchard

Dȳ-the-mis Hagen

230. fū-gax Hagen
231. ru-fi-nĕr-vis Burmeister
232. vē-lox Hagen

Er-ȳth-e-mis Hagen

233. ăt-ta-la Selys
234. col-lo-cā-ta Hagen
235. haē-mat-o-găs-tra Burmeister
236. pe-ru-vi-ā-na Rambur
237. ple-bē-ja Burmeister
238. sĭm-pli-ci-cŏl-lis Say

Er-ȳth-ro-dĭ-plax Brauer

239. ab-jĕc-ta Rambur
240. ba-sā-lis Kirby
241. be-re-nī-ce Drury
242. con-nā-ta Burmeister
243. fĕr-vi-da Erichson
244. fu-nē-re-a Hagen
245. fŭs-ca Rambur
246. jus-tin-i-ā-na Selys

Er-ȳth-ro-dĭ-plax *–Continued*
247. min-ŭs-cu-la Rambur
248. naē-va Hagen
249. um-brā-ta Linnaeus
250. ūn-i-ma-cu-lā-ta De Geer

Id-i-ăt-a-phe Cowley
251. cu-bĕn-sis Scudder

La-dō-na Needham
252. de-pla-nā-ta Rambur
253. e-xŭ-sta Say
254. jū-li-a Uhler

Lep-thē-mis Hagen
255. ves-ĭ-cu-lō-sa Fabricius

Leu-cor-rhīn-i-a Brittinger
256. bo-re-ā-lis Hagen
257. frĭg-i-da Hagen
258. gla-ci-ā-lis Hagen
259. hud-sŏn-i-ca Selys
260. in-tăc-ta Hagen
261. pa-trĭ-cia Walker
262. prŏ-xi-ma Calvert

Li-bĕl-lu-la Linnaeus
263. au-ri-pĕn-nis Burmeister
264. a-xi-lē-na Westwood
265. co-măn-che Calvert
266. com-pŏ-si-ta Hagen
267. cro-ce-i-pĕn-nis Selys
268. cy-ā-ne-a Fabricius
269. flā-vi-da Rambur
270. fo-rĕn-sis Hagen
271. in-cĕs-ta Hagen
272. luc-tu-ō-sa Burmeister
273. neēd-ham-i Westfall
274. no-di-stĭc-ta Hagen
275. o-di-ō-sa Hagen
276. pul-chĕl-la Drury
277. quăd-ri-mă-cu-lā-ta Linnaeus
278. sa-tu-rā-ta Uhler
279. sem-i-fas-ci-ā-ta Burmeister
280. vī-brans Fabricius

Ma-cro-dĭ-plax Brauer
281. bal-te-ā-ta Hagen

Ma-cro-thē-mis Hagen
282. ce-lē-no Selys
283. in-e-qui-ŭn-guis Calvert
284. leu-co-zō-na Ris
285. pseu-dĭ-mi-tans Calvert

Mi-a-thȳ-ri-a Kirby
286. mar-cĕl-la Selys
287. sĭm-plex Rambur

Mic-ra-thȳ-ri-a Kirby
288. ae-quā-lis Hagen
289. dĕ-bi-lis Hagen
290. dĭ-dy-ma Selys
291. dis-sō-ci-ans Calvert
292. hă-gen-i Kirby

Nan-nŏ-the-mis Brauer
293. bĕl-la Uhler

Ŏr-the-mis Hagen
294. fĕr-ru-gĭ-ne-a Fabricius

Pach-y-dĭ-plax Brauer
295. lon-gi-pĕn-nis Burmeister

Pal-tŏ-the-mis Karsch
296. li-ne-ă-ti-pes Karsch

Păn-ta-la Hagen
297. fla-vĕs-cens Fabricius
298. hy-mē-ne-a Say

Per-ĭ-the-mis Hagen
299. do-mĭ-tia Drury
300. in-tĕn-sa Kirby
301. moō-ma Kirby
302. sĕ-mi-nole Calvert
303. tĕn-e-ra Say

Plăn-i-plax Muttkowski
304. san-gui-ni-vĕn-tris Calvert

Plă-the-mis Hagen
305. lȳ-di-a Drury
306. su-bor-nā-ta Hagen

Pseŭ-do-lē-on Kirby
307. su-pĕr-bus Hagen

Sca-pā-ne-a Kirby
308. fron-tā-lis Burmeister

Sym-pē-trum Newman
309. am-bĭg-u-um Rambur
310. ă-tri-pes Hagen
311. cal-i-fōr-ni-cum Walker
312. cos-tĭ-fe-rum Hagen
313. dā-nae Sulzer
314. fas-ci-ā-tum Walker
315. in-tĕr-num Montgomery
316. mă-di-dum Hagen
317. ob-trū-sum Hagen

Sym-pĕ-trum—*Continued*

318. oc-ci-den-tăl-e Bartenev
319. păl-li-pes Hagen
320. ru-bi-cŭn-du-lum Say
321. se-mi-cĭnc-tum Say
322. vi-cī-num Hagen

Tar-nē-trum Needham & Fisher

323. cor-rŭp-tum Hagen
324. il-lō-tŭm Hagen

Tau-rĭph-i-la Kirby

325. aus-trā-lis Hagen

Thŏl-y-mis Hagen

326. ci-trĭ-na Hagen

Trā-me-a Hagen

327. ab-do-mi-nā-lis Rambur
328. bi-no-tā-ta Rambur
329. ca-ro-lī-na Linnaeus
330. co-phȳ-sa Hagen
331. la-ce-rā-ta Hagen
332. o-nŭs-ta Hagen

Part Two

Systematic Classification

Suborder ANISOPTERA Selys

The chief distinguishing characteristics of the suborder have already been stated (p. 7) in contrast with those of Zygoptera. Other characteristics having been detailed and illustrated at length in Part One, we may now proceed to a consideration of the components of Anisoptera. A classification of the major divisions of this suborder adequate for the purposes of this book disposes of them in five families:

PETALURIDAE
CORDULEGASTERIDAE
GOMPHIDAE
AESCHNIDAE
LIBELLULIDAE

KEY TO THE FAMILIES AND SUBFAMILIES

ADULTS

1—Triangle about equally distant from arculus in fore and hind wing, and similar in shape; two antenodal crossveins thickened (fig. 31, A) 2

—Triangle nearer arculus in hind wing than in fore wing and of different shape; no thickened antenodal crossveins (fig. 31, B) .... LIBELLULIDAE 5

2—Front margin of labium with a median cleft (fig. 32) 3

—Front margin of labium entire 4

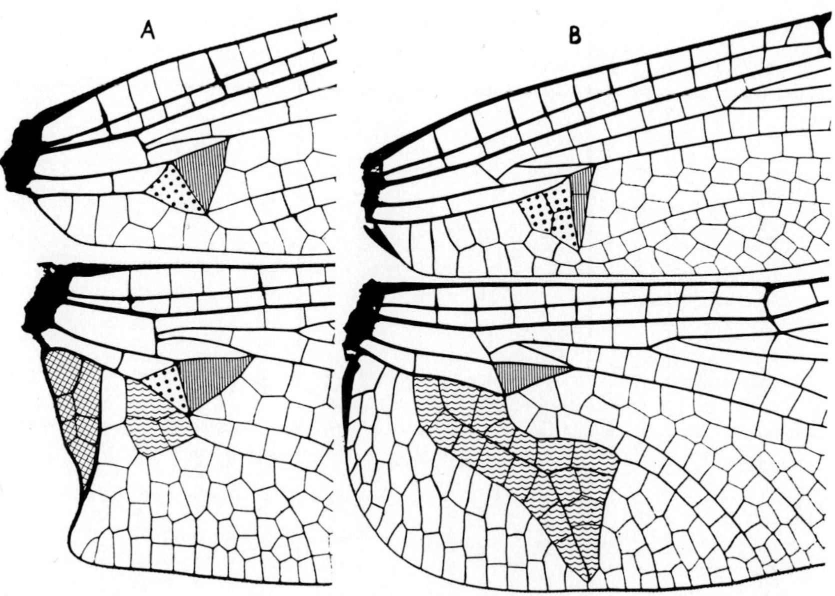

Fig. 31. Bases of dragonfly wings showing fundamentals of wing venation. A, *Ophiogomphus carolus* (family Gomphidae); B, *Erythemis simplicicollis* (family Libellulidae); both males. Triangles striated, subtriangles dotted, anal loops wavy lined, basal triangle cross-hatched.

3—Stigma long, linear, concave on posterior side; supported by a brace crossvein at or just before its inner end (fig. 37, p. 73) .... PETALURIDAE

—Stigma a little widened in middle, not concave on posterior side; no brace crossvein .... CORDULEGASTERIDAE

4—Eyes wide apart on top of head (fig. 33, A) .... GOMPHIDAE

—Eyes meeting in an eye-seam .... AESCHNIDAE

5—Triangle in hind wing about half as far from arculus as in fore wing; anal loop about as long as wide, its included cells not in two long rows (fig. 204, p. 332) .... MACROMINAE

—Triangle in hind wing at or very close to arculus; anal loop, if present, elongated, more or less foot-shaped, and divided lengthwise by a midrib into two rows of cells (fig. 31, B) 6

TABLE OF FAMILIES AND SUBFAMILIES

ADULTS

Families and subfamilies	*Labium cleft*[1]	*Compound eyes*[2]	*Wing triangles*[3]	*Hind wing triangle*[4]	*Thick. an. cvs.*[5]	*Midrib anal loop*	*Male auricles*[6]
PETALURIDAE	+	far apart	alike	far out	+	0	+
CORDULEGASTERIDAE	+	touching[7]	alike	far out	+	0	+
GOMPHIDAE	0	far apart	alike	far out	+	0	+
AESCHNIDAE	0	eye-seam	alike	far out	+	0	+[8]
LIBELLULIDAE							
MACROMINAE	0	eye-seam	unlike	nearer	0	0	+
CORDULINAE	0	eye-seam	unlike	opposite	0	+	+
LIBELLULINAE	0	eye-seam	unlike	opposite	0	+	0

[1]Cleft by a shallow median notch in tip: present (+), absent (0). (see fig. 32)

[2]Eyes widely separated, meeting at a point or broadly in an eye-seam.

[3]Triangles of fore and hind wing: form and direction of long axis.

[4]Triangle of hind wing: position in relation to arculus.

[5]Two antenodal crossveins much thicker than the others: present (+), absent (0).

[6]On sides of abdominal segment 2 in males only, and goes with presence of an anal triangle at base of hind wing.

[7]Or nearly so.

[8]Except in *Anax*.

6—Anal loop somewhat foot-shaped, but with little development of toe; males with a small auricle (fig. 4) on each side of abdominal segment 2, and an anal triangle in base of hind wing, to rear of which is an angulation of border (fig. 31, A) . . . . CORDULINAE

—Anal loop generally distinctly foot-shaped, with toe well developed; males without auricles, and inner angle (*tornus*) of hind wings rounded as in female (fig. 31, B) . . . . LIBELLULINAE

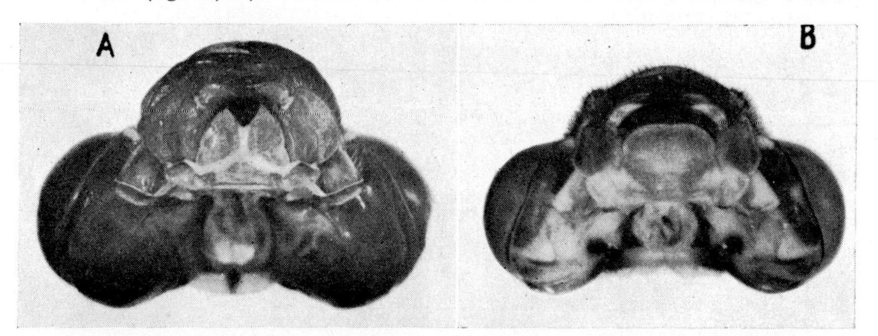

Fig. 32. Ventral view of heads of two adults. A, *Cordulegaster maculatus;* B, *Gomphus dilatatus;* especially for showing middle part of labium; cleft at its tip in A, entire in B.

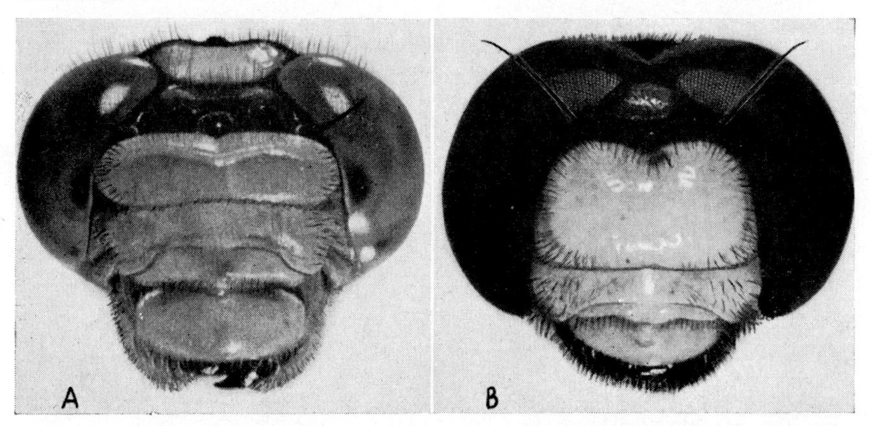

Fig. 33. Faces of two adult dragonflies to show compound eyes. Wide apart in A, *Gomphus villosipes;* meeting in an eye-seam in B, *Leucorrhinia intacta.*

NYMPHS

1—Mentum of labium flat or nearly so (fig. 15, C, p. 25) 2

—Mentum of labium spoon-shaped or mask-shaped, covering face up to eyes (fig. 17, p. 28) 4

2—Antennae 6- or 7-jointed; tarsi 3-3-3-jointed 3

—Antennae 4-jointed; tarsi 2-2-3-jointed . . . . GOMPHIDAE

3—Joints of antennae short, thick, hairy (fig. 36) . . . . PETALURIDAE

—Joints of antennae slender, bristle-like (fig. 169, p. 279) . . . . AESCHNIDAE

4—Lateral lobes of labium so deeply and irregularly cut on opposed margins as to form huge irregular teeth (fig. 17, p. 28) .... CORDULEGASTERIDAE

—Lateral lobes of labium less deeply cut, with smaller and more regular teeth LIBELLULIDAE 5

5—Prominent frontal horn on head; legs very long (fig. 205); abdomen strongly depressed, almost circular when seen from above; teeth on opposed margins of lateral lobes of labium large, with deep incisions between them . MACROMINAE

—No frontal horn on head (except in **Neurocordulia molesta**); legs shorter; abdomen less depressed, more cylindric; teeth on lateral lobes smaller, more numerous, with incisions shallower or even obsolete 6

6—Lateral caudal appendages generally more than half as long as inferiors; lateral spines of segment 9 longer than its dorsal length; middorsal hooks on abdomen often cultriform .... CORDULINAE*

TABLE OF FAMILIES AND SUBFAMILIES

NYMPHS

Families and subfamilies	*Joints*		*Labium*			*Ovp.*[6]
	Trs.[1]	*Ant.*[2]	*Shape*[3]	*Cleft*[4]	*Setae*[5]	
PETALURIDAE	3-3-3	6, 7	flat	+	0	+
CORDULEGASTERIDAE	3-3-3	7	mask	+	+	+
GOMPHIDAE	2-2-3	4	flat	0	0	0
AESCHNIDAE	3-3-3	7	flat	+	0[7]	+
LIBELLULIDAE						
MACROMINAE	3-3-3	7	mask	0	+	0
CORDULINAE	3-3-3	7	mask	0	+	±
LIBELLULINAE	3-3-3	7	mask	0	+	0

[1]Number of joints in front, middle, and hind tarsi.

[2]Last joint of antenna often much reduced.

[3]Shape of labium, reduced to two types: flat or nearly so, and mask-shaped or spoon-shaped; latter, when closed, covers face up to eyes.

[4]A median cleft or notch in margin of median lobe: present (+) or absent (0).

[5]Rows of strong raptorial bristles, each bristle set in a basal socket: present (+) or absent (0).

[6]Ovipositor — developed only in well-grown nymphs: present (+), absent (0), or variable (±).

[7]Present but weak in *Gynacantha* and *Triacanthagyna*.

* No single distinctive character can at present be named that will separate all the nymphs of Libellulinae from those of Cordulinae. When in doubt, try the keys to the genera in both subfamilies and compare specimens with the photographs of the nymphs of the two genera to which the keys lead.

—Lateral caudal appendages generally not more than half as long as inferiors; if lateral spines of 9 are longer than its middorsal length, then middorsal hooks on abdomen are not cultriform but more spinelike, or stubby, or altogether wanting .... LIBELLULINAE*

Family PETALURIDAE Needham

We begin with this ancient family of which, within our territorial limits, only two species are known to live today. Each species represents a different genus. Both are rather rare and very local in distribution: *Tanypteryx,* found in the mountains on the Pacific side of our continent; and *Tachopteryx,* found among the hills on the Atlantic side. Both seem to have been crowded out of the choicest places for dragonfly development and have made shift to live their long lives as nymphs in boggy woodland pools. Both are large dragonflies, stout of body and lacking in the finish of form and of coloration that goes with the specialization and increased numbers of the higher groups.

Other living genera of this family are found in the Antipodes, far-flung (along with primitive members of animals of other animal groups): *Petalura,* with three species in Australia; *Uropetala,* with one in New Zealand; and *Phenes,* with one in Chile. All these are but remnants of a great Petalurine fauna that flourished upon the earth in Solenhofen and Jurassic times. The abundance of their fossil remains shows that they were then the dominant dragonflies in the earth.

The Petaluridae are clear-winged dragonflies of rather rugged aspect. The eyes are wide apart on the top of the head. The labium is divided at the tip by a median cleft. The occiput is swollen in the rear. The thorax is compact, with its side plates only moderately aslant forward below. The abdomen is stout, with the last two segments of nearly equal length and very much shorter than the preceding segments. The genitalia of segment 2 in the male are little prominent, but the caudal appendages are large and strong. The female has a well-developed, stout, strongly upcurving ovipositor.

The venation of the wings, though strongly patterned, is very inconstant in minor details. There are two thickened antenodal crossveins; there is no basal subcostal crossvein. The stigma is long and narrowly linear, with its brace vein variously developed. There are two cubito-anal crossveins; occasionally, three. The triangles are similar in fore and hind wing (the latter, a little longer in the axis of the wing), and they are about equally remote from the arculus. The smaller subtriangle of the hind wing is generally without crossveins. The base of the hind wing is very broad. In the male there is a strongly developed anal triangle

* See note on p. 66.

that is generally three-celled. The anal crossing is situated at or beyond the base of vein A3. Their primitive character is indicated by the general coarseness of structure of the heavy bodies, by the widely separated eyes, by the cleft labium, by the moderate skewness of the thorax with the legs not shoved very far forward, by the upward inclination of the

Fig. 34. *Tanypteryx hageni,* adult male. (Collected at Swim, Oregon, by Mrs. Ruth C. Whitney.)

outer end of the triangles and subtriangles, by the narrowness and parallel-sidedness of the slender stigma, and by the general instability in number and position of the weak crossveins in the basal fourth of the wings.

The only considerable contribution to the knowledge of the adults of the North American species of this family is that of C. H. Kennedy (*Proc. U. S. Nat. Mus.,* 52:508–515, 1917).

The nymphs are sluggish creatures that wallow in the muck in seepage waters and in spring bogs. They have a rough-hewn exterior and short, somewhat twisted legs. The antennae are shorter than the head. They are six- or seven-jointed, and the joints are all thick, there being but slight differentiation between flagellum and base. The eyes are very prominent at the front angles of the squarish head. The labrum is short and stout, and has neither raptorial setae nor teeth. Its mentum is produced forward a little on the front border, where it is divided by a short, closed median cleft. The lateral lobes are stout and somewhat concave within; their blades are squarely truncate on the end, and the movable hook is strong and outstanding. The abdominal segments are all of about equal length, even the first and tenth being little shorter than the others.

KEY TO THE GENERA

Adults

1—Thorax black, spotted on front and sides with yellow; metathorax bearing a round hairy tubercle beneath; first and fourth or fifth antenodal cross-veins thickened; anal loop ill developed and very variable; subtriangle of fore wing one- or two-celled; anal angle of hind wing produced and rather sharply angulated in male .... **Tanypteryx**

—Thorax grayish in front, with carina narrowly black; sides lighter gray, with two oblique black stripes; metathorax with no round hairy tubercle beneath; first and sixth or seventh antenodals thickened; subtriangle of fore wing generally three-celled; anal angle of hind wing very obtusely angulated in male .... **Tachopteryx**

Nymphs

1—Antennae 6-jointed; lateral lobe of labium with minute extra hook on upper rim at base of movable hook .... **Tanypteryx**

—Antennae 7-jointed; no such extra hook present .... **Tachopteryx**

Genus TANYPTERYX Kennedy 1917

This genus includes one species in America and another in Japan. It is readily distinguishable by the presence, on the ventral side of the meta-thorax, of a large moundlike tubercle the rounded top of which is clothed with long whitish hair; also by the other characters stated in the key. In the wings the anal crossing is opposite or slightly beyond the base of vein A3. In the male the inferior caudal appendage is widely expanded at the tip, and distinctly three-lobed on its end.

The nymph of our species is still (1948) unknown; but the nymph of the Japanese species, *T. pryeri,* was described and figured by S. Asahina (*Mushi,* 19: 37, 38, 1949). Lacking the nymph of our American species, we copy Asahina's figures herewith. The nymph of *T. pryeri* is very much like the *Tachopteryx* nymph. Besides the characters stated in the key, it may be added that the postero-lateral angles of the middle abdominal segments are less outstanding, the front margin of the median lobe of the labium is less truncated, and the successive segments of the antenna taper a little more to the tip.

Tanypteryx hageni Selys

1879. Selys, C. R. Soc. Ent. Belg., 22:68 (in *Tachopteryx*).
1917. Kndy., Proc. U. S. Nat. Mus., 52:508 (figs.).
1929. N. & H., Handb., p. 54 (fig.).
1938. Ahrens, Ent. News, 49:10.
1938. Walk., Can. Ent., 70:145, 146, 149.
1941. Whts., Odon. of B. C., p. 508.
1947. Whitney, Ent. News, 58:103.

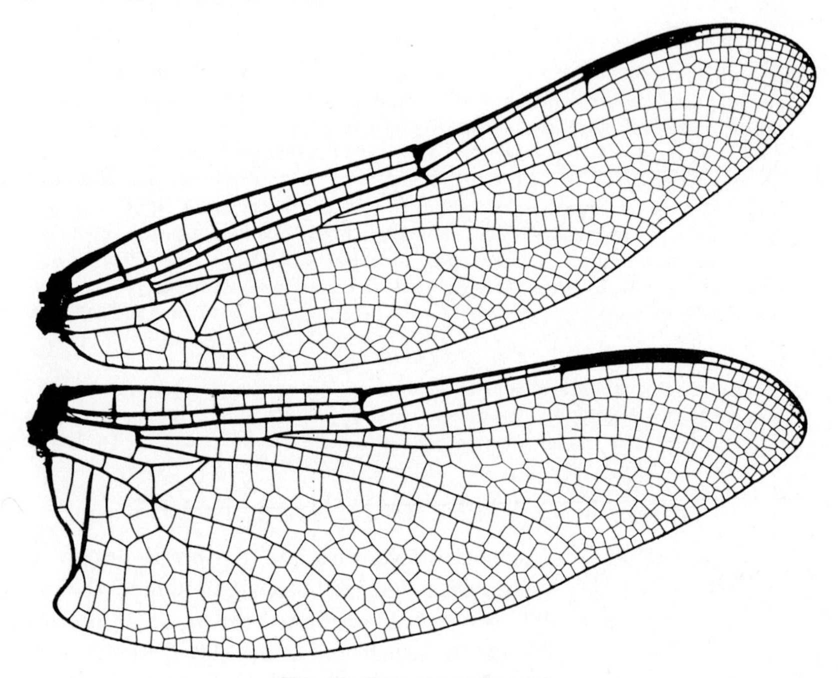

Fig. 35. *Tanypteryx hageni.*

Length 54–57 mm.; abdomen 40–44; hind wing 34–37.

A blackish dragonfly, marked on thorax with spots of light yellow and on abdomen with half rings of dull orange. Face all shining blackish below light yellow frons. A yellow spot on outer surface of each mandible. Top of head brown, beginning with a narrow base line across frons and extending rearward over whole of vertex and sloping front of occiput. On rear of slightly bilobed occiput are large twin spots of yellow.

Prothorax brown above, marked with two very unequal pairs of small spots of yellow. Synthorax blackish, with eight rather conspicuous spots in two longitudinal ranks, four spots on front and four on sides, those in lower rank being the larger; also, two pairs of yellow spots on dorsum between wing roots; all carinae black. Legs black.

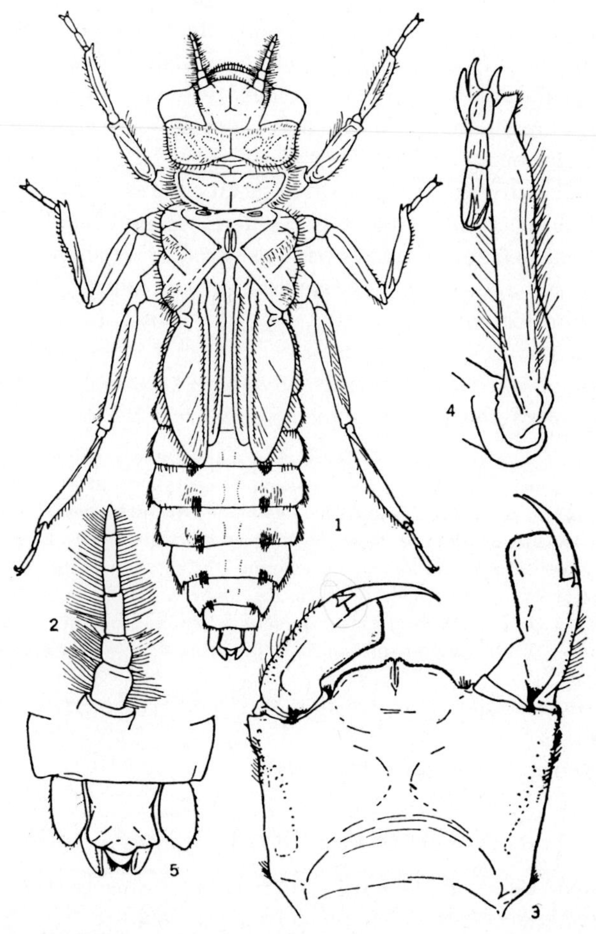

Fig. 36. Nymph of *Tanypteryx pryeri* (Japanese). 1, whole nymph; 2, an antenna; 3, mentum of labium (note short spur at base of movable hook); 4, a middle tibia with tarsus reflexed upon it to show its three terminal spurs; 5, caudal appendages in dorsal view. (From Asahina, in *Mushi*, vol. 19, pl. 7, 1949.)

Abdomen blackish, with paired orange spots on dorsum of segments 2 to 7; these spots, almost meeting on middorsal line, form more or less interrupted half rings. Yellow on segment 2 divided by a large basal triangle of black, and two small pear-shaped divergent streaks of black extend from it to rearward upon 3; also on 4 in female.

Venation of wings rather inconstant as to crossveins. Nodal crossveins 13–15:6–9/6–9:6–9, for example; second thickened antenodal varies in position from fourth to sixth, being oftenest 5/4; intermedian crossveins 6–7/3–4. Cells in triangle 1–2/1; in subtriangle 1–4/1; bordering outer side of triangle 2–4/2–3; anal loop weakly developed, of 2–5 cells; brace vein of stigma weakly developed, sometimes hardly recognizable.

Caudal appendages of male black, hardly longer than segment 10, or than inferior appendage; flattened to spatulate form, rather strongly divergent. Inferior appendage very broad, parallel-sided in its basal two-thirds, then broadly dilated to a three-lobed end, the outer lobes each upturned and abruptly tapering into a sharp terminal tooth. Viewed from beneath, apical margin of this appendage broadly W-shaped.

Caudal appendages (cerci) of female short and simple, close-laid, about as long as abbreviated segment 10. Ovipositor short, stout, and upcurving beyond tip of 10.

The habits of this rare dragonfly have been recorded by Mrs. Whitney in the paper cited above, from which we abstract the following:

> A number of *Tanypteryx* were to be seen in mating and feeding flights above the cat-tail swale at Swim [Oregon]. They seemed to be taking off from a "home base" on the warm sunny rocks that dotted the wet muddy trailway beside a white-painted post. Quite often one or more of them alighted on the sunny side of the post, and let me come close (within a foot or so) to examine them, when I did not have my net with me! One or two sat flattened down on the low rocks with wings and legs widely outspread. When they did take off, the flight was low and uneven. A height of perhaps twenty feet would be gained, and then the wing-flapping became gliding.

Distribution and dates.—CANADA: B. C.; UNITED STATES: Calif., Nev., Oreg., Wash.

June 21 (Oreg.) to August 30 (Oreg.).

Genus TACHOPTERYX Selys 1859

This genus consists of the single large Eastern species described below. It will be easily recognized by the characters stated in the key to genera, and by these additional: the inferior caudal appendage of the male is simply bilobed, with widely outspread and upturned tips; there are generally three distinct branches extending from the rear side of vein Cu2 to the wing margin; the anal crossing is generally situated about halfway from the base of vein A3 to the inner end of the subtriangle.

The nymph is readily recognized by its thick seven-jointed antennae; by the quadrate form of the mentum of the labium, with the peculiar closed cleft in its front margin; and by the strongly angulated side margins of the abdominal segments.

The nymph lives in bogs and other permanently wet spots buried in the mud of small pools that may contain very little water. It may climb perhaps two feet up a convenient tree trunk for transformation.

Tachopteryx thoreyi Hagen

1857. Hagen, in Selys, Mon. Gomph., p. 373 (figs.) (in *Uropetala*).
1861. Hagen, Syn. Neur. N. Amer., p. 117 (in *Petalura*).
1900. Wmsn., Ent. News, 11:398.
1901. Ndm., Bull. N. Y. State Mus., 47:472.
1901. Wmsn., Ent. News, 12:1–3 (figs., nymph).
1917. Kndy., Proc. U. S. Nat. Mus., 52:513 (figs.).
1927. Garm., Odon. of Conn., p. 116.
1929. N. & H., Handb., p. 55 (figs., nymph).
1930. Byers, Odon. of Fla., pp. 42 and 242.
1940. Fisher, Ent. News, 51:39 (habits).

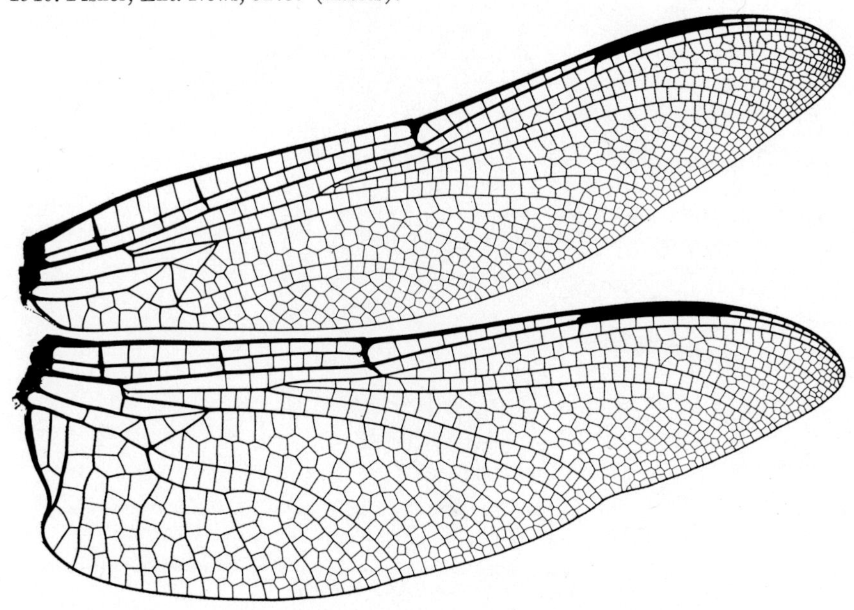

Fig. 37. *Tachopteryx thoreyi.*

Length 71–80 mm.; abdomen 50–61; hind wing 48–53.

Face pale yellowish, including top of frons, with black bands across clypeus and around border of labrum. Swollen occiput yellowish, with a black border around it, regularly convex on superior margin and with scattered black prickles close alongside free edge.

Thorax olivaceous, with black edgings on all carinae; blackish stripes cover humeral and third lateral sutures, which are conjoined below at intervening midlateral. A thin fringe of white hair about prothorax and around leg bases. Legs black. Wing veins blackish, with a line of brown on costa and stigma.

Venation of wings copious, with crossveins very variable. Nodal crossveins 17–19:10–12/10–12:8–12; second thickened antenodal 6–8/5–6; intermedian crossveins 8–10/4–6; brace vein of stigma strongly developed,

Fig. 38. *Tachopteryx thoreyi.*

but sometimes not touching stigma, removed for space of a cell or more proximally from it. Cells in triangles 2–3/1–2; in subtriangle 3–5/3–4; in space beyond, two rows increasing distally; anal loop of 3–7, often of 5 cells. Occasionally an extra cubito-anal crossvein in hind wing. Anal triangle of male of 3 cells, sometimes 5; behind its apex, inner wing margin gently curves, making hind angle less prominent than in *Tanypteryx*. Within this angle a tornal cell is ill defined, transversed by a few crossveins of irregular pattern and inconstant occurrence.

Abdomen little widened on segment 2, gradually tapering beyond, all the way to the end. Two tufts of long white hairs arise from dorsum of 1. Abdomen black, with wide markings of dull orange forming saddle marks on 1 and 2, and interrupted half rings on 3 to 7, diminishing in size to rearward.

Adults frequent sunny openings in woods, often alighting flat against a tree trunk or stone. Easy to approach, but swift of flight. Female oviposits among roots of dense grasses in wet and decaying vegetable matter above surface of water.

Distribution and dates.—UNITED STATES: Ala., Fla., Ind., Ky., Md., Mass., Mich., Mo., N. H., N. Y., N. C., Ohio, Pa., S. C., Tenn., Tex.

April 3 (Fla.) to August 18 (N. C.).

Family CORDULEGASTERIDAE Calvert

In this family we recognize but one genus occurring within our territorial limits, and in it eight species. These are far more generally distributed than the Petalurids, but still local. Their nymphs are more or less restricted to soft beds of muck and silt in flowing waters of woodland streams. The nymphs lie buried in the soft mud up to the tips of their high-peaked eyes, with their sensitive antennae laid out on the surface for contact with passing prey. They do not burrow with their fore feet, but kick the stuff out from under with their hind feet, and descend more or less by squirming the whole body. In such places there are few nymphs of other species to compete with them for food.

Fraser, in his monograph of this family (*Mem. Ind. Mus.*, 1929), recognizes four genera and forty-four species and subspecies. The family as a whole is almost cosmopolitan, but is not reported from the smaller islands nor from most of the African continent. The genus *Cordulegaster* is confined to the Northern Hemisphere, with thirteen known species and subspecies in the Old World and nine in North America (one of these in central Mexico). Some authors place *diastatops* in the genus *Zoraena;* and another genus, *Taeniogaster,* has been proposed for the two species *obliquus* and *fasciatus.*

Several other genera in the Old World are related to *Cordulegaster: Allogaster* and *Anotogaster* in the Orient; and the less closely related *Chlorogomphus* in China and the Indies. Like the Petalurids, these are rather large and hairy, and of rough and primitive finish.

The nymph is hairy and generally well smeared with mud when found. The head is very broad and flat on the upper side, with a rounded, shelf-like prominence between the bases of the rather stout antennae. The eyes are set well forward and a little elevated.

The legs are hairy, but become bare and heavily spined toward the more or less bent end of the tibiae and on the basal segment of the tarsi. The wings of the two pairs lie divergent upon the back.

The abdomen has no middorsal keel and has rather weak development of lateral keels. Lateral spines are small or wanting.

Genus CORDULEGASTER Leach 1815

Syn.: Taeniogaster Selys, Thecaphora Charpentier, Zoraena Kirby

Biddies, Flying Adders

The head is wide. The labium has a cleft in the center of its front margin. The compound eyes are a little separated or meet at a single

Fig. 39. *Cordulegaster maculatus*, female.

point on the top of the head; the face is mostly greenish or yellowish, but the anteclypeus in all our species is blackish, sometimes also the front of the frons. The occiput is usually flat or slightly convex, but may be produced upward into a conical eminence; its posterior surface is yellowish.

The thorax is brown or blackish, with a pair of oblique pale stripes on the front and two wide stripes on each side. The legs are blackish, brownish at the base in most species. The outer side of the second and third tibiae in the males is armed with a row of short, stubby uniform, closely set denticles, and the inner side with a row of long, fine, widely spaced spines.

The wings (see fig. 39) are hyaline, and are sometimes smoky, with costa narrowly yellow, and stigma without a brace vein. The triangles are generally two-celled, rarely of one or of three cells. The subtriangle is without crossveins and usually four-sided. There are two rows of cells in the trigonal interspace. There is no basal subcostal crossvein, and generally none in the midbasal space. There are two or three cubito-anal crossveins, rarely four. The anal loop is well developed, but not foot-shaped, is of two to ten cells, and generally is wider than long. The sectors of the arculus are well separated at their origin. The anal margin of the hind wing of the male is angular, and the anal triangle of the male generally of three cells, except in *fasciatus,* where it is usually four-celled.

The abdomen is long and cylindrical, enlarged dorso-ventrally at base and slightly constricted at segment 3, blackish and spotted or ringed with yellow. The anal appendages of the male are blackish and slightly shorter than 10; the superiors bear two ventral spines and the inferiors are quadrate and shorter than the superiors. The female has a long ovipositor.

The nymphs have the body stout, rough, hairy, and subcylindric, the abdomen beyond the middle tapering to a point. The slender antennae are seven-jointed. The eyes cap the antero-lateral angles of the head. The hind angles of the head are rounded and the posterior margin is not noticeably concave. The large spoon-shaped labium covers the face up to the antennae. The mentum is widened beyond the middle, and the median lobe, bearing a varying number of setae, ends in a prominent bifid median tooth. The broad lateral lobes are triangular and concave, bearing a row of short raptorial setae just within the outer margin, and a stout immovable end hook at the end of this row, with a series of coarse, irregularly interlocking teeth on the distal margin. A prominent frontal shelf projects forward, fringed on the outer edge with hairs which vary in thickness and length with the species; its disc is covered with a varying number of dark tubercles of different size in different species, and of some systematic value.

TABLE OF SPECIES

ADULTS

Species	*Hind wing*	*Occ. crest*	*Face*[1]	*Line labr.*[2]	*Super T cells*[3]	*Cu.A cvs.*	*Distr.*[4]
diadema	43-53	convex	black	+	1-2/1-2	2/2	SW
diastatops	36-42	convex	yellow	0	1/1	3/3	E
dorsalis	43-49	convex	yellow	0	1/1	2/2	W
erroneus	42-51	convex	black	+	2/2	3-4/3	E
fasciatus	53-60	conical	black	0	2-3/2	3-4/3	SE
maculatus	38-49	convex	yellow	+	1/1	3/2	E
obliquus	41-50	conical	yellow	±	2/1-2	3/3	E, C
sayi	37-42	convex	yellow	0	1/1	2-3/2	E

[1]Predominant color of anterior surface of frons.

[2]Median black line on labrum: present (+), absent (0), or variable (±).

[3]Number of cells in supertriangle. In this column and next, solidus (/) is placed between numbers and is to be read, "in fore and hind wing, respectively."

[4]Refers to United States only.

The prothorax is produced laterally on each side into a flattened transverse dorsal area, here termed an *epaulet*, which bears stiff marginal hairs; the shape of the epaulet and length and thickness of hairs vary with the species. The legs are short.

The nymphs are usually found in small woodland streams buried in the sand or silt of the bottom. The adults course slowly up and down these haunts.

KEY TO THE SPECIES

Adults

1—Abdominal segments 2 to 7 nearly encircled by bands of yellow 2
—Abdomen blackish with yellow spots 4

2—Frons yellow; eyes not meeting on top of head .... **sayi**
—Frons black, with a superior transverse oval yellow spot; eyes meeting at a single median point 3

3—Cubito-anal crossveins two; back of eyes yellow; segment 10 generally with at least a small yellow spot laterally .... **diadema**
—Cubito-anal crossveins three or four, rarely two; back of eyes black; segment 10 black (unmarked with yellow) .... **erroneus**

4—Occiput raised in a conical eminence; spots spear- or arrow-shaped on middle abdominal segments 5
—Occiput not so raised; spots not spear- or arrow-shaped on middle abdominal segments 6

5—Abdomen 64–72 mm.; hind wing 53–60 mm.; triangle usually three-celled; anal triangle of male of more than three cells .... **fasciatus**
—Abdomen 48–62 mm.; hind wing 41–50 mm.; triangle usually two-celled; anal triangle of male three-celled .... **obliquus**

6—Eyes not quite meeting; occiput yellow, usually fringed behind with short blackish hair; entire top of frons not darker than its anterior surface; ovipositor of female not extending beyond end of abdomen more than dorsal length of segment 10 .... **diastatops**
—Eyes meeting at one median point; occiput brown to black; ovipostor of female longer 7

7—Entire top of frons decidedly darker than its anterior surface; fore wing with three cubito-anal crossveins; abdomen usually with lateral spots **maculatus**
—Entire top of frons not darker than anterior surface (sometimes darker at base); fore wing with two cubito-anal crossveins; abdomen with middorsal line of subroundish or slightly bifid spots .... **dorsalis**

Nymphs

1—No lateral spines on segments 8 and 9; Western 2
—With lateral spines on 8 and 9; Eastern 3

2—Lateral setae 6 or 7, mental setae 5+3–4 .... **dorsalis**
—Lateral setae 5, mental setae 4+4–6 .... **diadema**

Fig. 40. *Cordulegaster erroneus.* A, frontal shelf; B, epaulet.

3—Lateral setae 4, mentals generally 5+5; hairs on margin of frontal shelf short, very much flattened and often wider at apex than at base; few large obsolescent brown dots (hair bases?) on frontal-shelf disc; hairs on epaulet less than twice as long as epaulet is wide .... **erroneus**
—Lateral setae 5–7, mentals generally not 5+5; brown dots on frontal-shelf disc smaller 4

4—Lateral setae 7 (distal seta often small and occasionally missing), mentals 8–9+4–5; lateral spines of segments 8 and 9 minute....................... 5
—Lateral setae 5, mentals 5–6+3–4; lateral spines of segments 8 and 9 larger.. 6

5—Anterior margin of frontal shelf broadly convex; femora with conspicuous dark spot on dorsal surface near apex .... **obliquus**
—Anterior margin of frontal shelf almost straight in middle; femora not conspicuously marked .... **fasciatus**

6—Median tooth of mentum secondarily bifid into two equal parts.......... **?sayi**
—Median tooth of mentum secondarily bifid into two unequal parts............ 7

7—Margin of frontal shelf usually with tuft of very long thin hair in center; long, usually black, hairs anterior to ocelli; obsolescent brown dots (hair bases?) on frontal-shelf disc generally few and irregular in outline; epaulets truncate on outer side .... **diastatops**
—Margin of frontal shelf usually without tuft of hair in center; brown dots on frontal-shelf disc generally numerous, small, and round; epaulets widely rounded at outer anterior edge .... **maculatus**

TABLE OF SPECIES

NYMPHS

Species	*Total length*[1]	*Setae of labium*		*Lateral spines on seg. 9*[2]
		Lateral	*Mental*	
diadema	45	5	4+5–6	0
diastatops	31	5	5+3–4	4
dorsalis	35	6-7	5+3–4	0
erroneus	38	4	5–6+4–5	3
fasciatus	45	6-7	8–9+4–5	2–
maculatus	40	5	6+4–5	2+
obliquus	39	6-7	8+4–5	2–
sayi ?	34	5	6+4–5	3

[1]Average length when grown.

[2]Length of spines of segment 9 expressed in tenths of entire length of lateral margin of that segment.

Cordulegaster diadema Selys

1868. Selys, C. R. Soc. Ent. Belg., 11:68.
1904. Ndm., Proc. U. S. Nat. Mus., 27:697 (fig., nymph).
1917. Kndy., Proc. U. S. Nat. Mus., 52:515 (figs.).
1929. Fraser, Mem. Ind. Mus., 9:127 (figs.).
1929. N. & H., Handb., p. 155.

Length 74–88 mm.; abdomen 57–66; hind wing 43–53.

Occiput yellowish to brown, fringed with yellowish (sometimes blackish) hairs. In a specimen reported from Mexico, occiput black, with whitish hairs. Thorax with the usual two wide lateral stripes, with a narrow yellow stripe between them, or a spot above to indicate position of stripe. Front coxae yellow.

Nodal crossveins* 15–20:9–13/11–14:10–13; second thickened antenodal 6–9/5–8; intermedian crossveins 6–10/5–6; anal loop of 3–9, usually 5 or 6, cells; usually 2, rarely 3, cubito-anal crossveins in hind wing, sometimes 3 cells in hind-wing triangle.

Yellow of abdomen in male as follows: segment 1 with one or two small lateral spots near ventral border; 2 with latero-basal spots including auricles and confluent over dorsum, enclosing a basal black spot on dorsum, with a narrow terminal half ring (sometimes narrowly interrupted on mid-line), and a large spot at junction of apical and lateral margins; 3 to 5 with median rings covering about one-fourth to one-third of dorsum of each, and a narrow terminal half ring (which may be interrupted on mid-line); rings of 6 and 7 cover about one-fourth to one-third of dorsum, that of 8 almost half; 3 to 8 with ring angulated on sides and bent forword at lower end; 9 with transverse basal band and an arm on each side projecting to rearward; 10 with yellow broken into a small basal spot and two larger lateral spots (these spots almost entirely lacking on some specimens). Abdomen of female similar, except that segment 1 is immaculate; 2 bears a pair of small apical rings and a narrow mid-dorsal stripe; 8 has a wide subbasal band covering almost half of length of segment; 9 with a small basal stripe; 10 without yellow markings.

Distribution and dates.—UNITED STATES: Ariz., Utah; also from Mexico. July 11 (Ariz.) to November (Mexico).

Cordulegaster diastatops Selys

Syn.: lateralis Scudder

1854. Selys, Bull. Acad. Belg., 21(2):101 (reprint, p. 81) (in *Thecaphora*).
1901. Ndm., Bull. N. Y. State Mus., 47:477 (nymph).

* For explanation of solidus see p. 52.

1927. Garm., Odon. of Conn., p. 119.
1929. Fraser, Mem. Ind. Mus., 9:137 (figs.).
1929. N. & H., Handb., p. 158.

Length 59–65 mm.; abdomen 42–49; hind wing 36–42.

Occiput yellow, with hind margin usually brown, and fringed with short stiff dark brown to black (rarely light tan) hairs which project dorsally. Eyes slightly separated and distinctly tumid behind.

Nodal crossveins 14–20:11–18/12–16:11–18; second thickened antenodal 5–7/6–7; intermedian crossveins 6–10/5–6; cubito-anal crossveins 3/2–3; anal loop of 4–6 cells in male, 7–10 in female; midbasal space and supertriangle only rarely crossed.

Yellow of abdomen in male as follows: segments 2 and 3 with dorso-lateral spots more or less confluent, forming a continuous stripe on each side; 3 to 8 with elongate dorso-lateral spots, usually pointed behind and well separated by middorsal carina, sometimes confluent over dorsum of 3; 9 with a subbasal spot on each side; 10 may or may not have a lateral spot.

Female similar, but yellow on sides of abdomen forms almost continuous stripe. Ovipositor shorter than in any of our other species except *sayi*.

Distribution and dates.—CANADA: N. B., N. S., Ont., Que.; UNITED STATES: Ala., Conn., Fla., Ga., Ind., Maine, Md., Mass., Miss., N. H., N. J., N. Y., N. C., Ohio, Pa., S. C., Vt., W. Va.

March 20 (Miss.) to August 29 (Que.).

Cordulegaster dorsalis Hagen

1857. Hagen, in Selys, Mon. Gomph., p. 347.
1904. Ndm., Proc. U. S. Nat. Mus., 27:686 (fig., nymph).
1917. Kndy., Proc. U. S. Nat. Mus., 52:515 (figs.).
1929. Fraser, Mem. Ind. Mus., 9:132 (figs.).
1929. N. & H., Handb., p. 156.

Length 70–85 mm.; abdomen 55–64; hind wing 43–49.

Occiput reddish brown fringed with stiff grayish brown or golden hairs. Antehumeral stripes not markedly wedge-shaped as in other North American species, but more or less parallel-sided. Thorax chocolate brown to black, with a small yellow spot on side just above spiracle, and usually another near crest between lateral stripes.

Nodal crossveins 14–19:8–16/10–14:10–15; second thickened antenodal 5–7/5–7; intermedian crossveins 7–10/4–6; anal loop of 3–6 cells in male, 7–8 in female.

Yellow of abdomen as follows: segment 2 with a lateral spot on auricles in male and in a similar position in female, a saddle-shaped spot on middorsum and two small apical spots; 3 to 7 with subcircular dorsal spots, posterior ones often quite deeply notched behind at middle line: 8 with a subbasal band; 9 with a narrower basal band, widened on sides in female; 10 in male variable, generally with a spot or two on each side, in female unmarked.

Distribution and dates.—Alaska; CANADA: B. C.; UNITED STATES: Calif., Nev., Utah.

May 15 (Calif.) to August 25 (B. C.).

Cordulegaster erroneus Hagen

1878. Hagen, in Selys, Bull. Acad. Belg., (2) 46:688.
1900. Wmsn., Odon. of Ind., p. 299.
1927. Garm., Odon. of Conn., p. 121.
1929. Fraser, Mem. Ind. Mus., 9:134.
1929. N. & H., Handb., p. 155.

Length 65–76 mm; abdomen 50–64; hind wing 42–51.

Occiput yellow, fringed with golden hair. Thoracic stripes greenish, a small spot above between the two wide lateral stripes.

Nodal crossveins 18–23:13–18/15–18:14–19; second thickened antenodal 6–9/7–9; intermedian crossveins 8–11/6–8; anal loop of 5–7 cells in male, 7–8 in female; cubito-anal crossveins in hind wing usually 3, rarely 2 or 4; supertriangle with 2 or 3 cells.

Yellow of abdomen as follows: segment 2 with a narrow transverse median band which becomes wider on sides and extends forward across lower part of 1; 2 also with a pair of apical middorsal spots, and a large spot at junction of apical and lateral margins; 3 to 8 with wider submedian rings, which become widened forward on ventral side and are only slightly separated by middorsal carina; 3 and 4 may also have small apical spot on each side; 8 with a large subbasal spot on each side; 9 in male with a small basal spot.

This species is unique in having the back of the eyes predominantly black, while in our other species this area is mostly yellow. Fraser in his monograph describes specimens from California in the collection of Selys, and notes that they differ from the type description. We believe his specimens to be incorrectly determined. They are probably *dorsalis.*

Distribution and dates.—UNITED STATES: Conn., Ky., N. Y., N. C., Ohio, Pa., S. C., W. Va.; Western records from Calif. and Utah need verification.

June 8 (Conn.) to September 3 (N. C.).

Cordulegaster fasciatus Rambur

1842. Rbr., Ins. Neur., p. 178.
1878. Selys, Bull. Acad. Belg., (2) 46:692 (reprint, p. 1001).
1929. Fraser, Mem. Ind. Mus., 9:131.
1929. N. & H., Handb., p. 158.
1930. Byers, Odon. of Fla., pp. 83–89, 257 (fig., nymph).

Length 83–88 mm.; abdomen 64–72; hind wing 53–60.

Occiput conical, yellow on sides, black at tip, fringed behind with blackish hairs. Thorax dark brown or black with usual yellow side stripes.

Nodal crossveins 24–27:19–22/16–21:17–21; second thickened antenodal 9/8; intermedian crossveins 11–12/9–10; anal loop of 4–9 cells; triangle usually of 3 cells, occasionally with 2 or 4 cells in one or more wings.

Yellow of abdomen as follows: segment 1 with a narrow basal band; 2 to 7 each with a spear-shaped mark on dorsum, those on posterior segments more dilated; 8 in male with a similar spot, much more dilated laterally to connect with a pale ventral area; 8 in female with an isolated basal spot on dorsum, this spot rounded or slightly widened posteriorly, and with a smaller spot on each side near ventral carina; 9 in male with a small dorso-basal spot, not present in female; 10 unmarked in both sexes. In addition, pale spots on under side of abdomen near ventral carina, these spots reduced on middle segments to narrow crescent-shaped marks.

Distribution and dates.—UNITED STATES: Fla., Ga., N. C.; Byers, 1930, lists a female from Mich.

June 20 (Fla.) to August (N. C.).

Cordulegaster maculatus Selys

1854. Selys, Bull. Acad. Belg., 21(2):105 (reprint, p. 86).
1901. Ndm., Bull. N. Y. State Mus., 47:476 (nymph).
1927. Garm., Odon. of Conn., p. 122.
1929. Fraser, Mem. Ind. Mus., 9:123 (figs.).
1929. N. & H., Handb., p. 159.
1930. Byers, Odon. of Fla., p. 84.

Length 64–76 mm.; abdomen 47–58; hind wing 38–49.

Labrum with at least median depression at base brownish, a black stripe sometimes extending to a point midway from base to apex; occiput brown to black, fringed behind with long light brown to whitish hair which projects backward. Two lateral yellow stripes of thorax with a narrow stripe between, often reduced to a line above and a spot below. Legs black, reddish brown at base.

Nodal crossveins 15–22:11–15/12–16:11–15; second thickened antenodal 6–9/6–8; intermedian crossveins 9–12/6–8; anal loop of 4–7 (more often 5–6) cells in male, 7–10 in female; supertriangle only rarely crossed; cubito-anal crossveins in front wing 3, rarely 2.

Yellow of abdomen in male as follows: segment 1 yellowish at sides; 2 with auricles and an area at junction of lateral and apical margins, and two median dorsal spots yellow: 3 to 7 with a pair of subconical spots slightly basal to middle of segment, and separated by middorsal carina; 8 with a pair of spots slightly elongate transversely and sometimes notched in hind margin; 9 generally with small basal yellow spot on each side; 10 unmarked; 2 to 5 usually with an additional pair of apical spots. Abdominal markings of female similar, but 9 has more extensive yellow markings.

Distribution and dates.—CANADA: N. B., N. S., Ont., Que.; UNITED STATES: Ala., Conn., Fla., Ga., Ind., Ky., La., Maine, Md., Mass., Mich., Miss.(?), N. H., N. J., N. Y., N. C., Ohio, Pa., S. C., Tex.(?), Vt., Va.

February 21 (Fla.) to October 5 (Conn.).

Cordulegaster obliquus Say

1839. Say, J. Acad. Phila., 8:15 (in *Aeschna*).
1900. Wmsn., Odon. of Ind., p. 300.
1905. Ndm., Ent. News, 16:3 (nymph).
1927. Garm., Odon. of Conn., p. 123.
1929. Fraser, Mem. Ind. Mus., 9:129 (figs.).
1929. N. & H., Handb., p. 158.

Length 72–81 mm.; abdomen 48–62; hind wing 41–50.

Occiput greenish yellow, conical, fringed behind with short stiff blackish hairs. Lateral thoracic stripes strongly margined with black.

Nodal crossveins 18–22:11–17/13–17:12–16; second thickened antenodal 6–7/6–8; intermedian crossveins 8–11/5–7; anal loop of 3–7 cells in male, 6–8 in female.

Yellow of abdomen as follows: segment 1 with a small spot at junction of apical and lateral margins, and a small middorsal spot, larger in female; 2 with a stripe across region of auricles, continuous with spot on 1, and a large diffuse spot on sides, also a middorsal stripe as long as segment; 3 with a stripe along antero-lateral margin; 3 to 7 with longitudinal middorsal stripes shaped like spearheads, with points directed to rear; 8 with a large spot reaching from base almost to apex, and sometimes continuous laterally with a large basal spot on sternum (7 often has a similar ventral spot); 9 in male with a small quadrangular basal spot, in female immaculate; 10 unmarked.

Distribution and dates.—CANADA: Ont., Que.; UNITED STATES: Ark., Conn., Ill., Ind., Kans., Ky., Maine, Mass., Mich., N. J., N. Y., Ohio, Okla., Pa., Va., W. Va., Wis.

May 22 (Ohio) to July 26 (Ind.).

Cordulegaster sayi Selys

Syn.: obliqua; var. A. Say

1854. Selys, Bull. Acad. Belg., 21(2):104 (reprint, p. 85).
1857. Selys, Mon. Gomph., p. 331.
1861. Hagen, Syn. Neur. N. Amer., p. 115.
1903. Ndm., Bull. N. Y. State Mus., 68:267 (fig., nymph supp.).
1908. Mtk., Odon. of Wis., p. 79.
1929. Fraser, Mem. Ind. Mus., 9:136 (figs.).
1929. N. & H., Handb., p. 155.

Length 60 mm.; abdomen 45–52; hind wing 37–42.

Occiput greenish yellow, fringed above with short stiff black or brown hairs. Eyes slightly separated and distinctly tumid behind. Thorax dark brown with two oblique yellow stripes on each side moderately narrow; between them a narrower, ill-defined or incomplete stripe of yellow.

Nodal crossveins 14–19:9–12/11–14:10–15; second thickened antenodal 5–7/5–6; intermedian crossveins 7–8/5–6; anal loop of 2–5 cells.

Dark abdomen of male marked with yellow as follows: segment 1 mostly pale; 2 with a narrow band just before middle, interrupted dorsally but widening downward on each side to cover auricles, also with an apical half ring which may be interrupted above; ventral margins of segment pale also; 3 to 6 with rings before middle, about one-fourth to one-third as long as segments, which extend obliquely forward almost to front margin, also with much narrower apical half rings which may be interrupted above; 7 with a wider ring nearer base, but no apical spots or rings; 8 with a still broader subbasal band; 9 with a basal spot on each side, extending apically below and sometimes enclosing a dark dorsal triangle with its base toward rear of segment; 10 with a pair of large lateral spots, sometimes much reduced in size.

Female similar to male, but with yellow rings broader and no markings on segment 9. Ovipositor projects but slightly from end of abdomen, and is shortest found in our species.

Closely related to *diastatops* in slight separation of eyes (which are markedly tumid behind) the shape of anal appendages, and short ovipositor.

The range is given in Muttkowski (*Catalog*, 1910) as Maine, Quebec, and Wisconsin, to Georgia. The type is from Georgia, and three males in the Cornell University collection are from Lake City, Florida. The northern records need further verification, as there has been confusion of the species. Fraser lists it from Georgia, and the White Mountains of New Hampshire, after Selys.

Distribution and dates.—UNITED STATES: Fla., Ga., Maine(?), Md.(?), N. H.(?), Wis.(?).

March 28 (Fla.) to August 15 (N. H.)

Family GOMPHIDAE Rambur

The dragonflies of this family are world-wide in distribution and very distinct from all others in both adult and nymphal stages. They are relatively primitive, as is shown by the wide distance between the eyes on the top of the head, by the very moderate skewness of the thorax (shift of wings backward and legs forward), and by the lack of planates and other specializations in the venation of the wings.

Gomphine dragonflies are mostly clear-winged. The legs are sprawling, as befits their low-perching habits. The caudal appendages are highly differentiated in the male, with the inferior one widely forked. The female has no ovipositor. In egg laying she is not aided by the male. She flies low over the water, descending to strike the surface at irregular intervals, liberating many eggs at each descent. In copulation the pairs seek the shelter of shore vegetation.

Wing venation is as shown in our table of families (p. 64) and in figure 41 of the base of the hind wing of *Gomphoides stigmatus*. The network of crossveins is rather dense and there are no planates developed (except for a trigonal planate in *Hagenius*). The triangles of fore and hind wing are of about the same shape and are regular, save that in the hind-wing subtriangles the two crossveins forming the inner and outer sides often do not meet at their contact with vein M4, and a short additional fourth side results.

In some of these characters this family is quite as primitive as are the *Petaluridae* and *Cordulegasteridae:* there is no fusion of the sectors of the arculus (M1–3 and M4) beyond the base; there is hardly more shifting of veins or reduction of crossveins than in those families. There are signs of advance, however, in the development of a brace vein to the stigma and in the reduction of the number of intermedian crossveins in more than half of the family. The development of a trigonal planate is rare. The females have small auricles on segment 2. The abdomen is swollen in its basal segments, clubbed on its apical segments, and narrow between, especially in males.

Gomphines do not intrude upon the collector's notice. They do not soar or hover in the open like the more familiar Libellulines, but spend most of their time at rest, squatting on stone or log or leaf, making occasional swift sallies from one resting place to another. Their bodies have a camouflage pattern of black and yellow or greenish* stripings that is so effective that they must be flushed to be seen; and then the eye must follow them to their next resting place or they are lost. One must go after Gomphines

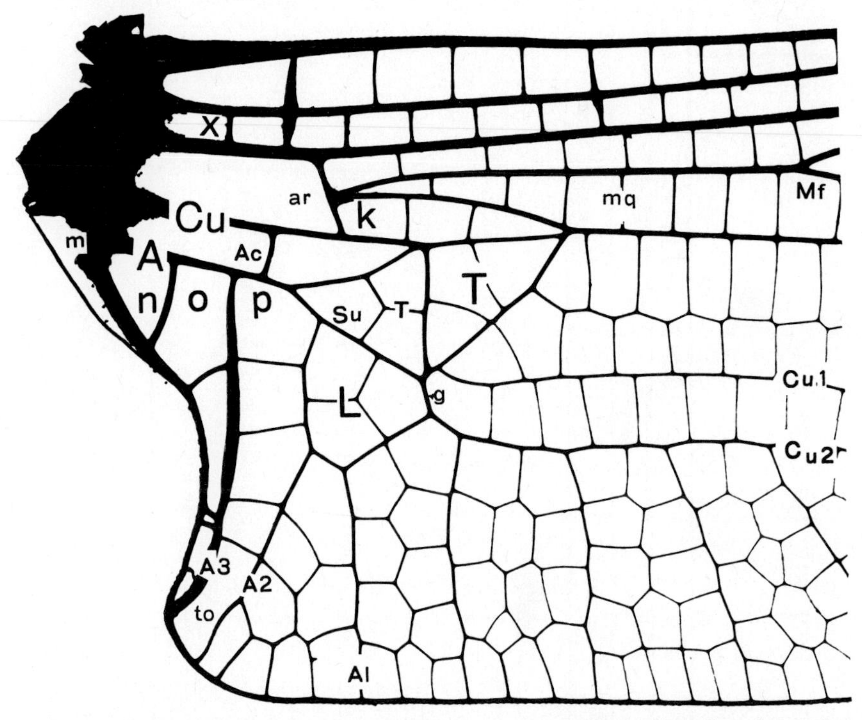

Fig. 41. Base of hind wing of *Gomphoides stigmatus,* illustrating Gomphine wing venation. Note presence of basal subcostal crossvein (x) and seven intermedian crossveins (mq). Note much-thickened first and fifth antenodal crossveins. Observe that triangle, supertriangle, and subtriangle are each 3-celled, also anal loop, and that each has a different arrangement of crossveins between cells; note extra long 5-celled basal triangle of male. Arculus (ar) and its sectors (k), anal crossing (Ac), gaff (g), and tornus (to) are as before, in introductory figure on page 17. Membranule (m) is outside framework of wing at extreme base. Of long veins only Cu and A are labeled here.

to find them, and one needs first of all to know where to go. They are for the most part stream-haunting. The larger species inhabit the larger muddy or sandy streams, where their abundance is sometimes evidenced by thousands of cast-off skins left behind on the shore at transformation. The smaller species inhabit brooks and cleaner streams, spring-fed rills and pools. A few species are pond dwellers.

The nymphs burrow shallowly in bottom mud, sand, and sediment. They are at once distinguishable by the depressed and wedge-shaped form

* The greens fade to yellow in old museum specimens.

TABLE OF GENERA

ADULTS

Genera	*Hind wing*	*Crossveins*		*Cells in hind wing*						*Distribution*
				Triangles[2]						
		2d thick.[1]	*Inter-median*	*Super*	*T*	*Sub*	*Male*	*Par-anal*[3]	*Post-anal*	
Aphylla	31-49	9/7	11/6	2/2	3/2	2/1	4	5	4-5	S
Dromogomphus	35-39	5/5	2-3/1	1/1	1/1	1/1	3	5	4	E,C
Erpetogomphus	24-35	5/5	2-3/1	1/1	1/1	1/1	3-5	5	3-5	S,W,E
Gomphoides	37-42	6/6	11/8	3/2	3/3	3/3	4-7	5	6	SW
Gomphus s.lat.[4]	20-45	5/5	2/1	1/1	1/1	1/1	3	3-7	4-5	G
Hagenius	47-58	8/7	5-6/2-3	1/1	2/2	1/1	3-5	5	5-6	E
Lanthus	20-26	5/5	2/1	1/1	1/1	1/1	3	4	4-5	E
Octogomphus	29-32	5/5	2/1	1/1	1/1	1/1	3	4-5	5	W
Ophiogomphus	19-38	5/5	2-3/1	1/1	1/1	1/1	4-6	5-6	5-7	N
Progomphus	24-35	5/5	6/4	1/1	3/3-4	2/2	3-6	5-7	4-6	S,W,E

[1]Serial number of second thickened antenodal: fore wing/hind wing.

[2]Prevailing number of cells composing supertriangle, triangle, and subtriangle; also basal triangle of ♂.

[3]Prevailing number of cells in paranal and postanal rows.

[4]This is the *Gomphus* complex. For further analysis of it see p. 163.

of the head; by the two-jointed front and middle tarsi; and by the thick four-jointed antennae, the fourth joint of which is minute or vestigial. The tips of the front and middle tarsi are more or less flattened, and hooked outward for burrowing. Their stiff legs being scarcely opposable, they do not often climb slender stems for transformation, but lie flat upon the sand of the shore, or sprawl over mats of grass, or climb a little way up on the broad rough surfaces of logs and stones.

KEY TO THE GENERA

Adults

1—Basal subcostal crossvein present; at least one crossvein in fore-wing subtriangle (fig. 41) 2
—Basal subcostal crossvein wanting; no crossvein in fore-wing subtriangle.... 4

2—One or more crossveins in supertriangles.......... 3
—No crossvein in supertriangles.......... **Progomphus**

3—Hind-wing subtriangle of two or more cells; anal loop of three to five cells formed by convergence of veins A1 and A2.......... **Gomphoides**
—Hind-wing subtriangle generally one-celled; veins A1 and A2 run direct to hind-wing margin and form no anal loop.......... **Aphylla**

4—Trigonal planate present in all wings; crossveins present in all triangles, none in subtriangles (fig. 52, p. 113).......... **Hagenius**
—Trigonal planate wanting or nearly so; crossveins absent from triangles and subtriangles 5

5—Hind wing with semicircular anal loop consisting generally of three cells (fig. 54, p. 117).......... **Ophiogomphus**
—Anal loop wanting or consisting of one or two weakly bordered cells......... 6

6—Long hind femur armed with five to seven strong spines, intermixed with numerous smaller ones .... **Dromogomphus**
—Femur with no outstandingly large spines.......... 7

7—Stigma of fore wing short and thick, at its widest about twice as long as wide; hind wing with five paranal cells; branches of inferior caudal appendage of male long, parallel full length and strongly hooked upward **Erpetogomphus**
—Stigma of fore wing generally more elongate, about three times as long as wide (except in **Octogomphus** and **Lanthus**); hind wing with four or five paranal cells; branches of inferior caudal appendage of male shorter and divergent 8

8—Usual middorsal dark stripe or pair of stripes absent from wide yellow front of synthorax; antehumeral black stripe very wide; inferior caudal appendage of male four-branched; Western.......... **Octogomphus**
—Middorsal thoracic dark stripe generally present; inferior caudal appendage of male two-branched 9

9—Crest of occiput (at rear of head, running crosswise from eye to eye) broadly rounded, thickly beset with coarse hairs sprinkled all over its surface; small species (hind wing 26 mm. or less); Eastern.......... **Lanthus**
—Crest of occiput sharper-edged, its hairs brought more in line to form a fringe along its summit .... **Gomphus s. lat.**

NYMPHS

1—Segment 10 cylindric, more than half as long as abdomen.............. **Aphylla**
—Segment 10 short, rarely as long as any single preceding segment............ 2

2—Naked antennal segment 4 generally about a fourth as long as big hairy 3; middle legs closer together at base than fore legs.................. **Progomphus**
—Segment 4 vestigial or nearly so; middle legs not closer at base than fore legs 3

3—Abdominal segment 10 a little longer than 9...................... **Gomphoides**
—Segment 10 shorter than 9.. 4

4—Wings strongly divergent.. 5
—Wings laid parallel along back... 6

5—Lateral caudal appendages in full-grown nymph about as long as inferiors **Erpetogomphus**
—These appendages about three-fourths as long as inferiors....... **Ophiogomphus**

6—Body very flat; abdomen nearly circular; paired tubercles on top of head; huge black nymphs .... **Hagenius**
—Body more nearly cylindric; no tubercles on head.......................... 7

7—Flattened antennal segment 3 nearly as wide as long...................... 8
—Long segment 3 more or less cylindric...................................... 9

8—Short lateral spines on abdominal segments 7 to 9............... **Octogomphus**
—Lateral spines on 8 and 9 only.. **Lanthus**

9—Dorsal hook on 9 is spinelike termination of middorsal ridge of 9. **Dromogomphus**
—Dorsal hook on 9, if present, rises above level of its rounded dorsum **Gomphus s. lat.**

Genus PROGOMPHUS Selys 1854

These short-legged dragonflies are brownish or blackish, varied with gray, green, and yellow. The fore-wing triangle is generally three-celled; the subtriangles of both wings are generally two-celled. The supertriangle is free from crossveins. A basal subcostal crossvein is present in all wings. The outer side of the triangle is bent at an angle in both wings, with a feeble planate springing from the angle, to extend zigzagged for a distance between the two rows of cells in the trigonal interspace. The three-celled anal loop of the male does not extend rearward to the hind angle of the wing as it does in *Aphylla* and *Gomphoides.* There is no anal loop.

The caudal appendages of the abdomen of the male are about as long as segment 9. The superiors are strongly flattened beyond their swollen base, where on the under side there are minute denticles. The superiors are obliquely truncated, with the apex at the outer side ending in a sharp point. The inferior appendage is divided to the base into two long incurving arms that are shorter and darker than the superiors. These also are flattened and widest near the end, where there is a large external upturned tooth. On the inner margin at the extreme apex are several upturned denticles. The hamules of abdominal segment 2 are very unequal in size:

the anterior hamule is small, hooked at the apex, and well concealed in lateral view by the broadly expanded base of the posterior hamule; the latter is incurved in its slender apical half, and more or less hairy before the tip. The deeply divided peduncle of the penis has something of the

TABLE OF GENERA

NYMPHS

Genera	*Total length*	*Labium*		*Wing cases*[3]	*Caudal append.*[4]		
		End hook[1]	*Teeth*[2]		*Lat.*	*Sup.*	*Inf.*
Aphylla	42-68	strong	4 huge	par.	8-9	10	10
Dromogomphus	32-38	var.	7-8	par.	9	10	10
Erpetogomphus	24-25	absent	many	div.	10	10	10
Gomphoides	33-35	long	8 small	par.	10–	10	10
Gomphus s.lat.[5]	24-45	var.	var.	par.	var.	var.	var.
Hagenius	36-41	absent	many	par.	4	9	10
Lanthus	20-23	minute	8 low	par.	5	10	10
Octogomphus	25-26	absent	8+ low	par.	5	10	10
Ophiogomphus	23-31	absent	many	div.	7-8	10	10
Progomphus	21-29	absent[6]	many	div.	5	10	10

[1]Portion of lateral lobe of labium that projects beyond level of line of teeth.
[2]Approximate number and size of teeth on inner margin of lateral lobe.
[3]Wing cases laid parallel (par.) along back or divergent (div.).
[4]Relative length of five appendages at end of abdomen in grown nymphs.
[5]For further analysis of *Gomphus* complex see p. 163.
[6]Except *serenus*, which has a minute end hook and no teeth.

appearance of a gaping clamshell; each deeply concave half provides a shelter in which the slender terminal filaments of the retracted penis lie coiled.

This genus includes three closely related large species in the United States, three in the West Indies, and about twenty additional in Mexico and Central and South America—a miscellaneous lot.

The nymphs live in the sandy beds of streams and lakes. Their color matches the sand in which they burrow. They are easily recognized by the cleanness of their trim bodies, the length and slenderness (in ours) of the caudal appendages, the unusual length (for a Gomphine) of the reclined fourth antennal segment, and the closely indrawn legs. Both front and middle legs are thickly beset on their most exposed surfaces with stiff outcurving bristles. Head and thorax are smoothed off on the dorsal side somewhat like a carapace, and slope forward and downward above the burrowing front feet.

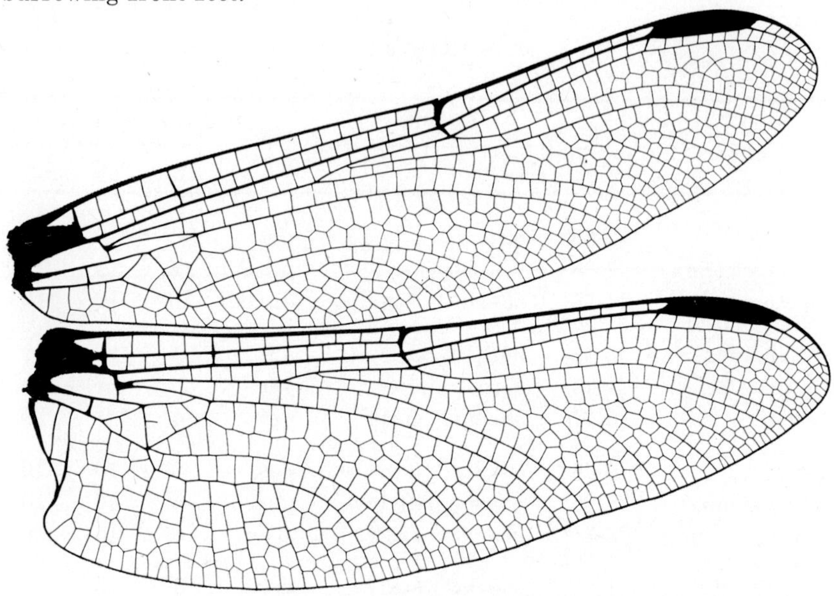

Fig. 42. *Progomphus obscurus.*

These are the marks of adept burrowers. They abound in shifting sand bars. The stream-dwelling species are among the most lotic of anisopterous dragonflies. The abdomen is smoothly tapered to rearward; segment 9 is longer than either 8 or 10, and the appendages are longer than 9.

The labium is very flat and rather narrow; the mentum is nearly or quite parallel-sided. The median lobe is prominently rounded in front, and fringed with flat microscopic scales in a double series, long and short. The slender lateral lobe bears a rather short movable hook, terminates in a bluntly rounded end, and is smooth along its inner margin. In the stout four-jointed antenna, segment 3 is about as long as all the others taken together, considerably depressed, and fringed with hair along its edges. Antennal segment 4 is a bare slender cylindric rudiment and, in position, is recurved upon the upper side of the broad segment 3. It is generally about a fourth as long as that segment.

Within this genus the nymphs are most easily distinguished by the relative development of dorsal hooks and lateral spines on the abdomen.

Fig. 43. *Progomphus obscurus.*

The most useful papers for reference in the further study of this genus are those of Byers (1927), and Needham (1941), cited under species below; also, for South American species (adults only), Ris (*Arch. Naturgesch.*, 82:139–144, 1916).

KEY TO THE SPECIES*

ADULTS

1—Costal edge of wing lined with yellow; superior appendages of male mainly yellow; larger species; in both sexes, a slender midventral tubercle or transverse ridge under abdominal segment 1 (sometimes not easily seen when laid flat); U. S. and Mexican 2

—Costa wholly black; appendages black; smaller species; no such tubercle under abdominal segment 1; Antillean 5

2—Midlateral stripe on thorax weak or undeveloped above spiracle and not wholly confluent with third lateral; male not reddish on abdominal segments 8 to 10 above; U. S. 3

—Midlateral and third lateral stripes on thorax complete and confluent throughout; male reddish on 8 to 10 above; hind wing about 30 mm.; Mexican .... **clendoni**

3—Extreme base of wings with a little brown patch extending outward only to basal subcostal crossvein; small carina under base of superior appendages of male armed with single line of denticles; western U. S. .... **borealis**

—Extreme base of wings with brown cloud extending outward to first thickened antenodal crossvein; eastern, central, and southern U. S. 4

4—Dorsal and antehumeral dark stripes on thorax generally confluent at both ends; small carina under base of superior appendages of male armed with single line of denticles .... **alachuensis**

—Dorsal and antehumeral dark stripes on thorax generally widely separated by yellow next to collar; denticles scattered .... **obscurus**

5—Labrum black, with a crescentic pale spot on each side .... **integer**

—Labrum greenish all around border, with a large basal black spot covering half of its surface .... **serenus**

NYMPHS

1—Lateral spines on abdominal segments 3 to 9 .... **borealis**

—Lateral spines on 5 to 9 2

—Lateral spines on 6 to 9 .... **integer**

2—Large species: length 26–29 mm. 3

—Small species: length 20–21 mm. 4

3—Abdomen with black markings mostly low on sides; dorsal hooks diminishing regularly to rearward from 2 to 9 .... **obscurus**

—Abdomen with black markings mostly submedian; dorsal hooks diminishing halfway, then increasing .... **alachuensis**

4—Antennal segment 3 very flat, little longer than wide; 4 cylindric, vestigial **serenus**

—Antennal segment 3 elongate, little flattened; 4 about as long as 3 is wide.... 5

5—Length of body 20 mm.; antennal segment 3 oval .... **zephyrus**

—Length of body 24 mm.; antennal segment 3 cylindric .... **clendoni**

* Key to the genera of Gomphines is on p. 91.

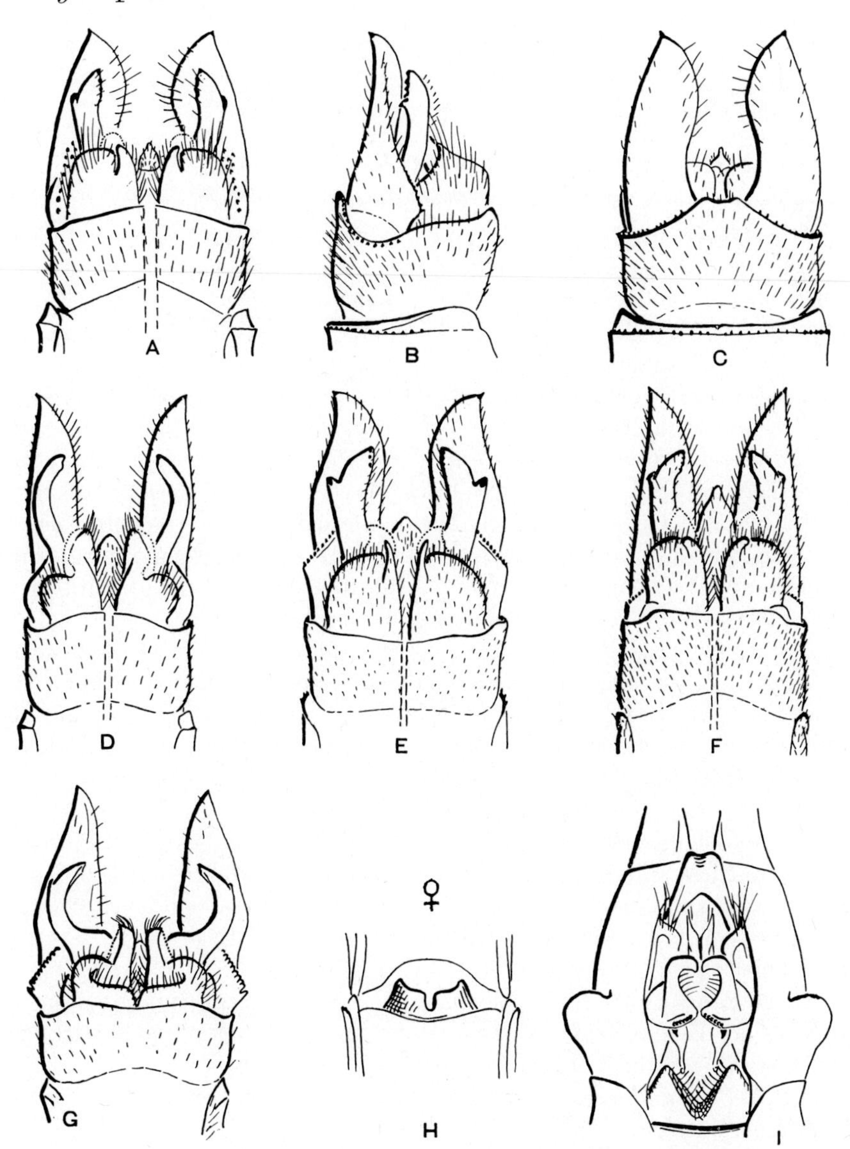

Fig. 44. A, B, and C, ventral, lateral, and dorsal views of *Progomphus obscurus*, male; D, ventral view of same parts in *P. integer;* E, same in *P. alachuensis;* F, same in *P. borealis;* G, same in *P. serenus;* H, subgenital plate of female in *P. serenus;* I, ventral view of genitalia of segment 2 of male in *P. serenus,* auricles projecting at sides. (Drawings by Dr. May Gyger Eltringham.)

Progomphus alachuensis Byers

1939. Byers, Proc. Fla. Acad. Sci., 4:50–56 (figs.).
1941. Ndm., Trans. Amer. Ent. Soc., 67:234 (nymph).

Length 52–57 mm.; abdomen 37–41; hind wing 30–35.

Largest species of the genus, brown marked with yellow. Face pallid, a little darker around margins and in sutures, and washed with yellow across top of frons. Blackish vertex paler on portion that slopes to rearward behind ocelli. Occiput yellow, except on side margins, which are blackish, straight-edged behind. Antennae black, with paler terminal rings on basal segments.

Mesothorax striped in front with brown and yellow. Carina narrowly yellow, and collar broadly so, except for a point at junction of the two. Middorsal and antehumeral brown stripes of thorax conjoined at both ends, leaving a narrow oblique streak of yellow between; humeral and antehumeral stripes wholly confluent, covering whole humeral area; humeral and midlateral joined only at ends with a broad intervening yellow band; this band and second lateral extensively confluent, with only a small faint yellowish spot above spiracle and a larger yellow one below it. Behind stripes, sides of thorax broadly yellow.

Wings hyaline, with a long stigma (6–7 mm.) and two rather conspicuous streaks of brown at wing base, subcostal one reaching out to arculus, cubital one shorter. Legs brown, touched with yellow at knees, more broadly yellow on ventral side of femora.

Abdomen black on sides, with yellow saddle marks on dorsum. These yellow marks on segments 3 to 7 cover basal third, and beyond that taper to rearward in a less distinctly yellow streak. Sides of 1 and 2 broadly yellow, of 8 and 9 narrowly so; 10 blackish. Superior appendages of male yellow except at extreme base within; inferior appendage brown. Stylets of female brown.

A lake-inhabiting species, common in some of clear sandy-bottomed lakes of northern Florida, where in early April it transforms abundantly, flat on sand, mostly within a foot of margin of water.

Distribution and dates.—UNITED STATES: Fla.

February to August 8.

Progomphus borealis McLachlan

1873. McL., in Selys, Bull. Acad. Belg., (2)35:764 (reprint, p. 36).
1905. Calv., B. C. A., p. 149.
1907. Calv., B. C. A., p. 398.

1917. Kndy., Proc. U. S. Nat. Mus., 52:524 (figs.).
1921. Kndy., Proc. U. S. Nat. Mus., 59:595 (figs.).
1929. N. & H., Handb., p. 62 fn.

Length 57 mm.; abdomen 42; hind wing 33.

A grayish brown and dull yellowish species of southwestern deserts. Face wholly pale grayish olive, with top of frons yellowish. Vertex black, with a broad transverse yellowish streak behind ocelli. Two low prominences in this streak thinly clad with whitish hairs. Occiput yellow, rear margin very slightly convex.

Prothorax grayish black, with a yellowish spot on each side. Synthorax striped with grayish brown and yellow. Carina yellow. Middorsal and antehumeral brown stripes more or less confluent at their upper ends below crest, but at lower ends widely separated by yellow of collar. Humeral and antehumeral stripes broadly conjoined at their lower ends but not at upper; often meet at two-thirds their height, interrupting there the narrow included strip of yellow. Midlateral stripe generally obsolete above spiracle, but conjoined with humeral at its lower end by a blackish band around lower edge of metepimeron. Third lateral stripe complete, ending below in a hook that curves to rearward.

Legs black beyond pale knee joints, wholly pale on short basal segments, and gray striped with blackish on all femora. Wings hyaline, a wash of brown in extreme base of both wings. Stigma brown, about 5 mm. long.

Abdomen indistinctly ringed with black and yellow, black on sides broadening to rearward on segments 1 to 7, almost separating lanceolate yellow spots of dorsum from narrow yellow lines of sides, except on 7, which is more than half yellow; 1 and 2 and half of 3 more broadly yellow on sides. Segment 8 black; 9 black, with a basal spot of yellow continued down side to join a submarginal (sometimes interrupted) yellow streak along lower margin; 9 and 10 black, each with a spot of yellow on side. Superior appendages of male yellow except at extreme base beneath; inferior appendage black; subanal plates yellow except at base. Stylets of female yellow, each ending in a minute black terminal spine.

Inhabits sandy shallows of desert streams.

Distribution and dates.—UNITED STATES: Ariz., Calif., Colo., N. Mex., Oreg., Tex., Utah; MEXICO: Baja Calif., Chihuahua, Sonora; also from Jalisco.

April 12 (Ariz.) to October 13 (Baja Calif.).

Progomphus clendoni Calvert

1905. Calv., B. C. A., p. 150.
1920. Wmsn., Occ. Pap. Mus. Zool. Univ. Mich., 77:12.
1941. Ndm., Trans. Amer. Ent. Soc., 67:235 (nymph).

Length 52 mm.; abdomen 35–41; hind wing 27–31.

This species near *obscurus,* differing in particulars as follows: obscure brownish markings on face below frons; middorsal and antehumeral brown stripes not confluent at their lower ends, or touch only at a point; midlateral brown stripe of thorax complete; abdominal segment 5 with something of a linear middorsal spot; basal fourth of superior abdominal appendage of male black, remainder pale green; tip of each arm of inferior appendage of male inequilaterally bifid.

Distribution and dates.—MEXICO: Tamaulipas; also from Costa Rica and Guatemala.

January 13 and June 19.

Progomphus integer Hagen

1878. Hagen, in Selys, Bull. Acad. Belg., (2)46:659 (reprint, p. 67).
1895. Calv., Proc. Calif. Acad. Sci., (2)6:500.
1941. Ndm., Trans. Amer. Ent. Soc., 67:222 (nymph).

Length 46–48 mm.; abdomen 35–37; hind wing 24–27.

A blackish species spotted with olivaceous gray. Face clothed with stiff black hair; sides of thorax below with long soft white hair. Face black, with three pairs of grayish spots at sides: one pair on labrum, one on postclypeus, and a pair closer together (sometimes conjoined into a cross stripe) on anteclypeus. Grayish top of frons somewhat invaded in low middle portion by the black extending forward from vertex; latter wholly black. Occiput yellowish, with black front and rear margins, and with a stiff fringe of long brownish hairs on its straight crest.

Prothorax black above, with an interrupted crossbar of yellow. Synthorax black, with a touch of yellow on carina; usual yellow collar stripe interrupted in middle and abbreviated at ends. About eight elongate pale gray spots on each side of thorax, vestiges of pale ground color of usual side stripes of brown remaining after fusion in their middle portion as well as at ends. At rear, a larger pale crescent covers most of metepimeron.

Legs black, but little paler basally. Wings subhyaline, with black veins and brown stigma.

Abdomen black, with only touches of yellowish or reddish brown on sides of segments 1, 2, 3, 8, and 9. Appendages of male black, including superiors except for their extreme tips.

Widely distributed in sand bars of small streams in Cuba.

Distribution and dates.—ANTILLES: Cuba, Dom. Rep., Jamaica.

Dates include January 7 (Jamaica) and April 30 and July 6 (Cuba).

Progomphus obscurus Rambur

1842. Rbr., Ins. Neur., p. 170 (in *Diastatomma*).
1854. Selys, Bull. Acad. Belg., 21(2):72 (reprint, p. 53).
1857. Selys, Mon. Gomph., p. 201.
1861. Hagen, Syn. Neur. N. Amer., p. 110.
1929. N. & H., Handb., p. 62 (figs.).
1939. Byers, Proc. Fla. Acad. Sci., 4:19.
1941. Ndm., Trans. Amer. Ent. Soc., 67:233 (figs., nymph).

Length 53 mm.; abdomen 40–41; hind wing 31–33.

A greenish yellow species, handsomely striped with brown. Face and top of head greenish, sometimes darkened in sutures, with a brown band across ocellar region from eye to eye and extending forward a little on top of frons.

Thorax with yellow and brown stripings, the two colors covering about equal areas. Middorsal thoracic brown stripes widened downward and isolated at yellow collar, but confluent with second stripe above along crest. Side stripes all broadly confluent at their lower ends, but narrowly conjoined above along subalar carina. Midlateral stripe continuous, though sometimes indistinct, widening to rearward at spiracle; third lateral stripe widest above, hooked to rearward at lower end.

Legs brown beyond knees and on anterior face of femora, with a large squarish yellow spot on dorsal side of femora just before knee. Wings subhyaline, with brown veins and a narrow line of yellow on costal margin. Each wing with a wash of brown that may extend outward from wing root to first thickened antenodal crossvein.

Abdomen brown on sides, also on dorsum of last four segments. Broadly marked with yellow on sides of segments 1 and 2; yellow diminishes to a narrow line on 3, continues so on 4 to 6, expanding again on 7 and 8. Yellow of dorsum forms basal saddle marks on 2 to 7, beyond which yellow diminishes in length on successive segments. Superior appendages of male yellow; inferiors black. Stylets of female yellow, ending in a minute black apical spine.

Inhabits sand bars in small streams, and shallows of wide lakes.

Distribution and dates.—UNITED STATES: Ala., Ark., Calif., Fla., Ga., Ill., Ind., Iowa, Kans., Ky., La., Mass., Mich., Miss., Mo., N. J., N. Y., N. C., Ohio, Okla., Oreg., Pa., S. C., Tenn., Tex., Va., W. Va., Wyo.; MEXICO: Baja Calif.

February (Fla.) to September 9 (Tex.).

Progomphus serenus Hagen

1878. Hagen, in Selys, Bull. Acad. Belg., (2)46:661 (reprint, p. 69).
1941. Ndm., Trans. Amer. Ent. Soc., 67:228 (figs., adult and nymph).

Length 42–45 mm.; abdomen 31–33; hind wing 24–26.

A slender blackish species, barred with gray on face and striped with greenish yellow on thorax. Blackish face marked with greenish gray in a wide border around front of labrum, a large spot on each side of postclypeus and on all prominences of frons. Side of mandibles white. Black of vertex broadly invades gray of upper surface of frons. Occiput grayish green, margined all around with black.

Front of black synthorax bears two opposed **7**-shaped yellow marks that are narrowly separated from each other at junction of carina and collar. Yellow line between humeral and antehumeral black stripes sometimes broken up into three spots. Inconstant side stripes of black are broad and diffuse, connected forward with humeral to enclose a band of yellow. Metepimeron broadly yellow, margined all around with black. Triangle of fore wings two- or three-celled; subtriangle one- or two-celled. Legs black beyond their pale bases, grayish on under side of front femora, tawny on outer side of hind femora.

Abdomen black; broadly overspread with yellow on sides of segment 2, on basal middorsal saddle marks on 2 to 7, and on sides of 8 and 9; 10 and caudal appendages wholly black. In female, extreme tips of appendages yellow.

Inhabits sand-and-gravel bars in swift streams of Dominican Republic. Adult commonly flushed from coarse rubble of aggrading shores.

Distribution and dates.—ANTILLES: Dom. Rep., Haiti.

March 15 (Haiti) to June 27 (Dom. Rep.).

Progomphus zephyrus Needham

1941. Ndm., Trans. Amer. Ent. Soc., 67:230 (figs., nymph).

Adult unknown. Characters of nymph as stated in key on page 96.

Distribution and dates.—ANTILLES: Dom. Rep.

June (nymph).

Genus GOMPHOIDES Selys 1850

This genus includes about a dozen species of large Neotropical dragonflies, two of which are found in Texas. They are stout-bodied, short-legged, clear-winged dragonflies. The face is pale, shining, with shadowy brown crossbands along its transverse sutures. The frons is low; its vertical height is less than that of the clypeus; its upper surface is bright yellow, with a straight-edged black crossband at its rear. The black of the vertex surrounds the ocelli and antennal bases, and extends downward along the eye border. The rear slope of the vertex and the front slope of the occiput are yellow or yellowish, with narrow black margins.

The thorax is about equally black and yellow. On the front a middorsal triangle of black is narrowly divided by the yellow edge of the carina. Two opposed **7**-shaped bright yellow marks surround that triangle; conjoined below, they cover the yellow collar. On the sides, four broad black or blackish stripes of about equal width are connected at both ends, narrowly along the subalar carina above, and more broadly at the leg level below. The third lateral (femoral) stripe is prolonged downward and rearward.

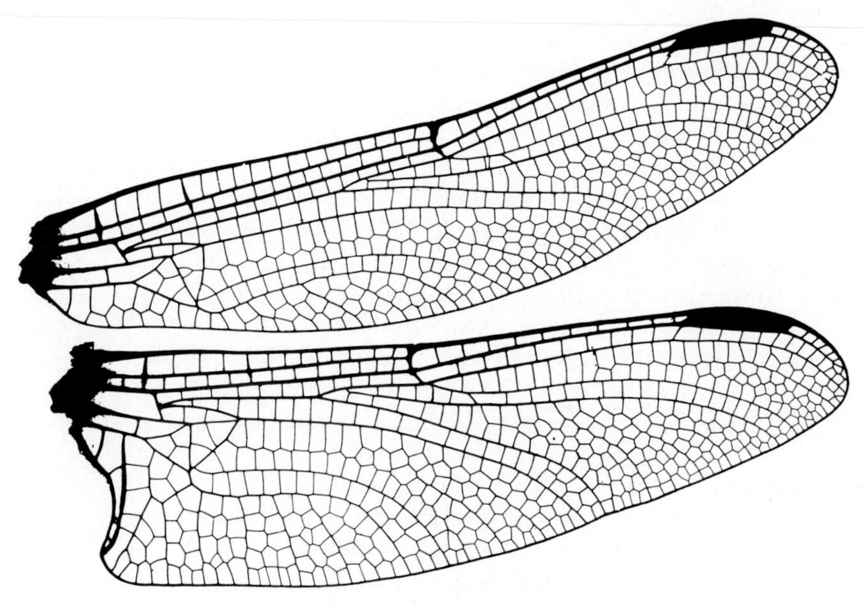

Fig. 45. *Gomphoides stigmatus.*

The legs are pale at the base and in streaks on the femora, with a little pale spot on the kneecaps; elsewhere they are black.

The wings are hyaline, with yellow costa and black veins and stigma. In their venation both triangles and subtriangles are unusually broad, and are divided by crossveins into a variable number of cells. A well-defined anal loop is present between the converging branches of the anal vein.

The abdomen is moderately enlarged on its basal segments, cylindric on its middle segments, especially in the male, and greatly expanded on 8 and 9, with 10 much smaller. It is broadly crossbanded with yellow on the slender middle segments. The sides of 1 to 3 are more broadly yellow; 4 to 7 are shining black behind the yellow crossbands; on 8 the yellow runs down into a yellow streak which then extends to rearward along the base of the leaflike lateral expansions of 8 and 9; 10 is dull yellowish.

The strong superior caudal appendages are forcipate, with a tooth above and one on the inner margin before the blunt tips. The branches of the inferior appendage are weak, flattened, and upcurved, and less than a third as long as the superiors. The copulatory appendages of segment 2 are conspicuously hairy along the midventral line. A transverse line of hairs borders the ventral edge of the anterior lamina. The hamules of the first pair are about as large as the second. They are bare and shining black, and widen upward to a blunt irregular summit, and are notched at their upper innermost angle. The second hamules are thick and hairy. Behind them a pair of densely hairy genital lobes continues the sheltering double line of hairs, and the end of the lane between them is occupied by the peduncle of the penis, which is covered by a round, paintbrush-like hair tuft. The peduncle is deeply divided by a long narrow cleft, in which the very long tapering tail of the penis tip lies concealed.

The adults frequent the streamside. They are found resting on stones in streams or on the tips of low vegetation. They are quick-starting, and difficult to approach as they sit always poised for flight.

The nymph is a burrower in muddy stream beds. It is of the usual Gomphine form, with depressed and wedge-shaped head, widespread hairy legs, and tapering abdomen. The burrowing hooks of the fore legs are large. The end hook on the lateral lobe of the labium is sharp, and before it the inner margin of that lobe is minutely denticulate. The median lobe is moderately prominent and fringed with hair. The tapering terminal segments of the abdomen increase progressively in length to rearward, segment 10 being about one and a third times as long as 9.

The nymph of *G. stigmatus* is in superficial appearance very like that of *Dromogomphus spinosus,* but differs in the uplifted spinelike dorsal hook of abdominal segment 9, in the greater length of 10, and in details of the structure of the labium.

KEY TO THE SPECIES

Adults

1—Third lateral stripe on thorax extends rearward and far upward at its lower end, forming a supernumerary thoracic side stripe on edge of metepimeron; segment 8 ends in a short apical declivity, the longest of its paired spinules not at its highest point.................................... **albrighti**

—Third lateral stripe on thorax with only a short upward turn at its lower end; segment 8 ends on middorsal line in a short, sharply ascending median ridge, with paired spinules of terminal cross ridge highest at its summit **stigmatus**

Nymphs

1—Dorsal hook on segment 3 about as high as those on 4 and 5; those on middle segments ridgelike .... **albrighti**

—Dorsal hook on 3 larger than on 2; those on middle segments moundlike **stigmatus**

Fig. 46. *Gomphoides albrighti.*

Gomphoides albrighti Needham

1950. Ndm., Trans. Amer. Ent. Soc., 76:1–3.

Length 60–63 mm.; abdomen 45–48; hind wing 37–40.

A recently described species, its pale face thickly besprinkled with minute black hairs and traversed by shadowy bands of pale brown overlying its transverse sutures. On labrum a short middorsal dash of brown extends half length from its basal suture, its outer margin fringed with reddish bristles. Opposed 7-shaped marks on front bright yellow with a tinge of green.

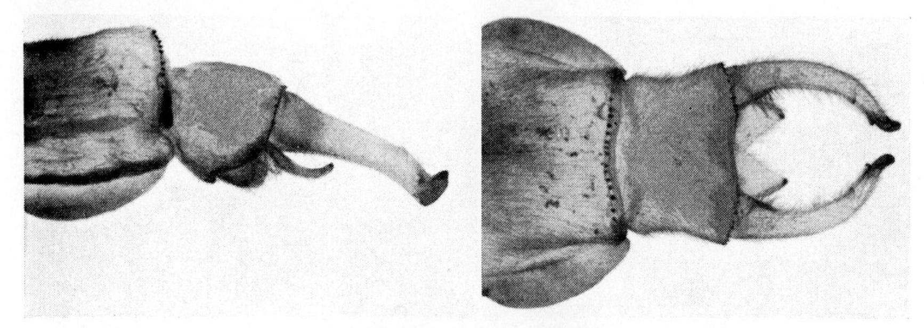

Fig. 47. *Gomphoides albrighti.*

Abdomen predominantly blackish, with yellow crossbands becoming darker, much broader, and more diffuse on end segments. Leaflike expansions of segments 8 and 9, when fully outspread, nearly twice as wide as segments; in female as broad as in male. Appendages of male yellow in middle, tinged with brown externally toward base, and shining black on tips of teeth and adjacent upcurving ends; those of female clear light yellow, with very long tapering tips.

In male, two teeth on each superior appendage of about equal size; inner tooth stands about midway between upper tooth and channeled and obliquely truncated tip. Inferior appendage thin, flat, upcurving; in drying it is subject to considerable distortion in appearance. Divided almost to base by a wide, straight-sided, V-shaped notch; edges of notch may fold in drying, and tip may curve far upward. Subgenital plate of female about a fifth as long as sternum of segment 9, deeply bifid by a U-shaped notch, the two parts unevenly truncated, partly uplifted, and set at an angle with each other.

At full maturity both male and female become more blackish. Yellow markings at base of leaflike expansions of 8 and 9 may wholly disappear, and a thin dusting of pruinosity may appear on under side of thorax.

Distribution and dates.—UNITED STATES: Tex.

June 12, June 14, September 18.

Gomphoides stigmatus Say

1839. Say, J. Acad. Phila., 8:17 (in *Aeschna*).
1854. Selys, Bull. Acad. Belg., 21(2):72 (reprint, p. 53) (in *Progomphus*).
1857. Selys, Mon. Gomph., pp. 205, 423.
1861. Hagen, Syn. Neur. N. Amer., p. 111.
1904. Ndm., Proc. U. S. Nat. Mus., 27:687 (figs., nymph).
1929. N. & H., Handb., p. 64 (figs.).
1940. Ndm., Trans. Amer. Ent. Soc., 65:368.

Length 65–70 mm.; abdomen 49–54; hind wing 39–42.

This long-known species differs from preceding one in characters stated in key; in its larger size; and in lighter yellow color, which inclines

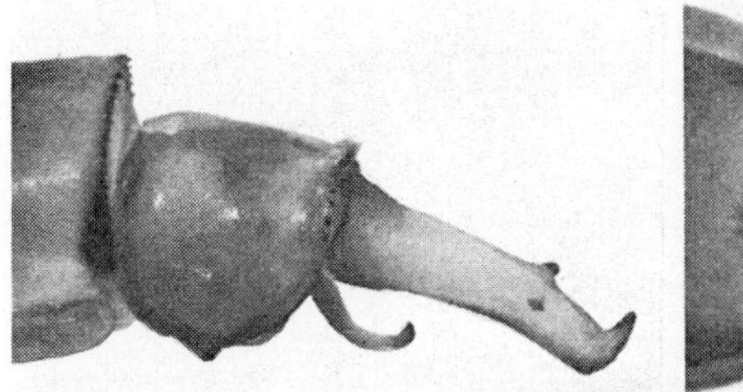
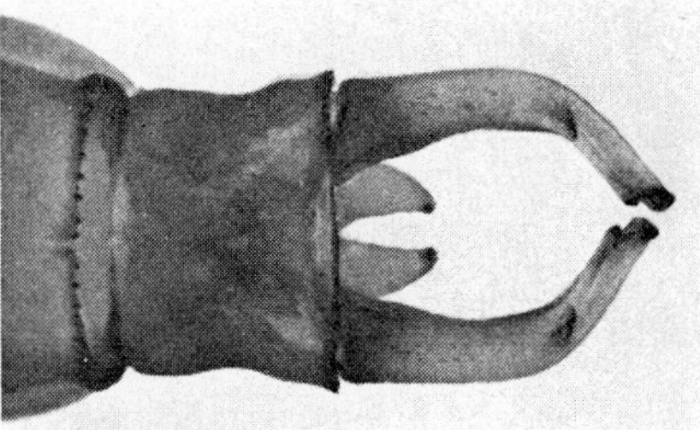

Fig. 48. *Gomphoides stigmatus.*

toward reddish on dorsum of terminal abdominal segments. Hairs fringing ridge of occiput tawny. Interrupted middorsal yellow line on abdominal segments 1, 2, and 3; 10 and appendages wholly yellow, except extreme tip of latter in male.

Female differs in having middle abdominal segments less slender, and lateral margins of 8 and 9 less expanded, hardly leaflike at all.

Distribution and dates.—UNITED STATES: streams of southern and western Tex.; MEXICO: Nuevo León, Tamaulipas.

July to August.

Genus APHYLLA Selys 1854

This Neotropical genus includes about a dozen species, four of which are ranged along our southern limits. They are large, clear-winged, short-legged dragonflies. The middorsal thoracic brown stripe is widened below, forming a broad triangle with a median yellow streak on the carina. The lateral stripes are all well developed, but vary in width and in continuity at their ends.

Venational characters of the genus, in addition to those shown in the table of genera, are: nodus beyond the middle of the fore wing; a very large stigma; a long anal triangle in the male, its slender tip outcurving and reaching to the hind angle of the wing, where there is very little development of a special tornal cell.

The lateral margins of abdominal segments 8 and 9 are very moderately dilated. The inferior caudal appendages of the male are so reduced as to appear to be wanting. Correlated with this lack of inferiors, the hind

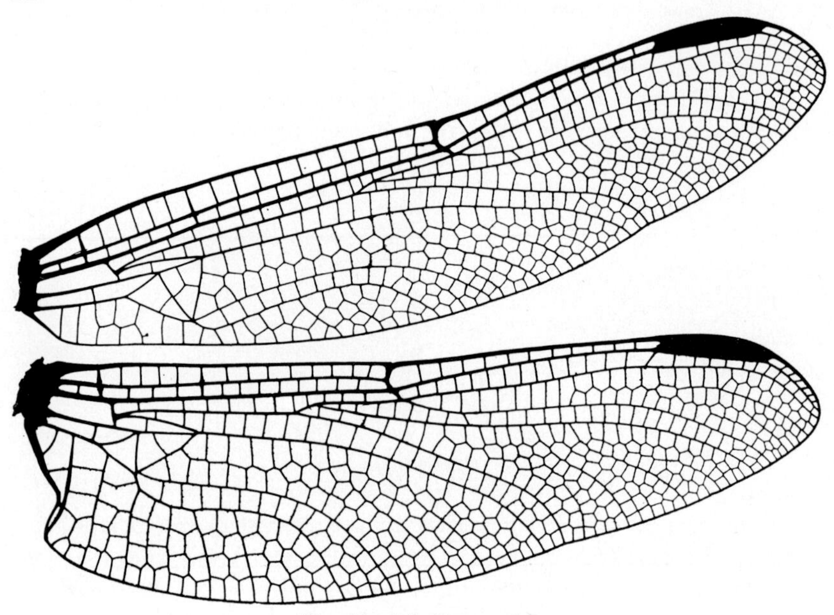

Fig. 49. *Aphylla caraiba.*

angle of 10 is prolonged into an acute point that is distinctive of the genus.

The nymph is unique in our fauna in having abdominal segment 10 greatly elongated, nearly as long as all the other segments taken together. It is further readily recognized by the entire absence of lateral spines on the abdomen, and by the huge teeth that margin the lateral lobe of the labium on the inner side.

The most useful reference papers for our species of this genus are those of Calvert (1905), Byers (1930), and Needham (1940).

KEY TO THE SPECIES

ADULTS

1—Sides of thorax with two wide brown bands formed by fusion of two pairs of stripes, humeral and antehumeral, and second and third laterals... **williamsoni**

—These stripes all separate, with pale areas between **2**

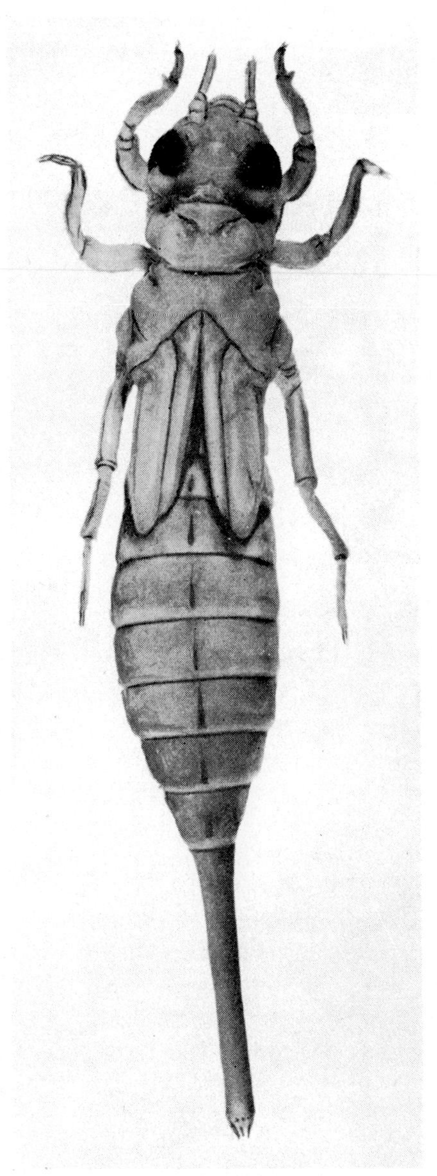

Fig. 50. *Aphylla caraiba.*

2—Labrum all pale .... **ambigua**
—Labrum with two pale areas separated by a triangular median blackish mark proceeding from base 3

3—Dorsum of segments 7 to 9 blackish; hind wing 31–39 mm.; Antillean.... **caraiba**
—Dorsum of 7 to 9 reddish; hind wing 36–42 mm. .... **protracta**

NYMPHS

1—Sides of mentum of labium parallel; two teeth before end hook of lateral lobe .... **caraiba**
—Sides of mentum of labium convergent toward base; three to five teeth before end hook 2

2—Three teeth before end hook of lateral lobe; median lobe evenly convex; segment 10 about as long as 5 to 9 inclusive.......................... **protracta**
—Four or five teeth before end hook of lateral lobe; median lobe similar but slightly truncate, with at least marginal fringe of scales interrupted there; segment 10 about as long as 4 to 9 inclusive..................... **williamsoni**

Nymph unknown: **ambigua.**

Aphylla ambigua Selys

1873. Selys, Bull. Acad. Belg., (2)36:505 (reprint, p. 61) (in *Gomphoides*).
1905. Calv., B. C. A., p. 157 (figs.); also p. 398 (in *Gomphoides*).
1940. Ndm., Trans. Amer. Ent. Soc., 65:364.

Length 64–66 mm.; abdomen 42–50; hind wing 35–49.

Greenish yellow face of this Mexican species rather strongly crossbarred with brown. Brown stripes of thorax rather wide, and intervening streaks of greenish yellow correspondingly narrow. Antehumeral stripe generally isolated below from middorsal stripe and above from midlateral stripe, but very variable. Abdomen faintly ringed with blackish at apex of slender middle segments; expanded apical segments broadly washed with reddish yellow. Deep middorsal notch in apical border of segment 10 and sinuate margin of lateral widening of 9 are characteristic of male.

Distribution.—MEXICO: Tamaulipas; also from Guatemala.

Aphylla caraiba Selys

1854. Selys, Bull. Acad. Belg., 21(2):79 (reprint, p. 60) (in *Gomphoides*).
1857. Selys, in Sagra, Hist. Cuba, Ins., p. 456.
1857. Selys, Mon. Gomph., p. 232.
1888. Gundlach, Contribucion a la Entomología Cubana, 2:234 (as *Gomphoides producta*).
1940. Ndm., Trans. Amer. Ent. Soc., 65:373–374 (figs., nymph).

Length 55–61 mm.; abdomen 41–45; hind wing 31–39.

Darker in color and smaller in stature than the other species. Two upper cross stripes of face broadly confluent in middle. Labrum bordered with brown and more or less completely divided on middle line by an invading basal triangle of black. Frons above half covered by a bilobed yellow crossband. Both vertex and occiput bordered with brown, former yellow only in a small central spot.

Middorsal and antehumeral stripes of thorax confluent at ends of yellow collar; latter isolated from humeral above by confluence of overarching yellowish green bands. Side stripes confluent below at level of spiracle.

Abdomen brown above, paler at sides. Each of middle abdominal segments bears weak encircling carina at a third its length, and a pair of subdorsal blackish marks near apex. In male, superior caudal appendage terminates in acute point that is distinctive of this species.

A not very common inhabitant of streams and ponds; adults perch on low vegetation at streamside. Nymphs when living in ponds burrow rather deeply in muddy bottoms.

Distribution and dates.—ANTILLES: Cuba, Dom. Rep., Haiti.

Apparently almost year-round; dates include April, June 16, June 28, December, and January 28.

Aphylla protracta Selys

1859. Selys, Bull. Acad. Belg., (2)7:546 (reprint, p. 20) (in *Cyclophylla*).
1861. Hagen, Syn. Neur. N. Amer., p. 113 (in *Gomphoides*).
1905. Calv., B. C. A., p. 157 (fig.) (in *Gomphoides*).
1929. N. & H., Handb., p. 65 (in *Cyclophylla*).
1940. Ndm., Trans. Amer. Ent. Soc., 65:372 (figs., nymph).

Length 62–68 mm.; abdomen 47–51; hind wing 36–42.

A handsome species (shown in our diagram on p. 11), marked in a rich pattern of brown and yellow. Two clypeal cross stripes of brown on face more or less confluent in middle. Labrum ringed and traversed with pale brown; occiput and rear of vertex broadly yellow. Middorsal and antehumeral thoracic dark stripes are separate at both collar and crest; latter abbreviated above, but meets humeral stripe below a narrow and intervening antehumeral line of yellow. Midlateral brown stripe broader than third lateral.

Slender middle abdominal segments yellowish, with a hairline of black on encircling carinae, and an apical middorsal triangle of black from which a wash of that color spreads diffusely forward and downward over sides, giving abdomen a ringed pattern that is conspicuous in males. Terminal segments reddish, with narrow apical rings of black. (See fig. 4.)

Inhabits pools of intermittent streams.

Distribution and dates.—UNITED STATES: Tex.; MEXICO: Tamaulipas; also south to Tabasco.

July 27 (Tex.).

Aphylla williamsoni Gloyd

1930. Byers, Odon. of Fla., pp. 46–48, 244–247 (fig., nymph) (as *Negomphoides ambigua*).
1936. Gloyd, Occ. Pap. Mus. Zool. Univ. Mich., 326:9 (figs.) (in *Gomphoides*).
1940. Ndm., Trans. Amer. Ent. Soc., 65:371 (figs., nymph).
1950. Bick and Aycock, Proc. Ent. Soc. Wash., 52(1):26.

Length 71–76 mm.; abdomen 52–62; hind wing 37–43.

Face obscurely brownish, brown merging into green on prominences of clypeus, labrum, and outer sides of mandibles. Head brownish dorsally; narrow occiput yellowish, with brown margins, its straight border densely

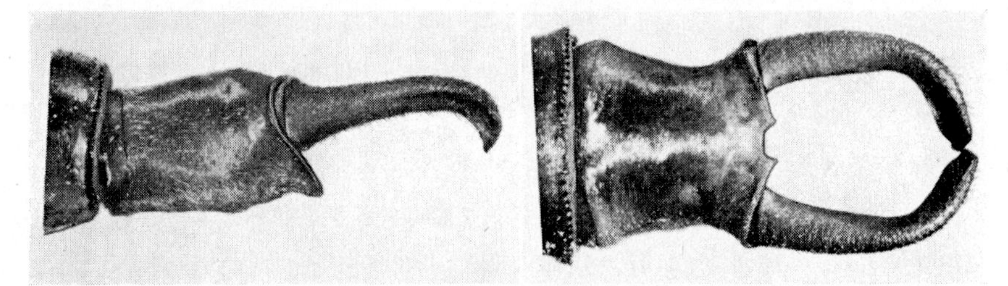

Fig. 51. *Aphylla williamsoni.*

fringed with short stiff brown hairs; longer bristles fringe top of a low postocellar ridge.

Thorax brown, with a pale T mark well down on front, formed by conjunction of yellow collar and carina; farther out on front, a pair of isolated yellow stripes are strongly divergent downward. Humeral and antehumeral stripes almost wholly fused into one wide brown band, behind which sides are yellowish green, traversed by another band of brown composed of the two fused lateral stripes. Wings clear, with brown veins and yellow stigma. Legs black beyond reddish femora.

Abdomen very long, with moderate enlargement of end segments. It is washed with black middorsally and on joinings of segments; sides paler, becoming yellowish on basal segments, rufous on apical segments. Narrow leaflike lateral expansions of segments 8 and 9 brighter reddish yellow. Superior caudal appendages of male black-tipped. Apical carinae of 8 and 9 have black nodules at hinge line; on 10 they are black around whole margin.

Female similarly patterned, with top of head and occiput less hairy; abdomen stouter. Leaflike expansions much smaller on segment 8, about same on 9 as in male.

A strong-flying species, difficult to approach for capture. Inhabits soft bottoms of shallow lakes, bayous, and streams. Transformation occurs on posts, on masses of floating vegetation, and on wet sandy shores.

Distribution and dates.—UNITED STATES: Ala., Fla., La., N. C.

April 14 (Fla.) to November 2 (Fla.).

Genus HAGENIUS Selys 1854

This genus includes only the single widely distributed North American species described below. Its nearest relatives are in Japan and India. In

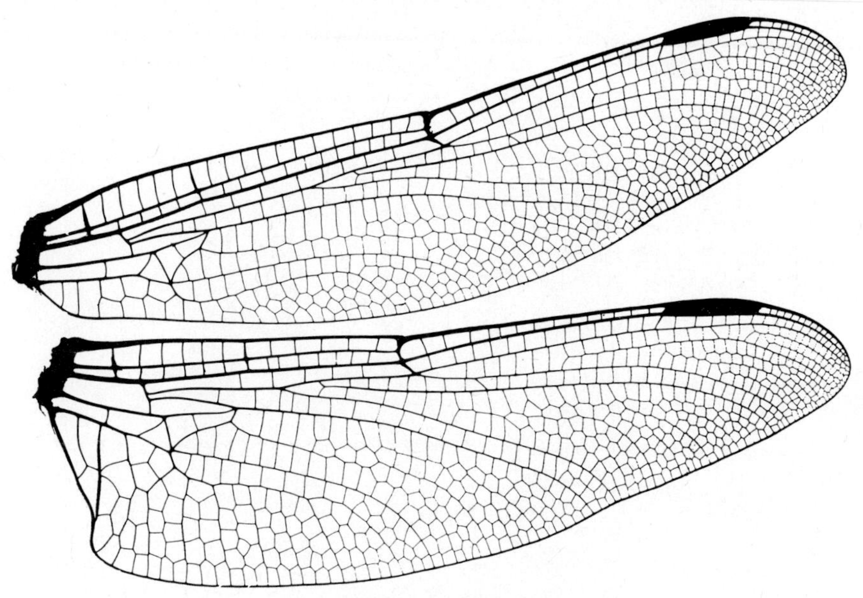

Fig. 52. *Hagenius brevistylus.*

our species both legs and wings are long and powerful. The stigma is rather narrow and well braced. The triangles are elongated, and angulated on the outer side, with a fairly well developed trigonal planate springing from the angle. The anal loop is normally four-celled and a little broader in the female than in the male. The tibiae are thickly beset with strong spines on the inner carinae, and sharply serrated on the outer ones; the femora are similarly armed within, but rounded, smooth, and shiny on the outer (dorsal) side.

The abdomen is very stout, moderately enlarged on the basal segments and nearly parallel-sided thereafter to the end. Its segments rapidly diminish in length from 6 onward, the last four being proportioned in length about as 18:13:10:6, with the appendages a little shorter than the last segment, and very stout.

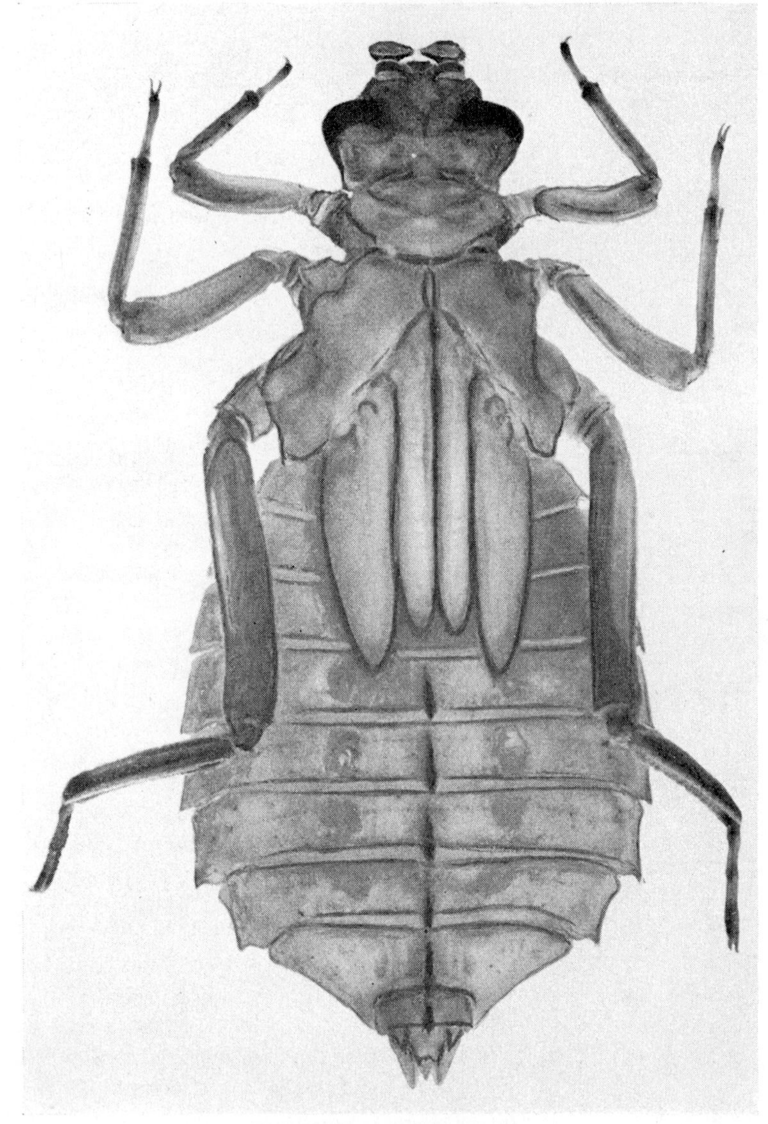

Fig. 53. *Hagenius brevistylus.*

The nymph is a very remarkable creature, flattened and widened, stiff-legged and slow. The synthorax widens to rearward until it is twice as wide as the head, and the nearly circular abdomen is still wider. Antennal segment 3 is likewise flattened; 4 is vestigial. On the rear of the head is a transverse row of blunt tubercles. Under the head a pair of heavy flanges support the very stout labium at the sides. There is a complete row of dorsal hooks on abdominal segments 1 to 10; the hooks are laterally flattened, blunt, low on the end segments, highest on 7. The lateral spines are hardly more than rearward-projecting angles of segments 4 to 8, largest on 7. Annular segment 10 is set in the end of 9. There are no burrowing hooks; the nymph is a sprawler, not a burrower.

The nymphs live along the edges of the larger woodland streams on the bottom, among chips of bark and leaves of their own dark coloration. The grotesque cast-off skins, left behind at transformation, may be found sticking to logs and trash within a foot of the water's edge. Several sizes of nymphs are usually found together, indicating that probably a number of years are required for growth.

Hagenius brevistylus Selys

1854. Selys, Bull. Acad. Belg., 21(2):82 (reprint, p. 63).
1857. Selys, Mon. Gomph., p. 241 (figs.).
1872. Cabot, Cat. M. C. Z., 5:9 (figs., nymph).
1901. Ndm., Bull. N. Y. State Mus., 47:440 (figs., nymph).
1929. N. & H., Handb., p. 66.

Length 73–90 mm.; abdomen 53–63; hind wing 47–58.

A giant black Gomphine, face crosslined with black on some or all sutures; occiput black. Front of synthorax black, with only summit of carina, sides of collar, and a pair of linear-oval stripes yellow; stripes entirely enclosed by confluence of broad middorsal and antehumeral black stripes at both ends. Often confluent also in middle, leaving intervening pale streaks above and below that meeting. Sides of thorax yellow, with lateral black stripes broad and continuous, and more or less fused. Legs black. Wing veins blackish, with edge of costa narrowly yellowish and stigma brown; nodal crossveins about 22:16/13:16. Usual middorsal pale stripe of abdomen abbreviated on segments 6, 7, and 8, and wanting farther back, 9 and 10 being wholly black above. Sides of 8 and 9 often broadly yellow.

Adults may be found foraging in open roadways through woods near streams; or may be seen swooping through sunny openings over streams pursuing butterflies and other insects, sometimes dive-bombing for a *Gomphus* half their own size and, if successful, flying away with it among the treetops. Females oviposit while flying, in a zigzag course, circling over a comparatively small area of open water, rising and descending and lightly striking surface with tip of abdomen to wash off a little cluster of eggs at each descent.

Hagenius was one of the few gomphines which could be seen almost every day all summer long. Despite its size and flying ability it was not very difficult to take, partly because of its fearlessness and partly because of its habit of flying a "beat" with definite perches along it. A female of this species was observed ovipositing late in the afternoon of July 14, 1945. She faced a bank of the stream which rose abruptly about four feet from the water. The bank was hung with a mass of roots and rootlets of a nearby silver maple. She would hover for an instant a few feet from the bank and about a foot above the water, then flying toward the bank and dropping, she would dip her abdomen at the water's edge, rise a few feet and return to her original position without turning around. This action was repeated several dozen times without interruption.—Donald C. Scott ("Notes on the Odonata of the Tippecanoe River State Park, Pulaski County, Indiana," *Proc. Ind. Acad. Sci.*, 55:200, 1946).

Distribution and dates.—CANADA: Man., N. B., N. S., Ont., Que.; UNITED STATES: Ala., Conn., Fla., Ga., Ill., Ind., Kans., Ky., La., Maine, Md., Mass., Mich., Minn., Miss., Mo., N. H., N. J., N. Y., N. C., Ohio, Okla., Pa., S. C., Tenn., Tex., Va., Wis.

May 13 (Fla.) to November 8 (Fla.).

Genus OPHIOGOMPHUS Selys 1854

These clear-winged stream-haunting dragonflies are of moderate size and of grayish green or yellowish green and black or brown coloration. The thorax is greenish, striped with black or brown, the stripe on the midlateral suture generally obsolete above the spiracle. The stout abdomen is yellowish, heavily overlaid with brown or black in a pattern that is repeated on successive segments, with a snaky effect that doubtless suggested the generic name (*ophis*, a serpent). The pale color of the abdomen is in three interrupted longitudinal bands: one middorsal band composed of spots that on each segment are broad basally, often incised at the sides, and narrowed and abbreviated and mostly isolated to rearward; the other two bands are often narrowed to a mere marginal line on the middle segments, broadened and more diffuse on the segments at front and rear ends, and generally heightened in yellow color in the males on the expanded lateral margins of segments 7 to 9.

The legs are rather short, and well armed with stout spines. The chief distinguishing feature of *Ophiogomphus* venation is the more or less semicircular form of the anal loop, shaped by the convergent angulation of veins A1 and A2, enclosing three (sometimes two or four) cells.

The abdomen is rather stout. Segment 8 is a little shorter than 9; 10 is a little more than half as long as 9. The superior appendages are about as long as 10; the inferiors are of very great diversity of form in the different species, furnishing the best of characters for specific determination. These and the hamules of segment 2 are, therefore, more trustworthy than the variable color pattern.

Spines on the occiput of the female are useful recognition characters, but they are very variable. One pair of spines occurs on the upper side, on or before the crest; the other pair, on the rear, wide apart and close to the hind border of the eye. The column in our table of species records merely their occurrence in the different species and tells nothing of their inconstancy of size and position. Largest and most conspicuous of them all is the forward-pointing approximated pair that stands just before the crest in *mainensis;* those of *arizonicus* are little truncated stumps far

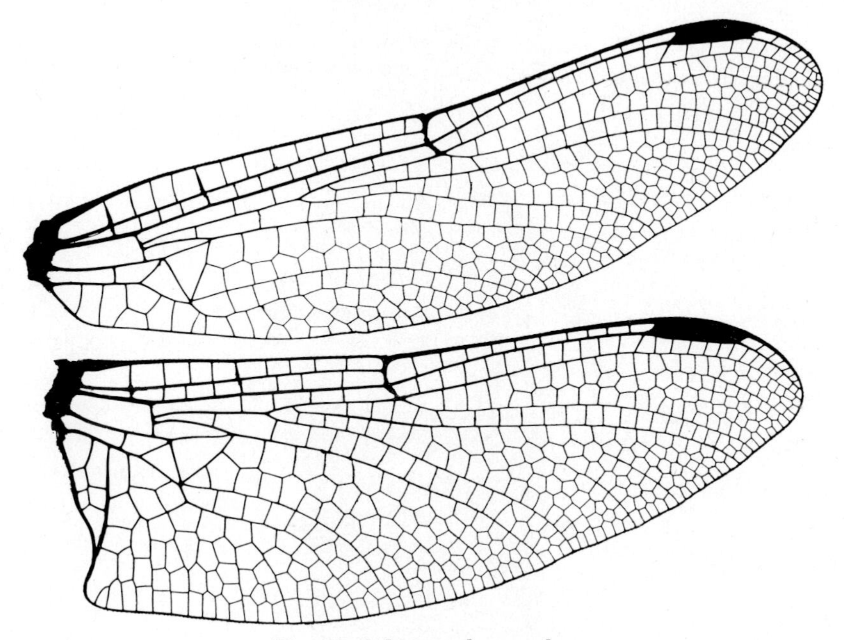

Fig. 54. *Ophiogomphus carolus.*

apart on the occipital crest. Largest of the rearward pairs are those of *occidentis* and *rupinsulensis.* They are short and thick, and look like crumpled horns. Those of *aspersus* resemble slender incurved thorns. Others both front and rear may be so very small that they are easily overlooked. Because of the fewness of good specimens for study, our listing is probably incomplete.

The nymphs of this genus are short and stout of body, with the developing wing cases strongly divergent. There are large burrowing hooks on the end of the fore and middle tibiae. The terminal (fourth) joint of the antenna is vestigial; the long third joint is more or less flattened and hairy on its edges. The labium is short and broad; the middle lobe of the mentum is slightly convex on its fore border, where it is fringed with a row of scalelike bristles and edged with many blunt, close-set microscopic teeth. The end of the lateral lobe is bluntly rounded; it is armed with microscopic teeth on its inner margin. There are rather stout and blunt dorsal hooks on abdominal segments 2 to 9,

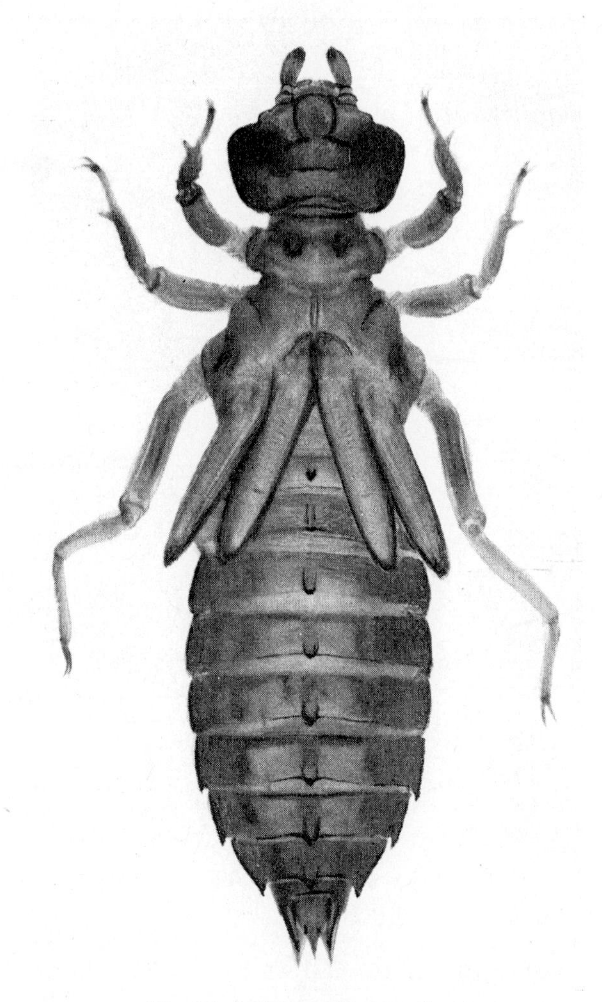

Fig. 55. *Ophiogomphus morrisoni.*

and lateral spines on 6 or 7 to 9. The caudal appendages are of unequal length, the lateral ones being much shorter than the superior. There is little color to the skin, which is rough and thinly hairy about the edgings.

This genus is of northward distribution. Sixteen species are known to occur within our limits—more than in Europe and Asia combined. The center of abundance for the genus is in the northeastern United States and Canada, with an outlying lesser group of larger species on our Pacific Coast. There is a greater density of venation in the smaller darker Eastern species than in the larger paler Western ones. This is seen, for example, in the number of crossveins under the stigma: not counting the brace vein at its front, there are generally four to six in the Eastern species; two or three in the Western. The density of venation is seen also in the number of cells in the anal triangle of the male: generally five or six in the Eastern species; four in the Western.

Papers most useful for further studies of *Ophiogomphus* are those of Needham (1899), Kennedy (1917), and Walker (1933).

I have never found one of this genus where there are alders [*Alnus*]; only where the sides [of the stream] are grassy and open, where they will be found on the grass or more frequently upon stones jutting from the water.... Their habits and the noticeably green thorax distinguishing them from anything else.—W. T. Davis (*Ent. News*, 51:63, 1940).

KEY TO THE SPECIES*

ADULTS

1—Face cross-striped with black 2
—Face not cross-striped with black 3

2—Hind wing 27–31 mm.; interclypeal cross stripe on face entire .... **colubrinus**
—Hind wing 24–27 mm.; interclypeal cross stripe interrupted .... **anomalus**

3—Tibiae yellow externally 4
—Tibiae black externally 11

4—Antehumeral thoracic brown stripe incomplete at both ends, or a mere spot, or wanting 5
—Antehumeral thoracic brown stripe continued downward to collar and there joining humeral 7

5—Middorsal thoracic brown stripe well developed .... **montanus**
—Middorsal thoracic brown stripe vestigial or wanting 6

6—Inferior appendage of male about four-fifths as long as superiors; no horns on occiput of female .... **severus**
—Inferior appendage of male about half as long as superiors; low blunt tubercles on occiput of female .... **arizonicus**

7—Middorsal brown stripe wanting .... **rupinsulensis**
—Middorsal brown stripe present 8

8—Antehumeral stripe sinuate on front margin .... **occidentis**
—Antehumeral stripe not sinuate on front margin 9

9—Inferior appendage of male with strong supero-lateral tooth on each side; Eastern .... **carolinus**
—Inferior appendage of male with no supero-lateral tooth; Western 10

10—Humeral and antehumeral stripes about three to four times as wide as interval between .... **morrisoni**
—Humeral and antehumeral stripes about as wide as interval between .... **nevadensis**

* Key to the genera of Gomphines is on p. 90.

TABLE OF SPECIES—ADULTS

Species	*Hind wing*	*Yellow on tibiae*[1]	♂ *Inf. append.*[2]	♂ *P.ham. shoulder*[3]	♀ *Spines on occiput*[4]	♀ *Sub-genit. plate*[5]	*Distri-bution*
anomalus	24-27	0	9	+	4	5	NE
arizonicus	33-37	+	6	0	2	8	Ariz.
aspersus	24-32	0	9	+	2	6	NE
bison	29-33	0	8	0	2	9	W
carolinus	25-27	+	10	+	2	7	N. Car.
carolus	24-28	0	9	+	2	10	NE
colubrinus	27-31	0	12	+	2	7	N
edmundo	24-29	0	10	0	2	7	E
howei	19-21	0	6	+	2	7	E
mainensis	25-31	0	9	+	4	8	N, E
montanus	30-32	+	8	0	0	6	W
morrisoni	28-33	+	10-11	0	0	6	W
nevadensis	30-38	+	10	0	0	6	W
occidentis	28-33	+	8	0	4	9+	W
rupinsulensis	27-32	+	9	±	4	8	N, E
severus	28-34	+	8	0	0	7	W

[1]Yellow on outer face of tibiae: present (+), absent (0), or variable (±).

[2]Length of inferior appendage of male, expressed in tenths of length of superior.

[3]Shoulder on front edge of posterior hamule of male: present (+) or absent (0).

[4]Number of spines on occiput of female.

[5]Length of subgenital plate of female, in tenths of length of ninth sternite.

11—Hind wing less than 22 mm.; inferior appendage of male about half as long as superiors .... **howei**
—Hind wing more than 23 mm.; inferior appendage of male more than half as long as superiors.. 12

12—Humeral and antehumeral stripes confluent or nearly so................. **bison**
—Humeral and antehumeral stripes well separated except at ends............ 13

13—Inferior appendage of male with a projecting lateral angle and with rectangular notch in its tip; female without horns on dorsal surface of occiput 14
—Inferior appendage of male with strong supero-lateral tooth on each side; female with horns on dorsal surface of occiput.......................... 15

14—Thoracic stripes faint; abdomen blackish; superior appendages of male much longer than inferior, with bulbous swelling in middle......... **aspersus**
—Thoracic stripes distinct; abdomen brownish; superior appendages of male equal in length to inferior, and not bulbous in middle.............. **edmundo**

TABLE OF SPECIES

NYMPHS

Species	*Total length*	*Lateral spines*[1]	*Appendages*[2]		
			Lat.	*Sup.*	*Inf.*
anomalus	23-25	7<8=9	7-8	10	10
aspersus	25-28	7<8>9	8	10	10
bison	28-29	6<7=8>9	9–	10	10
carolinus	24-28	7<8±9	8	10	10
carolus	23-26	7=8±9	8	10	10
colubrinus	26-30	7=8=9	8	10	10
mainensis	24-26	7<8>9	7	10	10
morrisoni	27-28	7<8=9	8	9+	10
nevadensis	30-31[3]	7<8>9	8	9+	10
occidentis	27-29	6<7<8>9	7	10	10
rupinsulensis	26-28	7<8<9	8	10	10
severus	25-28	7<8=9	8+	9+	10

[1]Relative size of spines on abdominal segments 6 to 9: increasing (<), decreasing on successive segments (>), equal (=), or subequal (±).

[2]Length of superior and lateral caudal appendages expressed in tenths of length of inferiors: plus (+), minus (–).

[3]Differs from *morrisoni* in having slightly larger dorsal hooks on 7, 8, and 9 also.

15—Tip of inferior appendage of male with a cleft about as deep as wide; subgenital plate of female reaching tip of segment 9.................. **carolus**
—Tip of inferior appendage of male with a cleft twice as deep as wide; subgenital plate of female covers about four-fifths of segment 9...... **mainensis**

NYMPHS

1—Lateral spines on abdominal segments 6 to 9.............................. **2**
—Lateral spines on 7 to 9.. **3**

2—Lateral appendages nearly as long as inferiors.......................... **bison**
—Lateral appendages two-thirds as long as inferiors................. **occidentis**

3—Antennal segment 3 broadly oval; 4 nearly as wide at base as 3, cut off obliquely .... **anomalus**
—Antennal segment 3 long and narrow; 4 a conical rudiment................. **4**

4—Antennal segment 3 narrowly oblong, densely scaly, twice as long as wide **mainensis**
—Antennal segment 3 two and a half to three times as long as wide.......... **5**

5—Superior caudal appendage a little shorter than inferiors; Western.......... **6**
—Superior appendage as long as inferiors; Eastern.......................... **8**

6—Dorsal hooks on 8 and 9 weak, slender, flattened; tips of those on 2 and 3 tapered, ascending .... **severus**
—Dorsal hooks on 8 and 9 stout, not flattened, ascending; those on 2 and 3 very blunt-tipped, and crooked to rearward............................ **7**

7—Side margins of 7 densely hairy.................................. **morrisoni**
—Side margins of 7 nearly bare.................................. **nevadensis**

8—Lateral appendage twice the middorsal length of 10...................... **9**
—Lateral appendage two and a half times as long as 10..................... **11**

9—Antennal segment 3 about three times as long as wide............ **rupinsulensis**
—Antennal segment 3 about two and a half times as long as wide............. **10**

10—Dorsal hooks on 7 to 9 sharp..................................... **carolinus**
—Dorsal hooks on 7 to 9 blunt... **carolus**

11—Dorsal hooks on 8 and 9, viewed from above, with acute tips........ **colubrinus**
—Dorsal hooks on 8 and 9 with obtuse tips........................... **aspersus**

Nymphs unknown: **arizonicus, edmundo, howei,** and **montanus.**

Ophiogomphus anomalus Harvey

1898. Harvey, Ent. News, 9:60 (figs.).
1901. Calv., Ent. News, 12:241.
1901. Harvey, Ent. News, 12:240.
1918. Howe, Odon. of N. Eng., p. 28.
1927. Garm., Odon. of Conn., p. 133 (figs.).
1929. N. & H., Handb., p. 70.
1933. Walk., Can. Ent., 65:221 (figs., nymph).

Length 39–44 mm.; abdomen 28–33; hind wing 24–27.

A slender greenish species, heavily striped with black. Face green, with full-length black cross stripes on frontal and labral sutures and an interrupted one on clypeus. Occiput green, with a touch of black on its outermost angles that is continued on a black ridge behind eye.

Narrow middorsal thoracic stripe brown, often with a median touch of green on upper end of carina. Humeral and antehumeral stripes blackish, fused together, more or less, with a narrow and very variable line of yellow below mid-length. Even when fused, a spot of green remains between them, near crest; or, if upper end of antehumeral be free, this spot is conjoined with larger green area of front of thorax. Midlateral brown stripe wanting above level of spiracle; from that level a black bar runs downward and backward to join lower end of complete second lateral stripe, forming a black N mark above leg bases. Legs blackish.

Abdomen mainly black, with usual interrupted longitudinal middorsal and lateral pale bands wide on base and reduced to rows of spots on middle segments: middorsal one, to half-length spearheads on 7 and 8, to a spot with a tail on 9, and to a larger round spot on 10. Sides of slightly enlarged end segments washed with yellow: indistinct forked streak on 7, a wider band (sometimes divided into two spots) on 8, and a C-shaped brighter yellow mark on 9; all these markings subject to considerable variation in both area and depth of coloration. On 10, an indistinct tinge of yellow. Caudal appendages brown.

In female, two pairs of nipple-shaped blackish horns on head. One pair rises from swellings on front of yellow occiput near its summit; other pair on rear, well over on edge of rear of eye. Horns minute and difficult to see. Low postocellar ridge nearly bare.

Distribution and dates.—CANADA: Ont., Que.; UNITED STATES: Maine. June 6 (Maine) to August 7 (Que.).

Ophiogomphus arizonicus Kennedy

1917. Kndy., Proc. U. S. Nat. Mus., 52:538 (figs.).
1929. N. & H., Handb., p. 73 (figs.).

Length 54 mm.; abdomen 40–42; hind wing 33–37.

A pale yellowish desert species, with dark markings reduced to a minimum for genus. Middorsal and midlateral stripes wanting on thorax; antehumeral stripe represented by a roundish or oval spot in space that stripe occupies in other species; humeral stripe narrow but complete; second lateral stripe appears only as a streak of brown in lower part of its suture. Legs yellow at base, becoming black on femora dorsally toward knees; tibiae pale externally; tarsi black.

Middorsal and lateral pale bands of abdomen very wide from end to end. Dorsal yellow full length on segments 1 and 2, progressively shortened on 2 to 7, reduced to spots on 8 and 9; 10 mostly diffuse yellow.

Lateral band of abdomen extends full length, narrowing only a little on middle segments. Between these bands, a stripe of black, conjoined with its fellow across dorsum at apical end of middle segments, gives appearance of half rings to abdomen.

This species nearest to *O. severus,* from which it is distinguished in male by having inferior caudal appendage only half as long as superiors; and in female by having a pair of short, obliquely truncated, widely spaced spines on ridge of occiput.

Distribution and dates.—UNITED STATES: Ariz.

June 14 to August.

Ophiogomphus aspersus Morse

1895. Morse, Psyche, 7:209.
1899. Ndm., Can. Ent., 31:236 (figs.).
1918. Howe, Odon. of N. Eng., p. 30.
1927. Garm., Odon. of Conn., p. 132 (figs.).
1929. N. & H., Handb., p. 71 (figs. as *occidentis*).
1933. Walk., Can. Ent., 65:228 (figs., nymph).
1939. Davis, Ent. News, 51:63.

Length 44–49 mm.; abdomen 30–35; hind wing 24–32.

A greenish species with narrow stripings of brown. Face and occiput green. Middorsal thoracic stripe weakly developed, sometimes apparently wanting, but generally narrow and parallel-sided and conjoined with crest above, but abbreviated below. Antehumeral and humeral stripes likewise variable, generally well separated by a line of green but inconstant in width, in joinings, and in depth of color. Midlateral stripe generally present and complete, but may be only a darkening in depths of its suture. Legs pale to near knees, then brown; tibiae only slightly yellow externally, tarsi black. Wings with brown veins and a line of yellow on front of costa; stigma tawny.

Abdomen blackish, with usual longitudinal bands of yellowish spots. Spots of middorsal band narrow to elongate-triangular form on base of segments 3 to 6, become about half length on 7 and 8, a short round spot with a tail on 9, and an oboval spot on 10. Appendages brownish. Lateral band of abdomen appears as yellowish spots on the little-expanded sides of end segments: a small diffuse spot on 7, and large, better-defined spots on 8 and 9, outside which the margin of segment is bordered with blackish.

Female with horns on rear of head only, minute and difficult to see, nipple-shaped, and blackish. Postocellar ridges very densely hairy.

Inhabits clear streams where shallow current ripples over sand. Males fly back and forth over stream, then rest on a prominent twig, preferably on a high bank, near by; not difficult to approach and to capture when at rest. Female makes a succession of sweeps back and forth over head of some little riffle, striking water again and again near same place, leaving her eggs in it. Nymphs sometimes very common in sandy bed. They trail along, burrowing rapidly at a slight depth through nearly clean sand under current. Often leave a faint trail, showing where tip of abdomen, upturned for respiration, has pushed sand grains aside.

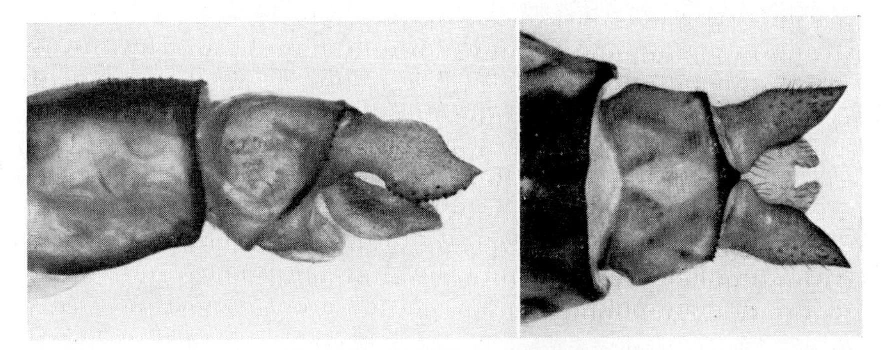

Fig. 56. *Ophiogomphus aspersus.*

Distribution and dates.—CANADA: N. B., N. S., Que.; UNITED STATES: Conn., Ky., Maine, Mass., Mich., N. H., N. Y., Vt.

May 26 (New England) to August 29 (New England).

Ophiogomphus bison Selys

Syn.: sequoiarum Butler

1873. Selys, Bull. Acad. Belg., (2)36:496 (reprint, p. 51).
1904. Ndm., Proc. U. S. Nat. Mus., 27:690 (figs., nymph).
1914. Butler, Can. Ent., 46:346 (as *sequoiarum* n. sp.).
1917. Kndy., Proc. U. S. Nat. Mus., 52:540 (figs.).
1929. N. & H., Handb., p. 77 (figs.).

Length 49–51 mm.; abdomen 33–38; hind wing 29–33.

A large and very handsome species, olive green with stripes of brown on thorax, and half-ringed black and yellow on abdomen. Face and occiput yellow. Middorsal thoracic brown stripe rather narrow, parallel-sided, constricted at collar to a narrow crossing. Its upper end produced laterally along crest to join antehumeral, which is then fused for a distance with humeral, then separated below by usual narrow intervening greenish line. This pair of stripes dark brown. Lateral stripes little developed; midlateral stripe practically wanting, the other a mere line in bottom of its suture.

Middorsal pale band of abdomen is narrowed as usual on the three basal segments to an elongated spot, with notch in each margin on middle segments, a half-length spearhead on 8, a band nearly full length on 9, and a half-length oboval spot on 10. Band of lateral margin appears as a narrow forked strip on 7, broad squarish half-height and more than half-length bands on 8 and 9, and a diffuse wash on 10. Yellow on 7 to 10 covers part of lateral margin of these moderately expanded segments. Caudal appendages yellow, tipped with brown.

In female, two large black-tipped horns project forward from ridge of yellow occiput, about as far from each other as from the eyes.

Inhabits small lowland West Coast streams.

Distribution and dates.—UNITED STATES: Calif., Nev.

May to August 16.

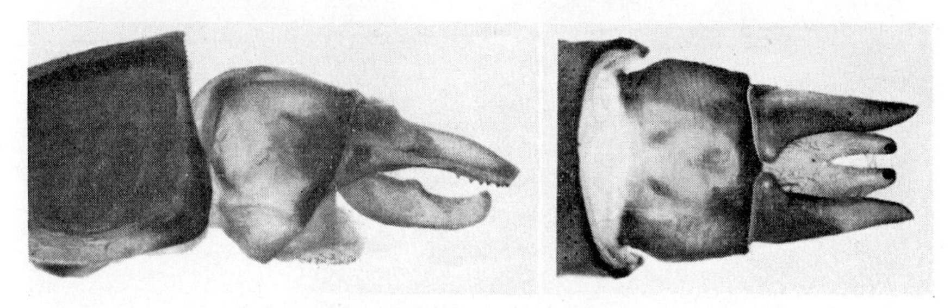

Fig. 57. *Ophiogomphus bison.*

Ophiogomphus carolinus Hagen

1885. Hagen, Trans. Amer. Ent. Soc., 12:259 (nymph only).
1899. Ndm., Can. Ent., 31:233–238 (figs., adult only).
1929. N. & H., Handb., p. 72 (figs.).

Length 44–46 mm.; abdomen 26–32; hind wing 25–27.

A pale greenish species, with stripings of brown. Face and most of top of frons yellow. On head in female, two minute yellow horns with black tips are laid flat, pointing forward. Middorsal thoracic stripe of brown wedge-shaped, widened forward and narrowly divided by yellow on carina; at its front end, suddenly narrows to a line.

Abdomen slender, moderately clubbed toward end and broadly marked with yellow at sides of end segments, and also of basal segments. Yellow on sides of segments 8 and 9 narrowly margined externally by black in both male and female. Interrupted band of yellow spots on back. Segment 10 and caudal appendages yellow in both sexes. Hamules of male black; anterior one wholly black; posterior, yellow on shoulder and tip.

This description is drawn from three specimens in the Museum of Comparative Zoölogy at Cambridge, Massachusetts, one male and two females, all labeled "types"; also labeled "North Carolina" and "C. U. Lot 35" (Cornell University Accessions, Lot Number 35): "a lot of specimens of insects . . . [including Odonata] from North Carolina purchased from H. K. Morrison." Doubtless these were among others presented by J. H. Comstock to H. A. Hagen. No further data are available.

Distribution.—UNITED STATES: N. C.

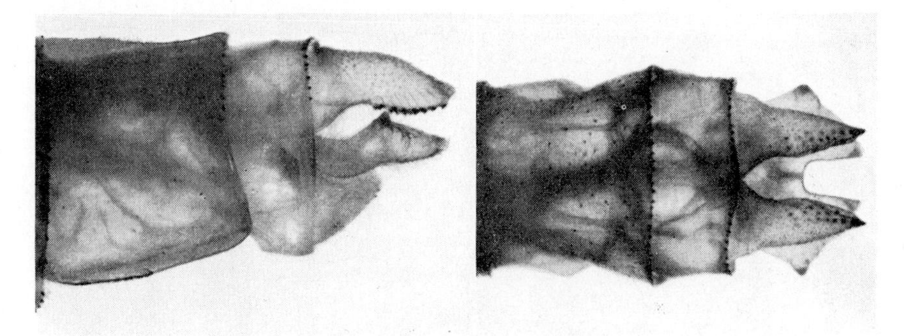

Fig. 58. *Ophiogomphus carolus.*

Ophiogomphus carolus Needham

1897. Ndm., Can. Ent., 29:183 (figs.).
1899. Ndm., Can. Ent., 31:235 (figs.).
1901. Ndm., Bull. N. Y. State Mus., 47:436 (figs., nymph).
1918. Howe, Odon. of N. Eng., p. 29 (figs.).
1927. Garm., Odon. of Conn., p. 134 (figs.).
1929. N. & H., Handb., p. 77 (figs.).
1933. Walk., Can. Ent., 65:223 (figs., nymph).

Length 40–45 mm.; abdomen 28–33; hind wing 24–28.

A small greenish species, rather smartly striped with brown. Face and occiput yellow. Middorsal thoracic stripe rather narrow and parallel-sided. Antehumeral stripe varies greatly in connections, from being entirely free from humeral at both ends or partially conjoined, to being wholly fused with it. Humeral at its lower end turns backward to form a **J** mark with upturned end beside spiracle. Above that level, midlateral is wanting. Second lateral generally complete but narrow. Legs blackish, with paler areas on sides of femora. Wings clear, with a fine yellow line on front of costa.

Middorsal pale band of abdomen broadest on segment 2, narrowed and pointed, notched on each side and abbreviated on 3 to 7, still shorter and quadrangular on 8, smaller and square or quadrangular (transversely placed) on 9, faint or wanting on 10. Side band of abdomen, reduced as usual on middle segments, appears as a yellowish band within the blackish borders of slightly expanded end segments: forked and dim on 7, much broader and yellower and nearer full length on 9, fading out on 10. Caudal appendages dull brown.

Female similar in coloration to male, with pale areas broader. Occiput generally bears on its hind border a pair of minute horns that vary greatly in size and degree of curvature; these horns sometimes lacking.

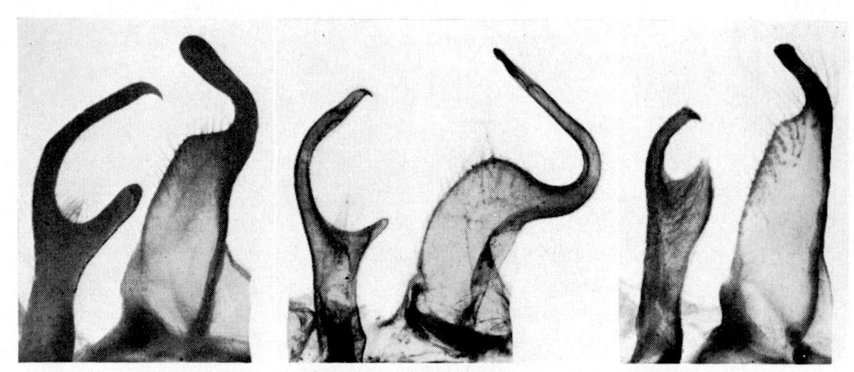

Fig. 59. *Ophiogomphus aspersus; O. carolus; O. colubrinus.*

Inhabits shallow stony riffles in woodland streams. Males perch on exposed tops of boulders; rather difficult to approach for capture. Nymphs burrow in silt beds and basins. On transformation they leave their cast skins flat on sloping banks at edge of water.

Distribution and dates.—CANADA: N. B., N. S., Ont., Que.; UNITED STATES: Maine, N. Y.

May 21 (N. Y.) to August 8 (Ont.).

Ophiogomphus colubrinus Selys

1854. Selys, Bull. Acad. Belg., 21(2):40 (reprint, p. 21).
1874. Hagen, Geol. Surv. Terr., p. 592.
1885. Hagen, Trans. Amer. Ent. Soc., 12:257 (nymph).
1899. Ndm., Can. Ent., 31:238 (figs.).
1918. Howe, Odon. of N. Eng., p. 29 (figs.).
1927. Garm., Odon. of Conn., p. 133.
1929. N. & H., Handb., p. 70 (figs.).
1933. Walk., Can. Ent., 65:227 (figs., nymph).

Length 41–48 mm.; abdomen 25–35; hind wing 27–31.

A stocky greenish species, heavily marked with black. Face green, heavily crosslined with black on all sutures and around front border of labrum. Occiput greenish. Strong middorsal stripe divided lengthwise by yellow edge of carina, and a little widened downward. Humeral and antehumeral stripes well developed, separated by usual narrow line of green. Occiput bright green. Midlateral stripe very short, ending below spiracle; second lateral complete and usually well developed. Legs black beyond base in male; also in female except for anterior face of femora, and a trace on antero-external edge of tibiae.

Abdomen with middorsal band of spots narrowed as usual on segment 3 and shortened thereafter, widened on 8 to quadrangular spot with a tail, and reduced to smaller oboval ones on 9 and 10, these also with

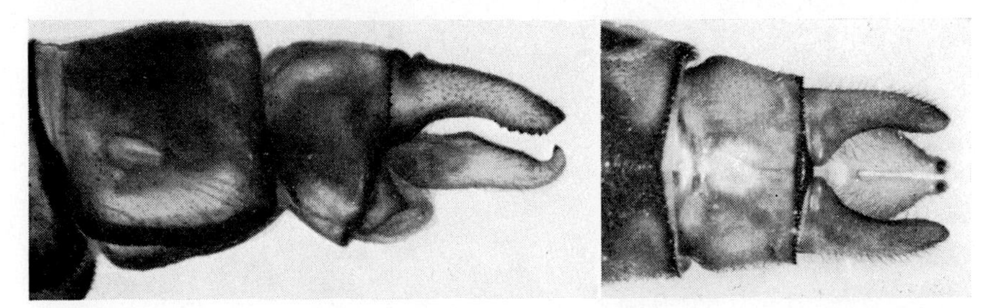

Fig. 60. *Ophiogomphus colubrinus.*

tails. Lateral band, obscured on middle segments, reappears in bright yellow as a half-length spot on 7, nearly full-length larger spots on 8 and 9, and a faint wash on 10. Caudal appendages of male dull olivaceous with black edgings.

Female with horns on both front and rear of occiput; latter minute. In corresponding position with rearward pair of female, male also has horns in this species—very small horns, low and truncated, and brown in color.

Distribution and dates.—CANADA: Alta., B. C., Man., Nfld., NW. Terr., Ont., Que., Sask.; UNITED STATES: Maine, Mich., N. H.

May 9 to September 3 (Ont.).

Ophiogomphus edmundo Needham

1950. Ndm., Trans. Amer. Ent. Soc., 76:1–12.

Length 45–48 mm.; abdomen 31–34; hind wing 24–29.

A greenish species, distinctly striped with brown on thorax, and with a blackish abdomen. Face pale greenish or yellowish green, with a narrow line of pale cinnamon brown on hair-fringed front margin of labrum. Top of head black, darkest across preocellar crossband, becoming paler brown behind postocellar ridge. Antennae black, stout basal segment tipped with a narrow ring of white. Occiput yellowish, its crest fringed with rather long blackish hair.

Middorsal stripe of synthorax well defined, divided full length by yellow of carina. In front it divides and spreads over crest. Humeral and antehumeral stripes well separated by a narrow pale line except at their ends. Midlateral stripe below level of spiracle; third lateral narrow, complete, well developed.

Legs pale basally, becoming black toward knees, and all black beyond. No yellow line on tibiae externally. Wings hyaline, with brown veins, a

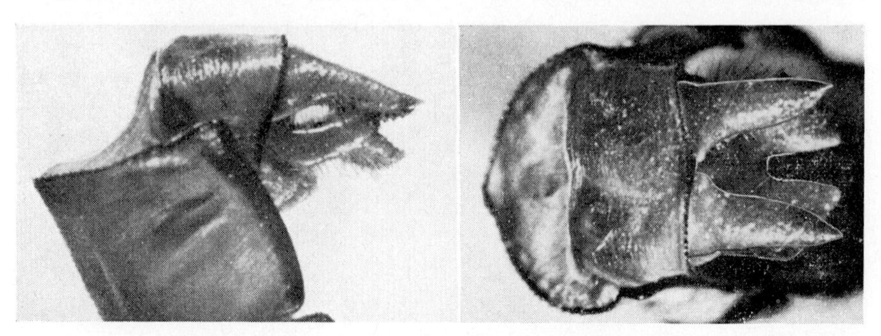

Fig. 61. *Ophiogomphus edmundo* (from the holotype).

white costa, and a tawny stigma. Hind wing with seven or eight antenodal crossveins; five cells in basal anal triangle of male.

Abdomen slender along middle segments. Beyond paler and hairy basal segments, it is smooth, blackish, and shining, with usual middorsal and lateral markings of yellow. Middorsal stripe broadly covers segments 1 and 2, becomes narrowed to a line on 3, does not reach apex on 4 to 7, broadens again and is reduced to half length on 8, and to a round basal spot on 9. Segment 10 mostly yellow; appendages yellow.

Superior appendages of male, viewed from above, broadly widened at base by a pair of opposed conical teeth that almost meet on middle line under a slight prolongation of dorsum of 10. Beyond these teeth they taper smoothly to rather sharp, diverging tips. Inferior appendage about as long as superiors. Viewed from beneath, its broad end is cleft by a U-shaped notch that is deeper than it is wide.

Female similar in color pattern, but with middorsal brown stripe of thorax less widely divided by yellow of carina. No horns on top of occiput, but a pair of very small black horns on rear side, wide apart and close to margin of eye; these horns blunt and crumpled in appearance.

Distribution and dates.—UNITED STATES: N. C., Pa.

June 3 (Pa.).

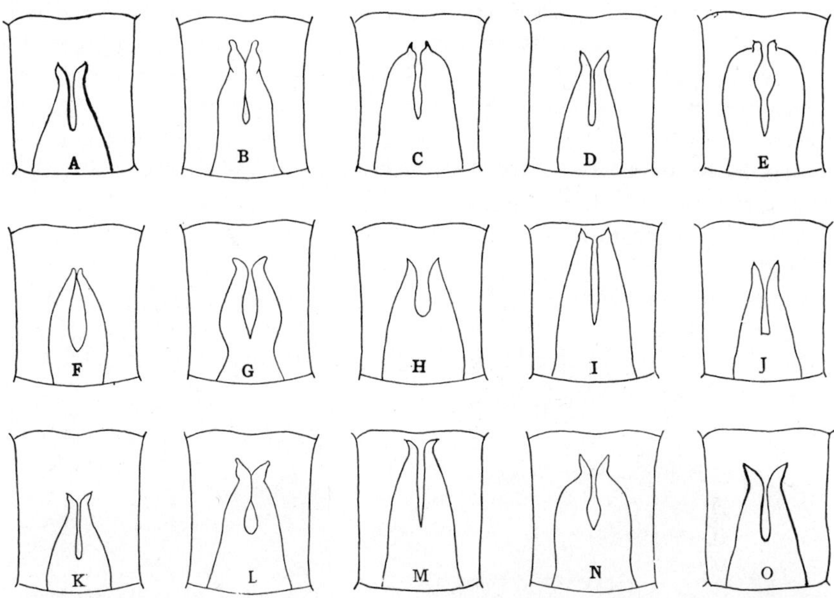

Fig. 62. Form of subgenital plate of female in fifteen species of genus *Ophiogomphus*. Plates as seen in ventral view; length indicated comparatively as they lie against sternum of abdominal segment 9. A, *anomalus;* B, *arizonicus;* C, *aspersus;* D, *bison;* E, *carolus;* F, *colubrinus;* G, *edmundo;* H, *howei;* I, *mainensis;* J, *montanus;* K, *morrisoni;* L, *nevadensis;* M, *occidentis;* N, *rupinsulensis;* O, *severus*. (Drawings by Dr. May Gyger Eltringham and Esther Coogle.)

Ophiogomphus howei Bromley

1924. Bromley, Ent. News, 35:343.
1924. Calv., Ent. News, 35:345 (figs.).
1927. Garm., Odon. of Conn., p. 135.
1929. N. & H., Handb., p. 76 (figs.).

Length 31–34 mm.; abdomen 22–24; hind wing 19–21.

Our smallest species, a very pretty one, clad in olive green, striped with brown and brightly marked with yellow on black abdomen. Face and greater part of top of frons and occiput greenish yellow; remainder of top of head black. Middorsal thoracic stripe much narrower than pale area at either side, and slightly widened downward. Humeral and antehumeral may be connected near upper end; separated by a pale greenish line below, where they are of about equal breadth; humeral stripe widens near top. Midlateral stripe wanting above level of spiracle; second lateral complete but very narrow. Legs black, with some streaks of paler on sides of femora. Wings tinged with yellow at middle of base; more deeply and much more, farther out in hind wings of female.

Abdomen short in female but little longer than hind wing. Two basal segments mainly yellow. Middorsal pale band of abdomen begins with a transverse bar on 1, a big trilobed spot covering most of dorsum of 2, then narrows to a long spear point on 3, reduced further to very narrow linear streaks on 4 to 7, widens again to a pointed spot on 8 and a smaller one on 9; 10 black above. Lateral pale band rather wide, covering sides of 1 and 2, reduced to a marginal line of ragged spots on 3 to 7, widens but is still ragged to rearward on 8 and 9, smaller and more evenly contoured on 10.

Female marked by a pair of short, sharp, widely separated horns on ridge of occiput.

Distribution and dates.—UNITED STATES: Mass., Pa.

May 20 (Pa.) to June 1 (Mass.).

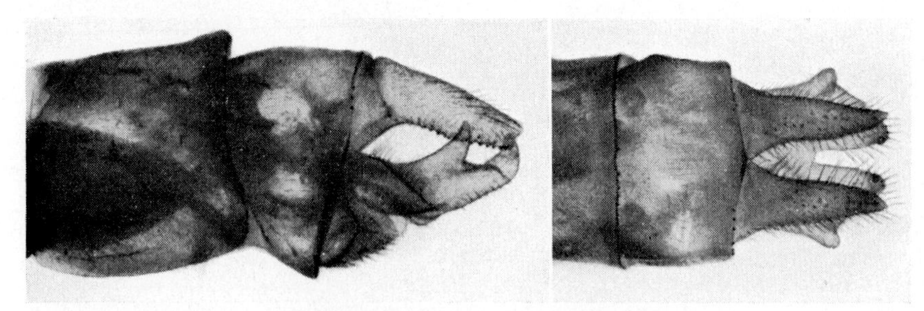

Fig. 63. *Ophiogomphus mainensis.*

Ophiogomphus mainensis Packard

Syn.: johannus Needham

1863. Packard, Proc. Ent. Soc. Phila., 2:255.
1874. Hagen, Geol. Surv. Terr., p. 595.
1899. Ndm., Can. Ent., 31:238 (figs.) (as *johannus* n. sp.).
1914. Woodruff, J. N. Y. Ent. Soc., 22:61 (figs., nymph).
1918. Howe, Odon. of N. Eng., p. 28.
1927. Garm., Odon. of Conn., p. 137.
1929. N. & H., Handb., p. 78 (figs.).
1933. Walk., Can. Ent., 65:222 (figs., nymph).

Length 42–46 mm.; abdomen 28–34; hind wing 25–31.

Another small yellowish green species. Face and occiput yellow. Middorsal thoracic stripe narrowly linear, divided by pale edge of carina. Rather narrow humeral and antehumeral stripes more or less separated by a narrow greenish line; generally conjoined for a short space near their upper ends. Lateral stripes weakly developed, midlateral stripe present only below spiracle, the humeral reduced to two dashes at ends of its suture. Legs blackish.

Middorsal pale band of abdomen, after narrowing to apex of segment 3, is reduced to abbreviated triangles on middle segments, becomes an oval spot on base of 8, a mere dot on 9, and another larger and more diffuse dot on 10. Yellow on sides of club a forked streak on 7, a broader, nearly full-length band on 8 and 9; does not overspread black margin on any of these three segments. On 10, two connected half-round spots of dull yellow cover sides next to terminal border. Segments 8 and 9 have slight expansion of side margins. Appendages blackish.

Female with two strong spines close together on ridge of occiput, with black tips proceeding from swollen bases. Also two very minute spines generally present behind, at edge of rear of eyes.

Distribution and dates.—CANADA: N. B., Que.; UNITED STATES: Ala., Conn., Maine, Md., Mass., N. H., N. J., N. Y., Pa., W. Va.

May 27 (New England) to July 17 (Que.).

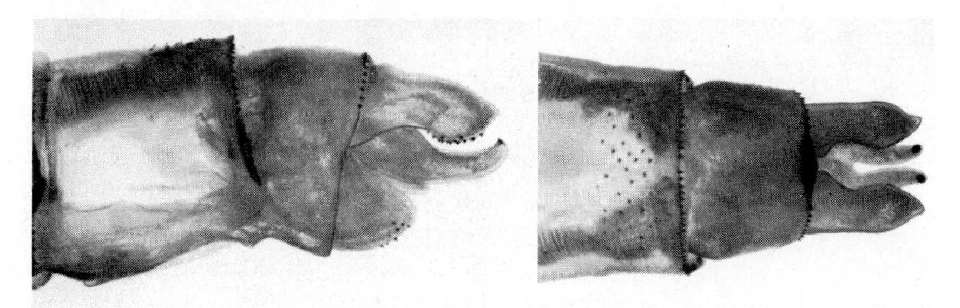

Fig. 64. *Ophiogomphus montanus.*

Ophiogomphus montanus Selys

1878. Selys, Bull. Acad. Belg., (2)46:432 (reprint, p. 27) (in *Herpetogomphus*).
1917. Kndy., Proc. U. S. Nat. Mus., 52:533 (figs.).
1929. N. & H., Handb., p. 75 (figs).

Length 50–51 mm.; abdomen 34–38; hind wing 30–32.

A large species, brightly patterned in yellow and black, with yellow caudal appendages. Face yellow up to basal third of frons on upper side. Middorsal thoracic stripe widens downward into triangle of moderate breadth. Antehumeral stripe reduced to half length, isolated island of black on a large field of yellow, well separated from humeral. Latter complete, narrow below, widened above at humeral pit, with a pale hairline of yellow in bottom of suture. Its lower end forms a J-shaped black mark with basal remnant of midlateral stripe. Second lateral stripe narrow, complete and forked at lower end. Legs yellow at base and well out on under side of femora and outer face of tibiae; elsewhere black.

Middorsal pale band of abdomen a complete row of yellow spots, divided transversely on segment 1, widest and constricted and then tapered and abbreviated on 2, shape of candle flame on 3 to 5, spear-pointed on 5 to 8, triangular on 9 and 10. Lateral pale band wide and continuous from end to end, only a little wider on larger end segments.

Perhaps merely a Northern variety of *severus,* with more black on oval spot that represents antehumeral stripe of thorax and on segment 10. Male with somewhat longer inferior caudal appendage; female with tips of bifid subgenital plate more bluntly pointed.

Distribution and dates.—CANADA: B. C.; UNITED STATES: Mont., Nev. June 14 (B. C.) to September 5 (B. C.).

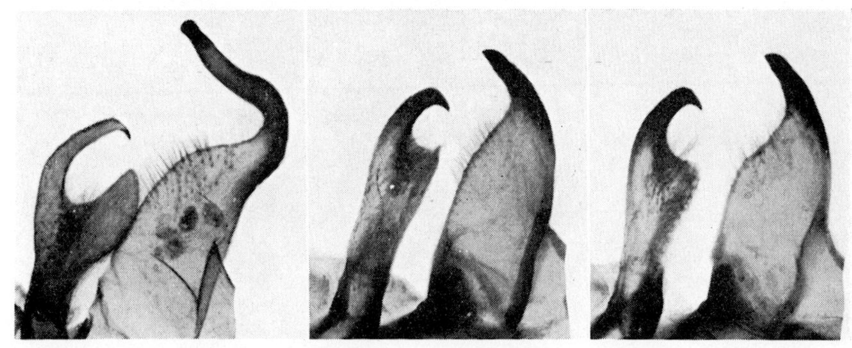

Fig. 65. *O. mainensis; O. montanus; O. morrisoni.*

Ophiogomphus morrisoni Selys

1879. Selys, C. R. Soc. Ent. Belg., 22:45.
1899. Ndm., Can. Ent., 31:238 (figs.).
1915. Kndy., Proc. U. S. Nat. Mus., 49:336.
1917. Kndy., Proc. U. S. Nat. Mus., 52:534 (figs.).
1929. N. & H., Handb., p. 75 (figs.).
1946. LaRivers, Ent. News, 57:209.

Length 50–52 mm.; abdomen 35–38; hind wing 28–33.

A large and handsome species, with an olive green thorax striped with black, and a chrome yellow abdomen both striped and ringed with black. Face greenish yellow, with a narrow black crossline at base of labrum and sometimes another on interclypeal suture. Frons yellow above, black on posterior half, with a slight forward prominence of middle of black portion. Vertex black except for an oval yellow spot behind ocelli.

Middorsal thoracic stripe narrow and slightly widened downward. Antehumeral ends freely above, wider than humeral and separated from it by a narrow green line. From narrowing lower end of humeral a rearward prolongation joins fragment of midlateral stripe to form a J-like mark, which then rises only to level of spiracle. Second lateral stripe narrow and complete. Femora yellowish beneath and on sides, black above toward knees; tibiae black, with a pale line on outer face; tarsi black. Wings clear; costa yellow; stigma brown.

Ground color of abdomen mainly yellow toward ends, whitish along middle segments. Middorsal pale band composed of large spots, wide and progressively abbreviated and somewhat trilobed at end of segments 2 to 9; 10 wholly yellow save for a basal black spot at side. Lateral pale band wide the full length of abdomen, only a little more so on 2, 8, and 9; on 2,

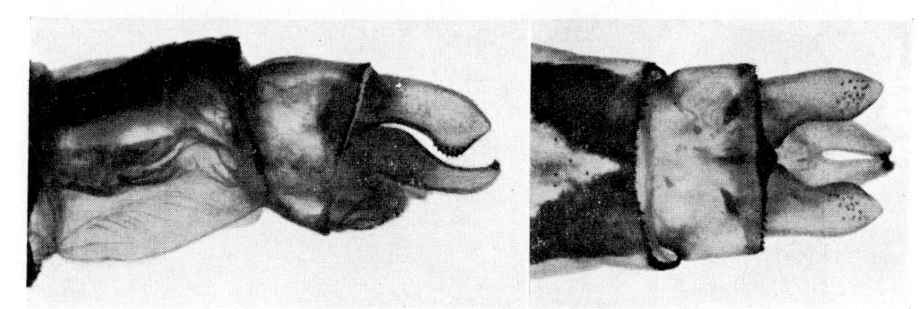

Fig. 66. *Ophiogomphus morrisoni.*

divided by a black mark descending from above; on 3 to 8, shortened a little by black descending from crossbands on joinings of segments, a recurrent branch from which makes a little emargination of the yellow below; only on 9 and 10 does yellow cover entire margin. Appendages yellow.

Female like male in coloration.

Distribution and dates.—UNITED STATES: Calif., Nev., Oreg.

June 19 (Nev.) to August 1 (Calif.).

Ophiogomphus nevadensis Kennedy

1917. Kndy., Proc. U. S. Nat. Mus., 52:536 (figs.).

Length 51–53 mm.; abdomen 37–41; hind wing 30–38.

A large pale desert species, its yellowish abdomen conspicuously patterned with dark brown. Face pale greenish gray, with a medially interrupted brown line across its base. Vertex black, with a transversely oval pale spot behind ocelli. Occiput yellow.

Thorax with exceedingly narrow middorsal stripe, very slightly widened downward. Antehumeral stripe, generally isolated only at upper end but sometimes interrupted below, is very narrow, isolated from equally narrow humeral by a pale line of its own width. Midlateral stripe very narrow, rising from below only to level of spiracle. Third lateral entire, narrow, forked below as in *morrisoni*. Legs yellowish almost to knees, on kneecap, and on a line on outer face of tibiae; elsewhere black.

Abdomen more yellow than brown. Its middorsal pale band is a row of elongate spots each contracted to a short point at posterior end. Lateral pale band correspondingly wide, edging upward to a basal confluence with yellow of dorsum, which is wide on 3 and narrow on 4 to 8; and to an apical confluence on 10. Wider yellow sides of 1 and 2 have a greenish tinge on 7 to 10; they are chrome yellow.

Species distinguished from *morrisoni*, its nearest ally, by its lighter coloration; also, in male, by having inferior caudal appendage only half as long as superiors.

Distribution and dates.—UNITED STATES: Calif., Nev.

June 25 (Calif.) to August 10 (Nev.).

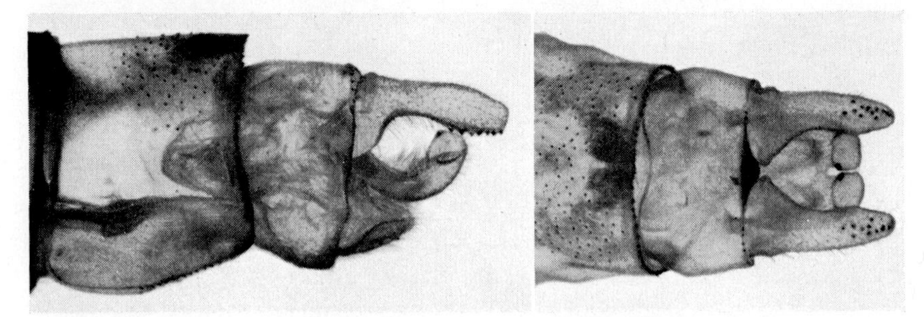

Fig. 67. *Ophiogomphus occidentis.*

Ophiogomphus occidentis Hagen

Syn: phaleratus Needham; var. californicus Kennedy

1882. Hagen, Nature, 27:173 (note on nymph).
1885. Hagen, Trans. Amer. Ent. Soc., 12:259 (nymph).
1899. Ndm., Can. Ent., 31:238 (figs.).
1902. Ndm., Can. Ent., 34:277 (as *phaleratus* n. sp.).
1917. Kndy., Proc. U. S. Nat. Mus., 52:542 (figs.).
1929. N. & H., Handb., p. 71 (figs. as *aspersus*).
1933. Walk., Can. Ent., 65:221 (figs., nymph).

Length 46–52 mm.; abdomen 34–38; hind wing 28–33.

A greenish species, striped with dark brown on thorax; abdomen largely yellow. Face greenish yellow; vertex black in front and yellow behind, at least on postocellar ridge. Middorsal stripe of thorax very narrow and a little widened downward. Humeral and antehumeral stripes consolidated, or with only vestiges of usual intervening pale line; front margin undulate. Midlateral stripe vestigial, but extends upward a little above level of spiracle. Second lateral stripe slender but complete, ending below in a forward-reaching streak to base of hind coxa. Legs yellow at base, becoming black toward knees on dorsal side; tibiae black with a line of yellow on their outer faces; tarsi black.

Middorsal pale band of abdomen beyond segment 3 becomes a row of broad abbreviated yellow spots, rounded and more or less pointed to rearward; on 10 expanded to cover whole segment except for a pair of basal brown dots. Lateral band of abdomen wide and continuous from

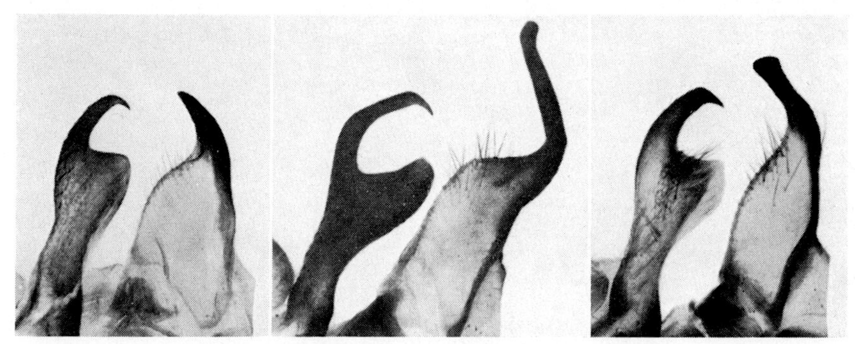

Fig. 68. *O. occidentis; O. rupinsulensis; O. severus.*

end to end, little widened on end segments; white on middle segments, and partly surrounds a small brown spot at lower apical corner.

Female similar in pattern, but with lighter areas somewhat more extensive. Occiput bears two pairs of horns: a smaller, blunt, spinelike pair near front margin directed forward; and a larger, thicker pair at outmost corners, directed backward and then crooked laterally.

Inhabits larger western mountain streams.

Distribution and dates.—CANADA: B. C.; UNITED STATES: Calif., Nev., Oreg., Utah, Wash.

June 6 (Oreg.) to October 1 (B. C.).

Ophiogomphus rupinsulensis Walsh

Syn.: pictus Needham

1862. Walsh, Proc. Acad. Phila., p. 388 (in *Erpetogomphus*).
1874. Hagen, Geol. Surv. Terr., p. 594.
1897. Ndm., Can. Ent., 29:181 (as *Herpetogomphus pictus* n. sp.).
1900. Wmsn., Odon. of Ind., p. 298.
1927. Garm., Odon. of Conn., p. 138.
1929. N. & H., Handb., p. 73 (figs.).
1933. Walk., Can. Ent., 65:225 (figs., nymph).

Length 45–54 mm.; abdomen 32–39; hind wing 27–32.

A fine greenish species, with brown shoulder stripes, slightly amber-tinted wings, and abdomen tipped with rusty red. Face and occiput yellow. Middorsal and midlateral stripes of thorax wanting. Antehumeral stripe abbreviated above, not reaching crest, and rather widely separated from humeral stripe by a line of yellowish green. Complete but narrow humeral stripe much widened toward its upper end. Second lateral stripe very little or not at all developed. Brown legs pale to near yellow knees and on outer face of tibiae. Clear but lightly tinted wings have costal margin yellow and a slightly rufous stigma.

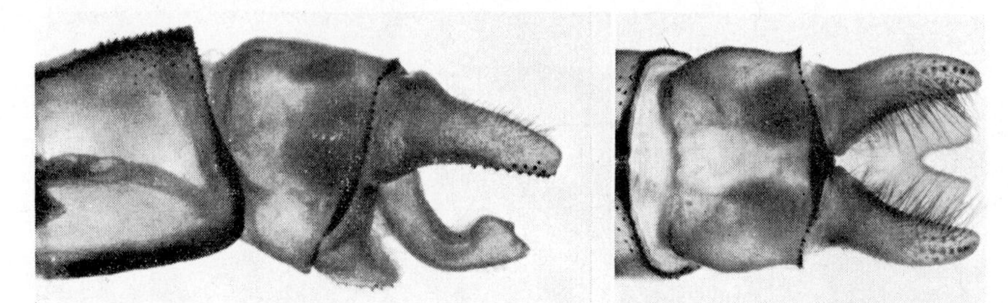

Fig. 69. *Ophiogomphus rupinsulensis.*

Middorsal pale line of abdomen, wide and continuous on basal segments and narrowed on 3, becomes diffuse and obscure on 5 and 6, and has a russet tinge on remaining segments. Segments 7 to 9 blackish in apical half; narrowly divided by rufous on middle line in male. Caudal appendages tinged with russet externally.

Female similarly but less brightly colored, with a pair of rather large blunt and wrinkled horns at rear of occiput, and two less conspicuous spines on front.

Distribution and dates.—CANADA: Man., N. B., Ont., Que., Sask.; UNITED STATES: Conn., Ill., Ind., Ky., Maine, Md., Mich., Mo., N. H., N. J., N. Y., N. Dak., Ohio, Pa., Tenn., Wis.

May 1 (Ind.) to September 23 (Ont.).

Ophiogomphus severus Hagen

1874. Hagen, Geol. Surv. Terr., p. 591.
1917. Kndy., Proc. U. S. Nat. Mus., 52:531 (figs.).
1929. N. & H., Handb., p. 74 (figs.).
1933. Walk., Can. Ent., 65:227 (figs., nymph).

Length 49–52 mm.; abdomen 34–37; hind wing 28–34.

Very handsome desert species; colors in life are said to be: on thorax green, on abdomen yellow, with markings on both dark brown. Face yellow; vertex and a narrow strip across rear of frons black; occiput yellow. Usual stripes of thorax almost wanting, save for an oval spot (antehumeral) and a parallel separate thickening at the humeral pit continued as a hairline down suture; also, a hairline in depths of second lateral suture, slightly widened at its lower end. Legs yellow at bases, becoming black toward knees on dorsal side of femora; also on knees. Tibiae black, yellow on outer faces; tarsi black. Wings slightly flavescent; costa yellow; stigma grayish brown.

Abdomen more yellow than black, with usual middorsal and lateral pale bands greatly expanded. Middorsal band is a row of large spots on

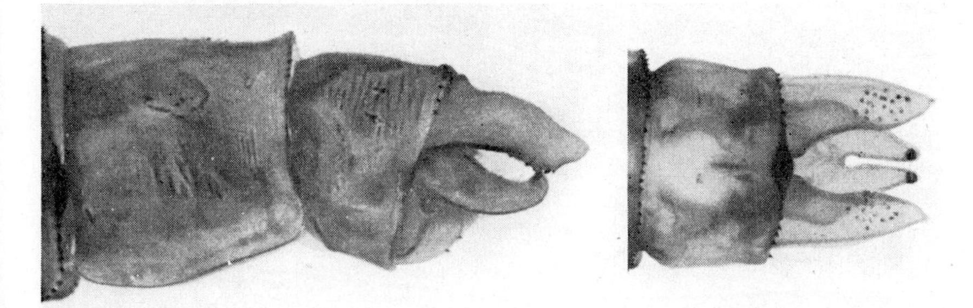

Fig. 70. *Ophiogomphus severus.* Dots near tip of superior appendages are denticles on under side showing through by transparency.

segments 2 to 9, each spot trilobed, with middle lobe pointed to rearward; small spot on 10 triangular. Lateral band wide and continuous, narrowed a little and white on middle segments. Caudal appendages yellow.

Female similar in coloration to male except that legs are more yellow, and on lower edges of middle abdominal segments there is more black.

Inhabits coarse sand and gravel beaches of lakes and streams.

Distribution and dates.—CANADA: Alta., Sask.; UNITED STATES: Calif., Colo., Idaho, Nebr., Oreg., Utah, Wash., Wyo.

June 15 (Wash.) to September 10 (Alta.).

Genus ERPETOGOMPHUS Selys 1858

This genus is mainly Neotropical, with about a dozen species in Mexico and Central America, and a few farther south. Four species are recorded from the United States, three of which are restricted to our southwestern states. Three additional are considered here because of their proximity in bordering states of Mexico.

These are Gomphines of medium size, short-legged and clear-winged, with sometimes a tint of yellow in the wing membrane.

The face is yellow or green. The middorsal stripe of the thorax (wanting in *crotalinus*) is widened downward into a rather broad triangle of brown, which is more or less divided lengthwise by a narrow line of yellow on the carina. The legs are pale at the base and blackish beyond the knees, with the black spreading inward over the femora in a narrowing dorsal streak. In the wings the first anal interspace (x) is greater than the

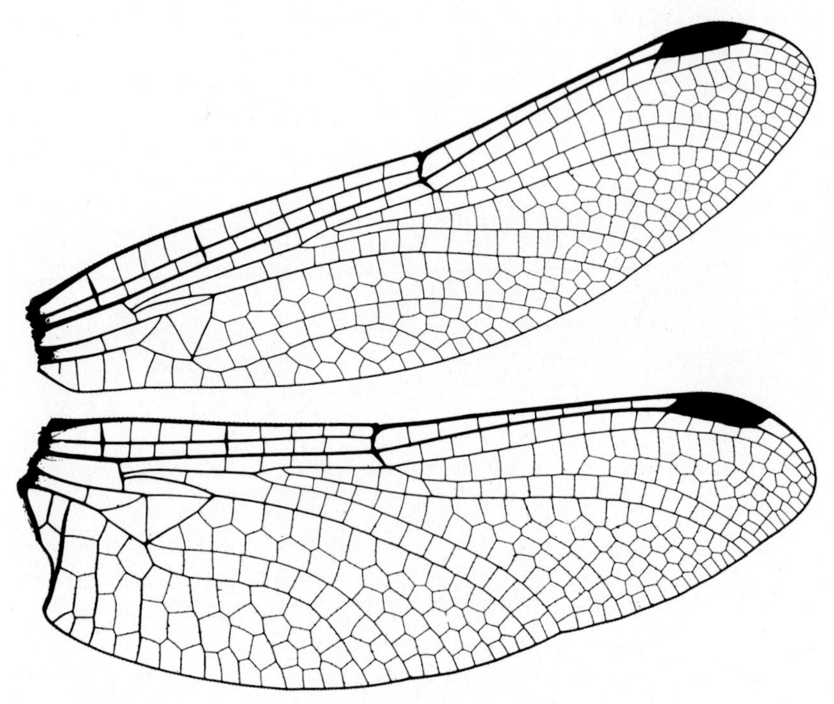

Fig. 71. *Erpetogomphus coluber.*

second (y), and is filled with two rows of cells from the triangle to the wing margin; the second interspace (y) generally begins with the two full-width cells next to the anal vein. Other venational characters are as shown in the table of genera.

The ringed abdomen is clubbed only in the male, and segment 10, as if enlarged to carry its heavy caudal appendages, is as long as 9 or longer. There is little development of leaflike free lateral margins on 8 and 9; none at all in the female. There are no spines on the occiput of the female.

The nymph in this genus is thickset, depressed of body, and rather abruptly narrowed at the rear. The wings are broadly divergent on the back. Antennal segment 3 is long and linear; the median lobe of the labium is moderately convex and edged with small, close-set brown teeth

Fig. 72. *Erpetogomphus designatus.*

and fringed with scales as in *Ophiogomphus;* the lateral lobe is similar, but with the movable hook set about midway of its length; the part beyond the base of that hook is shorter and broader than in *Ophiogomphus.*

On the abdomen there are dorsal hooks (at least, vestiges of hooks) on segments 2 to 9, a little higher and blunter on the basal segments; and lateral spines (in ours) on 6 to 9. The caudal appendages are all about the same length. This last character will distinguish these nymphs from those of *Ophiogomphus,* in which the laterals are distinctly shorter than the superior.

Two very useful reference papers for special study of this American genus are those of Calvert (B. C. A., 1905), and Williamson (1930).

KEY TO THE SPECIES

Adults*

MALES

1—Middle of occiput with large dome-shaped bulge covering more than half of upper surface .... **designatus**
—No such elevation present 2

2—No middorsal stripe on front of thorax .... **crotalinus**
—Middorsal thoracic stripe present 3

3—Superior abdominal appendages of male, viewed from side, distinctly angulate above 4
—These appendages not distinctly angulate above; smoothly downcurved.......... 5

4—First and second lateral thoracic stripes usually joined; inferior dilated edge of segments 8 and 9 black.......... **lampropeltis**
—These stripes not joined; inferior dilated edge of 8 and 9 light brown, like sides .... **natrix**

5—Hind wing 24–26 mm.; usually a single row of paranal cells in fore wing . **coluber**
—Hind wing 30–35 mm.; paranals with one or more added marginal cells in fore wing .... **compositus**

Nymphs

1—Dorsal hooks well developed on abdominal segments 2 to 9.......... **designatus**
—Dorsal hooks rudimentary or wanting on some posterior segments 2

2—Dorsal hooks prominent only on the more basal segments, vestigial on 7 to 9; lateral spines about equal in size on 6 to 9.......... **compositus**
—Dorsal hooks rudimentary or wanting on 4 to 9; lateral spines on 6 and 9 smaller than those on 7 and 8 .... **lampropeltis**

Nymphs unknown: **coluber, crotalinus, diadophis,** and **natrix.**

* Not included in key: **diadophis** (only two females known).

Erpetogomphus coluber Williamson

1930. Wmsn., Occ. Pap. Mus. Zool. Univ. Mich., 216:14–19 (figs.).

Length 44–45 mm.; abdomen 32–33; hind wing 24–26.

This species similar to better-known Texan *compositus,* but smaller. Face greenish white, with some pale brownish stripings on sutures. Occiput pale grayish green, with side margins next to eyes washed with brown.

Middorsal thoracic dull brown stripe broadly widened downward quite to collar and divided on middle line by pale carina. Humeral and antehumeral generally complete and regular, but latter may be abbreviated next to crest; they are generally separated by a complete pale line, but may be conjoined near upper end, isolating a pale spot above junction. Sides of thorax faintly yellowish; two lateral brown stripes well defined but variable, sometimes conjoined either above or below middle. Legs black beyond femoral knee stripe.

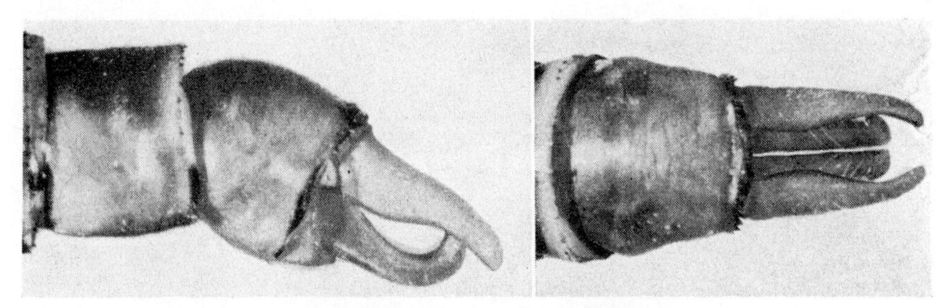

Fig. 73. *Erpetogomphus coluber.*

Abdomen on middle segments black, with a wide pale band across base of each; joinings of segments deep black. Pattern obscure on two paler basal segments. Segments 7, 8, and 9 moderately enlarged at sides in male; not at all in female. They are black above in male; paler in female. Segment 10 yellow, with a wash of black across base above, and with thickened transverse apical carina jet black. Yellow on small leaflike lateral expansions of end segments scanty on 7, broad on 8 and 9, bordered externally by black on all. Caudal appendages yellow.

Usually differs in wing venation from other five species in having no marginal cells behind paranals of fore wing. Frequently, *natrix* also has no marginal cells.

Distribution and dates.—MEXICO: Baja Calif.

October 10.

Erpetogomphus compositus Hagen

1857. Hagen, in Selys, Mon. Gomph., p. 400 (fig.).
1861. Hagen, Syn. Neur. N. Amer., p. 99.
1873. Hagen, in Selys, Bull. Acad. Belg., (2)35:740 (reprint, p. 12).
1885. Hagen, Trans. Amer. Ent. Soc., 12:256 (nymph).
1905. Calv., B. C. A., p. 166 (figs.).
1929. N. & H., Handb., p. 80.
1930. Wmsn., Occ. Pap. Mus. Zool. Univ. Mich., 216:1–4 (figs.).
1934. Tinkham, Can. Ent., 66:215.

Length 46–55 mm.; abdomen 36–41; hind wing 30–35.

Great variety in ground coloration of this handsome species: white on face, front of thorax, and base of abdomen; yellow on occiput and sides

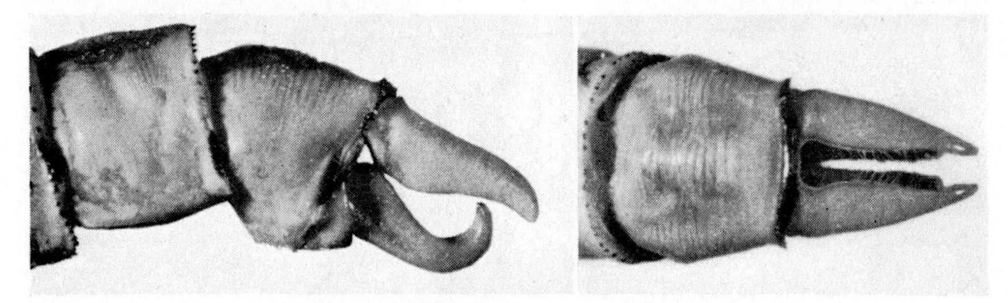

Fig. 74. *Erpetogomphus compositus.*

of thorax; fawn on club of abdomen, with markings of brown and black on whatever ground color. Face and occiput wholly pale.

Middorsal thoracic brown stripe widens downward into a broad triangle that rests against a wholly white collar; it is divided on middle line by a white carina. Antehumeral brown stripe, isolated at upper end, runs down to outer end of collar. It is widest in middle; separate and parallel humeral stripe widens toward upper end. Lateral stripes well developed and separate, with area between them lighter yellow than that at either side. Legs whitish basally, with femora bearing a cap streak of black; tibiae and all spines black. Wings clear, with only a trace of yellow in membrane at base, which extends farther out in female; stigma black.

Abdomen heavily ringed with black on middle segments. End segments yellowish in ground color, with an overwash in male of reddish fawn, and in female of black. In female, dorsum of 8 and 9 also broadly black. Middle segments ringed with dull yellow about their basal fourth; on each side, ring is followed by a more or less isolated spot of brown. In female, pattern of middle segments continues rearward on 7; 8 and 9 wholly blackish above; 10 wholly yellow.

Distribution and dates.—UNITED STATES: Ariz., Calif., Nev., N. Mex., Oreg., Tex., Utah, Wyo.; also from Mexico.

April 18 (Tex.) to September 22 (Ariz.).

Erpetogomphus crotalinus Hagen

1854. Hagen, in Selys, Bull. Acad. Belg., (2)21:40 (reprint, p. 21) (in *Ophiogomphus*).
1857. Hagen, in Selys, Mon. Gomph., p. 72 (figs.).
1861. Hagen, Syn. Neur. N. Amer., p. 101.
1905. Calv., B. C. A., p. 165.

Length 45–49 mm.; abdomen 33–37; hind wing 29–35.

A dull yellowish species, almost lacking usual brown stripes of thorax and with more or less interrupted black stripes on side of abdo-

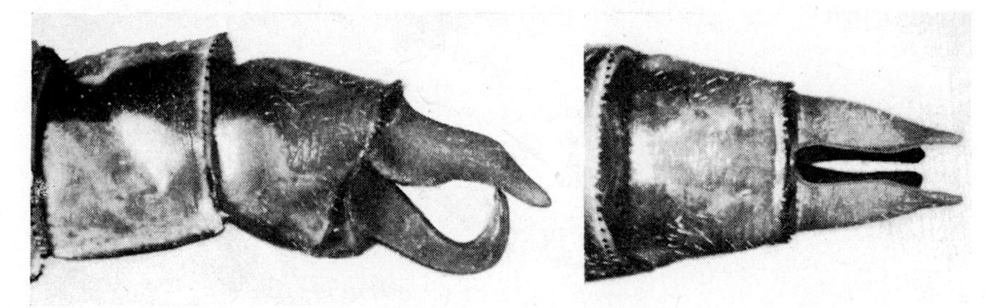

Fig. 75. *Erpetogomphus crotalinus.*

men. Face and frons wholly dull yellow. Vertex brown, paler behind ocelli. Occiput yellow, with a brown hairline on crest.

Thorax yellow, practically without stripes on front or sides or with only vestiges of some of them. Legs yellow as far out as knees, with only a streak of black on upper side of femora; yellow also on kneecap and on outer face of tibiae; inner side of tibiae and all tarsi and spines blackish.

Abdomen mainly yellow on two basal segments, but with spot of brown on each side of dorsum of segment 2, and a streak of it runs down beside yellow auricle. Segments 4 to 6 indistinctly ringed with black; their paler basal two-thirds is traversed by a narrow black crossline that is interrupted middorsally; 7 becomes blacker on sides where enlargement of segment begins. A diffuse middorsal pale band on 7 to 10; oval in outline on 8. Little expansion of lateral margins of 7, 8, and 9. Pale band edged with black; just within black are usual yellowish spots, narrow on 7, wide on 8 and 9, but dull and not conspicuous. This species easily recognized by lack of brown stripes on thorax.

Distribution and dates.—MEXICO: Chihuahua; also southward in Mexico.

May to October.

Erpetogomphus designatus Hagen

1857. Hagen, in Selys, Mon. Gomph., p. 401 (figs.).
1861. Hagen, Syn. Neur. N. Amer., p. 99.
1885. Hagen, Trans. Amer. Ent. Soc., 12:255 (nymph).
1901. Wmsn., Proc. Ind. Acad. Sci. (figs.).
1901. Wmsn., Ent. News, 14:226.
1927. Tinkham, Can. Ent., 66:216.
1929. N. & H:, Handb., p. 80 (figs.).

Length 49–55 mm.; abdomen 34–39; hind wing 30–35.

A yellowish species, marked with bright brown and black and faintly tinted with saffron, especially on wing bases. Face yellow, tinged with

Fig. 76. *Erpetogomphus designatus.*

greenish on clypeus and labrum, latter with a narrow straw-colored border. Occiput yellow.

Middorsal stripe of thorax broadens downward to all-yellow collar and is divided by yellow carina. Antehumeral brown stripe elongate oval in form, abbreviated at both ends and well separated from humeral, which is narrower but widens upward. Midlateral stripe weak and inconstant; second lateral narrow but well defined. Legs yellowish at base and out to knees except on upper side of femora. Tibiae reddish brown, with black spines and edgings. Saffron color of wing bases usually ends abruptly at first crossveins, with a little touch of it between bases of sectors of arculus. Stigma yellow.

Abdomen, as seen from side, pale on the two obscure basal segments, broadly ringed with black on joinings of middle segments, rusty yellow on clubbed end segments. Pale rings encircle these segments at base. Black rings extend forward from joinings increasingly to rearward.

Species easily recognized by little quadrangles of saffron yellow in wing base and by moundlike eminence on upper surface of occiput.

Distribution and dates.—UNITED STATES: Ala., Ark., Fla., Ga., Ind., Kans., Ky., Md., Mo., Nev., Ohio, Okla., S. C., Tenn., Tex.; also from Mexico.

May 27 (Tex.) to October 7 (S. C.).

Erpetogomphus diadophis Calvert

1905. Calv., B. C. A., p. 167.

Length ? mm.; abdomen 33; hind wing 28.

Female (male as yet unknown): A greenish species, striped with brown on thorax; blackish abdomen with pale bands completely encircling middle segments. Face greenish, with touches of brown on clypeal sutures. Top of frons with pale brown line across base next to brown vertex.

Middorsal thoracic stripe, widened to a broad triangle below, is completely divided lengthwise by pale carina, and is connected above with brown of crest. Humeral and antehumeral brown stripes well developed, latter not reaching up to crest and being a little wider than humeral. Side stripes of thorax narrow lines of brown. Legs pale at base and on under side of femora; dorsal side of femora overspread with black increasingly out to knee; more black on fore and middle femora than on those of hind legs. Wings clear. Costa brown, with a yellow line on its front margin; stigma dark brown.

Middorsal pale band on abdomen wide on segments 1 and 2, narrows and becomes a row of elongate abbreviated spots on 3 to 7, becomes obscure on 8 and 9, which are paler on sides; 10 and appendages obscure brown. Acutely pointed appendages longer than 10 and shorter than 9. (Description abstracted from Dr. Calvert's original.)

Distribution.—UNITED STATES: Tex.

Erpetogomphus lampropeltis Kennedy

1918. Kndy., Can. Ent., 50:297 (figs.).
1929. N. & H., Handb., p. 79 (figs.).
1934. Tinkham, Can. Ent., 66:215.
1934. Wmsn., Occ. Pap. Mus. Zool. Univ. Mich., 216:19–26.

Length 46–52 mm.; abdomen 34–40; hind wing 29–30.

A handsomely decorated species, showing great variety of tints in color pattern. Face, frons, and occiput white or pale yellowish. Middorsal thoracic brown stripe completely divided by white carina; collar wholly white. Humeral and antehumeral stripes well developed, latter generally interrupted at upper end, and the two well separated by a line of green. Lateral brown stripes well developed and generally conjoined. Legs whitish at base, streaked with brown toward knee on upper side; tibiae and tarsi wholly black. A faint tinge of yellow in extreme base of wings.

Abdomen very black on wide encircling bands of middle segments; shorter intervening pale bands whitish, as is also ground color of short basal segments. In male, clubbed end segments tinged with rusty red.

In female, ringed pattern of middle segments continues to 7, is brightest on 8 and 9, and merges into yellow on 10, with only a thickening of transverse apical carina jet black. Segments 8 and 9 black above; 10 and appendages wholly yellow.

Distribution and dates.—UNITED STATES: Calif., Tex.

July 17 (Calif.) to October 5 (Tex.).

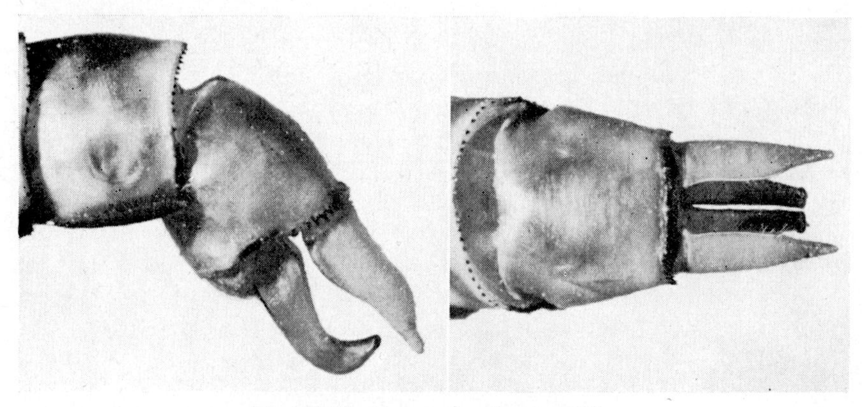

Fig. 77. *Erpetogomphus lampropeltis.*

Erpetogomphus natrix Williamson

1930. Wmsn., Occ. Pap. Mus. Zool. Univ. Mich., 216:19–26 (figs.).

Length 49–53 mm.; abdomen 35–38; hind wing 27–31.

A green and yellow species, striped on thorax with brown. Face light bluish green, a little more yellowish on labrum, with a pale brown stripe across base of labrum and another across frons. Vertex black; occiput light brown, with a black edging on crest.

Middorsal thoracic stripe broadens downward into wide triangular form, narrowly divided lengthwise by a greenish carina. Humeral and antehumeral stripes of brown well developed, more or less conjoined near their upper ends; antehumeral rounded at upper end and not reaching crest. Elsewhere between the two is a narrow line of green. Lateral stripes complete, a little irregular in contour, tending toward contact near upper end. Legs pale at base on short basal segments and on under surface and anterior face of femora; elsewhere black.

Abdomen mainly brown or blackish, usual middorsal pale band reduced to interrupted line of narrow elongate spots that cease on segment 6.

Middle segments mostly brown, with black rings on joining of segments continuous with pale brown of sides. Yellow rings encircling base of these segments connect with pale band on dorsum. Superior appendages of male yellow; inferior darker, tinged with reddish.

Distribution and dates.—MEXICO: Baja Calif.

October 6–13.

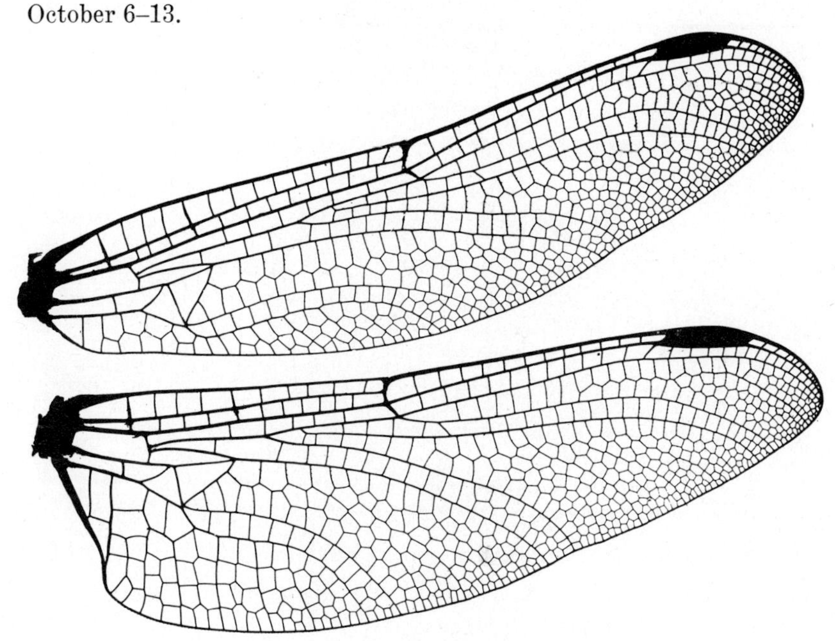

Fig. 78. *Dromogomphus spinosus.*

Genus DROMOGOMPHUS Selys 1854

Dragonflies of this genus are easily recognized by their extremely long hind femora that are armed with an inner row of four to eight very strong spines. The claws of the hind tarsi are long and nearly straight, with the inferior tooth situated just before their middle point. The face and occiput are yellow or green. Carina and collar are broadly bordered with yellow, and humeral and antehumeral brown stripes are much broader than the other stripes of the thorax, and more or less confluent, at least at their ends; but the breadth of these brown stripings is very variable. The wings are hyaline, with at least a trace of yellow on the costal veins.

In wing venation an anal loop of one, two, or rarely three cells is often recognizable though weakly developed, and very inconstant in its shape. Vein A1 is strongly angulated at its base, being bent proximally to the

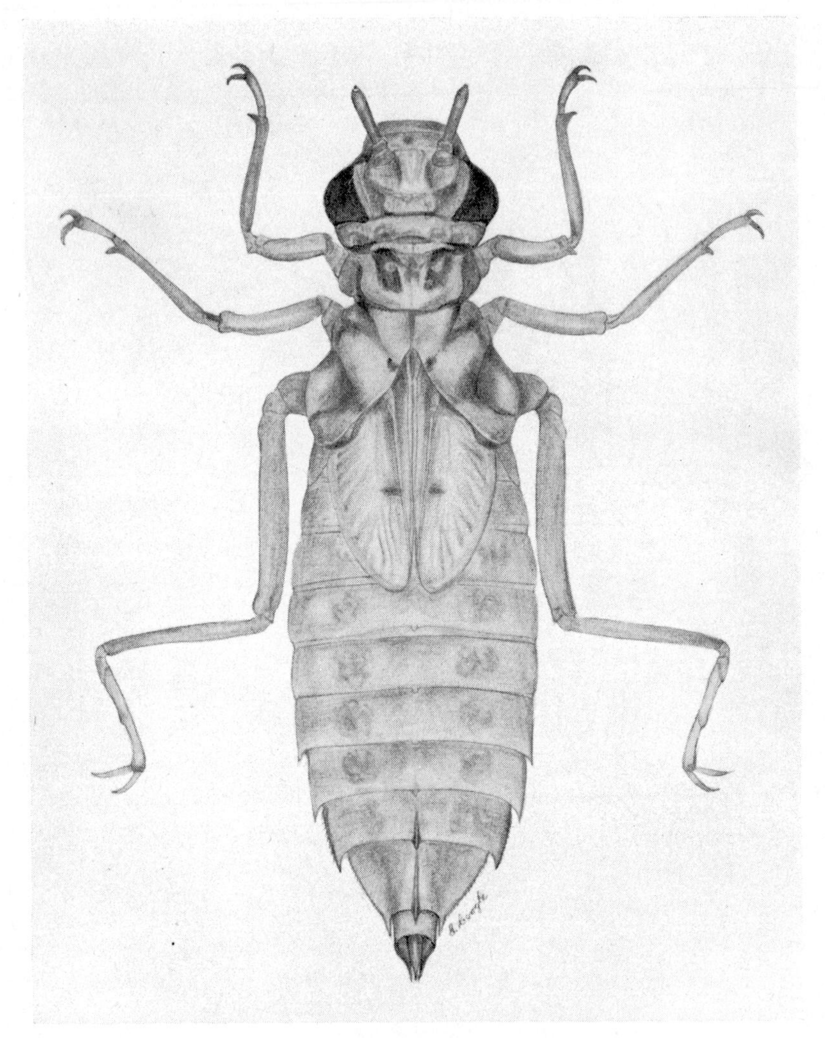

Fig. 79. *Dromogomphus spinosus.*

first crossvein, then posteriorly to run parallel with the rather straight A2 to the hind margin. The dividing crossveins of the anal triangle of the male are exceedingly inconstant, forming two to five cells.

The abdomen is distinctly clubbed in the male; less so in the female, with segments 7 to 9 moderately dilated on the lateral margins. The auricles of 2 are weakly developed, almost unarmed with denticles. The usual interrupted middorsal line of yellow markings is present, at least basally. The sides of the moderately swollen basal segments and the laterally expanded sides of 7 to 9 are broadly marked with yellow. The superior appendages of the male and the broadly divaricate branches of the inferior are about of equal length. The subgenital plate of the female is triangular and rather deeply bifid at the apex with acute tips.

The three known species of this genus are mainly found in the southern portion of the eastern United States.

The nymphs of this genus are elongate in form, with a moderately depressed abdomen that becomes triquetral toward the tapering posterior end, especially on segment 9. The antennae are more slender than in related genera, and the long segment 3 surpasses the tip of the labrum by more than half of its own length; 4 is a minute spherical rudiment. The labium is rather short, its median lobe cut off straight or even slightly concave in front. The hooked lateral lobe is armed with minute teeth on its inner margin. The burrowing hooks of the fore and middle tibiae are well developed.

There are dorsal hooks on the abdomen on segments 3 to 9, highest on the middle segments, mere humps on 3 and 4, declined to sharp prolongation of the middorsal ridge on 9. The middorsal length of abdominal segments 7, 8, 9, and 10 is as 7:8:10:5, with the caudal appendages, in this ratio, 7.

KEY TO THE SPECIES

Adults

1—Costa mainly yellow; caudal appendages yellow; lateral expansion of segment 8 and 9 broad, extended to rearward much beyond general level of apical margin of segment .... **spoliatus**

—Costa mainly black; caudal appendages black; lateral expansion of 8 and 9 narrow, not produced backward conspicuously beyond general level of apical margin of segment 2

2—Face generally with a cross stripe of black on fronto-clypeal suture ...... **armatus**

—Face all yellow .... **spinosus**

Nymphs

1—Tip of lateral lobe of labium roundly incurved to terminal tooth on its inner margin .... **spoliatus**

—Tip of lateral lobe of labium prolonged in a very distinct end hook that projects inward well beyond line of teeth 2

2—Lateral spines of segments 7 and 8 outarching from sides of abdomen; front border of median lobe of labium with no minute tooth in middle ...... **armatus**

—Lateral spines of 7 and 8 more slender and scarcely outarched at all; front border of median lobe of labium with a very minute tooth in middle... **spinosus**

Dromogomphus armatus Selys

1854. Selys, Bull. Acad. Belg., 21(2):59 (reprint, p. 40).
1857. Selys, Mon. Gomph., p. 122 (in *Gomphus*).
1861. Hagen, Syn. Neur. N. Amer., p. 102 (in *Gomphus*).
1873. Selys, Bull. Acad. Belg., (2)36:499 (reprint, p. 54).
1878. Selys, Bull. Acad. Belg., (2)46:467 (reprint, p. 62).
1929. N. & H., Handb., p. 117.

Length 59–63 mm.; abdomen 44–48; hind wing 35–39.

A blackish species, usually recognizable by a black cross stripe on fronto-clypeal suture of face. Top of frons and occiput yellow. Vertex black. Thoracic stripes of first pair narrow, prolonged laterally at front end to meet those of second pair. An inverted Y-shaped yellow area entirely covers carina and collar, narrow dorsal brown stripes being widely divergent downward. Humeral and antehumeral stripes broad and fused to more or less eliminate usual intervening antehumeral line of yellow. Sides yellow, with traces of brown in sutures. Legs black. Wings hyaline, with brown veins and stigma, and a hairline of yellow on front of costa.

Middorsal yellow line of abdomen broad on segment 1, trilobed on 2, progressively narrowed on 3 to 7, widened again on 8 and 9, usually absent on 10. Sides of 1, 2, and 3 broadly yellow; of 7, 8, and 9 with yellow areas of variable extent within black-bordered, expanded lateral margins.

Female similar to male in coloration. The border of occiput doubly convex, with a slight emargination in middle; no spines on vertex.

Distribution and dates.—UNITED STATES: Ala., Fla., Ga.

June 7 (Ga.) to October 3 (Ga.).

Dromogomphus spinosus Selys

1854. Selys, Bull. Acad. Belg., 21(2):59 (reprint, p. 40).
1857. Selys, Mon. Gomph., p. 120 (figs.) (in *Gomphus*).
1861. Hagen, Syn. Neur. N. Amer., p. 102 (in *Gomphus*).
1893. Calv., Trans. Amer. Ent. Soc., 20:245.
1901. Ndm., Bull. N. Y. State Mus., 47:462 (figs., nymph).
1927. Garm., Odon. of Conn., p. 168 (figs.)
1929. N. & H., Handb., p. 118 (figs.).

Length 60 mm.; abdomen 45; hind wing 36.

Very similar to preceding species, but less overspread with blackish color. Inverted Y-shaped yellow mark on crest and collar wider; so, also, adjacent stripes of yellow. Dorsal yellow stripes very narrow but they widen below, where strongly divergent. Sides of thorax yellow. Middorsal yellow line of abdomen nearly continuous, being interrupted only on segments 8 and 9, and broadened again to a square middorsal patch on 10. Genitalia practically identical with preceding species. Occiput of female slightly convex and notched in middle, with a minute triangular tooth in bottom of notch; from ends of elevated ridge behind ocelli arise a pair of sharp thornlike black spines.

Inhabits clear streams and clean lake shores.

Distribution and dates.—CANADA: Ont., Que.; UNITED STATES: Ala., Conn., Fla., Ga., Ill., Ind., Kans., Ky., La., Maine, Mass., Mich., Miss., Mo., N. H., N. J., N. Y., N. C., Ohio, Okla., Pa., S. C., Tenn., Tex., Vt., W. Va., Wis.

April 14 (Fla.) to November 11 (Fla.).

Dromogomphus spoliatus Hagen

1857. Hagen, in Selys, Mon. Gomph., p. 409 (figs.) (in *Gomphus*).
1859. Hagen, in Selys, Bull. Acad. Belg., (2)7:543 (reprint, p. 17).
1861. Hagen, Syn. Neur. N. Amer., p. 103 (in *Gomphus*).
1920. Wmsn., Proc. Ind. Acad. Sci., p. 101 (habits).
1929. N. & H., Handb., p. 117 (figs.).

Length 60–61 mm.; abdomen 43–46; hind wing 35–38.

A fine big yellowish species, striped with brown. Face and occiput yellow. Inverted Y-shaped yellow mark on front of synthorax covering carina and collar is abbreviated above where dorsal brown stripes coalesce. Dorsal and antehumeral stripes meet at ends of collar, leaving a rather wide yellow stripe included between them. Humeral and antehumeral about equal in width, well separated by yellow. Side stripes of brown, with third lateral stripe and midlateral stripe linear but variable. Legs black, with streaks of yellow from base invading sides of femora. Long claws of hind tarsus reddish in middle. Wings hyaline; costa yellow; stigma tawny.

Abdomen slightly swollen on basal segments, cylindric on 3 to 6, greatly swollen on terminal segments. Segment 1 bears a tuft of brown hairs on each side. Obscure yellow markings invade the brown on dorsum of 1 to 6; sides of basal segments broadly yellow. Swollen terminal segments yellow, tinged with reddish.

Female similar to male, but a little lighter in color.

Distribution and dates.—UNITED STATES: Ala., Fla., Ind., Kans., Ohio, Okla., S. C., Tenn., Tex., Wis.

June 21 (Ohio) to October 3 (S. C.).

Genus LANTHUS Needham 1897

This genus includes the smallest of our Gomphines, two clear-winged blackish species, smartly marked with narrow lines of yellow. The face is cross-striped with black. The top of the frons is half yellow; and black covers the entire remaining top of the head, also both vertex and occiput, save for a little median yellow spot that lies close behind the postocellar

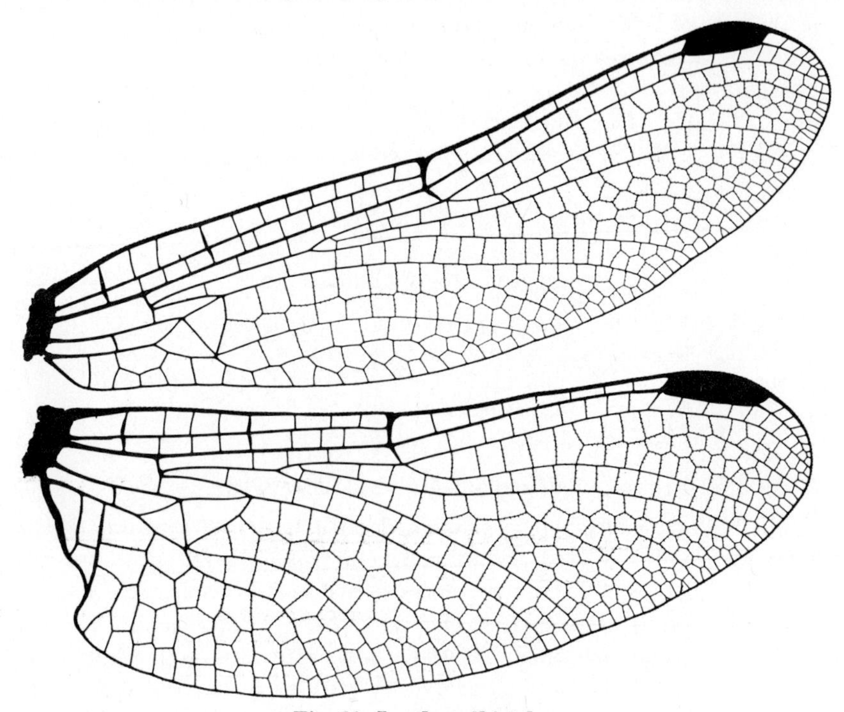

Fig. 80. *Lanthus albistylus.*

ridge. The front of the synthorax is much more black than yellow; the abdomen is mainly black. The legs are black, except for the under side of the front femur. The wings are clear, with brown veins and stigma.

The abdomen is slender on the middle segments, and black, with a minimum development of the usual middorsal and lateral pale bands. The remaining segments diminish regularly in length to the end. The superior appendages of the male are longer than segment 10. The posterior hamules of the male are strongly inclined to rearward.

This genus is restricted in distribution to the eastern United States and Canada.

The nymphs of this genus are stocky little fellows with short, abruptly

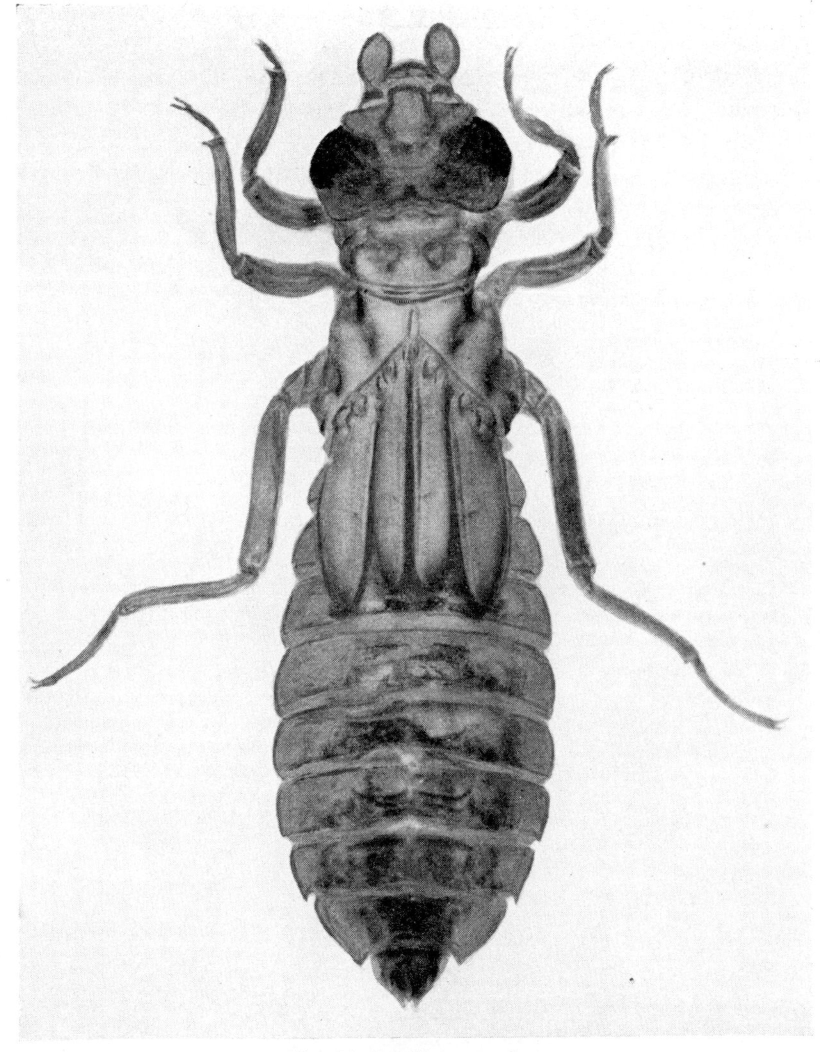

Fig. 81. *Lanthus parvulus.*

pointed, depressed abdomen, and with the flat, broadly oval antennal segment 3 extended before the face. The labium is short; the front border of its median lobe is straight or very slightly convex and scale-fringed, with a few low chitinous teeth in the middle. The lateral lobe is rounded apically to the first of the series of coarse teeth on its inner margin, and there is thus no distinct end hook. The teeth diminish in size proximally.

There are very short lateral spines on abdominal segments 8 and 9, and no dorsal hooks at all.

These nymphs inhabit sandy places in the beds of rocky spring-fed brooks where they burrow shallowly and whence they are easily obtained by sifting. They feign death for some minutes after being taken from the water and are apt to be thrown away with the trash, undiscovered by the careless collector.

Most useful for reference of the papers that deal with this genus are those of F. L. Harvey on adults (*Ent. News*, 9:63, 85, 1898) and J. G. Needham on nymphs (*Can. Ent.*, 29:165, 1897).

KEY TO THE SPECIES

ADULTS

1—Outer side of fore-wing triangle distinctly angulated near middle; superior caudal appendages of male mostly yellow or white .... **albistylus**

—Outer side straight or nearly so; appendages black.................. **parvulus**

NYMPHS

1—Wide antennal segment 3 inequilateral, straight on its inner margin; denticles on front border of median lobe of labium generally three; sides of mentum of labium somewhat convergent toward base; lateral spines of 9 wide, short, and stout .... **albistylus**

—Wide antennal segment 3 more nearly oval and regularly convex; denticles on front border of median lobe of labium generally four; sides of mentum of labium parallel toward base; lateral spines of 9 more slender, longer, and more clawlike .... **parvulus**

Lanthus albistylus Hagen

Syn.: naevius Hagen

1878. Hagen, in Selys, Bull. Acad. Belg., (2)46:460 (reprint, p. 55) (in *Gomphus*).
1898. Harvey, Ent. News, 9:63 (figs.) (as *G. naevius*); *ibid.*, p. 85 (*naevius* = *albistylus*).
1901. Ndm., Bull. N. Y. State Mus., 47:443.
1927. Garm., Odon. of Conn., p. 139 (figs.).
1929. N. & H., Handb., p. 120 (figs.).
1940. Leonard, Occ. Pap. Mus. Zool. Univ. Mich., 414:1–6 (figs.).

Length 31–36 mm.; abdomen 21–26; hind wing 20–23.

Smallest and daintiest of our Gomphines. Heavy cross stripes of black on sutures of face, conjoined by convergent diagonal lines. Occiput blackish. Front of synthorax blackish, with a touch of pale yellow on middorsal carina and two pairs of conspicuous yellow streaks: one pair on sides of collar; the other, more divergent pair in place of usual wide pale stripes of front. Sides pale, with both lateral stripes well developed, but very variable, sometimes fused into one, sometimes wholly apart; ground color changes from green before stripes to yellow on sides of basal segments of abdomen.

Abdomen mainly black, with a rapidly dwindling middorsal yellow line on basal segments, some faint reddish yellow spots on sides of middle segments, and larger but still diffuse markings of same color on sides of slightly enlarged segments 7 to 10: on 7, extending across dor-

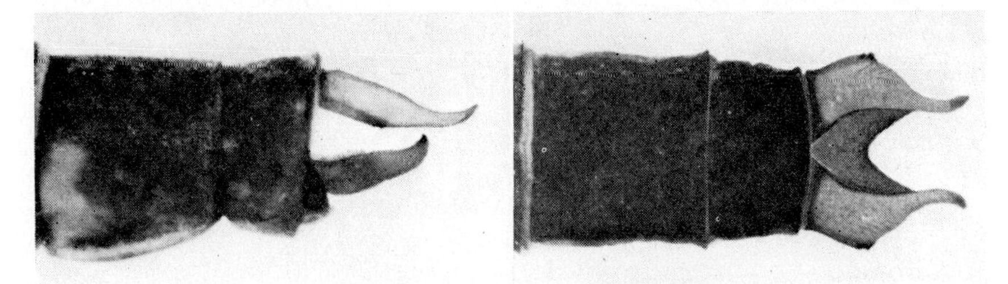

Fig. 82. *Lanthus albistylus.*

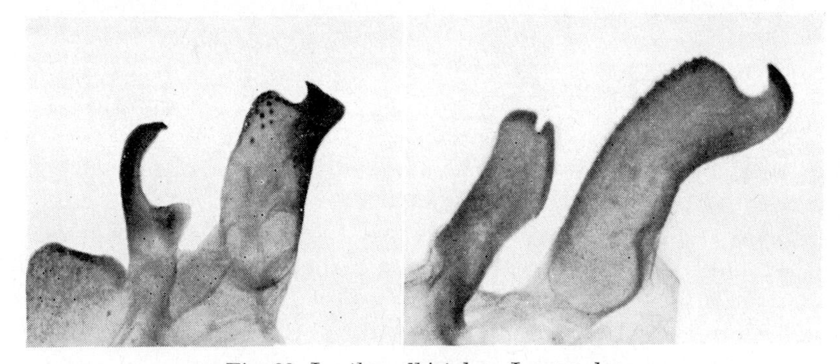

Fig. 83. *Lanthus albistylus; L. parvulus.*

sum as a basal band; on 8, along black lower margin, full length and much broadened at base; on 9, smaller and restricted to base; on 10, wider but more diffuse. Caudal appendages of male yellowish or whitish, with base black; inferior black.

Female paler, black markings less extended, especially on fronto-clypeal suture of face and midlateral suture of thorax. Pale markings of sides of middle abdominal segments may extend upward and meet to form pale basal rings on middle abdominal segments.

Distribution and dates.—CANADA: N. B., N. S., Ont., Que.; UNITED STATES: Ala., Conn., Ky., Maine, Md., Mass., Mich., Mo., N. H., N. J., N. Y., N. C., Ohio, Pa., Tenn., Va.

May 30 (Pa.) to August 17.

Lanthus parvulus Selys

1854. Selys, Bull. Acad. Belg., 21(2):56 (reprint, p. 37) (in *Gomphus*).
1857. Selys, Mon. Gomph., p. 157 (figs.).
1885. Hagen, Trans. Amer. Ent. Soc., 12:281 (nymph as *Uropetala thoreyi?*).
1893. Calv., Trans. Amer. Ent. Soc., 20:242.
1897. Ndm., Can. Ent., 29:165.
1927. Garm., Odon. of Conn., p. 140 (figs.).
1929. N. & H., Handb., p. 120 (figs.).

Length 35–40 mm.; abdomen 25–30; hind wing 22–26.

Another small blackish species, with sides more broadly yellowish. Face greenish yellow; heavy middle cross stripe generally covers entire

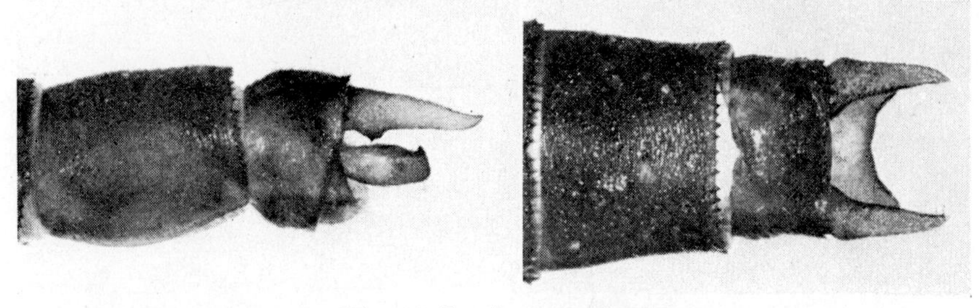

Fig. 84. *Lanthus parvulus.*

clypeus. Collar broadly yellow, constricted in middle, but connected to rearward with top edge of yellow carina.

Front of synthorax broadly black, with two pairs of divergent yellow streaks, much as in *albistylus,* but with the two on collar generally slightly connected, and a vestige of a yellow line between fused black stripes on humeral and antehumeral sutures. Sides of thorax yellowish green, with two brown stripes on lateral sutures broad and variously conjoined, or first lateral one interrupted.

Abdomen mainly black, tricolored on slightly enlarged basal segments, with a yellow crossbar and a half-length cross dash covering auricle on segment 2; a fawn-colored lateral stripe and margin on 1 and 2 and sometimes on 3; remainder black. Middorsal yellow band wide at base, narrows to rearward to end on 3 (male) or 4 (female). End segments of abdomen black, including caudal appendages; in female, with lateral yellow spots on 8 and sometimes also on 9.

Species easily distinguished from *albistylus* by having superior caudal appendages of male wholly black. In female, yellow markings more extensive, and there are diminishing side spots on abdominal segments 3 to 8; only 9, 10, and appendages are black. In *albistylus,* outer side of triangle distinctly angulated near middle; in *parvulus* straight or nearly so. Arculus differs also: in *albistylus*, portion of it above base of sectors longer than part below them; in *parvulus*, the reverse.

Inhabits small rapid woodland brooks; perches on leafy sprays at waterside, sometimes on big stones beside a waterfall. Flies in swift short sallies with quick return to rest.

Distribution and dates.—CANADA: N. S., Que.; UNITED STATES: Conn., Ky., Maine, Mass., N. H., N. J., N. Y., N. C., Pa., S. C.

May 13 (Pa.) to August 17 (Que.).

Genus OCTOGOMPHUS Selys 1873

This genus consists of a single species that is restricted in its distribution to the Pacific slope of the continent. It is similar to the East

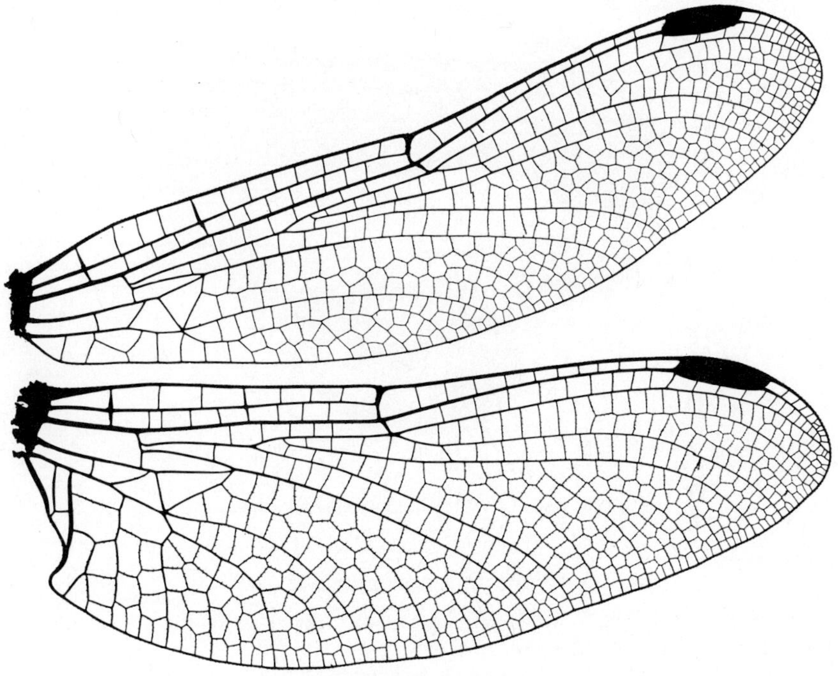

Fig. 85. *Octogomphus specularis.*

Coast genus *Lanthus*, but differs in having a wider hind wing with a rather marked widening of the hind border at the level of the middle fork; also in having the fourth paranal cell in the hind wing nearly as long as the third; that cell is much shorter in *Lanthus*. Two correlated sex characters distinguish it at once from all our other Gomphines: a deeply four-branched inferior appendage in the male, and a four-humped postocellar ridge on the top of the head in the female.

Lanthus and *Octogomphus* agree in having a relatively thick stigma; a widely branched vein Cu2; a very long third paranal cell (in the male), with but one cell between it and the tornus (*to*); and a four-sided subtriangle, the crossvein that forms its inner side not quite reaching the inner angle of the triangle.

The nymphs are very similar to those of *Lanthus,* depressed of body, oval in outline of abdomen, with the same type of labium, and a similarly flattened but narrower antennal segment 3. There are on the abdomen no dorsal hooks and very short lateral spines. They are readily distinguished by their larger size and by the presence of lateral spines on abdominal segment 7.

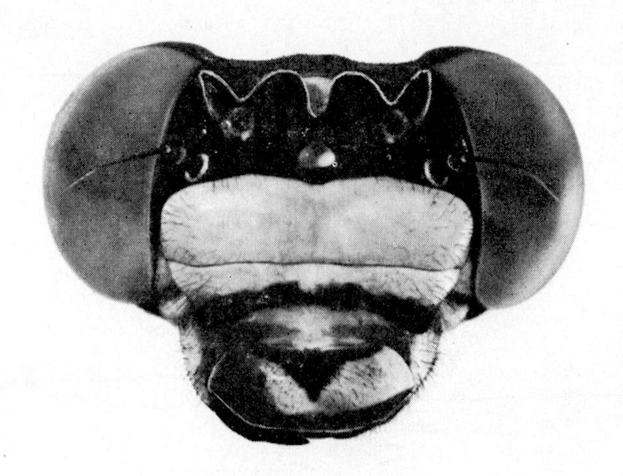

Fig. 86. Face of *Octogomphus specularis,* female.

Octogomphus specularis Hagen

1859. Hagen, in Selys, Bull. Acad. Belg., (2)7:544 (reprint, p. 18) (in *Neogomphus?*).
1861. Hagen, Syn. Neur. N. Amer., p. 110.
1873. Selys, Bull. Acad. Belg., (2)35:759 (reprint, p. 32).
1895. Calv., Proc. Calif. Acad. Sci., (2)4:470 (figs.).
1917. Kndy., Proc. U. S. Nat. Mus., 52:574–581 (figs.).
1929. N. & H., Handb., p. 121 (figs.).

Length 51–53 mm.; abdomen 35–39; hind wing 29–32.

A very pretty little Western Gomphine with a yellow mark, shaped like an inverted urn, occupying front of thorax, bounded laterally by wide shoulder stripes of black. Face yellowish, crosslined with black on sutures, lower stripe running down as a triangular projection on middle

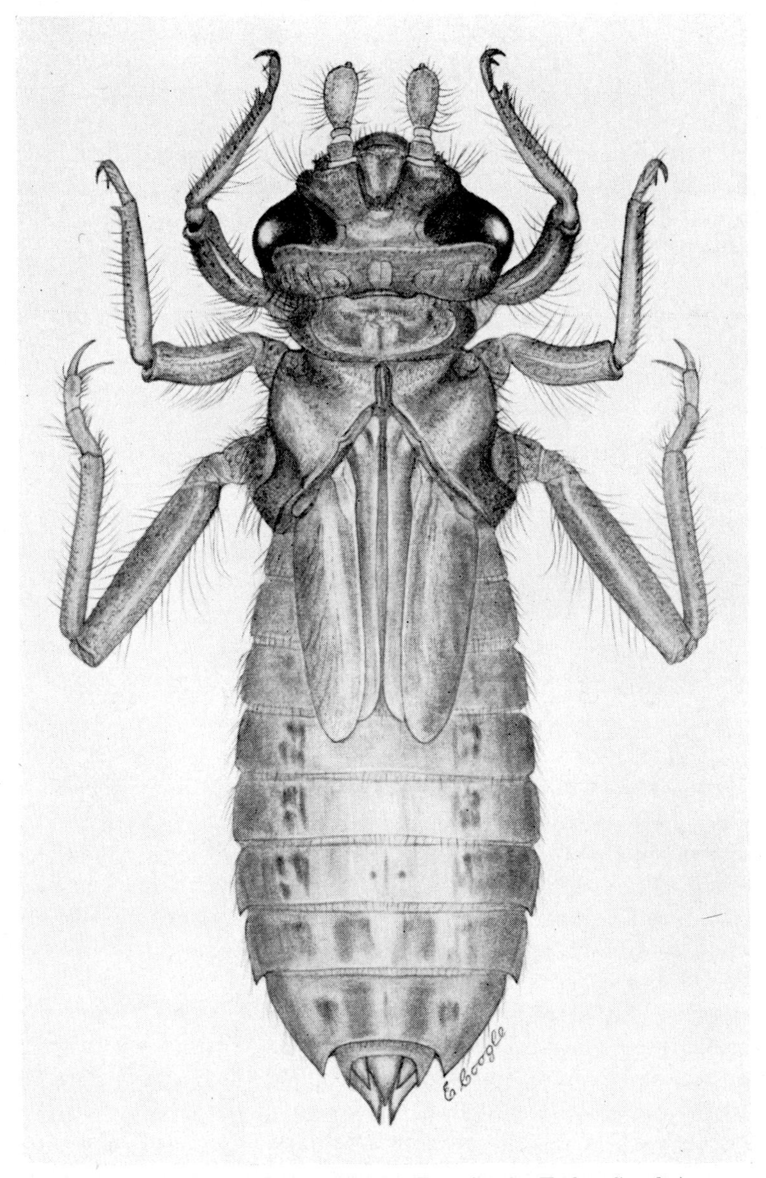

Fig. 87. *Octogomphus specularis*. (Drawing by Esther Coogle.)

of labrum; latter margined with black. Yellow ring about peduncle of black antenna. Vertex black, with a round yellow spot behind low (male) postocellar ridge. Occiput black, yellow at and behind ridge.

The absence of middorsal stripe leaves a very wide yellow dorsal area. Antehumeral and humeral black stripes almost completely fused to form one broad shoulder stripe. Midlateral wanting above spiracle; femoral continuous but narrow, conjoined with black of lower margin of thorax. Legs black.

Fig. 88. *Octogomphus specularis.*

Abdomen black, marked with yellow at ends. Middorsal pale band becomes a mere hairline on segments 3 and 4, continues so to 8 or 9, and widens to an oval yellow spot on 10. Yellow of sides covers most of 1 and about half of 2, where it is divided by a bar of black, a small basal triangle on 3; a faint hairline on lower border of 3 continues faintly on 4 to 7; on 9 yellow reappears as two spots, which may fuse and become a C-shaped figure. Caudal appendages of male yellow on upper surface of superiors; elsewhere black.

Female differs markedly in form of vertex. Postocellar ridge arises in a submedian pair of conspicuous rounded humps that stand stiffly erect. At its outer ends is another more slender pair that curve inward; there is a streak of yellow on front side of each of these lateral prominences. Kennedy has shown (1917, p. 578) that hooked outer tips of four-pronged inferior appendage of male hook around and clasp these lateral prominences in copulation. Behind humps, both vertex and occiput amber brown.

Yellow areas on female much more extensive, with interrupted line of spots all the way down sides to and including 9, but are full length of segment only on 1 and 2. Caudal appendages and parts between them yellow.

Inhabits small streams of Pacific Coast that are not fed in summer by snow water, and only in swifter passages. Oviposits in riffles. Nymphs live mainly in leafy trash that gathers in edges, and crawl up any solid object at shoreline to transform.

Distribution and dates.—CANADA: B. C.; UNITED STATES: Calif., Nev.; MEXICO: Baja Calif.

April 20 (Calif.) to August 18 (B. C.).

Genus GOMPHUS Leach 1815

The *Gomphus* Complex

GOMPHUS s. lat.

This genus as introduced in our key to the genera of Gomphidae (p. 91) is rather a complex of lesser genera. In our fauna it is more numerous in species than all the other nine genera put together. Its further analysis, to which we now come, calls for the use of other characters than those found in that key and table.

The dragonflies of the *Gomphus* complex are singularly uniform in wing venation. They have no crossvein in triangle, supertriangle, or subtriangle of either fore or hind wing; no anal loop, or one of a single cell only. The stigma has a brace vein; the first and fifth antenodal crossveins are thickened. The middle fork is symmetrical; the basal triangle of the hind wing in the male is generally three-celled; intermedian crossveins 2/1 or 3/1. All these characters are well shown in our introductory figure of wings of *Gomphus cavillaris* (p. 17). Any departure from these numbers is exceptional.

Notwithstanding this venational uniformity, there is great diversity in size and appearance among the many species.

In all, the thorax is yellow, olive, or green, striped with brown or black. On the abdomen the darker colors prevail. The middorsal pale band is broad and continuous on the three basal segments, variously reduced or interrupted thereafter; when broken into spots these are shortened and spear-pointed to rearward. The lateral band also is wide and continuous on the basal segments, then greatly reduced on the middle segments, often greatly enlarged again on 7, 8, and 9. The latter spots are largest and brightest yellow on the end segments in those species which have the greatest expansion of the lateral margins. There is often an edging of yellow on the costal vein of the wings. There is generally a small median twin spot of yellow on the disc of the pro-

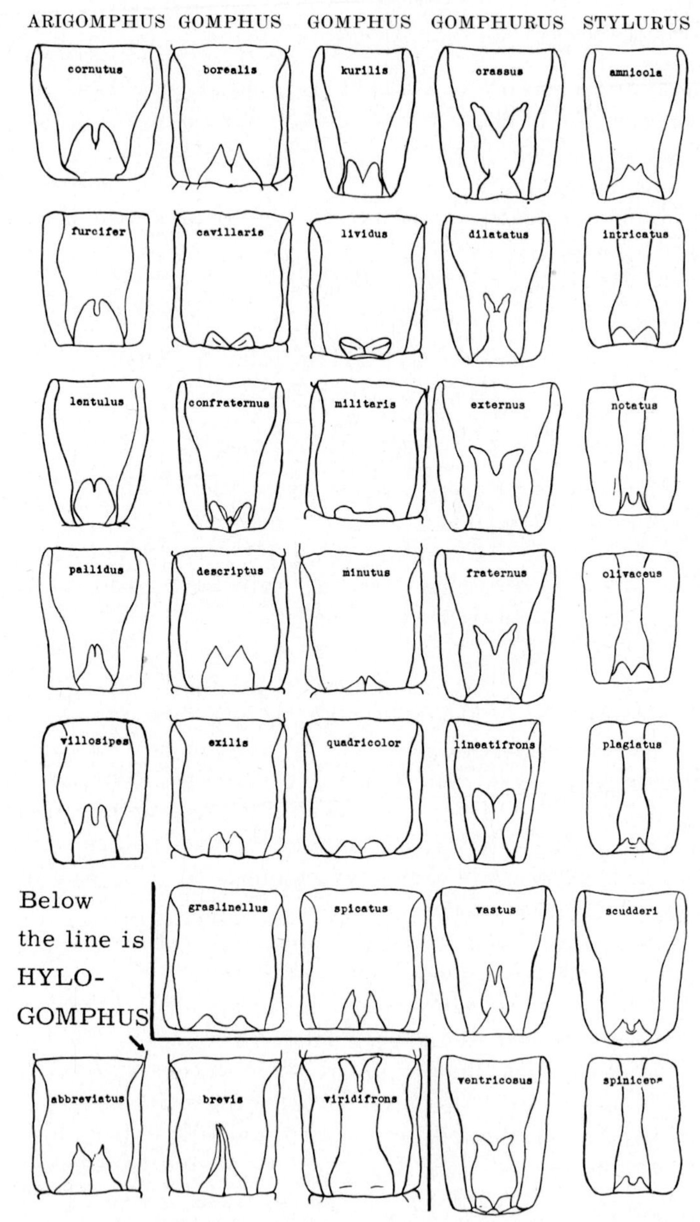

Fig. 89. Form of subgenital plate of female in thirty-four species of genus *Gomphus;* species arranged by subgenera in alphabetic order. Plates as seen in ventral view; their length indicated comparatively, as they lie against sternum of abdominal segment 9. (Figures rearranged from Needham and Heywood's *Handbook,* and are from drawings made by Dr. C. F. Byers.)

thorax, with a larger spot at each side. The spines on the hind legs are much longer and stronger in the female than in the male. The subgenital plate of abdominal segment 8 of the female lies against the sternum of 9. It varies greatly in length and is always divided more or less deeply by a median cleft.

We now offer, with some trepidation, keys for the further analysis and placement of the species in this very difficult group. When determining isolated females it may be helpful to remember that in *Gomphurus* and *Hylogomphus* the subgenital plate is generally more than half as long as the sternum of abdominal segment 9, and of varied outline; in *Arigomphus,* about a third as long as 9, and of one deeply cleft, round-tipped pattern; in *Gomphus* and *Stylurus,* generally less than a fourth as long as 9, and divided into a pair of triangular or low, rounded lobes.

The nymphs are all the color of the silt in which they shallowly burrow. All have developed burrowing hooks on fore and middle tibiae, and more or less hairiness about the leg bases. The mentum of the labium is nearly parallel-sided, with a slightly convex middle lobe, and the lateral lobe generally ends in a well-developed hook, before which are small teeth on the inner margin.

For further studies on the North American species of this complex the most useful single paper is that of the senior author (*Trans. Amer. Ent. Soc.,* 73:307–339, 1898), with its bibliographic list of others.

KEY TO THE SUBGENERA OF THE GOMPHUS COMPLEX*

Adults

1—Vein A1 angulated or kinked at outer end of gaff; in blackish species of **Gomphurus** sometimes straight 2
—Vein A1 in hind wing runs straight or in an open curve from gaff to wing margin 3

2—Front side of fore-wing triangle shorter, or not longer, than inner side† **Gomphurus**
—Front side of fore-wing triangle distinctly longer than inner side...... **Stylurus**

3—Gaff longer, or not shorter, than inner side of triangle; short, stocky; Eastern .... **Hylogomphus**
—Gaff shorter (except in **exilis**) than inner side of triangle 4

4—Stripes on front of thorax, bordering middorsal carina, faint or wanting; general coloration pallid .... **Arigomphus**
—Those stripes, two, dark, well defined, often fused into a wide middorsal band .... **Gomphus**

* For verification of this key it may be necessary to consult the descriptions and figures of appendages of males of the species in the group to which the key leads.

† Except in **crassus.**

TABLE OF SUBGENERA

NYMPHS

Subgenera	*Total length*	*Labium*			*Abdomen*	
		Med. lobe convexity[1]	*End hook*[2]	*Teeth*[3]	*Dorsal groove*[4]	*Lat. spines*
Arigomphus	33-41	strong	absent	5-8	absent	7, 8-9
Gomphurus	28-40	scant	variable	4-9	present	6-9
Gomphus	26-39	variable	variable	6-10	absent	6, 7-9
Hylogomphus	24-26	none	short	6-8	absent	6-9
Stylurus	27-45	variable	variable	2-4	present	6-9

[1]Convexity of front margin of median lobe of labium.

[2]End hook of lateral lobe.

[3]Number of teeth on inner margin of lateral lobe.

[4]A faint longitudinal middorsal groove on middle abdominal segments.

NYMPHS

1—Abdomen ends to rearward in a long tapering point; a low wide median ridge but no median groove on middle segments................**Arigomphus**

—Abdomen ends to rearward more abruptly (except in **Gomphus cavillaris** and **G. australis**); there may be low dorsal hooks or a median groove on middle segments 2

2—Abdomen lanceolate (moderately pointed to rearward); small middorsal hooks on middle segments; no median groove.....................**Gomphus** s. str.

—Abdomen wider than head, flattened, ending more bluntly, narrowed abruptly on segment 9 (where lateral spines are spinulose-serrate on outer edge)...... 3

—Abdomen elongate, narrower than head (except in **S. scudderi**); regularly tapering all the way to rearward; lateral spines of 9 less flattened and merely hairy on outer edge ....**Stylurus**

3—Small species: length less than 27 mm...........................**Hylogomphus**

—Larger species: length 28–40 mm................................**Gomphurus**

Subgenus ARIGOMPHUS Needham 1897

Syn.: Orcus Needham

In this group of pale species the general tone of coloration is greenish gray. The face and occiput are pale green. The front margin of the costa is yellow. The thorax is predominantly pale, with stripes of obscure brown reduced or wanting. The hind femora of the male are very hairy; of the female, only spiny.

There are generally nine antenodal crossveins in the hind wings. The triangles of the wing are rather large, and that of the hind wing tends to have a sagging outer side, and A3 beyond it runs straight to the wing margin. Crossveins are considerably reduced, especially toward the wing base, where the intervals between them are rather wide. In the fore wing the front side of the triangle is one and one-tenth, and in the hind wing one and one-fifth, times as long as the inner side. The gaff is more than half as long as the inner side of the triangle. The paranal cells in the fore wing are five or six, with very few marginals behind them or with no marginals at all. In the hind wing the paranals are generally five; when but four, the last one is elongated in the axis of the wing, and it may simulate an anal loop. The anal triangle in the male is long, extending well toward the hind margin of the wing, and vein A3 runs straight out to the margin; it is not recurved as in other subgenera, and forms a less prominent angle at the tornus. The postanal cells are four or five.

The abdomen is little swollen on the basal segments and still less on the apical segments.

The tips of the superior caudal appendages of the male are angulated, or tend to become forked, with the long arm of the fork on the inner side, and the sharp tips of the pair convergent. There are no teeth on them. The outer arm or angle is short and blunt, or even reduced to a black tubercle. The fork of the inferior appendage is wide, its tips extending laterally farther than the tips of the superiors (except in *furcifer*).

The appendages of the genital pocket are distinctive. The anterior hamule is rather stout, simple, more or less dilated above the middle,

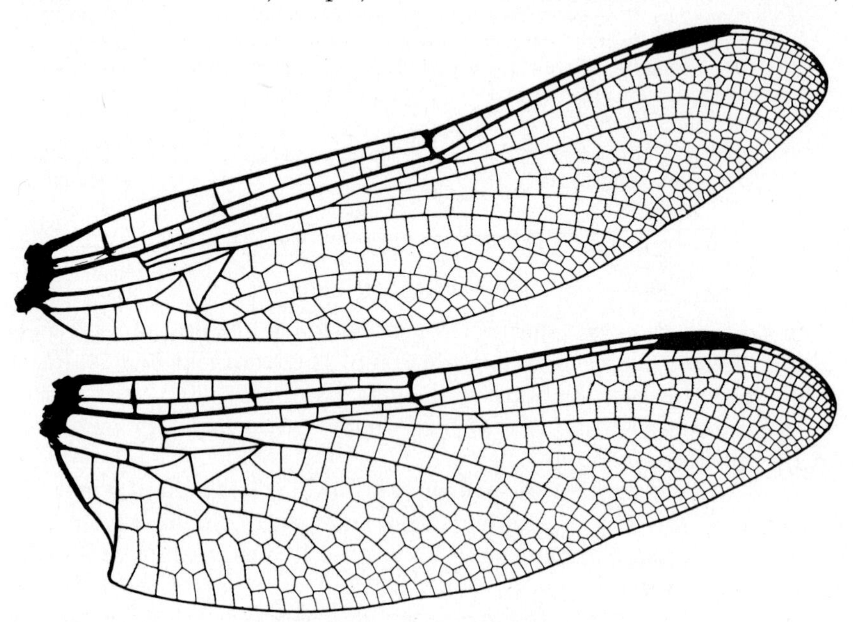

Fig. 90. *Gomphus villosipes.*

and contracted at the end to a single strong clawlike inturned hook. The posterior hamule is large and strongly angulated in *furcifer,* less so in *cornutus,* low and scarcely angulate in the other species. Beyond the shoulder this hamule declines and tapers into a long gooseneck-like inturned hook, the recurved end of which is obliquely truncated and chisel-edged on the inner side.

The peduncle of the penis is deeply divided in front and a little way along its crest, and very hairy within the edges of the cleft. Those edges may be more or less thickened or outrolled like a collar, even presenting in lateral view a backwardly projecting rim. The fourth joint of the penis is of unusual length. It tapers very gradually outward from its thick base into two rather stout lashlike tails of very variable length. The tails are longest in *submedianus* and *villosipes.* In *furcifer* only, the tails are short, and sharply delimited at their base from the body of penial segment 4 by a notch and an angulation.

The subgenital plate of the female is generally about a third as long as the sternum of 9, and deeply bilobed at its tip.

The nymph of this subgenus is recognizable at a glance by the long taper of the hind end of the abdomen. A low middorsal ridge ends at a very minute vestigial middorsal hook that is generally present on 9; there is no lateral spine on 6. The middle lobe of the labium is convex and generally armed with a minute median tooth. The lateral lobe lacks a distinct end hook, and on the inner margin (except in *maxwelli*) bears coarse, irregular, obliquely truncated teeth.

The form of the nymph is as shown in figure 91, and its structural characters are as stated in the table of species which follows. The coarse and deeply cut teeth on the opposed edges of the lateral lobes of the labium, together with the lack of development of an end hook on these lobes, are distinctive characters. The lateral spines of abdominal segment 9, besides being often very much longer than those of the other segments, are different in form. They are laterally flattened and closely appressed to the sides of 10.

The nymph of *furcifer* stands apart from the nymphs of the other species in that it has developed an end hook on the lateral lobe of the labium, and has the middle lobe less prominent.

KEY TO THE SPECIES

Adults

1—Tibiae wholly black 2
—Tibiae with at least an external line of yellow 3

2—Crest of occiput in both sexes high, strongly convex and notched in middle; spread of inferior appendage of male twice the middorsal length of segment 10; female with large horns behind lateral ocelli; hind wing 32–37 mm. **cornutus**
—Occiput less strongly convex and not notched in middle; spread of inferior appendage of male hardly more than middorsal length of 10; female without horns behind lateral ocelli; hind wing 27–31 mm. .... **furcifer**

3—Occiput with sharp elevation or spine at middle; edge usually black .. **villosipes**
—Occiput without sharp elevation or spine at middle; edge usually not black .. 4

4—Thorax with black side stripes well developed; hind wing about 29–32 mm. **maxwelli**
—Thoracic stripes not well developed; hind wing about 29–38 mm. 5

5—Tibiae dull yellow .... **pallidus**
—Tibiae black and yellow, in sharp contrast 6

6—Antehumeral and humeral black stripes subequal in width .... **lentulus**
—Humeral stripe reduced to a line, much narrower than antehumeral **submedianus**

Nymphs

1—Lateral spines of segment 9 a fifth as long as middorsal length of 10, or less .. 2
—Lateral spines of 9 two-fifths as long as 10, or longer 3

TABLE OF SPECIES

NYMPHS

Species	*Total length*	*Labium teeth*[1]		*Abdomen*					
				Dorsal hooks[2]	*Lat. spines*[3]	*Spine of 9*[4]	*Appendages*[5]		
		No.	*Size*				*Lat.*	*Sup.*	*Inf.*
cornutus	41	7-8	med.	absent	8-9	2	8	9	10
furcifer	33	5-6	lrg.	absent	8-9	1	8	10	10
lentulus	40	7-8	lrg.	8-9	7, 8-9	5	9	10	10
maxwelli[6]	33	5-8	vst.	9	7, 8-9	?	10	10	10
pallidus	40	7-8	med.	8-9	7-8-9	3	8	10	10
submedianus	40	5-7	lrg.	8-9	8-9	9	9	10	10
villosipes	37	5-8	lrg.	absent	8-9	2	9	10	10

[1]Number and size of teeth on inner margin of lateral lobe: medium (med.), large (lrg.), vestigial (vst.).

[2]Dorsal hooks (always vestigial) on designated segments of abdomen.

[3]Lateral spines (generally vestigial except on segment 9).

[4]Length of lateral spines on segment 9 in tenths of length of segment 10.

[5]Relative length of superior and lateral appendages compared with inferior taken as 10.

[6]Supposition: not reared.

2—Length of body about 41 mm.; superior abdominal appendage shorter than inferiors .... **cornutus**
—Length of body about 33 mm.; superior abdominal appendage as long as inferiors .... **furcifer**
3—Lateral lobe of labium rounded at end; teeth obsolescent............ **maxwelli**
—Lateral lobe ending in a tooth; other teeth deeply cut........................ 4
4—Lateral spines of segment 9 almost equal to middorsal length of 10..**submedianus**
—Lateral spines of 9 not more than half as long as 10 5
5—No dorsal hooks .... **villosipes**
—Rudimentary dorsal hooks on 8 and 9...6
6—Lateral spines of 9 about half of middorsal length of 10.............. **lentulus**
—Lateral spines of 9 about a fourth to a third as long as 10............ **pallidus**

So close is the likeness of these species in all essential characters that we have been unable to construct a table for adults that would be of much aid in their determination. There are, however, some single distinctive characters that will aid: *pallidus* alone lacks the divided middorsal stripe on the thorax; *cornutus* alone has a notch in the middle of the occipital crest in both sexes, and a pair of horns on the occiput of the female; *villosipes* alone has a tooth or teeth on the middle of the occipital crest.

For adults of other species the figures of the genitalia will suffice.

Gomphus cornutus Tough

Syn.: whedoni Muttkowski

1900. Tough, Mem. Chicago Ent. Soc., 1:17.
1908. Mtk., Odon. of Wis., p. 89.
1908. Walk., Ottawa Natural., 22:52 (fig.).
1914. Whedon, Minn. State Ent. Rep., p. 95.
1915. Mtk. & Whedon, Bull. Wis. Nat. Hist. Soc., (2)13:88, 90 (nymph).
1929. N. & H., Handb., p. 115 (figs.).

Length 55–57 mm.; abdomen 40–44; hind wing 32–37.

A greenish species, striped with brown or black on thorax, and mainly blackish on abdomen. In female, on the rear of vertex behind lateral ocelli, is a pair of horns, short, blunt, finger-like, and incurving. Pale green occiput raised in a bilobed border, higher and more deeply emarginate in middle in female.

Short middorsal thoracic brown stripe an inverted **Y**, with arms narrowly separated by pale green carina. Darker antehumeral stripe abbreviated above, not confluent with narrower humeral except at lower end. Latter stripe distinctly widened at humeral pit; at lower end is conjoined with stub of a midlateral stripe that does not rise above level of spiracle. Narrow, irregular, and sometimes interrupted second lateral ends in an inverted **Y** below. Legs black beyond their pale bases. Wings clear; veins black; costa pale; stigma yellow.

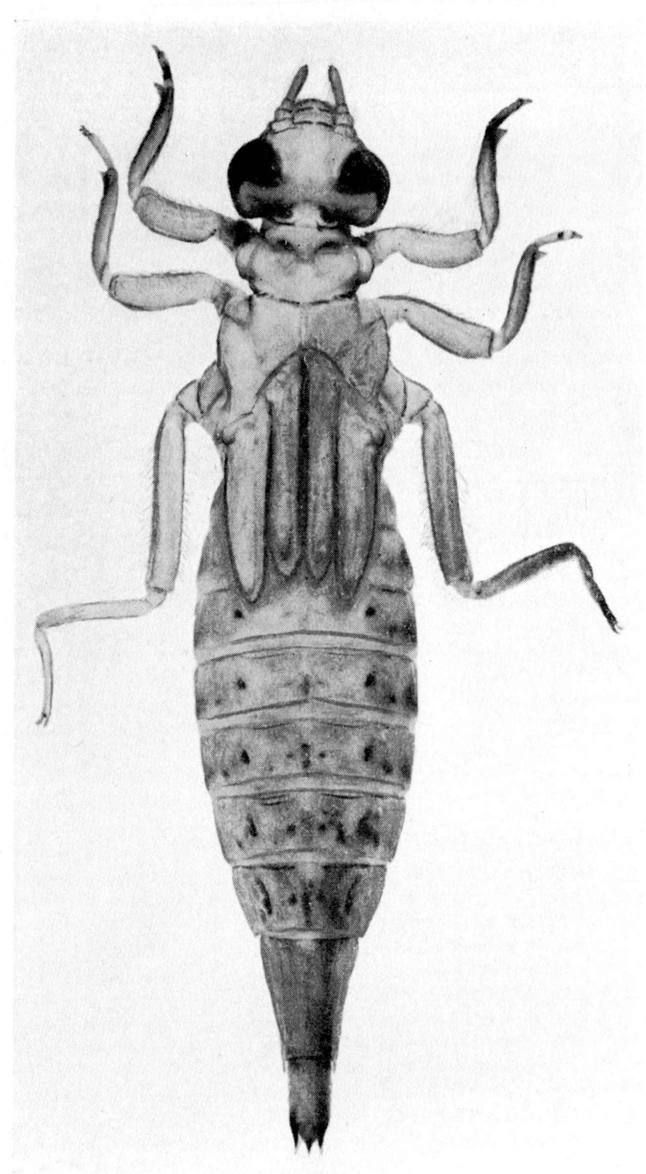

Fig. 91. *Gomphus villosipes.*

Abdomen blackish, with middorsal pale band on segments 1 to 8, wide as usual on basal segments, narrowed and more or less abbreviated on 3 to 8, very small and basal on 9, and large, covering most of dorsum of 10. Lateral pale band of abdomen covers lower half of sides on 1 and 2, is reduced to a narrow edging on 3 to 6, then is progressively widened and heightened in color on 7 to 10. Caudal appendages olivaceous, with blackish tips. Little development of a club out of four end segments, and little expansion of their lateral margins. Instead, viewed from above, abdomen seems to widen a little all the way from 7 to 10. Relative

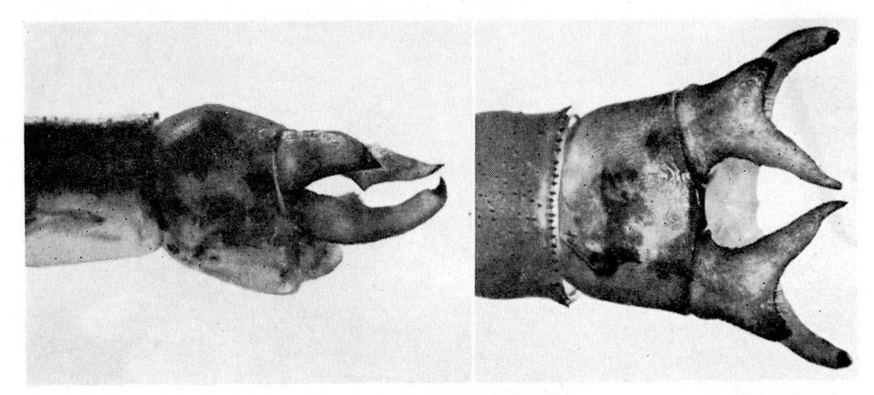

Fig. 92. *Gomphus cornutus.*

length of last four segments as 11:9:10:7, with appendages 6 on same scale in male; or as 12:11:10:6 and appendages 7 in female.

Easily recognized by high occipital margin, notched in middle in both sexes; by wide forking of superior caudal appendages in male; and by horns on vertex in female.

Frequents lake shores.

Distribution and dates.—CANADA: Man., Ont., Que.; UNITED STATES: Colo., Ill., Ind., Mich., Minn., Nebr., N. Dak., Wis.

May 30 (Ill.) to July 21 (Ont.).

Gomphus furcifer Hagen

1878. Hagen, in Selys, Bull. Acad. Belg., (2)46:458 (reprint, p. 53).
1900. Wmsn., Odon. of Ind., p. 292 (figs.).
1902. Hine, Ohio Natural., 2:61 (figs.).
1904. Walk., Can. Ent., 36:358 (figs.).
1917. Howe, Odon. of N. Eng., p. 34 (figs.).
1917. Kndy., Bull. Univ. Kansas, 18:137 (figs.).
1927. Garm., Odon. of Conn., p. 158 (figs.).
1929. N. & H., Handb., p. 115 (figs. as *villosipes*).

Length 46–54 mm.; abdomen 34–39; hind wing 27–31.

A dark green species striped with black. Top edge of occiput straight, fringed with long black hair. Abbreviated middorsal thoracic stripe cleft into an open Y-shaped figure, much as in *cornutus;* but with arms of Y widened downward, and farther outspread. Antehumeral black stripe much abbreviated at upper end and well separated from antehumeral; it becomes more slender below where it curves rearward into short

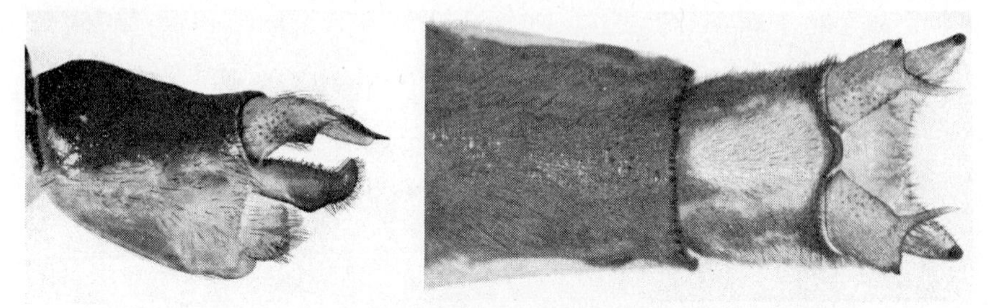

Fig. 93. *Gomphus furcifer.*

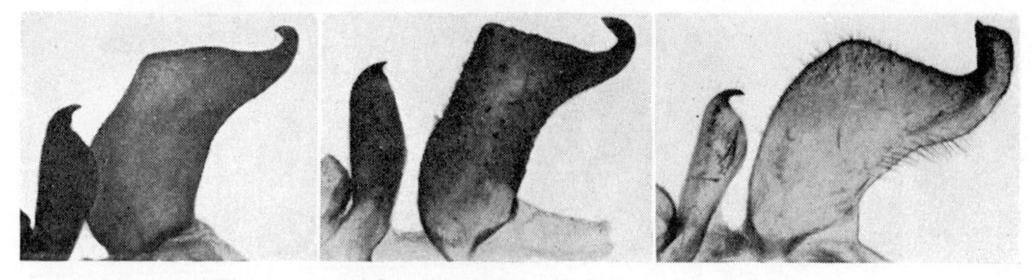

Fig. 94. *Gomphus cornutus; G. furcifer; G. lentulus.*

remnant of midlateral stripe that rises below spiracle. Second lateral weak and ill defined, especially near lower end, where it may form stem of an inverted Y-shaped black mark. Legs pale at base and out to middle of femora; black beyond. Stigma of wings tawny.

On abdomen, middorsal pale band is reduced on middle segments to narrow spearheads that become obsolete on 8; 9 wholly black above; 10 black, with a large middorsal yellow spot. Lateral pale band wide on sides of 1 and 2, narrows on 3 as usual, to continue as a narrow strip along lower margin, with heightening yellow color on 7, 8, 9, and 10, especially on lower half of 10. Caudal appendages yellow in both sexes.

Recognizable by extreme length and slenderness of inner branch of superior appendages of male. Female has no horns on vertex.

Distribution and dates.—CANADA: Ont., Que.; UNITED STATES: Conn., Ind., Iowa, Mass., Mich., N. J., N. Y., Ohio.

May 18 (N. Y.) to August 1 (Mich.).

Gomphus lentulus Needham

Syn.: subapicalis Williamson

1902. Ndm., Can. Ent., 34:275.
1908. Mtk., Odon. of Wis., p. 82.
1911. Mtk., Bull. Wis. Nat. Hist. Soc., 9:36 (figs.).
1914. Wmsn., Ent. News, 25:54 (figs.).
1929. N. & H., Handb., p. 113 (figs.).
1941. Gloyd, Bull. Chicago Acad. Sci., 6:127.

Length 48–58 mm.; abdomen 34–41; hind wing 29–36.

An olive green species, striped faintly with brown. Pale green occiput convex. Low postocellar ridge as viewed from above is bow-shaped,

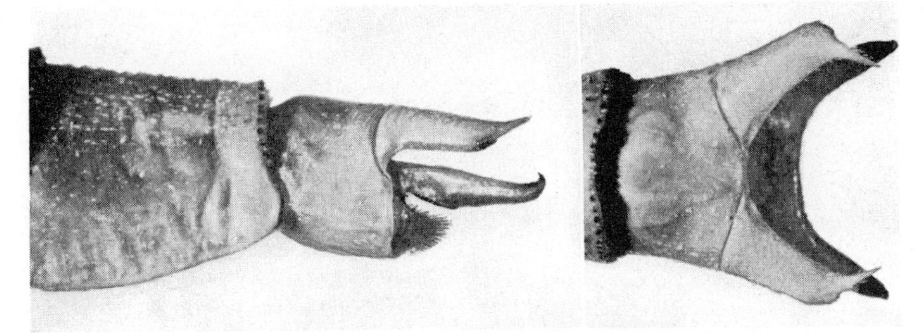

Fig. 95. *Gomphus lentulus.*

with a wide median depression. Middorsal thoracic brown divided completely by carina and fork of crest into two stripes, which may be very faint, and which end squarely below, well above collar. Humeral and antehumeral about equally developed and well separated by a line of green. Antehumeral abbreviated at upper end. Lateral stripes undeveloped except sometimes at their ends.

Usual longitudinal pale bands of abdomen scarcely recognizable except on basal segments. Diffuse brown color of middle segments has taken on a ringed appearance, darkest across apices of segments, and with a secondary crossline at about a fourth the length of each. Dorsum of 7 and 8 black; 9 paler; 10 yellowish; caudal appendages yellow.

Female differs as usual in having a generally greater extent of pale areas and a marked narrowing of end segments of abdomen.

This species nearly like *submedianus,* from which it will be best distinguished by long needle-sharp point on end of superior caudal appendage in male; also by the longer and more decurved subgenital plate of female.

Distribution and dates.—UNITED STATES: Ill., Ind., La., Tex.

May 8 (Tex.) to July 9 (Ind.).

Gomphus maxwelli Ferguson

1950. Ferguson, Field and Lab., 18(2):93 (figs.).

Length 50–54 mm.; abdomen 35–40; hind wing 29–32.

A pale greenish olive species, striped on thorax and ringed on abdomen with brown. Face yellow; top of head black, including basal third of frons. Occiput rises in a transverse nearly hairless ridge that is squarely truncated on its summit.

Thorax greenish olive, yellowish beneath, with middorsal brown stripe divided by yellow carina into a pair of slender, rather widely separated

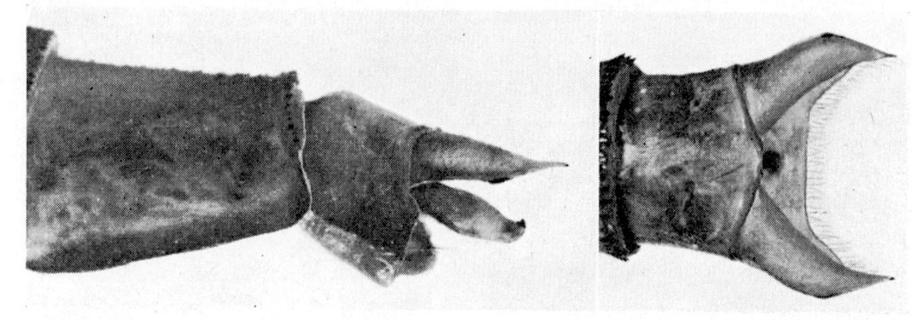

Fig. 96. *Gomphus maxwelli.*

submedian stripes of brown that diverge downward and are isolated at both ends. All edgings of crest deep black. Antehumeral brown stripe wider, abbreviated above; humeral narrow below, widening upward to humeral pit. Other two lateral stripes present, but indistinct and variable. Pale color of legs at base extends far out on sides of hairy femora and along outer side of tibiae. Outer edges beset with black prickles. Tibiae elsewhere and tarsi black beneath; all spines black. Black lines on hind femur coalesce outward to wholly blacken it at knee. Wings hyaline, with yellow costa and stigma and light brown veins.

Abdomen dull green, on middle segments doubly ringed with brown: a wide ring on joinings of segments and a narrowly linear obscure one just before it. End segments darker, with diffusely reddish undertone.

Female differs in having top of head olivaceous, general coloration paler, abdomen little swollen basally and increasingly slender all the way to rear end, where segment 10 and appendages are yellow.

Distribution and dates.—UNITED STATES: Ala., Tex.

May 11 (Tex.) to June 14 (Ala.).

Gomphus pallidus Rambur

Syn.: pilipes Hagen

1842. Rbr., Ins. Neur., p. 163.
1861. Hagen, Syn. Neur. N. Amer., p. 105.
1897. Ndm., Can. Ent., 29:157.
1900. Wmsn., Odon. of Ind., p. 291 (figs.).
1908. Mtk., Odon. of Wis., p. 82.
1914. Wmsn., Ent. News, 25:54.
1927. Garm., Odon. of Conn., p. 158 (figs.).
1929. N. & H., Handb., p. 114 (figs.).
1930. Byers, Odon. of Fla., pp. 62, 250 (figs.).

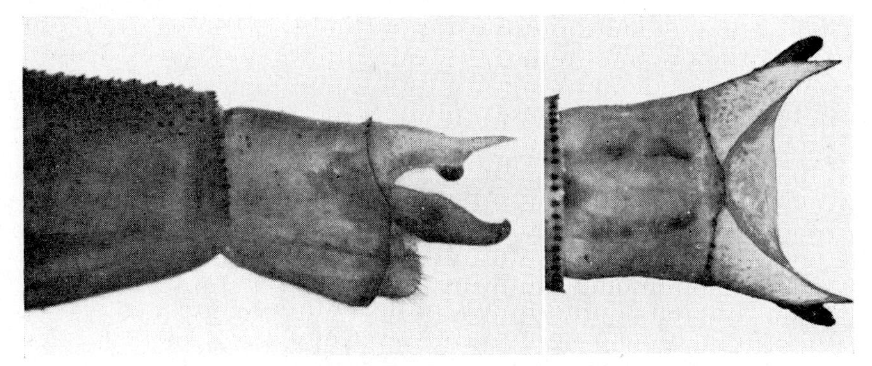

Fig. 97. *Gomphus pallidus.*

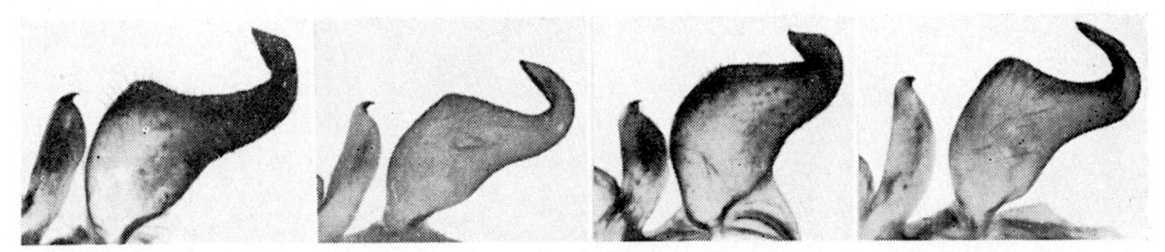

Fig. 98. *Gomphus maxwelli; G. pallidus; G. submedianus; G. villosipes.*

Length 60–62 mm.; abdomen 45–47; hind wing 37–38.

An olivaceous or grayish green species, almost without thoracic brown stripings. Postocellar ridge, as viewed from above, about straight in its middle third. Thorax heavily clad with brown hairs. Usual thoracic stripes represented by indistinct tracings, the one on humeral suture being generally best developed. Legs mostly pale, blackened on tip of tibiae and on tarsi; first and second tarsal segments of middle and hind legs yellow on dorsal side; on front tarsi black.

Middle abdominal segments have a somewhat ringed appearance, being darkened across apices and down sides. Segments 7 to 9 rusty brown; 10 and caudal appendages dull yellow. Claws tawny, with black tips.

This species much like *lentulus* and *submedianus*, but easily distinguished from them in male by having black nodule at outer angle of superior caudal appendage projecting stumplike; in other species it is a low black edging. In female, subgenital plate shorter, narrower, straight, scooplike, not decurved.

Distribution and dates.—UNITED STATES: Ala., Fla., Ga., Ky., La., S. C., Tenn.

March 20 (Fla.) to July 23 (Ky.).

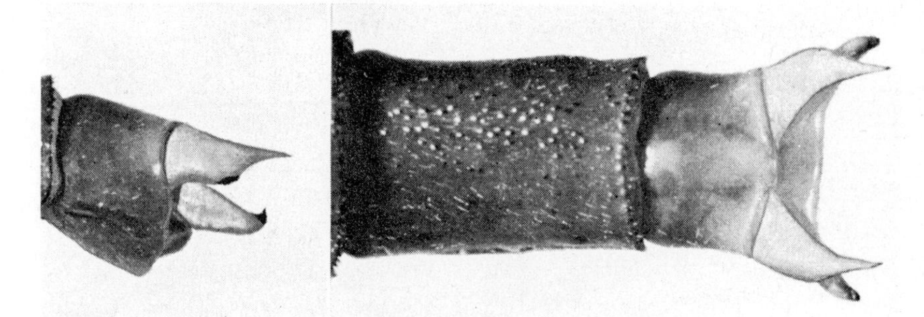

Fig. 99. *Gomphus submedianus.*

Gomphus submedianus Williamson

1914. Wmsn., Ent. News, 25:54 (figs.).
1929. N. & H., Handb., p. 113 (figs.).
1941. Gloyd, Bull. Chicago Acad. Sci., 6:127.

Length 51–55 mm.; abdomen 37–40; hind wing 34–36.

This species very similar to *lentulus*, having same pattern of faint brown stripings on thorax, and same suffusion of blackish on abdomen, darkest about segment 7, tending toward crossbands on middle segments, and lightening toward yellow on end segments. It will therefore suffice to point out some minor differences.

Occiput similar to that of *lentulus*, but larger and evenly convex. Postocellar ridge, as viewed from above, bilobed by a shallow median notch. Humeral and antehumeral brown stripes generally quite unequally developed, humeral being much fainter. In male, tip of superior caudal appendages shorter and stouter. In female, subgenital plate shorter and less strongly decurved.

Distribution and dates.—UNITED STATES: Ill., Ind., La., Mass., Mich., Miss., Okla., Tex.

May 21 (Ind.) to July 31.

Gomphus villosipes Selys

1854. Selys, Bull. Acad. Belg., 21(2):53 (reprint, p. 34).
1861. Hagen, Syn. Neur. N. Amer., p. 105.
1893. Calv., Trans. Amer. Ent. Soc., 20:244.
1899. Klct., Odon. of Ohio, p. 63 (figs.).
1900. Wmsn., Odon. of Ind., p. 291 (figs.).
1901. Ndm., Bull. N. Y. State Mus., 47:460 (nymph).
1902. Hine, Ohio Natural., 1:61.
1927. Garm., Odon. of Conn., p. 159 (figs.).
1929. N. & H., Handb., p. 116 (figs. as *furcifer*).

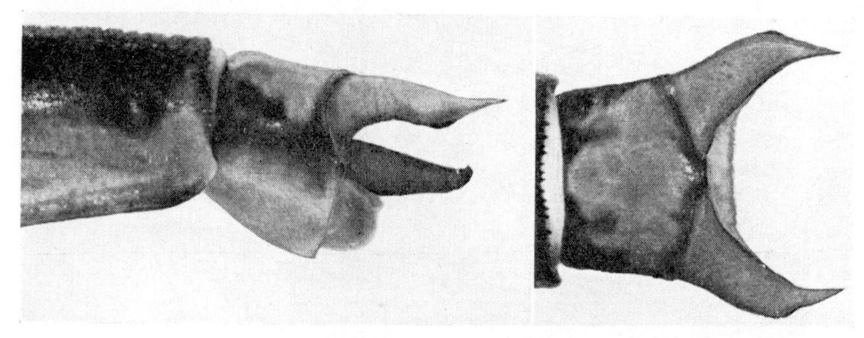

Fig. 100. *Gomphus villosipes.*

Length 50–58 mm.; abdomen 37–41; hind wing 29–36.

A large greenish species, well striped with brown on thorax, and with a blackish abdomen. Green occiput strongly convex on upper margin, with a median marginal tooth in both sexes. Middorsal thoracic stripe widens downward, cleft longitudinally below fork of crest along carina and well separated from collar. Humeral brown stripe abbreviated above and separate from humeral. Side stripes of brown obsolete except at ends. Legs black except at base. Hind femora of male densely clad with long soft hairs (whence the specific name), commingled with numerous slender spines.

Abdomen stout. Middorsal pale band reduced beyond basal segments to short interrupted lines on 3 to 7; 8 and 9 wholly black above; 10 and superior appendages yellow, blackish at sides.

Distribution and dates.—CANADA: Ont.; UNITED STATES: Conn., Ill., Ind., Ky., Md., Mass., Mich., Minn., N. J., N. Y., N. C., Ohio, Pa.

May 10 (Ohio) to July 24 (Ont.).

Subgenus GOMPHURUS Needham 1901

The members of this subgenus are the heavyweights of the *Gomphus* complex. They are characterized by great widening of abdominal segments 7, 8, and 9. These segments are generally successively a little shortened, and 10 is about half as long as 9; the appendages are a little longer than 10. The blackish stripings of the sides of the thorax are

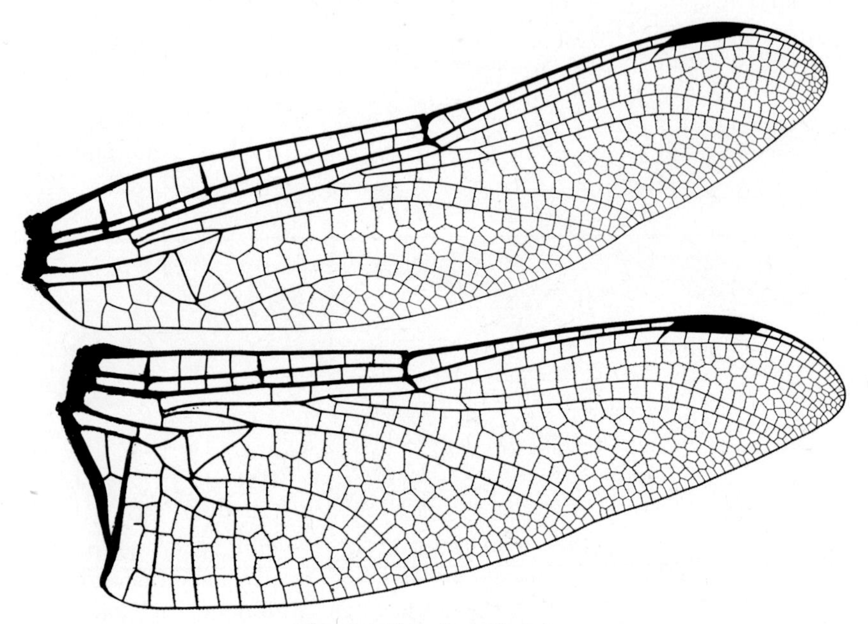

Fig. 101. *Gomphus dilatatus.*

consistently present, only the midlateral stripe being interrupted or sometimes wanting. Venational characters are as stated for the *Gomphus* complex (p. 163), with these added. The front side of the triangle is not longer, often a little shorter, than the inner side in the fore wing, and is at least one and a fifth times as long in the hind wing; the gaff is from one-half to nine-tenths as long as the inner side. The paranal cells of the fore wing are five, six, or seven, generally six, with one to six marginals; postanals are three, four, or five. In the hind wing the paranal cells are generally four. The first and second anal interspaces (x and y) generally start with one, two, or three full-width cells. The single tornal cell is two or three times the height of the cells of the marginal row beyond it.

The superior caudal appendages of the male are stout at the base and quickly narrowed and arched thereafter, variously carinate, and toothed more or less before the acute tip. The inferior appendage is widely forked, the tips of the fork reaching laterally beyond the tips of the superiors.

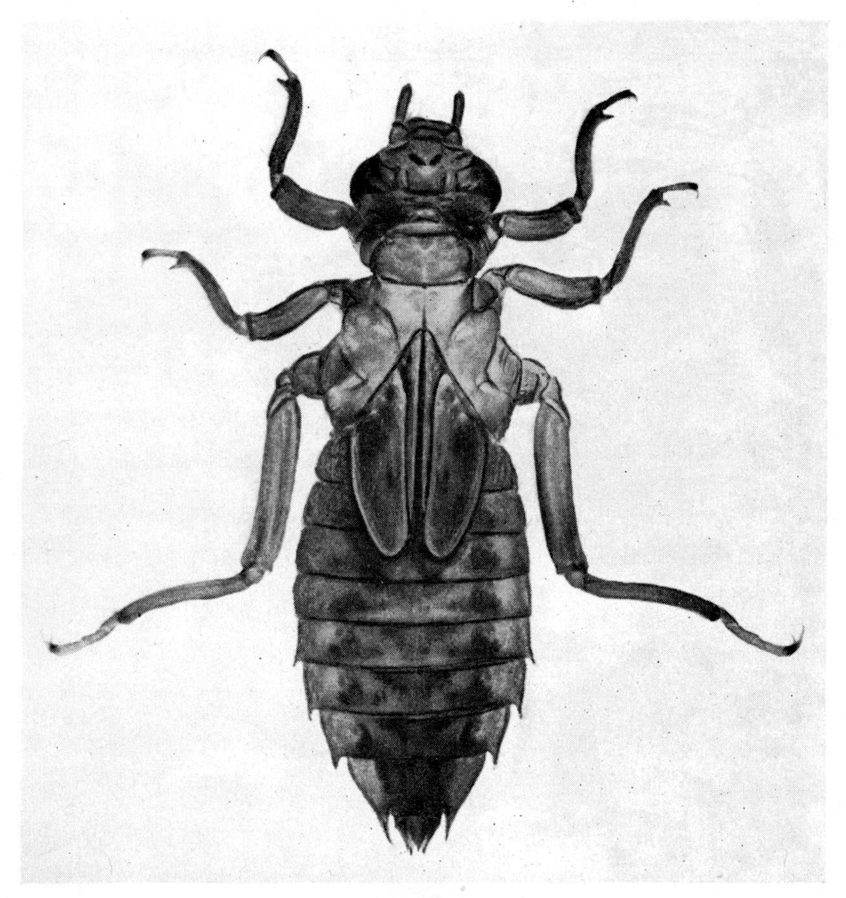

Fig. 102. *Gomphus externus.*

The anterior hamule is short, with minute inflexed teeth. The posterior one is prominent, stout, and hairy, with a shoulder on its front margin (low in *ventricosus*).

In *Gomphurus* there are two minor groups of more closely associated species, and two strays that seem to stand a little apart from both (the strays being the little-known species *adelphus* and *consanguis*). In the *Dilatatus* group fall *lineatifrons, modestus, vastus,* and *ventricosus.* In the *Fraternus* group fall *crassus, externus,* and *hybridus.* The two groups differ in form of body, in certain minor features of venation and genitalia, and in depth of pigmentation in the coloration of the body. The *Dilatatus* group we regard as the more specialized, having more open network in venation, more broadly clubbed abdomen, deeper pigmentation; also less similarity to other groups of the great *Gomphus* complex.

This group parallels *Stylurus* in many ways, and with that group some of its species may easily be confused. In form of body the species of *Gomphurus* are shorter and stockier, with a more heavily clubbed abdomen, as the subgeneric name indicates. There is a great range in depth of coloration, but all the species have well-developed thoracic side stripes of brown or black.

The triangles of the wing are somewhat smaller than in *Stylurus,* and somewhat more nearly equal in size in fore and hind wing.

The nymphs of this subgenus, so far as known, corroborate the foregoing grouping of the adults. They are more strongly depressed than in other subgenera, with the abdomen broader, much wider than the head, and more abruptly narrowed to the caudal appendages. The mentum of the labium is little longer than wide, parallel-sided or nearly so, with small teeth on the inner margin of the lateral lobe diminishing in size proximally. The tibial burrowing hooks are large and strong. There are low, flattened dorsal hooks on segments 8 and 9, sometimes a very minute vestige of a hook on 7. The lateral spines on 6 to 9 are large, strong, and conspicuous; on 6, sometimes much smaller than on 7; on 9, generally larger than the others, varying in length from one to three times as long as segment 10. Segment 9 is generally a little longer than 8 and more than twice as long as 10; its lateral margins are finely serrated. The caudal appendages are blunt; superior and inferiors are subequal in length, laterals from four-fifths to nine-tenths as long.

KEY TO THE SPECIES

ADULTS

1—Dorsal surface of occiput wholly black .... **consanguis**
—Dorsal surface of occiput almost wholly yellow 2
2—Abdominal segments 8 and 9 almost wholly black .... **adelphus**
—Abdominal segments 8 and 9 conspicuously yellow on sides 3
3—Face yellow or green 4
—Face cross-striped with black 8
4—Dilation at end of abdomen in male wider than thorax; third lateral brown stripe of thorax wanting .... **ventricosus**
—Dilation not wider than thorax; third lateral stripe more or less present 5
5—Midlateral suture with dark stripe complete; outer surface of tibiae generally with pale yellow stripe; third femora of female without this yellow stripe 6
—Midlateral suture with dark stripe incomplete; outer surface of tibiae generally without yellow stripe; third femora of female with this yellow stripe 7

6—Dorsum of segment 9 with a broad yellow stripe .... **externus**
—Dorsum of 9 blackish, with little if any yellow .... **hybridus**

7—Inferior appendage of male straight-edged between forks .... **fraternus**
—Inferior appendage of male semicircular between forks .... **crassus**

8—Hind wing 34 mm. or more 9
—Hind wing 35 mm. or less 10

9—Face with black line on fronto-clypeal suture; middorsal black stripe generally not wider than bordering pale stripe .... **lineatifrons**
—Face with wide black band on fronto-clypeal suture; middorsal black stripe as wide as, or wider than, bordering pale stripe .... **dilatatus**

10—Humeral and antehumeral brown stripes separated by long stripe of yellow **modestus**
—Humeral and antehumeral stripes in contact near their upper ends, or separated by very narrow line of yellow .... **vastus**

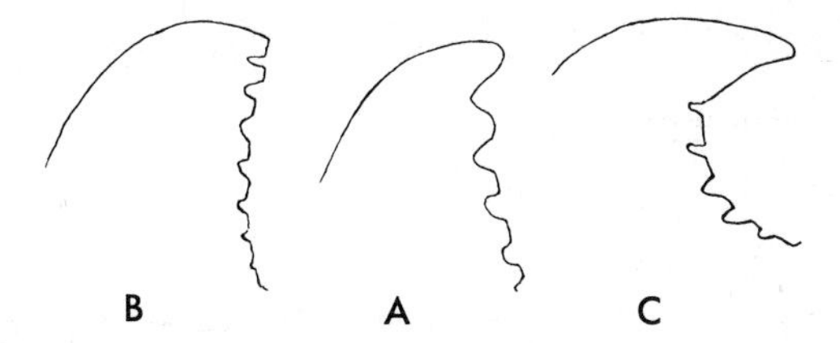

Fig. 103. End of lateral labial lobe. B, *Gomphus hybridus;* A, *G. fraternus;* C, *G. vastus.*

NYMPHS

1—Front margin of median lobe of labium deeply concave .... **lineatifrons**
—Front margin of median lobe of labium straight or a little convex 2

2—Lateral spines of segment 9 equal to middorsal length of 10 .... **?consanguis**
—Lateral spines of segment 9 about one and one-half to twice the middorsal length of 10 3

3—End hook on lateral lobe of labium short or wanting; teeth small, 7–9; tips more or less in straight line (fig. 103, A, B) 4
—End hook longer, more incurved, line of teeth curving with it; teeth larger, 4–6 (fig. 103, C) 7

4—Lateral spines larger, more outstanding, spine on segment 6 almost as large as that on 7 5
—Lateral spines smaller, less outstanding, spine on 6 small, usually less than half as large as that on 7 6

5—Tip of lateral lobe of labium very blunt, terminal tooth not longer than preceding teeth (fig. 103, B) .... **hybridus**
—Tip of lateral lobe of labium less blunt, terminal tooth distinctly longer than others (fig. 103, A); lateral spine of 6 very outstanding .... **crassus**

TABLE OF SPECIES

ADULTS

Species	Hind wing	Facial stripe[1]	Thoracic stripes		Yellow markings[4]			Distr.
			Midd.[2]	Humerals[3]	On tib.	On 8	On 9	
adelphus	25-29	+	△	separ.	+	0	0	NE
consanguis	32-37	+	△	joined	0	+	+	E
crassus	31-36	0	=	joined	0	+	+	C
dilatatus	34-43	+	△	separ.	0	0	0	SE
externus	30-33	0	=	±	+	+	+	C,W,S
fraternus	28-33	0	=	joined	0	+	±	E,C
hybridus	27-29	0	=	separ.	+	±	±	SE
lineatifrons	38-45	+	=	separ.	0	0	0	E,C,S
modestus	34-35	+	△	separ.	0	+	0	S
vastus	28-34	+	△	±	0	+	+	E,C,S
ventricosus	24-33	0	=	joined	0	+	+	E,C

[1]Cross stripe of black or brown on face: present (+), absent (0).

[2]Thoracic stripes: middorsal pair widened forward to form a dark triangle (△), approximately parallel (=).

[3]Humeral and antehumeral stripes meeting above, near crest: (joined), separate (separ.), variable (±).

[4]Yellow on outer face of tibiae and on sides of abdominal segments 8 and 9: present (+), absent (0).

TABLE OF SPECIES

NYMPHS

Species	Total length	Labium			Abdomen			
		Median lobe[1]	Lat. lobe		Lat. spines of seg. 9[3]	Appendages[4]		
			Hook[2]	Teeth		Lat.	Sup.	Inf.
consanguis?	29	conv.	A	8	10	9	10	10
crassus	34	stra.	A	8-9	20	9	10	10
dilatatus	40	stra.	C	4-6	20	8	10	10
externus	33	conv.	A	8-9	20	8	10	10
fraternus	31	stra.	A	7	20	8	10	10
hybridus	28	stra.	B	8-9	20	8	10	10
lineatifrons	38	conc.	C	6	20	8	10	10
vastus	31	stra.	C	4-6	15	9	10	10

[1]Front margin of median lobe: straight (stra.), slightly convex (conv.), concave (conc.).

[2]End hook of lateral lobe of types most like three shapes A, B, and C shown in figure 103.

[3]Length of lateral spines of segment 9 in tenths of length of segment 10, alongside which they are extended (this segment may be telescoped in 9, or pulled out, with a stretch of intersegmental membrane showing).

[4]Relative length of lateral and superior caudal appendages compared with inferior taken as 10.

6—Fringed front border of median lobe slightly convex; lateral lobe generally ending in two large teeth.......................................**externus**

—Fringed border straight; lateral lobe generally ending in one large tooth **fraternus**

7—Length of body about 40 mm.; lateral spines of 9 reaching well beyond level of tip of 10...**dilatatus**

—Length of body about 31 mm.; lateral spines of 9 not reaching tip of 10..**vastus**

Nymphs unknown: **adelphus, modestus,** and **ventricosus.**

Gomphus adelphus Selys

1857. Selys, Mon. Gomph., p. 413.
1861. Hagen, Syn. Neur. N. Amer., p. 104.
1878. Selys, Bull. Acad. Belg., (2)46:457 (reprint, p. 52).
1917. Howe, Odon. of N. Eng., p. 33 (fig.).
1929. N. & H., Handb., p. 94.

Length 43–48 mm.; abdomen 32–37; hind wing 25–29.

This species strikingly patterned in yellow and black. Face yellow, heavily crossbanded with black on borders of labrum, on sutures, and on sides of postclypeus. Anteclypeus also blackish. Top of head all black except front of occiput. Crest of occiput fringed with long black hairs.

Synthorax yellow, heavily striped with black in front. Middorsal band divided by a yellow carina into two stripes which widen forward to yellow collar, and are prolonged upward laterally beneath crest to a junction with humeral stripe. Antehumeral black stripe isolated at upper end, where obliquely truncated. Midlateral and femoral stripes almost wanting: former wholly so above level of spiracle; latter reduced to a narrow line.

Legs wholly black beyond their short basal segments except for a small yellow spot on knee joint. Wings clear. A few details of venation may be noted: antenodal crossveins in hind wing nine; first (x) and second (y) anal interspaces begin with two full-width cells (cells elongated in axis of wing); basal triangle of male three-celled.

Abdomen strongly club-shaped on end segments; black beyond, with usual middorsal pale line reduced progressively to abbreviated spear-pointed spots on 3 to 7; 8, 9, 10, and appendages all black, save for faint traces on broadly expanded lateral margins of 8 and 9.

Female differs from male in some details. On top of head a pair of short conical black spines stand behind lateral ocelli. Occipital crest, viewed from front, low and emarginate in middle. Under side of front femur yellow. Leaflike lateral expansion of 8 and 9 more broadly and vividly yellow. Under side of hind femur bears strong spines (half as long as femur is wide), where male has merely bristles. Third anal interspace (z), equivalent of basal triangle in male, contains four cells.

Subgenital plate half as long as sternum of 9, against which it lies; cleft in its apical third into two pointed tips that lie parallel.

Species appears to be rare. Nymph unknown.

Distribution and dates.—UNITED STATES: Mass., N. Y.

June 4 (N. Y.) to September (N. Y.).

Gomphus consanguis Selys

Syn.: rogersi Gloyd

1879. Selys, C. R. Soc. Ent. Belg., 22:66.
1929. N. & H., Handb., p. 94.
1936. Gloyd, Occ. Pap. Mus. Zool. Univ. Mich., 326:1–5 (figs.) (as *rogersi* n. sp.).
1944. Klots, Amer. Mus. Nov., 1258:1–5 (figs.).

Length 48–50 mm.; abdomen 35–38; hind wing 32–37.

An olive green species, very heavily marked with black on face, front of thorax, and dorsum of abdomen. Labium black on middle lobe, greenish elsewhere. Labrum cross-striped with black on front and hind margins, with a median black spot or connecting crossbar joining stripes. Mandibles green at exposed sides, black at tip. Anteclypeus black. A wide black crossband covers fronto-clypeal suture, variously widened downward for middle half of its length and down again at its ends following lower margin of compound eye. Vertex wholly black. Occiput black in front, with a yellow spot in rear, very narrow and slightly concave on its nearly hairless margin.

Thorax black in front, green on sides. Prothorax black on dorsum, with its front lobe greenish; three large green spots in line across its top, with middle spot single, not geminate. Behind humeral stripe, sides mainly greenish. All black stripes joined to wholly black crest. Mid-lateral stripe generally interrupted in middle. Its lower piece narrows upward and then widens to a triangular tip at spiracle. Femoral stripe complete but narrow and irregular. A large spot of black in green on each supra-coxal plate. Black stripings on front of synthorax so broadened and confluent at ends that little is left of pale ground color: only a thin line on carina, an isolated green stripe on each side (the pair divergent downward), and a round spot and narrow strip below it, still partly separating humeral and antehumeral stripes.

Legs long and thin, wholly black except for a pale streak on under side of front femur. Claws unusually long, with a sharp tooth underneath, midway of their length. Wings hyaline, with veins black, costa black, and stigma brown. A faint wash of brown beyond and behind stigma in membrane. Nine or ten antenodal crossveins in hind wing. Six paranal cells in fore wing (with no added marginals) and four in hind wing. First two cells in both first (x) and second (y) anal interspaces full width. Five or six crossveins under stigma.

Abdomen black beyond three basal segments, with usual middorsal and lateral pale stripes greatly reduced. Dorsal pale stripe on segment 1 very wide, narrowing to a full-length spear point on 3, and to a third of segment's length on 4; may be wanting on 5, and reappear as a trace on 6 and 7. In all this the extent of pale color is very variable. Beyond 7 all black on dorsum, including caudal appendages. Lateral pale stripe covers lower half of sides of 1 and 2 (partly enclosing auricle on 2), narrows

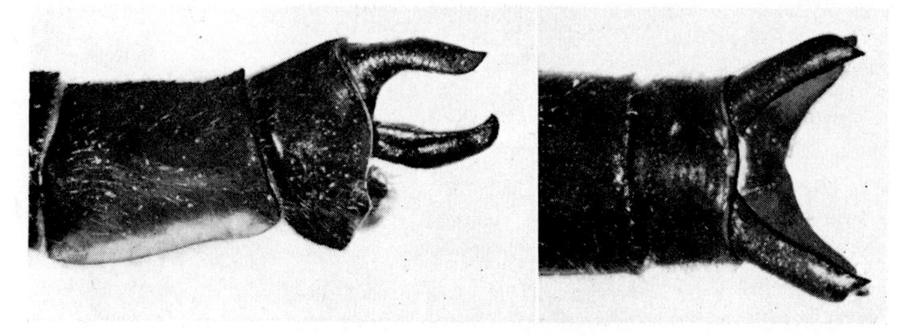

Fig. 104. *Gomphus consanguis.*

down to a trace on 3, reappears as a trace on 6 and 7, widens on end of 7, still more on 8 and 9, becoming brighter yellow on expanded margins of 8 and 9.

Genitalia of segment 2 black. Anterior lamina hood-shaped. Anterior hamule, black and shining, rises erect and is widened to top, where one long corner is folded over end of channel. Posterior hamule sway-backed and flat, tapering to a sharp recurved end hook, and hairy on both edges. Caudal appendages about equal in length, their four ends with about equal lateral spread.

Female similar to male but somewhat less broadly marked with black. No horns on vertex. Abdomen stouter, with less enlargement of ends, and none at all on 8 and 9. Subgenital plate about half as long as sternum of 9. Midway of its length it is divided by a V-shaped cleft; divisions are regularly tapered to round points and directed straight backward.

Distribution and dates.—UNITED STATES: Ala., N. C., Pa., Tenn.

May 12 (Ala.) to June 25 (N. C.).

Gomphus crassus Hagen

Syn.: walshii Kellicott

1878. Hagen, in Selys, Bull. Acad. Belg., (2)46:453 (reprint, p. 48).
1900. Wmsn., Odon. of Ind., p. 288.
1901. Calv., Ent. News, 12:65 (figs.).
1927. Garm., Odon. of Conn., p. 162 (figs.).
1929. N. & H., Handb., p. 92 (figs.).

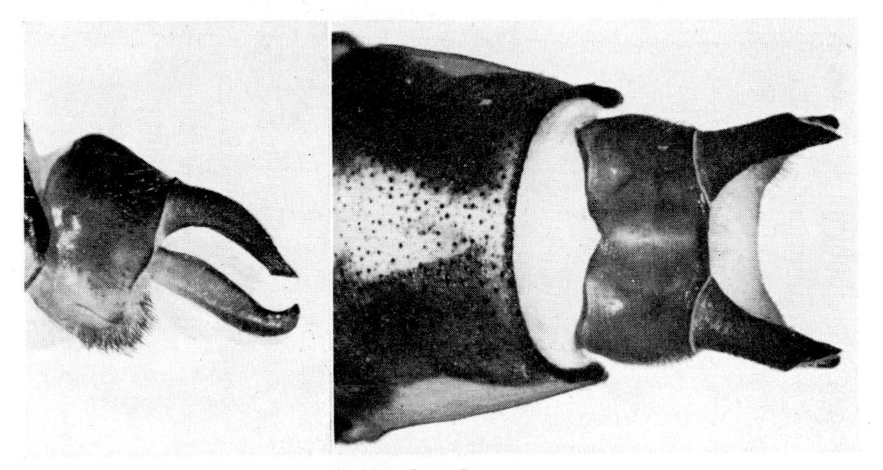

Fig. 105. *Gomphus crassus.*

Length 54–59 mm.; abdomen 41–43; hind wing 31–36.

A greenish yellow species, heavily striped with black. Middorsal thoracic stripe divided by yellow carina, each half of it much narrower than adjacent bordering yellow stripe. Antehumeral black stripe isolated at its upper end, then conjoined with humeral for a little space, then separated again below by a narrow line of yellow. Midlateral stripe reduced to a short tapering stub that rises hardly above level of spiracle. Third lateral stripe wanting.

Legs black, including hind femur of female, with an external yellow line of variable length on tibiae.

Middorsal line of yellowish on abdomen wide at base, narrowing as usual to a slender tip on segment 3, still more on 4 to 6, becoming a short triangle on base of 7, and again wider on 8 and 9, oval in contour on 9. It takes on a much brighter yellow on 7, 8, and 9. Lateral pale band of abdomen practically wanting on middle segments; a trace of yellow on lower part of sides of 7, a big blotch of yellow, sometimes divided, on 8, and a better delimited similar area on 9, where it reaches base of 10. Segment 10 in both sexes paler brown than 8 and 9, with obscure roundish middorsal spot of yellow. Caudal appendages and hamules of male black.

Distribution and dates.—UNITED STATES: Ind., Iowa, Ky., Minn., Ohio, Tenn.

May 11 (Ind.) to July 31 (Ind.).

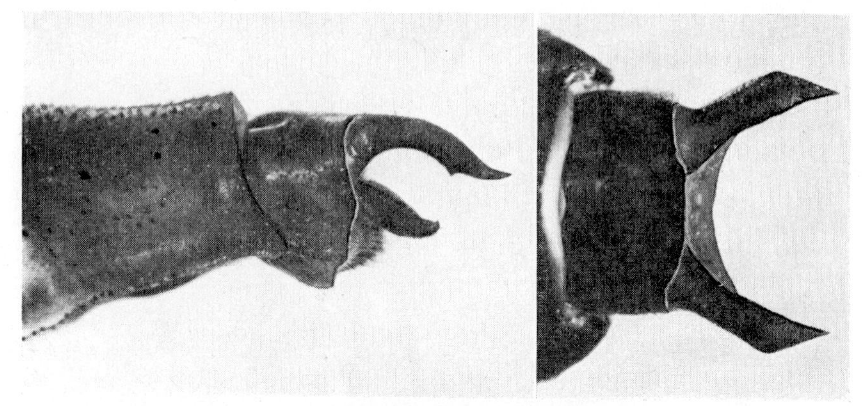

Fig. 106. *Gomphus dilatatus.*

Gomphus dilatatus Rambur

1842. Rbr., Ins. Neur., p. 155.
1861. Hagen, Syn. Neur. N. Amer., p. 103.
1903. Ndm., Proc. U. S. Nat. Mus., 26:710 (figs.).
1921. Calv., Trans. Amer. Ent. Soc., 47:224.
1929. N. & H., Handb., p. 97 (figs.).

Length 67–73 mm.; abdomen 46–53; hind wing 34–43.

A very large species, one of handsomest of genus. Two broad black cross stripes on face and a black front border on labrum. Yellow occiput narrowly margined with black. Middorsal thoracic stripe broadens downward into a wide triangle, lower angles of which may or may not be connected laterally with lower end of antehumeral; wide humeral and antehumeral separated by a narrow line of yellow, latter free at upper end; second and third laterals generally complete and of moderate width. Legs black, or sides of hind femora with a streak of pale yellow.

Middorsal stripe on abdomen, narrowed and interrupted or obsolete to rearward, ends in a half-length spearhead on segment 7 (longer in female); 7 to 9 broadly yellow on widely dilated side margins; 8, 9, and 10 black above.

From the senior author's field notes:

This species is common on the lower Chipola River in west Florida. The adult goes steaming along in steady horizontal flight two or three feet above the open river with tail aloft and wings scarcely showing vibration. It is a striking figure. The slender middle part of the abdomen is inclined upward, and the broadly dilated end segments are held parallel with the course of flight, but at a higher level than that of the bulky striped thorax. Back and forth it goes, steadily, easily, as ruler of the lesser life over the open stream. I saw one carrying a black-wing (*Agrion maculatum*), a long limp captive, perhaps on the way to a feeding perch among the willows.

The nymphs live in the muddy banks of the river, and clamber several feet up the swollen bases of the tupelo trees to transform.

Distribution and dates.—UNITED STATES: Ala., Fla., Ga., La., S. C. March 6 (Fla.) to June 23 (Ga.).

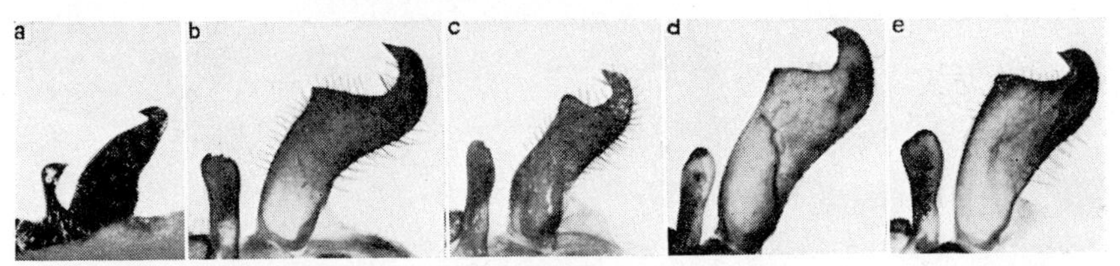

Fig. 107. A, *Gomphus consanguis;* B, *G. crassus;* C, *G. dilatatus;* D, *G. externus;* E, *G. fraternus.*

Gomphus externus Hagen

Syn.: consobrinus Walsh, fraternus Walsh

1857. Hagen, in Selys, Mon. Gomph., p. 411.
1861. Hagen, Syn. Neur. N. Amer., p. 104.
1900. Wmsn., Odon. of Ind., p. 174.
1901. Calv., Ent. News, 12:65 (figs.).
1901. Ndm., Bull. N. Y. State Mus., 47:451 (figs.).
1917. Kndy., Bull. Univ. Kansas, 18:137 (figs.).
1927. Garm., Odon. of Conn. (figs., pp. 146–152).
1929. N. & H., Handb., p. 90 (figs.).

Length 52–59 mm.; abdomen 37–42; hind wing 30–33.

A stout light yellowish green species, handsomely striped with brown. Middorsal stripe wholly divided, narrow, parallel-sided, abruptly truncated a short distance above collar, with each half much narrower than bordering pale bands. Antehumeral, as in *crassus,* isolated above, then joined to humeral near its upper end, then separated from it again by a narrow yellow line for most of its length. Midlateral stripe ill defined but complete; tends to be conjoined by a wash of brown with stripe on third lateral suture for its whole length. Legs black; tibiae externally yellow. Wings faintly tinged with yellow in membrane at base.

Middorsal pale line of abdomen as usual broad on basal segments, reduced on middle segments, long and spear-pointed on 7, reduced to an elongate basal triangle on 8, expanded to a wide yellow band on entire length of 9, and reduced to a little oval basal dash of yellow on 10. Species may generally be distinguished from its nearest allies by wide dorsal band of yellow on 9.

Inhabits larger streams; males fly in long regular sweeps chasing each other; rests on bare sand or on logs or boards, tail elevated, wings declined to touch the sand, in an attitude of great alertness. Females oviposit by dipping tip of abdomen to wash off eggs (5,200 eggs once obtained

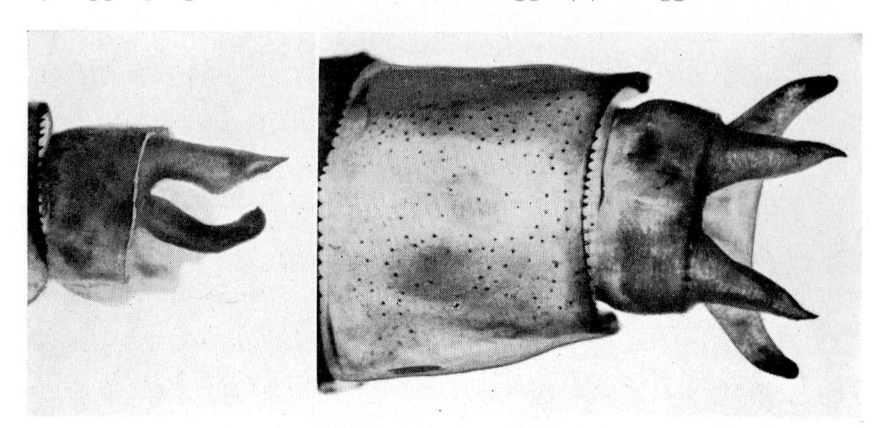

Fig. 108. *Gomphus externus.*

from a single female). Transformations occur on piers or stumps, a foot or two above the water.

Distribution and dates.—CANADA: Man.; UNITED STATES: Ill., Ind., Iowa, Kans., Ky., Mich., Minn., Mo., Nebr., N. Mex., Tex., Utah, Wis.

May 1 (Tex.) to August 10 (Mo.).

Gomphus fraternus Say

1839. Say, J. Acad. Phila., 8:16 (in *Aeschna*).
1861. Hagen, Syn. Neur. N. Amer., p. 104.
1900. Wmsn., Odon. of Ind., p. 289.
1901. Calv., Ent. News, 12:65 (figs.).
1917. Howe, Odon. of N. Eng., p. 33 (figs.).
1927. Garm., Odon. of Conn., p. 163 (figs.).
1929. N. & H., Handb., p. 90 (figs.).

Length 48–55 mm.; abdomen 34–40; hind wing 28–33.

A greenish yellow species, striped on thorax with brown. Thorax and swollen basal segments of abdomen rather densely clothed with fine soft black hair. Sides of narrow middorsal stripe parallel, somewhat distant from collar at lower end. Antehumeral brown stripe free at its upper end, then joined more or less with humeral, then free again and separated by a narrow greenish yellow line down to level of collar. Upper third more or less of midlateral stripe ill defined or wanting; third lateral generally complete and well defined, but may be entirely wanting.

Middorsal abdominal stripe, interrupted and reduced as usual on middle segments, is expanded to form triangular basal spots on 7 and 8, with often a little spot on 9; 10 blackish. Yellow on lateral margins of expanded end segments a streak on 7, a wide patch on 8 and 9; 10 yellow beneath. Legs blackish; the sides of hind femora of female yellowish.

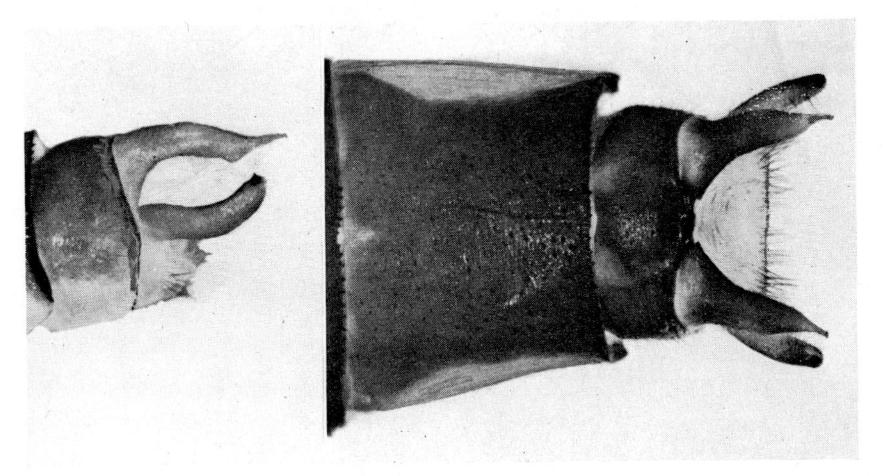

Fig. 109. *Gomphus fraternus.*

"Very often taken in pastures and open woodlands back from the rivers. About rapids they dart swiftly here and there above the turbulent water, dash in and out of the leafy arches along the banks, or rest tightly flattened against the boulders in midstream."—Whedon (*Minn. State Ent. Rep.*, p. 94, 1914).

Distribution and dates.—CANADA: Man., Ont., Que.; UNITED STATES: Ala., Ark., Fla., Ill., Ind., Iowa, Ky., Mass., Mich., Minn., Miss., N. H., N. Y., Ohio, Pa., Tex., Va., W. Va., Wis.

April 9 (Fla.) to July 23 (Wis.).

Gomphus hybridus Williamson

1902. Wmsn., Ent. News, 13:47.
1929. N. & H., Handb., p. 90 (figs.).

Length 50–52 mm.; abdomen 35–37; hind wing 27–29.

Another slightly smaller, stockier species, like the three preceding in having face and occiput green, most like *fraternus* in coloration. Middorsal brown stripe of synthorax much widened downward to yellow collar, more or less divided on median line by yellow on carina. Antehumeral stripe free at its upper end, close to humeral and lightly in contact with it at one or two points, the intervening yellow being narrow and inconstant. Midlateral and third lateral stripes complete, or former may be interrupted above spiracle. Legs mostly blackish beyond their pale basal segments. No yellow stripe on hind femur. Tibia black, with a yellow line on its outer side. Tarsi black, with more or less yellow on back

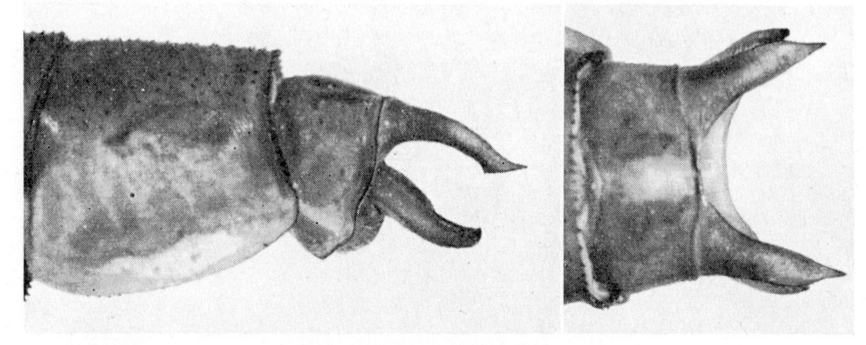

Fig. 110. *Gomphus hybridus.*

of first and second joints, least marked in hind tarsi. Wings hyaline, with black veins and stigma.

Abdomen black, with usual middorsal line of paler markings wide and continuous on swollen basal segments, narrowed and reduced to spots beyond them. Spot on 7 about half as long as segment or a little longer. Widened segments 7 to 10 darker than middle ones and varied with brown and yellow; 10 obscure brownish, with a more or less constant round spot of yellow on dorsum.

Superior caudal appendages of male regularly convex on dorsal side. Small tubercle on under side. Widely extended branches of inferior appendage lie outside superiors and end in a semicircular curve.

Female differs from male in having dark colors less extended. A short, erect spine on vertex at each end of postocellar ridge. Subgenital plate parallel-sided beyond its spreading base, and ends in a very wide V-shaped excision; plate slightly less than half as long as sternum of segment 9, against which it lies.

Distribution and dates.—UNITED STATES: Fla., Ind. (?), Tenn.
March 24 (Fla.) to June 7 (Tenn.).

Gomphus lineatifrons Calvert

1921. Calv., Trans. Amer. Ent. Soc., 47:222 (figs.).
1929. N. & H., Handb., p. 98 (figs.).
1937. Borror, Ohio J. Sci., 37:187.

Length 67–69 mm.; abdomen 46–52; hind wing 38–45.

This species much like *dilatatus,* about as large, a little less extensively black. Instead of black band across frons in that species, only a sutural line of black. Middorsal thoracic stripe much narrower, being scarcely widened downward; midlateral stripe interrupted in middle.

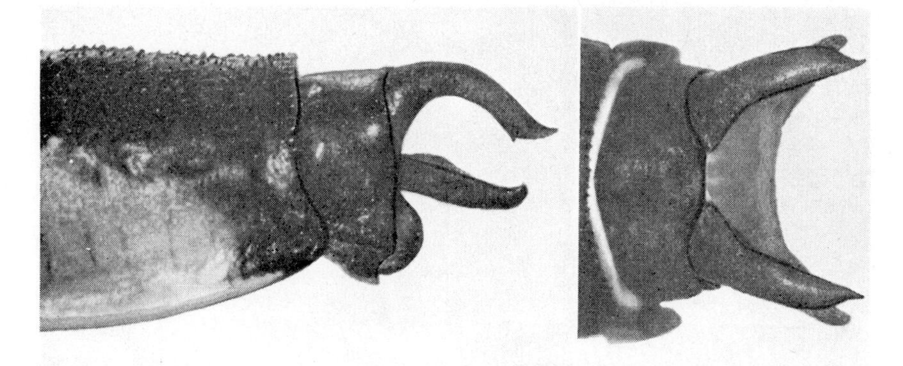

Fig. 111. *Gomphus lineatifrons.*

Middorsal pale line of abdomen exceedingly narrowed on middle segments, but expands again on 7 to a basal triangle that may be half as long as dorsum of that segment; 8, 9, and 10 black above. Yellow on expanded lateral margins of end segments a mere brush streak on 7, a quadrate blotch on 8, and a wider full-length area on 9; 10, caudal appendages, and hamules black.

Distribution and dates.—UNITED STATES: Ala., Ill., Ind., Ky., Mich., Minn., N. Y., Ohio, Pa., Tenn., Va.

May 17 (Ky.) to July 4 (Ind.).

Gomphus modestus Needham

1914. Wmsn., Ent. News, 25:447 (as *consanguis?*).
1942. Ndm., Can. Ent., 74:72 (figs.).

Length 62 mm.; abdomen 45; hind wing 34–35.

This species much like *vastus,* but larger. Cross stripes on face similar, but black on crest of occiput, which covers only line of roots of fringing hairs in that species, in *modestus* is wider, extending down below margin.

Middorsal thoracic stripe less widened downward, and forms a narrower triangle. Antehumeral brown stripe entirely separated from humeral by a wide yellow stripe. Midlateral and third lateral stripes well developed.

Middorsal pale line of abdomen runs out to a hairline on segment 3, continues so on 4 to 7, widens to a short basal triangle of yellow on 7 and again on 8; 9 and 10 black above. Yellow on expanded lateral margins of end segments very scant on 7, a quadrangular spot on 8, and a large full-length marginal field on 9; 10 blackish.

Distribution and dates.—UNITED STATES: Ala., Miss., Tex.

April 25 (Miss.) to May 27 (Tex.).

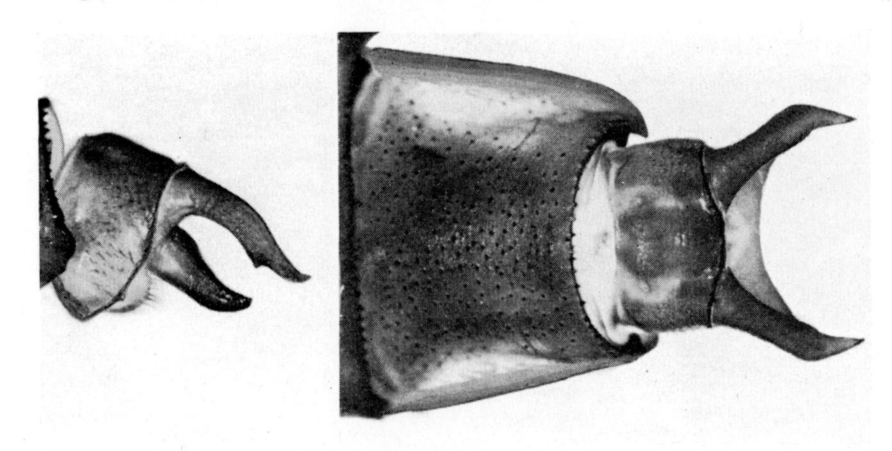

Fig. 112. *Gomphus modestus.*

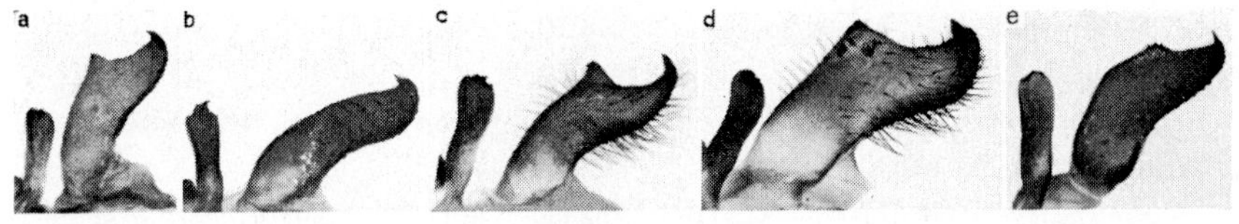

Fig. 113. A, *Gomphus hybridus;* B, *G. lineatifrons;* C, *G. modestus;* D, *G. vastus;* E, *G. ventricosus.*

Gomphus vastus Walsh

1862. Walsh, Proc. Acad. Phila., p. 391.
1872. Cabot, Mem. M. C. Z., 5:3 (figs., nymph).
1900. Wmsn., Odon. of Ind., p. 287.
1903. Wmsn., Ent. News, 14:226.
1917. Wlsn., Proc. U. S. Nat. Mus., 43:190.
1927. Garm., Odon. of Conn., p. 165 (figs.).
1929. N. & H., Handb., p. 95 (figs.).

Length 47–57 mm.; abdomen 33–41; hind wing 28–34.

A pretty yellowish green species; thorax striped with rich brown, and a black abdomen brightly marked with yellow at both ends. Labrum, heavily bordered and crossed with black, more black than greenish. Occiput yellow, with a narrow black line on its summit that covers hardly more than bases of long, ciliate, fringing hairs. Middorsal thoracic stripe widens below to a triangular form. Antehumeral stripe free above, but often touches humeral near top and is separated from it thereafter by an inconstant and very narrow yellow line; midlateral and third lateral stripes well developed.

Fig. 114. *Gomphus vastus.*

Middorsal pale yellow line of abdomen ends in half-length basal yellow spot on segment 7; 8, 9, and 10 black above. Yellow of expanded lateral margins of end segments faint on 7, a half-length included basal spot on 8, and a wider, full-length marginal area of brighter yellow on 9.

Inhabits river banks, thickets, and lake shores where there are alternating stretches of sand and gravel; frequents brush near water's edge.

Distribution and dates.—CANADA: Ont., Que.; UNITED STATES: Ala., Conn., D. C., Ga., Ill., Ind., Iowa, Kans., Ky., Md., Mass., Mich., Minn., Miss., Mo., N. H., N. Y., N. C., Ohio, Pa., S. C., Tenn., Va., W. Va., Wis.

April 8 (Miss.) to September 15 (Ohio).

Gomphus ventricosus Walsh

1863. Walsh, Proc. Ent. Soc. Phila., 2:249.
1900. Wmsn., Odon. of Ind., p. 287.
1908. Mtk., Odon. of Wis., p. 90.
1917. Howe, Odon. of N. Eng., p. 33 (figs.).
1927. Garm., Odon. of Conn., p. 166 (figs.).
1929. N. & H., Handb., p. 97 (figs.).

Length 48–53 mm.; abdomen 32–39; hind wing 24–33.

A very dainty and attractive little species, greenish yellow in body, brightly striped with dark brown. Yellow occiput slightly bilobed on its crest and tilted forward, black-margined only at sides of middle notch.

Middorsal stripe of thorax parallel-sided, enclosing a yellow dash on carina that widens below to a confluence with yellow collar. Antehumeral stripe darker, free above, then confluent with humeral for a space, then free again below for a space and separated by a narrow greenish line.

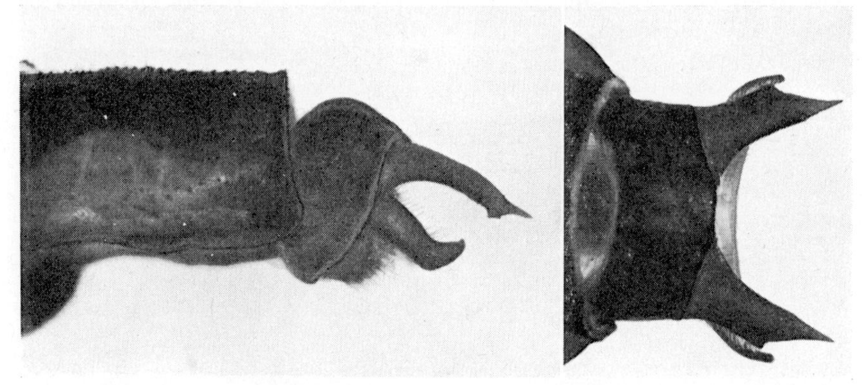

Fig. 115. *Gomphus ventricosus.*

Midlateral stripe extends from below only a little way above level of spiracle. Third lateral stripe faintly developed.

This species has widest flare to sides of end abdominal segments. They are considerably wider than thorax. This, on end of a very slender mid-abdomen, gives it a striking appearance. Middorsal pale line on abdomen narrows almost to disappearance on segment 3, widens again to a basal spearhead on 7; 8, 9, and 10 black above. Yellow of expanded lateral margins of end segments scarcely visible on 7, forms enclosed subquadrate spot on 8, and a wide marginal area that narrows to rearward on 9; 10 below obscure.

Distribution and dates.—CANADA: N. B., N. S., Ont., Que.; UNITED STATES: Conn., Ill., Ind., Mass., Mich., N. Y., Ohio, Pa., Va.

May 17 (Mich.) to July 24 (Mich.).

Subgenus GOMPHUS Leach 1815

Under this name were once included all the species of what we now consider a family, Gomphidae; but the known species were few, and systematic entomology was then in its infancy. With the vast increase in the number of known species, the group has been repeatedly divided and subdivided into tribes and genera of lesser scope, each with its own name. Under the original name *Gomphus* only eighteen American species hereinafter appear. The type species of the genus is the Old World *Libellula vulgatissima* of Linnaeus (1758).

The dark stripings of the thorax are well developed. The front margin of the costa is yellow or yellowish. The face is yellow or pale greenish, with no cross stripes of brown except in *australis*.

The front side of the triangle of the fore wing is generally about equal in length to the inner side (fig. 7, p. 17); in the hind wing, longer. The paranal cells of the fore wing are five, six, or seven, with accompanying marginals none to seven. The paranals of the hind wing are five, two of which (or sometimes one wide cell) may fill the base of the first anal interspace (x). The first anal vein runs direct to the hind margin of the wing, without angulation below the gaff. The gaff is from four-tenths to nine-tenths as long as the inner side of the triangle.

The superior appendages of the male, viewed from above, are strongly divergent at base, develop more or less of a projecting angle midway of the outer side, beyond which they straighten out and extend their acute tips to rearward. On the outer angle, and on the inner margin as well, there may be a low downwardly directed tooth. In the genitalia of segment 2 the form of the interior hamule is extremely varied.

Our eighteen species still form a rather heterogeneous group. Perhaps their most distinctive character in wing venation is the rather wide first anal interspace (x), with its bordering veins, A1 and A2, running directly to rearward, and generally with two complete rows of cells between them extending from the hind angle of the triangle to the wing margin (fig. 7, p. 17). This character they have in common with *Hylogomphus,* but that subgenus has a longer gaff. In *Gomphus* the gaff is generally shorter than the inner side of the triangle. The species of *Hylogomphus* are mostly smaller, with stockier form of body, stubbier superior caudal appendages in the male, and longer subgenital plate in the female (fig. 89, p. 164).

Unique in this group by reason of their scanty venation are two small Southern species, *cavillaris* and *brimleyi,* in both of which the cells in the trigonal interspace are reduced to a single row for a distance of from two to six cells, with only one or two crossveins under the stigma, and in the hind wing four postanal cells; the venation is correspondingly sparse throughout.

Two other species, *borealis* and *descriptus,* stand somewhat apart because of their more abundant venation: mostly six or seven bridge crossveins, five to seven crossveins under the stigma, six or seven postanal cells, intermedian crossveins often 3/1. In both nymphal and adult stages

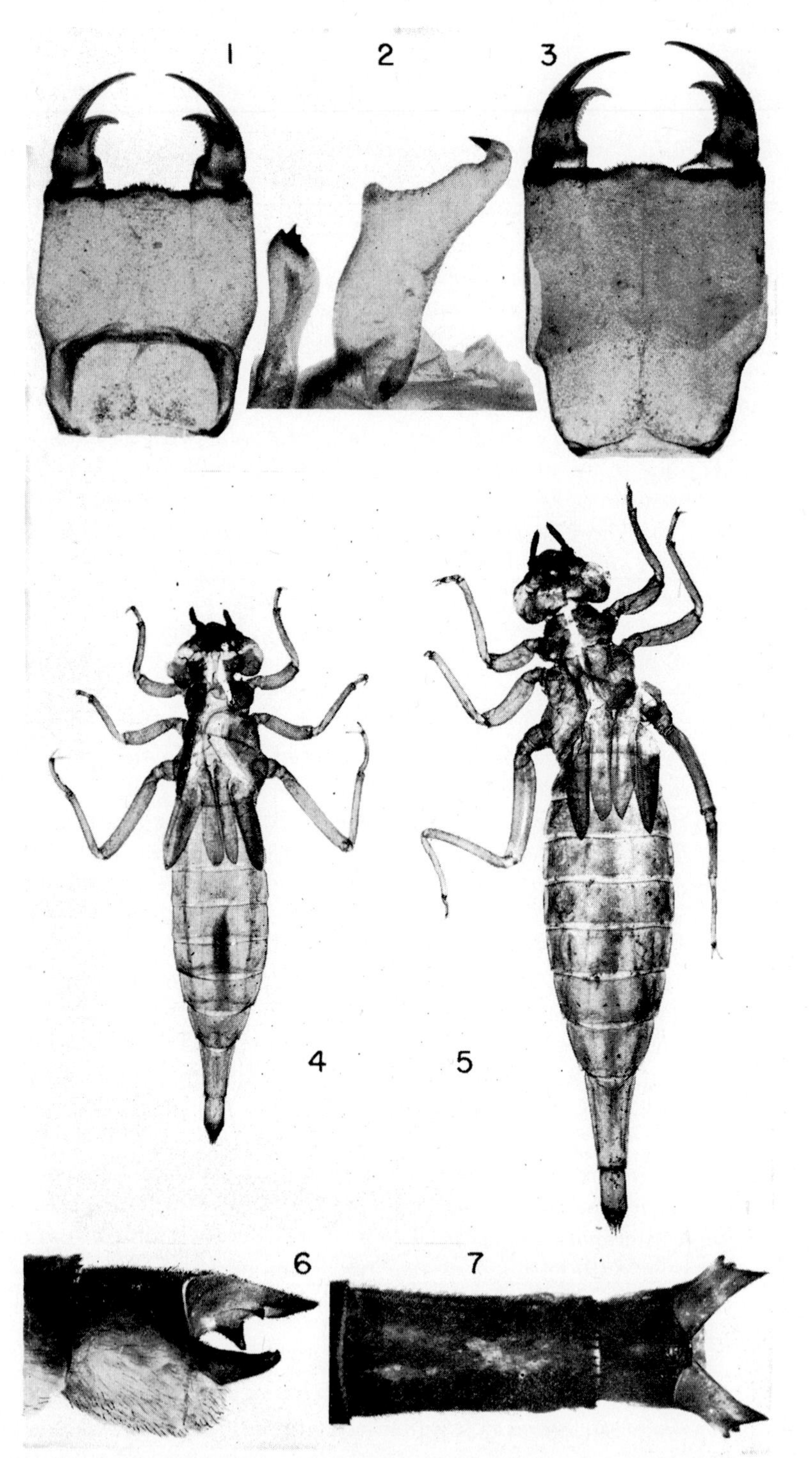

Fig. 116. Comparative figures (to same scale) of *Gomphus australis* and *G. cavillaris.* 1, labium of nymph of *G. cavillaris;* 2, hamules of adult male of *G. australis;* 3, labium of nymph of *G. australis;* 4, last nymphal skin (*exuvia*) of *G. cavillaris,* 5, same of *G. australis;* 6 and 7, lateral and dorsal views of adult male of *G. australis.* (From a paper by the junior author, *Florida Entomologist,* 33:35, 1950.)

the two species are similar in aspect. In the genitalia of the second segment they agree in some particulars and differ strongly in others. The fourth segment of the penis has very long tails in both, the longest of any in this subgenus; but the basal segment (peduncle) is very different: split loaf in *borealis,* and uniquely high-backed in *descriptus.* And *borealis* is like *vulgatissimus* in frequently having five cells in the anal triangle of the male.

The nymphs of this group are less consistently of one form of abdomen than are the nymphs of the other subgenera. They are less flattened and

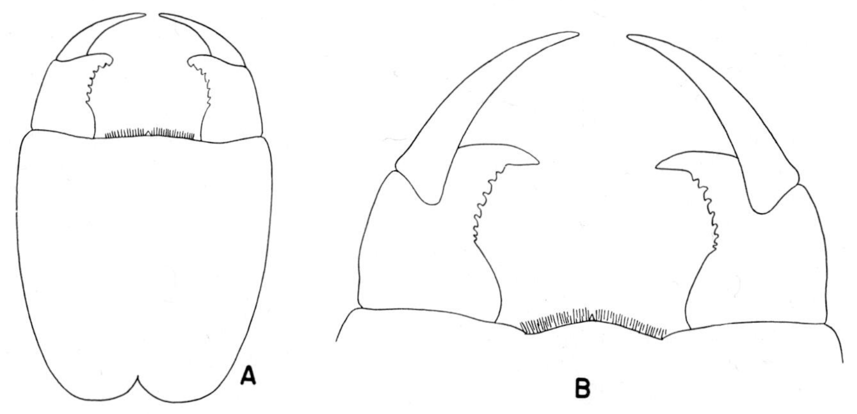

Fig. 116*a*. A, *Gomphus descriptus;* B. *G. cavillaris,* showing two types of end hook of lateral labial lobe in nymphs of the subgenus *Gomphus.*

more pointed to rearward than in *Gomphurus* and *Hylogomphus;* less narrowed than in typical *Stylurus.* Two species, *australis* and *cavillaris,* parallel *Arigomphus* in having narrow elongated end segments, and in having the teeth on the lateral lobes of the labium obliquely truncated, with points directed to rearward; but, unlike *Arigomphus,* they have a long end hook beyond the teeth.

The species that seem most like *Gomphurus* are *graslinellus* and *quadricolor;* these are perhaps as close to *G. vulgatissimus* (the type species) as any other in form of nymph and in details of parts in both adult and nymphal stages.

The nymph has no middorsal groove on the back of the abdomen, but has a row of low dorsal hooks in its place. They are never hooklike in form, but are often mere stubs. The range of variability in these and other nymphal characters is shown in the preceding table of the subgenera of nymphs. The teeth on the inner margin of the lateral lobes of the labium are smaller. There are lateral spines on abdominal segments 6 to 9, with those on 8 and 9 about equally well developed. The abdomen is simply lanceolate to rearward.

KEY TO THE SPECIES*

Adults

1—Abdominal segment 9 much longer than either 8 or 7; face cross-striped with dark brown... **australis**
—Abdominal segment 9 not distinctly longer than either 8 or 7; face pale or only faintly darkened along sutures.................................... 2

2—Hind tibiae wholly dark (sometimes with pale external line in **spicatus**).... 3
—Hind tibiae with yellow line externally or wholly yellow................. 6

3—Fore wing generally with no marginal cells behind paranals; hind wing 27 mm. or less.. **quadricolor**
—Fore wing with one or more marginal cells behind paranals; hind wing 28 mm. or more except in some **borealis**................................ 4

4—Slender blackish species; occipital crest slightly concave in female......... 5
—Stouter species, brightly marked with yellow; occipital crest straight or slightly convex in middle third in female...............................11

5—Legs all black; hood of peduncle of penis higher than wide......... **descriptus**
—Base of hind femora yellow; hood of peduncle wider than high........ **borealis**

6—Fore wing with no marginal cells behind paranals, and trigonal interspace usually with one or more through-cells; small; Southeastern............... 7
—Fore wing generally with one or more marginal cells behind paranals, and trigonal interspace with no through-cells.................................10

7—Midlateral and femoral brown stripes of thorax and area between covered by single wide brown band; first and second joints of tarsi all brown **diminutus**
—These brown stripes separated by pale area; first and second joints of tarsi partly yellow on upper side.. 8

8—Middorsal yellow stripe on abdomen well defined, on black background...... 9
—Middorsal stripe vague, ill defined, on brownish background......... **cavillaris**

9—Trigonal interspace of fore wing generally with one or more through-cells **brimleyi**
—Trigonal interspace with no single through-cells; two rows, unbroken... **hodgesi**

10—Far Western 11
—Not Far Western (except **graslinellus**, Northwestern)....................12

11—Antehumeral and humeral dark stripes on thorax more or less separated by pale line; abdominal segment 9 partly yellow................. **confraternus**
—These stripes entirely fused; 9 almost wholly black................... **kurilis**

12—Dorsum of 9 generally with no yellow; body densely hairy; a dark-hued species .... **spicatus**
—Dorsum of 9 with some yellow..13

13—Middorsal thoracic dark stripe parallel-sided..............................14
—This stripe widened downward, forming triangle of brown.................15

14—Tibia with yellow line ending at tarsus.......................... **graslinellus**
—Tibia with yellow line running down on tarsus................. **oklahomensis**

15—Caudal appendages yellow; peduncle of penis warty externally....... **militaris**
—Caudal appendages brown; peduncle smooth externally....................16

* **williamsoni** omitted; see p. 223.

16—Hind wing 29 mm. or more; antenodal crossveins generally nine........**lividus**
—Hind wing 29 mm. or less; antenodals generally eight....................17

17—Tarsi uniformly light-colored....................................**minutus**
—Tarsi blackish 18

18—Abdominal segment 8 dorsally more yellow than black..........**flavocaudatus**
—Segment 8 more black than yellow above.............................**exilis**

TABLE OF SPECIES

ADULTS

Species	*Hind wing*	*Yel. on tibia*[1]	*Mid-dorsal*[2]	*At crest*[3]	*Distr.*
australis	27-29	0	par.	free	Fla.
borealis	25-29	0	par.	free	NE
brimleyi	20-23	+	tri.	free	SE
cavillaris	23-25	+	tri.	fused	Fla.
confraternus	31-34	0	par.	fused±	W
descriptus	29-32	0	diff.	free	E, C
diminutus	23-24	+	tri.	free	N. Car.
exilis	23-27	+	tri.	fused±	E, C
flavocaudatus	25-26	+	tri.	free	S
graslinellus	29-34	+	par.±	free	N, C
hodgesi	26	+	tri.	free	S
kurilis	28-33	0	par.	fused	W
lividus	29-34	+	tri.	free±	E, C
militaris	28-33	+	tri.±	free±	SW
minutus	26-29	+	tri.	free±	SE
oklahomensis	26-30	+	par.±	free	S, C
quadricolor	25-27	0	tri.	fused	E
spicatus	26-30	+	tri.	fused	NE, C

[1]Yellow stripe on outer face of middle and hind tibiae: present (+) or absent (0).

[2]Dark middle stripes (often fused) tend to form a band, lateral margins parallel (par.), form a triangle, their margins strongly divergent forward (tri.), diffuse, or obscure (diff.).

[3]Upper end of dark antehumeral stripe fused with black of crest, or free from it.

TABLE OF SPECIES—NYMPHS

Species	Total length	Labium			Abdomen					
		Mentum[1]	*Lat. lobe*		*D. hooks*[4]	*Lat. spines*	*Spine*[6]	*Caudal append.*[7]		
		W. x L.	*Hook*[2]	*Teeth*[3]	*on segs.*	*on segs.*[5]	*of 9*	*Lat.*	*Sup.*	*Inf.*
australis	39	10:12	B	8-9	0	7-9	1	9	10	10
borealis	30	10:10	B	8-9	0	7-9	2	9	10	10
cavillaris	29	10:11	B	7-8	0	7-9	2	8	10	10
confraternus	34	10:10	A	6-7	8-9	6-9	3	8	10	10
descriptus	30	10:11	A	10-11	6-9	6-9	3	9	10	10
exilis	26	10:10	B	4-7	8-9	6-9	6	8	10	10
graslinellus	30	10:12	A	6-9	4-9	6-9	3	8	10	10
kurilis	29	10:12	A	6-8	2-9	6-9	3	8	9	10
lividus	31	10:11	A	6-8	7-9	6-9	7	9	10	10
militaris	34	10:13	A	6-7	9	9	4+	7	9	10
minutus	30	10:12	B	6-7	0	6-9	4	9	10	10
oklahomensis	30	10:12	A	7-8	4-9	6-9	4			10
quadricolor	27	10:12	A	6-7	6-9	6-9	5	8	9	10
spicatus	29	10:12	B	6-8	9	7-9	5	9	10	10

[1]Length of mentum of labium in relation to width taken as 10.

[2]Shape of end hook of lateral labial lobe most nearly resembling A or B of figure 116a.

[3]Number of teeth on inner margin of same.

[4]Dorsal hooks (mostly very small) on middle line of abdominal segments.

[5]Lateral spines on abdominal segments.

[6]Length of lateral spine of segment 9 in terms of tenths of length of segment 10 alongside which it lies.

NYMPHS

1—Lateral spines on abdominal segment 9 only .... **militaris**
—Lateral spines on 7 to 9; very minute if present on 6 2
—Lateral spines of 6 to 9 well developed 5

2—Lateral spines of 9 about a tenth as long as 10 3
—Lateral spines of 9 about half as long as 10 .... **spicatus**

3—Segment 9 but little longer than width at base .... **borealis**
—Segment 9 about twice as long as width at base 4

4—Length of body 39 mm.; mentum of labium one and a fifth times as long as its greatest width .... **australis**
—Length of body 28–31 mm.; mentum of labium one and a tenth times as long as its greatest width .... **cavillaris**

5—Dorsum of 9 rounded in cross section 6
—Dorsum of 9 sharply ridged in cross section; Southern .... **minutus**

6—Dorsal hooks well developed, upstanding, present on 4 and 5 7
—Dorsal hooks low, often vestigial, generally not present on 4 and 5 8

7—Dorsal hooks on 8 and 9 of about equal size .... **graslinellus**
—Dorsal hook on 9 smaller than on 8, spinelike .... **oklahomensis**

8—Superior caudal appendage shorter than inferiors; teeth on lateral lobes of labium obsolete or poorly developed 9
—Superior caudal appendage as long as inferiors; teeth on lateral lobes of labium well developed 10

9—Lateral spines of 9 about one-third as long as 10, or less; Western ...... **kurilis**
—Lateral spines of 9 about half as long as 10; Eastern .... **quadricolor**

10—Median lobe of labium straight-edged .... **lividus**
—Median lobe of labium convex-edged 11

11—Length of body less than 27 mm.; end hook on lateral lobe projecting well beyond level of tips of teeth .... **exilis**
—Length of body more than 28 mm.; end hook more or less confluent with last row of teeth, hardly projecting beyond tip 12

12—Median lobe strongly convex; Pacific Coast .... **confraternus**
—Median lobe only slightly convex; Eastern .... **descriptus**

Nymphs unknown: **brimleyi, diminutus, flavocaudatus,** and **hodgesi.**

Gomphus australis Needham

1897. Ndm., Can. Ent., 29:184.
1929. N. & H., Handb., p. 109 (figs.).
1930. Byers, Odon. of Fla., p. 56.
1950. Wstf., Fla. Ent., 33:33 (figs.).

Length 52–54 mm.; abdomen 39–42; hind wing 27–29.

An olivaceous species striped with brown. Face yellowish, clad with short black pubescence, and marked with two brownish cross stripes, one on base of labrum and other across front of frons; upper half of frons yellow. Vertex black; occiput yellow, elevated in middle and fringed with long black hairs.

Middorsal brown of synthorax widens forward, does not extend quite to collar, and is cleft by yellow of carina and surrounded at sides and in front by a pair of opposed **7**-shaped yellow marks. In front of crest, median brown stripe is confluent with antehumeral brown stripe around base of each yellow **7**. Three lateral stripes distinct, entire, narrow. All stripes of thorax conjoined by their ends. Legs black; hind femora of male thinly clothed with long pale hairs; under side of thorax white-woolly. Wings hyaline; costa yellow; stigma brown or tawny. Fore-wing triangle short, its front side generally a little shorter than inner side.

Abdomen black on slender middle segments, with usual middorsal line of pale spearheads obsolete there, though present and bright yellow on 7 and 8. Sides of basal segments largely pale; lower half of sides of 7 to 9 bright yellow; 9 and 10 wholly blackish; appendages brown (fig. 116, p. 200).

This is one of three species that approach *Arigomphus* in several respects: *australis, brimleyi,* and *cavillaris.* Nymphs of *cavillaris* have long slender bodies that taper to rearward as in *Arigomphus;* no lateral spines on segments 6 and 7. Adult male has posterior hamule long and sway-backed, but with much less of a smooth, graceful, gooseneck-like curve; in other characters there is less likeness. The likeness is probably a mere parallelism, for *australis* stands well apart in other characters, notably its extremely long abdominal segment 9, one and a fourth times as long as 8 in male. Relative length of last four segments about as 7:8:10:4.

Distribution and dates.—UNITED STATES: Fla.

March 28 (reared) to April 21.

Gomphus borealis Needham

1900. Ndm., Bull. N. Y. State Mus., 47:453.
1903. Ndm., Bull. N. Y. State Mus., 68:265 (nymph).
1908. Mtk., Odon. of Wis., p. 86.
1917. Howe, Odon. of N. Eng., p. 33 (figs.).
1927. Garm., Odon. of Conn., p. 157 (figs.).
1929. N. & H., Handb., p. 110 (figs.).

Length 44–49 mm.; abdomen 32–35; hind wing 25–29.

A blackish species, shining black on slender abdomen. Face and occiput greenish yellow. Vertex wholly blackish. Rather narrow middorsal thoracic stripe ends bluntly at collar. Humeral and antehumeral stripes darkest of stripes in color, separated by only a narrow, irregular, often interrupted yellow line. Sides of thorax mainly greenish, with narrow dark stripes on lateral sutures ill developed; area between them often washed with brown.

Abdomen mostly black, with very little widening of end segments. Usual middorsal yellow stripe starts wide on segment 1, and is narrowed between two black hair tufts toward its end; continues wide on 2; is reduced to narrow spear points on 3 to 7; 9, 10, and appendages wholly black dorsally; 8 also black, except occasionally for a minute basal spot of yellow. Wings hyaline, with fawn-colored stigma and a very narrow pale line on front edge of costal vein.

This species resembles *descriptus* in stature and coloration, but is generally a little darker. Best distinguished by sex characters, shape of superior caudal appendages of male, and bilobed contour of ridge of occiput in female.

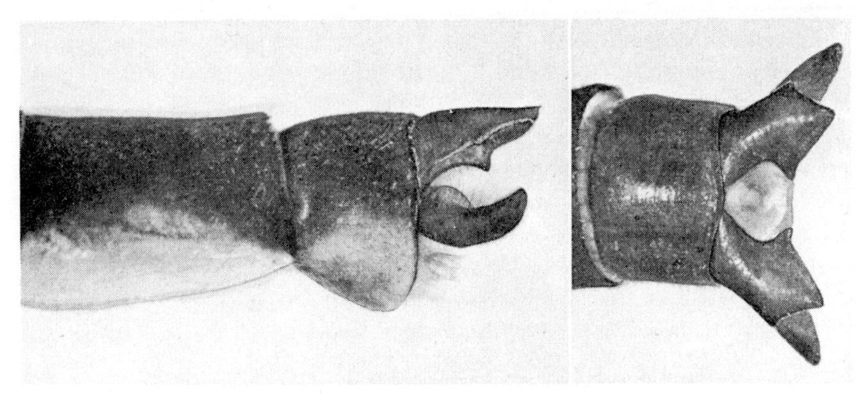

Fig. 117. *Gomphus borealis.*

Distribution and dates.—CANADA: Ont., Que.; UNITED STATES: Conn., Mass. (?), N. H., N. Y., N. C.

May (N. H.) to July 24 (N. C.).

Gomphus brimleyi Muttkowski

1911. Mtk., Ent. News, 22:221 (figs.).
1929. N. & H., Handb., p. 106 (figs.).
1950. Ndm., Trans. Amer. Ent. Soc., 76:6–8.

Length 37–41 mm.; abdomen 24–28; hind wing 20–23.

Smallest species of the genus. Face greenish, with a blackish cross stripe on fronto-clypeal suture. Top of frons green; at its base a triangle of black projects forward in median furrow. Top of vertex black save for touches of paler on rather prominent lobes of postocellar ridge; occiput yellow on both front and rear sides.

Thorax greenish yellow, brightly marked with narrow brown stripes. Middorsal stripe strongly widened forward, narrowly divided by yellow carina; humeral stripes of about equal width, antehumeral not reaching crest; midlateral and femoral stripes about equally well developed. Behind latter is a short strip of brown at edge of intersternum.

Legs brown, yellow on under side of femora and outer side of tibiae. Wings hyaline, with brown veins; tawny stigma, rimmed by heavily pigmented veins. In trigonal interspace of fore wing, usual two rows reduced by two to four through-cells a little beyond triangle. Hind wing of male with four paranal cells and three postanals.

Abdomen black and yellow, mostly black beyond swollen basal segments, with a narrow middorsal line of yellow that widens and ends on 8, and with obscure lateral edgings of yellow that become widened and brighter on lower half of 7 and 8; 9, 10, and appendages black.

Very similar to *cavillaris* but smaller and more brightly striped; humeral stripes more nearly equal in breadth and femorals much better developed, antehumeral being isolated at its upper end; tarsi blackish (reddish in *cavillaris*). Hamules of male scarcely distinguishable in the two, but smaller in *brimleyi;* superior appendages of male more suddenly contracted and more slender beyond their external tooth in *brimleyi,* which by some may be considered a race of *cavillaris.*

Distribution and dates.—UNITED STATES: Fla., N. C.

March 12 (Fla.) to May 15 (N. C.).

Gomphus cavillaris Needham

1902. Ndm., Can. Ent., 34:276.
1908. Mtk., Odon. of Wis., p. 82.
1911. Mtk., Bull. Wis. Nat. Hist. Soc., 9:37 (figs.).
1929. N. & H., Handb., p. 105 (figs.).
1930. Byers, Odon. of Fla., pp. 57, 249.
1950. Wstf., Fla. Ent., 33:33 (figs., nymph).

Length 41–45 mm.; abdomen 29–31; hind wing 23–25.

A small olive green species, obscurely striped with brown. Face olive green, with narrow pale brownish lines in sutures. Rearward slope of postocellar ridge almost as green as occiput; touches of brown on outer corners of occiput.

Middorsal thoracic brown stripe divided lengthwise by greenish carina, and each half of brown reaches outward from its upper end almost to conjunction with antehumeral brown stripe. Between that stripe and humeral is a pale line that widens a little at each end. Sides of thorax greenish, with narrow and complete brown stripes on lateral sutures. Costal vein broadly yellow. Caudal appendages yellowish brown.

Venation of wings more scant than in all other species but *brimleyi.* Usually only five paranal cells in fore wing, with no marginals, and three postanals in hind wings; oftenest but two crossveins under stigma.

Reduction of cells to a single row for a short distance in trigonal interspace of these two species is almost distinctive in family Gomphidae.

Abdomen greenish, with brown bands along sides. Usual middorsal pale stripe continues wide from base outward on middle segments; on 7 to 10 it spreads laterally to cover entire segments.

Inhabits reedy shoals in sand-bottomed lakes; very common in southern Florida.

Distribution and dates.—UNITED STATES: Fla.

February 12 to May 19.

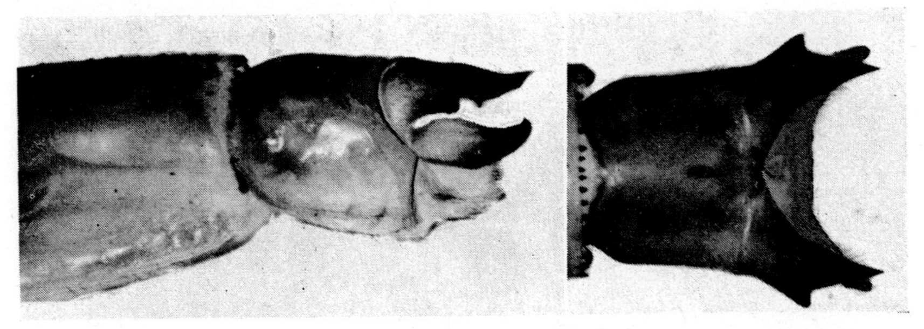

Fig. 118. *Gomphus cavillaris.*

Gomphus confraternus Selys

Syn.: sobrinus Selys

1873. Selys, Bull. Acad. Belg., (2)35:744 (reprint, p. 16).
1904. Ndm., Proc. U. S. Nat. Mus., 27:291 (nymph).
1917. Kndy., Proc. U. S. Nat. Mus., 52:558 (as *sobrinus*?).
1927. Seemann, J. Ent. Zool. Claremont, Calif., 19:21 (figs.).
1929. N. & H., Handb., p. 92 (figs.).

Length 50–57 mm.; abdomen 36–41; hind wing 31–34.

A black-legged West Coast species, with well-defined green and brown stripes. Considerable enlargement of end segments of abdomen in male. Parallel-sided middorsal stripe narrower than yellowish area on either side of it, and runs down close against collar. Antehumeral and humeral stripes of dark brown mostly conjoined, with only traces of usual intervening pale line: a roundish spot at top (sometimes not completely enclosed) and a variable narrow streak farther down. Midlateral brown stripe incomplete, not continued above level of spiracle. Second lateral stripe complete, with lower end hooked to rearward around metepimeron.

Markings of abdomen bright yellow. Usual middorsal line of these, broad at base as usual, is reduced to basal spearheads pointed to rear on segments 4 to 7, to basal spots on 8 and 9; 10 above and appendages black. Sides of three basal segments yellow below as usual, with small enclosed yellow spots on 4 to 6. Large and conspicuous yellow spots on sides of 8 and 9 cover lower edges of these segments but do not reach their apical margin. Male and female very similar in depth of coloration.

Nymphs live in silted beds of rapid streams; when in a pool that dries up, they are quite capable of crawling along mud-encrusted bed to find another pool.

Distribution and dates.—UNITED STATES: Calif., Oreg., Wash.

April (Calif.) to July 15 (Wash.).

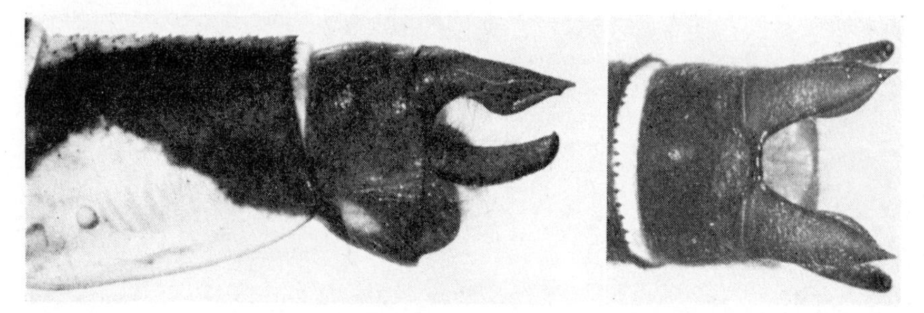

Fig. 119. *Gomphus confraternus.*

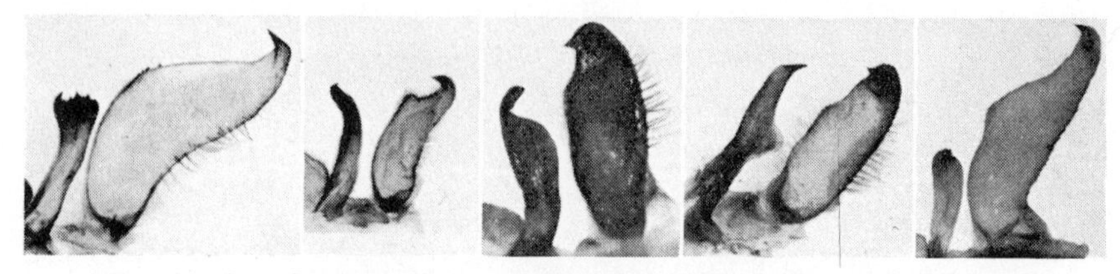

Fig. 120. *Gomphus borealis; G. cavillaris; G. confraternus; G. descriptus; G. diminutus.*

Gomphus descriptus Banks

Syn: argus Needham, mortimer Needham

1896. Banks, J. N. Y. Ent. Soc., 4:194.
1901. Ndm., Bull. N. Y. State Mus., 47:452 (figs.).
1901. Ndm. & Hart, Bull. Ill. State Lab., 6:70.
1927. Garm., Odon. of Conn., p. 151 (figs.).
1929. N. & H., Handb., p. 109 (figs.).

Length 48–52 mm.; abdomen 33–38; hind wing 29–32.

Another blackish species that has a greenish yellow thorax striped with brown. Hair on front of thorax brownish and less dense than in *borealis*. Sides of middorsal thoracic stripe nearly parallel to collar, where contracted and narrowly connected with black below collar. Antehumeral brown stripe may be free at upper end or conjoined to humeral, and to black of crest. Between the three, a minute green spot is usually left at top, below which is a narrow and variable fine line of greenish yellow.

Middorsal yellow stripe of abdomen full length on segments 1 and 2, narrowed to a spear point on 3, and to points that shorten successively on 4 to 7; 8, 9, 10, and appendages dorsally black. Paler yellow covers lower half of 1, 2, and 3. Middle abdominal segments have small round

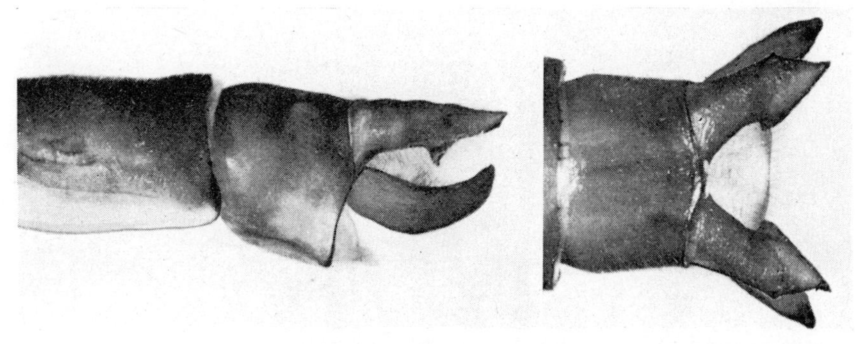

Fig. 121. *Gomphus descriptus.*

included basal spots and a very narrow and inconspicuous infero-lateral edging of yellow that widens on 7, and rises to half the height of segments on 8, 9, and 10.

This species nearest *borealis;* see distinctions under that species.

Distribution and dates.—CANADA: Ont., Que.; UNITED STATES: Fla., Iowa, Ky., Mass., Mich., N. Y., N. C., Pa.

April 6 (Fla.) to August 3–6 (Iowa).

Gomphus diminutus Needham

1950. Ndm., Trans. Amer. Ent. Soc., 56:6–8 (figs.).

Length 40 mm.; abdomen 29; hind wing 23–24.

A very small olive green species, rather faintly striped with brown. Head and thorax thinly clothed with brownish hairs. Rather smooth blackish abdomen, mostly bare. Face and mouth parts olivaceous, with faint lines of brown around margin of labium and in transverse sutures. Base of frons above and all vertex black; antennae black save for a very narrow pale ring around base of each. Occiput yellow, with upper edge very slightly convex and fringed with long black hairs.

Prothorax brownish above, with yellowish front lobe; lateral spots and a submedian twin spot on middle lobe. Synthorax yellow, striped with brown. Rather broad middorsal stripe widened forward to collar, and narrowly divided by yellow of middle of carina. Antehumeral brown stripe isolated above and narrower than yellow area before it, and separated from humeral brown stripe by a narrower stripe of yellow. Behind rather wide humeral brown stripe, sides of thorax dull yellow, with a single wide band of brownish that covers second and third lateral sutures and all the area between them.

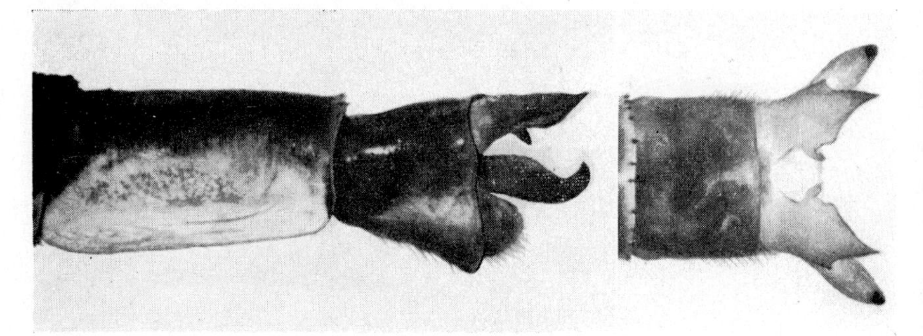

Fig. 122. *Gomphus diminutus.*

Legs blackish, pale basally and on sides of femora. Hind femora long. Tibiae paler on outer side. Wings hyaline, with brown veins and a tawny stigma. A very faint pale line in costa.

Abdomen black, with usual middorsal line of yellow spots. At front, yellow covers most of dorsum of segment 1, wide and full length on 2, thereafter narrower and progressively reduced in length on 3 to 7; yellow reappears on hind end of lateral margin of 8, and diffusely covers most of 8 and 9; 10 obscure pale brownish. Sides of 1 to 3 mostly yellow, including auricles on 2. A conspicuous tuft of black hair at each side of dorsum of 1.

Superior appendages of male moderately divergent. Branches of inferior much more widely outspread.

Distribution and dates.—UNITED STATES: N. C.

April 14.

Gomphus exilis Selys

1854. Selys, Bull. Acad. Belg., 21(2):55 (reprint, p. 36).
1893. Calv., Odon. of Phila., p. 243.
1900. Wmsn., Odon. of Ind., p. 293 (figs.).
1901. Ndm., Bull. N. Y. State Mus., 47:455 (nymph).
1929. N. & H., Handb., p. 108 (figs.).

Length 39–48 mm.; abdomen 28–35; hind wing 23–27.

A small greenish yellow species, diffusely striped with pale brown. Vertex brown, paler on rearward slope of ocellar ridge. Middorsal thoracic stripe widens downward into a broad triangle. Antehumeral brown stripe free at upper end and well separated from humeral by a more or less complete line of greenish yellow. Sides of thorax yellowish, with complete brownish stripes on second and third lateral sutures. Area between these stripes more or less brownish. Femora brownish, paler beneath; wide stripes of yellow on outer side of tibia run down on top of tarsal segments.

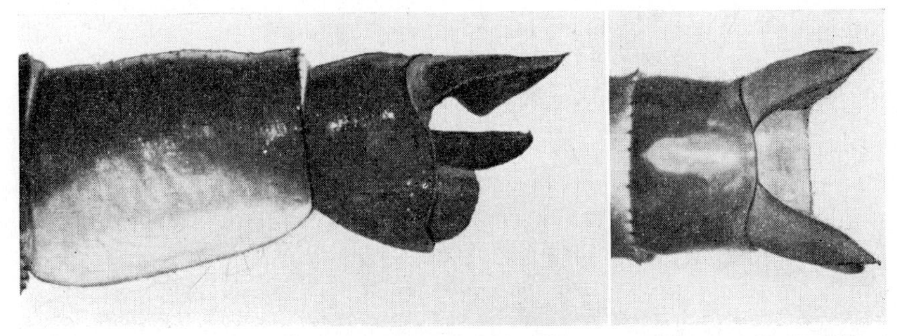

Fig. 123. *Gomphus exilis.*

Abdomen predominately brownish. Its middorsal pale line nearly continuous, narrowed on middle segments, narrowed most on 8, broadened and diffused on 9 and 10. Slightly dilated lateral margins of 7 to 9 dull yellow; appendages brown. Posterior hamule of male yellow.

Slender nymph a shallow burrower in edges of streams and ponds.

Distribution and dates.—CANADA: Man., N. B., N. S., Ont., Que.; UNITED STATES: Ala., Conn., Ill., Ind., Ky., Maine, Md., Mass., Mich., N. H., N. J., N. Y., N. C., Ohio, Pa., R. I., S. C., Vt., Va., Wis.

April 7 (N. C.) to September (N. Y.).

Gomphus flavocaudatus Walker

1940. Walk., Ent. News, 51:194–196 (figs.).

Length 45–47 mm.; abdomen 34–36; hind wing 25–26.

This species very like *exilis* in most particulars, but more reddish brown in markings, and male has somewhat shorter caudal appendages. Probably to be considered a variety of *exilis*. Dark stripes on thorax reddish brown. All conjoined by their lower ends, and about equal in area to intervening stripes of yellowish green.

Abdomen blackish on slender middle segments, where middorsal line of interrupted yellow spearheads is well defined. Spread of branches of inferior caudal appendages of male not greater than that of superiors. Angulation of inferior margin on superiors starts somewhat farther out from base than in *exilis.*

Distribution and dates.—UNITED STATES: La., Miss.

March 26 (La.) to April 17 (Miss.).

Gomphus graslinellus Walsh

1862. Walsh, Proc. Acad. Phila., p. 394.
1885. Hagen, Trans. Amer. Ent. Soc., 12:264 (nymph).
1900. Wmsn., Odon. of Ind., p. 290 (figs.).
1927. Garm., Odon. of Conn., p. 164 (figs.).
1929. N. & H., Handb., p. 106 (figs.).

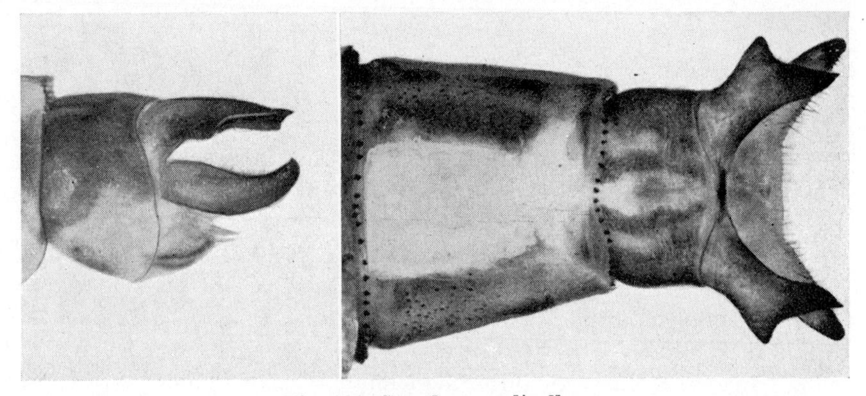

Fig. 124. *Gomphus graslinellus.*

Length 47–53 mm.; abdomen 33–39; hind wing 29–34.

A handsome species; greenish thorax heavily striped with dark brown, and blackish abdomen marked with bright yellow; brightness and extent of colors about same in both sexes. Face and occiput bright yellow. Parallel-sided middorsal brown stripe about as wide as green on either side of it. Antehumeral stripe ends free above, then joins humeral, and again separated from it by a narrow greenish line. Lateral stripes broad and complete, their margins ill defined. Legs brown, but outer face of tibiae yellow.

Abdomen considerably enlarged at both ends and not as slender in middle as in *oklahomensis,* its nearest ally. Middorsal line of pale spots wide at base and reduced to rather large yellow spear points on all middle segments, with spot on 8 much shortened, but widened again to cover most of dorsum of 9 and 10. Sides of 8, 9, and 10 brightly marked with yellow, that on 9 covering nearly half of side surface. Caudal appendages brownish, with darker tips. Large peduncle clad with rather long sparse hair.

Distribution and dates.—CANADA: B. C., Man., Ont.; UNITED STATES: Ill., Ind., Iowa, Kans., Mich., Minn., Mo., Ohio, Okla., Wash., Wis.

May 11 (Ind.) to August 14 (Ont.).

Gomphus hodgesi Needham

1950. Ndm., Trans. Amer. Ent. Soc., 56:1–12.

Length 44 mm.; abdomen 33; hind wing 26.

A very slender olive green species, with black abdomen. Face all pale greenish, shining, thinly beset with short sparse black hairs at sides and across base of labrum. Margins of labrum and lower edges of facial lobes whitish. Top of frons shining green except for a low narrow quadrangle of black thrust forward upon its base from black of vertex. Frons high and unusually prominent, bearing a well-defined transverse crest. Vertex black except for a somewhat brownish rearward slope behind sharp ocellar carina. Occiput greenish, slightly convex on its crest, where it is beset with thin scattering hairs.

Middorsal stripe of synthorax widens forward into a broad triangle of black, and ends above dark gray green collar. At upper end this stripe spreads out laterally to cover entire thoracic crest. Antehumeral dark stripe incomplete, ending in green at its upper end. Below it runs straight down across supra-coxal plate. Humeral dark stripe wide at crest, narrow and parallel-sided below humeral pit. It runs down into supra-coxal furrow and around a bulge in mesepimeron, to join midlateral stripe below bulge. Pale streak between humerals narrow and irregular. Midlateral and third lateral stripes complete, though not sharply defined, and are joined above by shiny black subalar carina. Behind these stripes, metepimeron broadly greenish white. Venter of synthorax wholly pale, thinly clad with long white hairs.

Legs pale at base, on under side of femora, on upper side of tibiae, and on upper side of first two joints of tarsi. Hind femur strikingly long (6 mm.). Wings hyaline, with a faint tinge of brown in membrane beyond nodus. Indistinct white line on costa. Veins black, stigma brown. Five paranal cells in fore wing, with no added marginals; five in hind wing also, two of them in first anal interspace. Tornal cell nearly half the height of its interspace; above it are two full-width cells.

Slender abdomen very little clubbed on end segments. It is black, with a dull greenish yellow middorsal stripe continuous from base to end of segment 8: wide on 1 and 2, narrowing on 3, parallel-sided on 4, 5, and 6, narrowing again on 7, and fading out on 8. Usual lateral stripe green: wide on 1 to 3, covering about lower half of 1 and 2, including auricle; narrows on 3, and is reduced to round basal spots on 4, 5, and 6; widens greatly for full length on 7, 8, and 9, becoming yellow, bright yellow on slightly flaring lateral expansion of these segments, but is clouded above yellow on 7 and 8. Segment 10 dull black above and dingy yellow below. Caudal appendages dull olive washed with black, with very black tips.

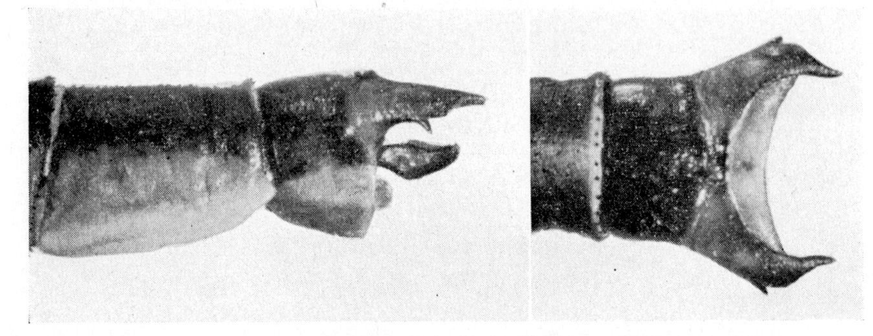

Fig. 125. *Gomphus hodgesi.*

This species differs from *cavillaris* in being more slender and blacker, and in having a higher frons. Known from three male specimens.

Distribution and dates.—UNITED STATES: Ala., Miss.

March 1 (Miss.) to May 7 (Miss.).

Gomphus kurilis Hagen

Syn.: donneri Kennedy

1857. Hagen, in Selys, Mon. Gomph., p. 132.
1917. Kndy., Proc. U. S. Nat. Mus., 52:562 (figs.).
1929. N. & H., Handb., p. 94 (figs.) (as *donneri*).
1941. Gloyd, Bull. Chicago Acad. Sci., 6:131.

Length 48–53 mm.; abdomen 35–39; hind wing 28–33.

Another West Coast black and yellow species, so similar to *confraternus* that it will suffice here to state discoverable differences.

It differs in having antehumeral and humeral blackish stripes wholly fused into one solid band, blackish color of abdomen more extended, and pale areas more restricted. In addition are the slight differences in form shown in our figures of genitalia.

It can hardly be considered more than a variety of *confraternus*.

Distribution and dates.—UNITED STATES: Calif., Nev., Oreg. July 23–24 (Calif.).

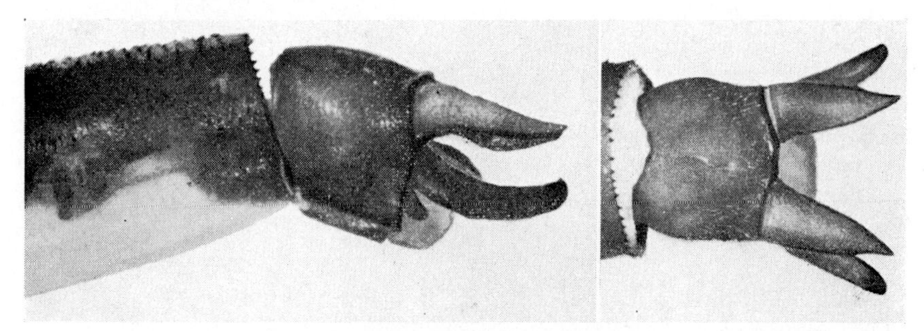

Fig. 126. *Gomphus kurilis.*

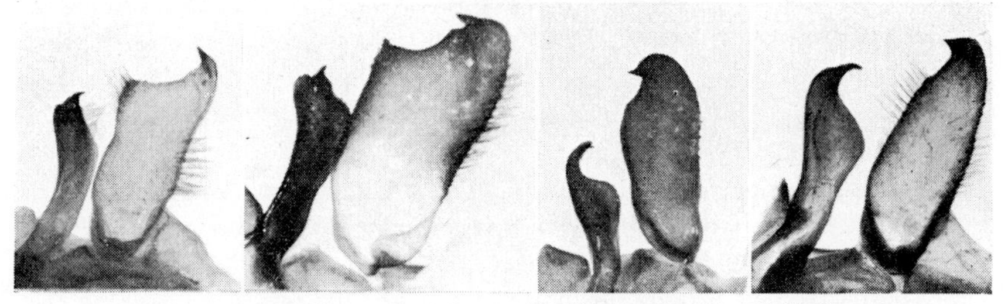

Fig. 127. *Gomphus exilis; G. graslinellus; G. kurilis; G. lividus.*

Gomphus lividus Selys

Syn.: sordidus Hagen, umbratus Needham

1854. Selys, Bull. Acad. Belg., 21(2):53 (reprint, p. 34).
1900. Wmsn., Odon. of Ind., p. 292 (figs.).
1927. Garm., Odon. of Conn., p. 154 (figs.).
1929. N. & H., Handb., p. 104 (figs.).

Length 48–56 mm.; abdomen 34–40; hind wing 29–34.

A brownish species of rather dull coloration. Vertex brown; antennae narrowly ringed about basal segments with paler. Slightly convex border of occiput sparsely fringed with brown hairs. Middorsal brown stripe of thorax slightly widened downward to collar, and about as wide as bordering greenish yellow. Antehumeral brown stripe sometimes free at upper end and sometimes fused there with humeral; variable yellow line between them farther down is very narrow and often interrupted in middle or wanting altogether. Brown stripes on lateral sutures of thorax present but diffuse, and entire area between them pale brown. Femora brown above, yellow beneath.

Abdomen slender in middle and but little enlarged on segments 8 and 9. Middorsal line of yellowish spearheads nearly continuous along middle segments. Lower edgings on middle segments pale yellow, unusually narrow and ill defined, but widened on 7, 8, and 9; 10 obscure yellowish brown, paler dorsally and toward lateral margins. Caudal appendages blackish. Hamules of male yellowish, peduncle more or less black.

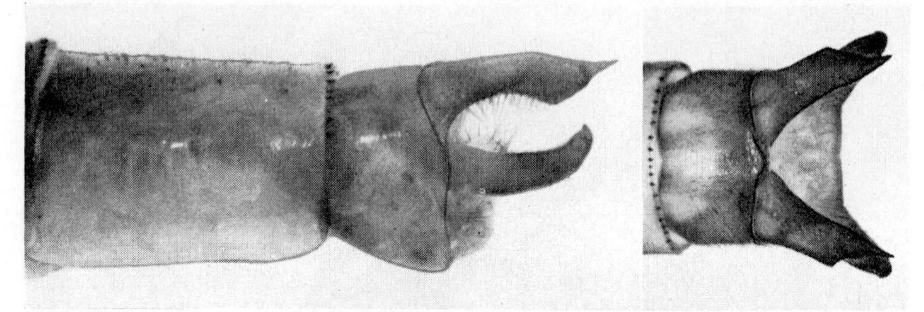

Fig. 128. *Gomphus lividus.*

Distribution and dates.—CANADA: Ont., Que.; UNITED STATES: Ala., Ark., Conn., Fla., Ga., Ind., Ky., Mass., Mich., Miss., Mo., Nebr., N. J., N. Y., N. C., Ohio, Pa., S. C., Va., Wis.

March (N. C.) to August 30 (Pa.).

Gomphus militaris Hagen

1857. Hagen, in Selys, Mon. Gomph., p. 416.
1861. Hagen, Syn. Neur. N. Amer., p. 107.
1917. Kndy., Bull. Univ. Kansas, 18:138 (figs.).
1929. N. & H., Handb., p. 104 (figs.).
1934. Bird, Ent. News, 45:44 (fig., nymph).

Length 47–53 mm.; abdomen 33–39; hind wing 28–33.

Yellowest member of the group, a handsome bright saffron yellow and brown species, with a rusty redness overspreading expanded tip of abdomen in fully mature males. Front of thorax more yellow than brown. Middorsal brown stripe considerably widened at collar, and divided its entire length by a wholly yellow carina. Antehumeral and humeral stripes well separated by a continuous yellow line. Midlateral stripe may be interrupted above level of spiracle, but stripe on third suture is complete and well developed. Wings slightly flavescent toward base. Legs yellow and black; sides of femora and outer face of tibiae yellow.

Abdomen slender on middle segments and considerably enlarged at both ends. Middorsal yellow band rather broad and nearly continuous on middle segments, with 8, 9, and 10 mainly, and diffusely, yellow. Sides of middle segments deepen in color to blackish with age. Caudal appendages yellow. Anterior hamule of male black; posterior yellow with black tip. Surface of peduncle warty, or at least pimply, especially in rear.

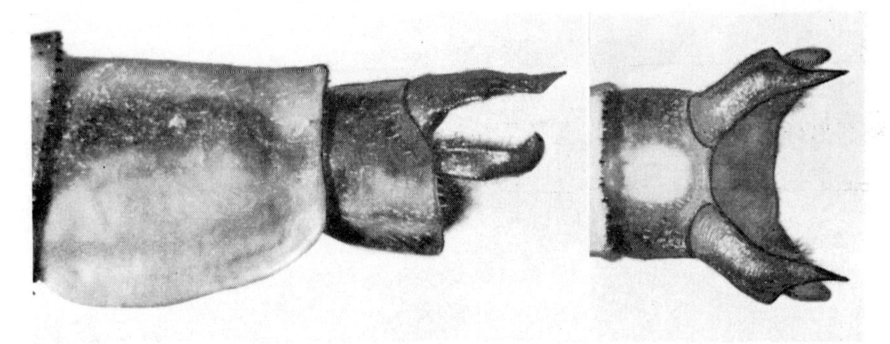

Fig. 129. *Gomphus militaris.*

A well-marked species. Warty peduncle of penis of male distinctly larger than in related species of *Gomphus*. Head of female with a pair of very minute black spines, rising from near ends of outarching ridge of vertex, each about midway between lateral ocellus and eye.

Distribution and dates.—UNITED STATES: Kans., N. Mex., Okla., Tex. April 1 (Tex.) to August 7 (Tex.).

Gomphus minutus Rambur

1842. Rbr., Ins. Neur., p. 161.
1893. Calv., Trans. Amer. Ent. Soc., 20:244.
1904. Ndm., Proc. U. S. Nat. Mus., 27:690 (fig., nymph).
1929. N. & H., Handb., p. 107 (figs.).
1930. Byers, Odon. of Fla., pp. 61, 250.
1950. Wstf., Fla. Ent., 33:33 (nymph).

Length 43–50 mm.; abdomen 32–35; hind wing 26–29.

A greenish yellow species, striped with brown and rather dull in coloration, being much infuscated. Middorsal thoracic stripes much widened downward, wider than yellow on either side. Antehumeral stripe isolated at upper end and separated from humeral by a variable narrow yellow line. Two hinder brown stripes on side of thorax complete, but sometimes hidden in general infuscation of their area. Legs brown, with sides of femora as well as outer face of tibiae and tarsi extensively yellow. Membrane of wings sometimes lightly tinged with yellow.

Abdomen moderately widened to front and rear ends. Its middorsal yellow line obscure but continuous, becoming diffuse on segments 9 and 10. Sides of 1 to 3 and of 8 to 10 yellowish; caudal appendages pale brown. Hamules of male yellow with black tips. Peduncle low, broad, and thinly hairy.

Distribution and dates.—UNITED STATES: Fla., Ga., La., Mass., Miss. February 21 (Fla.) to May 12 (Fla.).

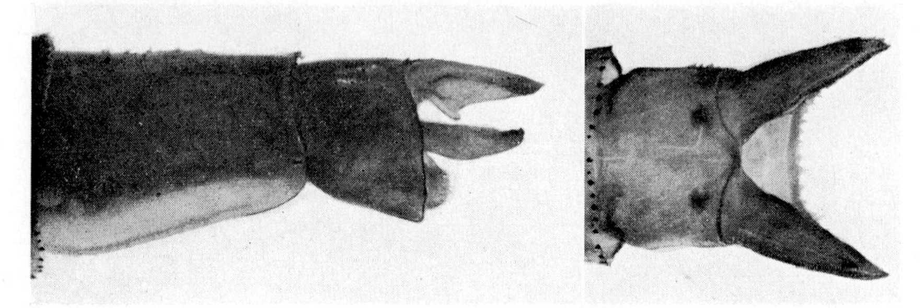

Fig. 130. *Gomphus minutus.*

Gomphus oklahomensis Pritchard

1935. Pritchard, Occ. Pap. Mus. Zool. Univ. Mich., 319:1.

Length 46–49 mm.; abdomen 34–36; hind wing 26–30.

A rather slender greenish yellow species, striped with brown on thorax, suffused with brown on abdomen. Anteclypeus whitish on a greenish yellow face. On dark brown vertex is a pair of pale spots, or a crossbar behind lateral ocelli. Parallel margins of middorsal thoracic stripe swing outward abruptly as they approach collar. Antehumeral and humeral stripes have ill-defined margins, and are separated by a complete greenish yellow line. Hair on front of thorax blackish; sides of thorax yellowish, with complete diffuse stripes of brown on lateral sutures. Front edge of costal vein narrowly whitish yellow. Legs light reddish brown; femora paler beneath, and outer side of tibiae whitish its full length; white extends to basal segments of tarsi.

Abdomen with usual middorsal yellowish stripe continuous on segments 1 and 2, reduced to diminishing spear points on 3 to 7, broken into a basal spot followed by a fine line on 8, greatly widened to cover a large part of dorsum of 9, and reduced to a small elongate spot on 10. Sides of 1 to 3 yellowish below, followed by quadrangular basal spots of yellow on 3 to 7, by wide lateral stripes of bright yellow on 8 and 9, and by a spot on lower margin of 10. Caudal appendages dark brown; genitalia of 2 reddish brown.

Distribution and dates.—UNITED STATES: Okla., Tex.

April 25 (Okla.) to July 10 (Tex.).

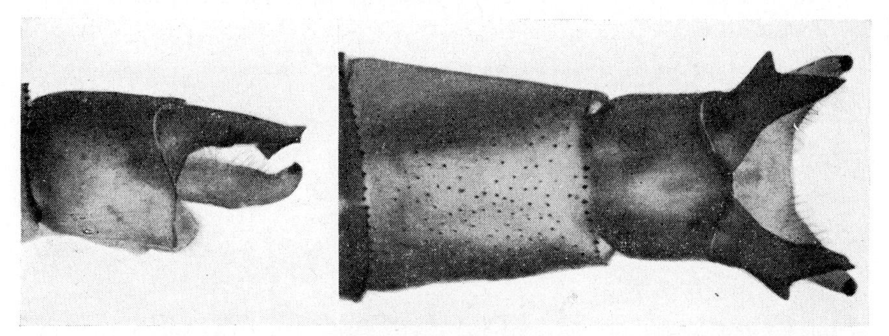

Fig. 131. *Gomphus oklahomensis.*

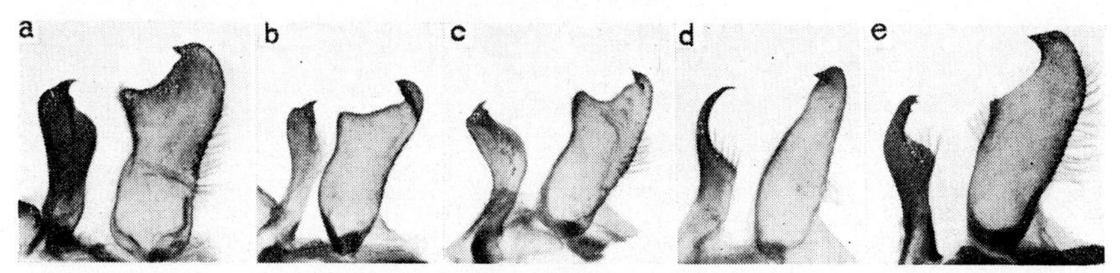

Fig. 132. A, *Gomphus militaris;* B, *G. minutus;* C, *G. oklahomensis;* D, *G. quadricolor;* E, *G. spicatus.*

Gomphus quadricolor Walsh

Syn.: alleni Howe

1862. Walsh, Proc. Acad. Phila., p. 394.
1900. Wmsn., Odon. of Ind., p. 288 (figs.).
1908. Mtk., Odon. of Wis., p. 85.
1929. N. & H., Handb., p. 111 (figs.).
1932. Walk., Can. Ent., 64:273 (figs., nymph).

Length 42–45 mm.; abdomen 31–34; hind wing 25–27.

A dainty little greenish species, striped with blackish brown, and with a blackish abdomen. Rear of eyes black, with a midlateral pale spot. Middorsal brown stripe of thorax not as wide as bordering green, and truncated below, where narrowly connected with brown stripe of front of collar. Antehumeral and humeral stripes both conjoined with black of crest, and between them lies a little triangular spot near crest, and a thin green line below that extends downward to level of collar. Two hinder brown stripes complete, narrow and conjoined at upper end. Legs black.

Abdomen little widened at ends. Its middorsal pale band, wide on basal segments, is reduced to distant basal triangles on middle segments; 8, 9, and 10 wholly black above. Sides of 1, 2, and 3 yellowish below as usual; in black on sides of 4 to 7 are small roundish included basal spots, obscure yellowish brown. Sides at lower margin of 8, 9, and 10 brighter yellow. Caudal appendages black. Anterior hamules of male black; posterior yellow.

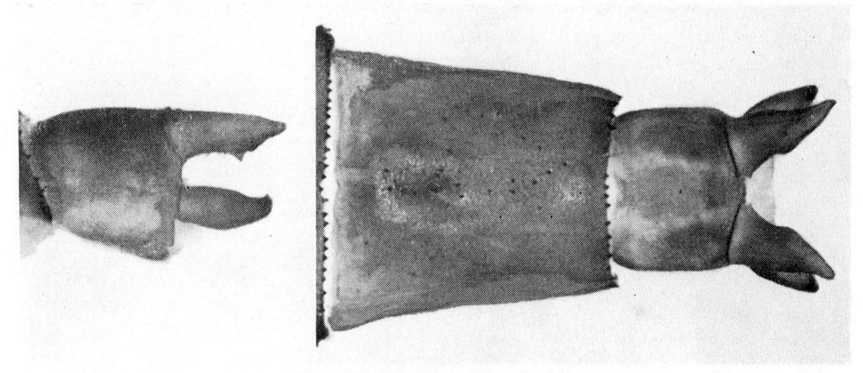

Fig. 133. *Gomphus quadricolor.*

Distribution and dates.—CANADA: Ont.; UNITED STATES: Ala., Ill., Ind., Ky., Mass., Mich., N. H., N. Y., Ohio, Pa., Tenn., Va., Wis.

May 20 (Ohio) to July 2 (Ont.).

Gomphus spicatus Hagen

1854. Hagen, in Selys, Bull. Acad. Belg., 21(2):54 (reprint, p. 35).
1900. Wmsn., Odon. of Ind., p. 292 (figs.).
1901. Ndm., Bull. N. Y. State Mus., 47:459 (nymph).
1917. Howe, Odon. of N. Eng., p. 34 (figs.).
1918. Stout, Ent. News, 29:68 (figs., nymph).
1927. Garm., Odon. of Conn., p. 156 (figs.).
1929. N. & H., Handb., p. 108 (figs.).

Length 46–50 mm.; abdomen 34–38; hind wing 26–30.

A dull-colored greenish yellow species. Hairy thorax striped with brown; abdomen suffused with brown. Middorsal thoracic stripe rounded to collar, where it is narrowly connected forward to brown below. Antehumeral brown stripe free above, then joins humeral, then is separated again by a narrow green line. Humeral sometimes narrowly divided by pale greenish on upper part of suture. Two rearward stripes complete, angulated near middle, variable in width, the brown more or less overspreading area between them; rear stripe may end below in a reversed J that is hooked around border of metepimeron to rear. Legs blackish.

Abdomen little dilated at ends. Its middorsal band a line of spots, wide and connected at base, shortened to spearheads on segments 6 and 7; 8, 9, and 10 blackish on dorsum. On lower part of sides of enlarged segments at front and rear ends of abdomen are usual pale yellowish markings, somewhat broader in female than in male. Caudal appendages blackish. Anterior hamule blackish; posterior one paler. Peduncle bulbous, smaller than in other species.

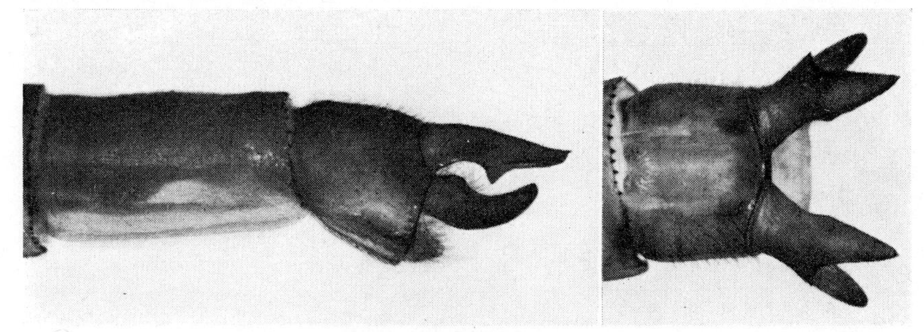

Fig. 134. *Gomphus spicatus.*

Distribution and dates.—CANADA: N. S., Ont., Que.; UNITED STATES: Conn., Ill., Ind., Maine, Mass., Mich., Minn., N. H., N. J., N. Y., Ohio, Wis.
May 4 (Mass.) to July 31 (Ont.).

Gomphus williamsoni Muttkowski

1903. Wmsn., Ent. News, 14:253 (figs.) (no name).
1910. Mtk., Cat. Odon. N. Amer., p. 98 (named; nothing more).
1929. N. & H., Handb., p. 107 (figs. from Wmsn.).

This species, if it be one, was described in detail by Williamson in 1903 from a single male specimen taken near Bluffton, Indiana. He compared it with *graslinellus* and *lividus,* its nearest allies, and thought it might possibly be a hybrid between these two species. He gave it no name. Muttkowski named it and treated it as a distinct species in his *Catalog.* The specimen is in the Zoological Museum at Ann Arbor; the senior author studied it there in 1944 and concluded that it is more like *lividus* than *graslinellus,* but differs from *lividus* in these particulars: (1) it has more of a "shoulder" on the front margin of the posterior hamule than has *lividus;* (2) the terminal hook on the anterior hamule is shorter. Williamson, Montgomery, and others have collected Odonata diligently in Indiana, and in forty-five years no other such specimen has been found. We doubt whether *williamsoni* is a valid species.

Subgenus HYLOGOMPHUS Needham 1951

This is a group of four small Eastern species (hind wing less than 30 mm.) that have short, stocky greenish yellow bodies, heavily marked with black. Humeral and antehumeral stripes are well developed, but midlateral and femoral stripes are weak or wanting. The stubby abdomen is mostly black, wholly so on the dorsum of segments 9 and 10. The

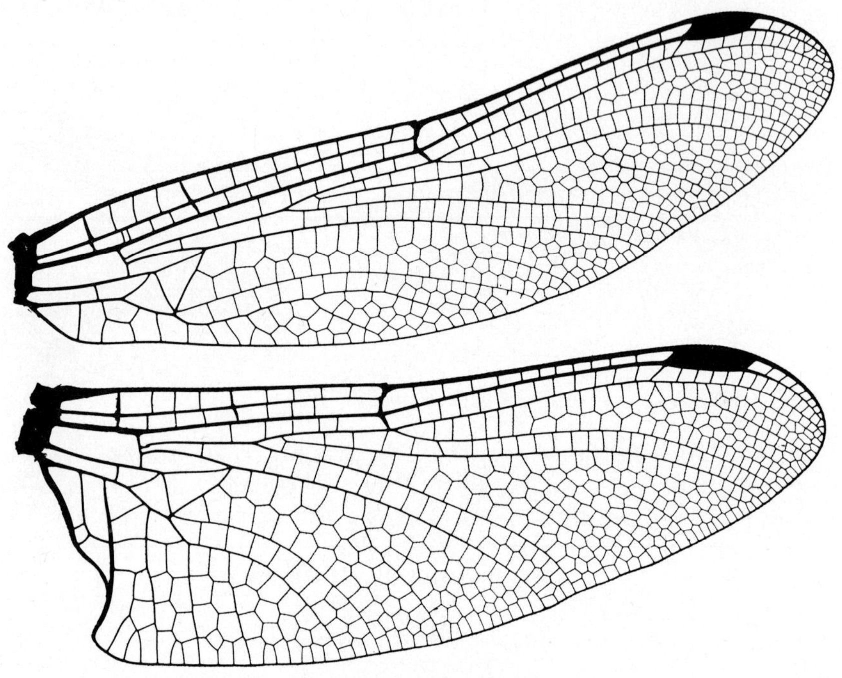

Fig. 135. *Gomphus brevis.*

yellow-marked sides of 7, 8, and 9 are little dilated; 10 is hardly more than half as long as 9. The females have a pair of spines on the top of the head behind the lateral ocelli.

The venation of the wings is chiefly distinguished by the long gaff: as long as, generally longer than, the inner side of the triangle. The front side of the triangle is in the fore wing about one and a tenth, in the hind wing about one and a fifth, times as long as the inner side. In the hind wing, veins A1 and A2 generally run parallel and straight to the wing margin, and there is no anal loop. In the hind wing are four paranal cells and four to six postanals. Vein A2 arises at about the middle of the subtriangle. There are two full-width cells in the base of anal space *y*.

The superior appendages of the male are very short and stout, with a sharp inferior tooth far out near the tip; the branches of the inferior appendage are separated by a wide and very shallow U-shaped, almost

Fig. 136. *Gomphus brevis.*

semicircular, notch. Superior and inferior branches are very widely and about equally divaricate.

The two best papers on the adults of this group are those of James S. Hine (1901) and Bertha P. Currie (1917).

The nymph is as shown in figure 136. It is quite like the nymph of *Gomphurus,* but differs in having no distinct middorsal groove on the middle abdominal segments and in being of smaller size.

KEY TO THE SPECIES

Adults

1—Face heavily cross-striped and heavily margined with black; front of occiput narrow, its crest blunt and nearly straight, not rising as high as level of top of compound eyes; labrum traversed by a median black line that divides its surface into two large pale spots .... **brevis**

—Face pale, not conspicuously cross-striped with black; front of occiput wider, its hair-fringed crest rising in a curve to level of top of compound eyes.... 2

2—Face green, shining .... **viridifrons**

—Face yellowish 3

3—Hind wing 22–25 mm.; stigma yellow; shoulder on anterior edge of posterior hamule of male, viewed from side, rising in an acute prominence; triangular subgenital plate of female not more than half as long as sternum of segment 9 .... **abbreviatus**

—Hind wing 24–25 mm.; stigma brown or blackish; shoulder on anterior edge of posterior hamule of male, viewed from side, rising in a rounded knob **parvidens**

Nymphs

1—Small vestiges of middorsal hooks on abdominal segments 8 and 9........... 2

—No vestiges of dorsal hooks on any abdominal segments.................... 3

2—Length of grown nymph 24 mm.; end hook on lateral lobe of labium hardly projecting beyond tip of tooth before it on inner margin......... **abbreviatus**

—Length of grown nymph 26 mm.; end hook better developed, its tip distinctly projecting beyond tip of tooth before it...................... **brevis**

3—Length of grown nymph 24 mm. .... **parvidens**

—Length of grown nymph 27 mm. .... **viridifrons**

Gomphus abbreviatus Hagen

1878. Hagen, in Selys, Bull. Acad. Belg., (2)46:464 (reprint, p. 59).
1901. Hine, Ohio Natural., 1:61 (figs.).
1901. Ndm., Bull. N. Y. State Mus., 47:444 (nymph).
1917. Currie, Proc. U. S. Nat. Mus., 63:223–226 (figs.).
1917. Howe, Odon. of N. Eng., p. 31 (figs.).
1927. Garm., Odon. of Conn., p. 149 (figs.).
1929. N. & H., Handb., p. 89 (figs.).

Length 34–35 mm.; abdomen 26–30; hind wing 22–25.

Face and frons yellowish green above, with faint lines of brownish in sutures. Occiput yellow, with narrow blackish side margins; its hair-fringed convex hind border rises above level of eyes; middorsal thoracic stripe ill defined at lower end, contracted to a mere line where it crosses yellow collar stripe; humeral and antehumeral stripes a little confluent near upper end, more broadly confluent below. Midlateral stripe wanting above level of spiracle; second lateral weak or wanting.

Abdomen stout, with sides of basal segments broadly yellow up to and including auricles; dorsum of these segments yellow, the yellow narrowing to rearward into an interrupted line of lanceolate spots that end on 8. Segments 9, 10, and appendages black.

Distinguished from next species by greater convexity of occipital margin. Spines on vertex of female short, very slender, bristle-like, and strongly divergent.

Distribution and dates.—UNITED STATES: Conn., Maine, Mass., N. H., N. J., N. Y., N. C., Ohio, S. C.

April 27 (S. C.) to July 3 (Maine).

Gomphus brevis Hagen

1878. Hagen, in Selys, Bull. Acad. Belg., (2)46:462 (reprint, p. 57).
1901. Hine, Ohio Natural., 1:61 (figs.).
1901. Ndm., Bull. N. Y. State Mus., 47:449–450 (figs., nymph).
1917. Currie, Proc. U. S. Nat. Mus., 53:223–226 (figs.).
1917. Howe, Odon. of N. Eng., p. 32 (figs.).
1927. Garm., Odon. of Conn., p. 149 (figs.).
1929. N. & H., Handb., p. 88 (figs.).

Length 44–46 mm.; abdomen 26–33; hind wing 25–27.

Frons broadly yellow, entirely encircled by a band of black. Facial lobes of postclypeus black. Greenish labrum has a heavy black border from which, on basal side, proceeds a black triangle that more or less divides the green into two large spots. Occiput yellow with black outer corners. Middorsal blackish stripe of thorax parallel-sided, or only slightly widened downward, narrowly divided by yellow on middle of carina, and abruptly constricted to crossing of collar. Humeral and antehumeral stripes conjoined at both ends, and again near upper end where they enclose a small triangular yellow spot; midlateral stripe incomplete above spiracle; third lateral well developed, forked below, connected above with humeral along black subalar carina.

Abdomen mostly black except for swollen end segments. Band of spots on sides greatly narrowed, as usual, on middle segments, but widens again on 7, 8, and 9 to cover lower half of sides, including their lower margin. Middorsal pale line also greatly reduced on middle segments to end in a very small one, generally, on 8; 9, 10, and caudal appendages, and genital hamules of male, black.

Female similar in pattern, but more extensively yellowish. A pair of stout black spines on vertex stand nearly erect behind lateral ocelli. Subgenital plate about half as long as sternum of segment 9, cleft into two slender curving lobes that rise from a broad base, and are upcurving, appressed, and carinate.

Species may be recognized by pattern of face; male by short stout appendages, with superiors a little shorter than branches of inferior.

Found near clear-flowing streams that have alternating riffles and pools, mostly near riffles; sometimes on sandy open shores of lakes; spends most of its time resting on low vegetation, and is not difficult to approach or to capture.

Nymphs are shallow burrowers in silt and sand, mostly below riffles. Place of transformation close to water's edge on any solid support having not too smooth a surface.

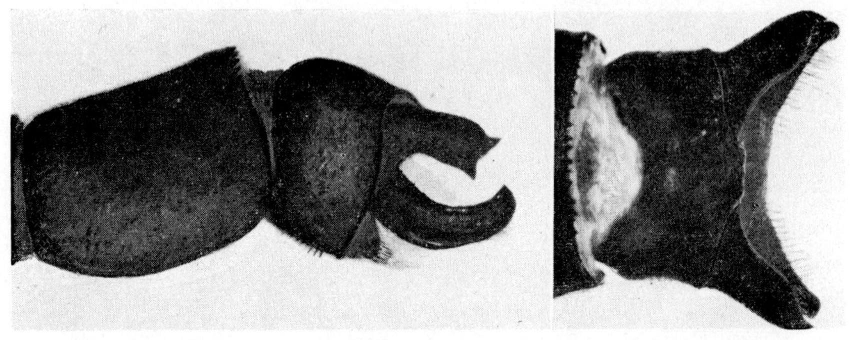

Fig. 137. *Gomphus brevis.*

Distribution and dates.—CANADA: N. B., N. S., Ont., Que.; UNITED STATES: Conn., Maine, Mass., Mich., N. H., N. J., N. Y., Pa., Vt., Wis.

May 21 (N. J.) to August 20 (Ont.).

Gomphus parvidens Currie

1917. Currie, Proc. U. S. Nat. Mus., 53:223 (figs.).
1929. N. & H., Handb., p. 89 (figs.).
1940. Mtgm., J. Elisha Mitchell Sci. Soc., 56:289.
1942. Wstf., Ent. News, 53:98.

Length 37 mm.; abdomen 27–29; hind wing 24–25.

A pretty greenish yellow species, heavily striped with blackish at front of thorax, but little touched with that color at sides; abdomen very black dorsally. Rather wide middorsal stripe of thorax divided narrowly by greenish yellow of carina. Humeral and antehumeral stripes broad and well separated by a narrow pale greenish yellow line.

Middorsal stripe of abdomen wide, as usual, at base, narrows sharply on segment 2, is reduced to a line on 3, and to short basal triangles on 4 to 6; 8, 9, and 10 black above, as are caudal appendages. Sides of 1 and 2 and base of 3 yellow, then that color almost disappears low down on middle segments, comes out again on 7, more broadly but diffusely on 8 and 9, and extends downward to cover entire lower margin; 10 black above.

In fore wings of this species are five or six paranal cells with, generally, no added marginals (as in *viridifrons*) ; hind wings have two full-width cells in second anal interspace.

Abdomen brightly marked with clear yellow. Superior caudal appendages of male more widely divaricate than in other three species. End

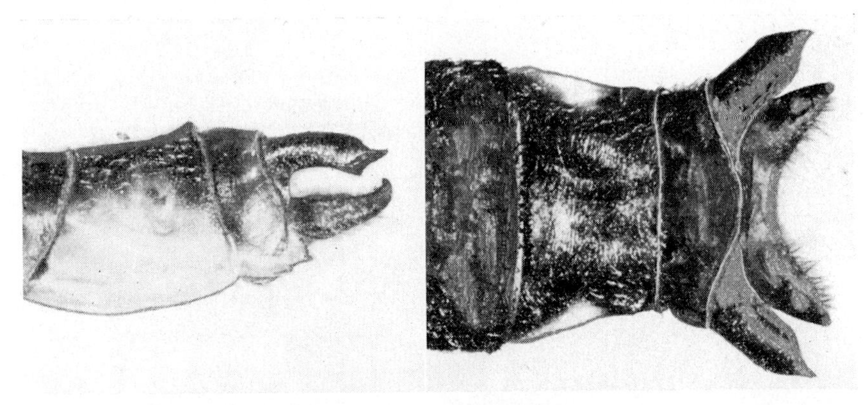

Fig. 138. *Gomphus parvidens.*

Fig. 139. *Gomphus brevis; G. parvidens* (photographed from dry specimen, also showing peduncle) ; *G. viridifrons.*

notch in inferior caudal appendage, as viewed from side, concave to rearward in *parvidens;* convex in *viridifrons.* Female may be distinguished from *viridifrons* by shorter subgenital plate, which is hardly more than half as long as sternum of segment 9.

Found beside clear streams, especially near rapids. It perches much of the time out in the open on leafy sprays, sometimes on stones and logs at the bank.

Distribution and dates.—UNITED STATES: Ala., Md., N. C., S. C.

April 19 (S. C.) to June 30 (S. C.).

Gomphus viridifrons Hine

1900. Wmsn., Odon. of Ind., p. 294 (figs.) (no name).
1901. Hine, Ohio Natural., 1:60 (figs.).
1917. Currie, Proc. U. S. Nat. Mus., 53:226 (figs.).
1929. N. & H., Handb., p. 89 (figs.).

Length 45–46 mm.; abdomen 33–35; hind wing 27–28.

This species mainly green and black. Face green, with only sutures slightly darkened, and often a pair of impressed dots on postclypeus. Black crossband of vertex extends forward to cover basal third or fourth of upper surface of frons.

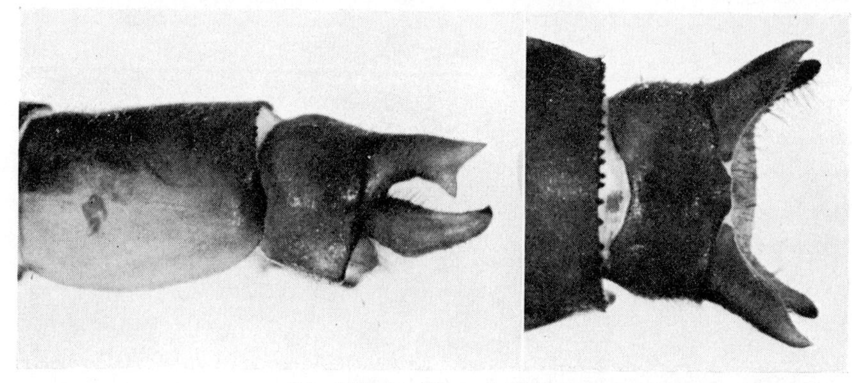

Fig. 140. *Gomphus viridifrons.*

Pale green line of carina divides middorsal black stripe of thorax into bands, each of which is widened to collar and narrower than one of the bordering **7**-shaped green marks. Humeral and antehumeral black stripes well separated by a long greenish stripe that is somewhat widened at its upper end; incomplete midlateral black stripe rises but little above level of spiracle. Humeral stripe feebly developed.

Abdomen spotted with yellow at ends, but mainly blackish. Usual middorsal and lateral pale bands yellowish green on segments 1 and 2, tapering off to a line on 3, more or less interrupted on middle segments. Dorsal band reduced to short basal triangles on 4 to 6, widened again on 7, almost obliterated on 8. Bright yellow side spots on 7, 8, and 9 extend high up on sides, but do not cover lower edge of segments. Caudal appendages black.

Distribution and dates.—CANADA: Ont.; UNITED STATES: Ala., Ind., N. Y., Ohio, Pa., Tenn.

May 3 (Pa.) to July 23 (N. Y.).

Subgenus STYLURUS Needham 1897

These rather large Gomphines exhibit considerable variety in form and coloration. The blackish stripings on the front of the synthorax are so widened and confluent at the ends that often they leave only a pair of isolated pale stripes on the front and a divided pale cross stripe on the collar. The abdomen is for the most part little clubbed (except in *scudderi*). The wings are clear and the venation rather open. The front side

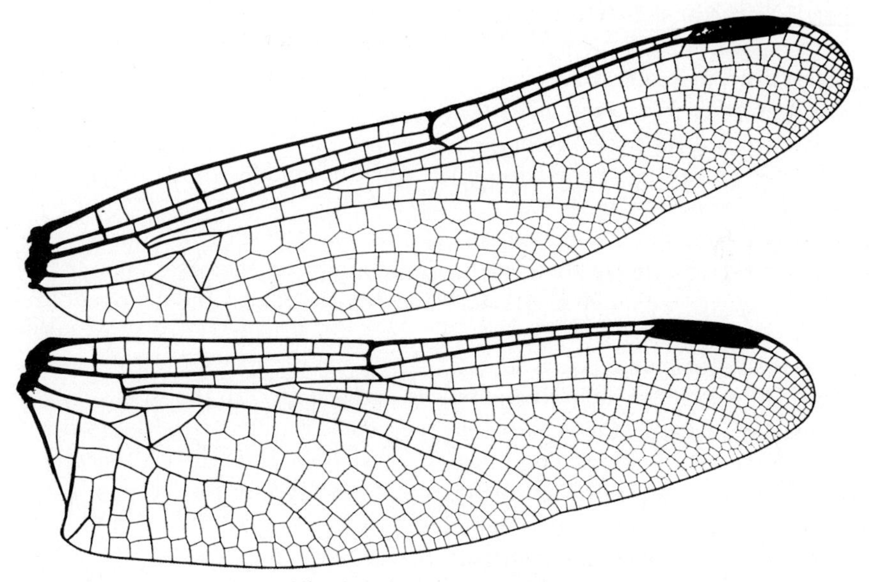

Fig. 141. *Gomphus plagiatus.*

of the triangle in the fore wing is about one and a tenth times as long as the inner side; in the hind, about one and a third times as long. The gaff varies from about half to about nine-tenths as long as the inner side of the triangle. The paranal cells of the fore wing are usually six, sometimes five or seven; of the hind wing, four or five. The first and second anal interspaces of the hind wing (x and y) generally begin, each with two full-width cells, sometimes with one or three (oftenest three in *scudderi*); postanal cells oftenest four or five, sometimes three (in *potulentus*) or six (in *scudderi*). Vein A1 generally has a kink in it a little beyond the gaff, more pronounced in males than in females. Vein A2 arises a little before the middle of the rear side of the subtriangle.

The superior caudal appendages of the male are rather simple. They lack forks or distinct teeth, but may bear low ridges near their upwardly beveled tips. They point more or less directly rearward. The arms of the fork of the inferior appendage have about the same lateral spread as the superiors.

The anterior hamule is small and simple; a thin, flattened, more or less linear, unarmed piece, often hidden between the strongly developed posteriors. The latter are almost equally simple, but they are large and prominent, and end in a sharp inturned tooth. They stand erect (or very slightly inclined backward in *scudderi*) or are strongly inclined forward in several species. They are strongly *bent* forward edgewise in a full-length curve in *olivaceus*.

The small fourth segment of the penis also is distinctive. It is very short, hardly longer than the third segment is wide; and its tip-tilted tail is only about as long as its body.

The subgenital plate of the female is very short, hardly ever reaching a fourth of the length of the ninth sternite, and is rather widely and deeply notched at the tip.

The subgenus as here delimited is best distinguished by characters of sex: by the simplicity of form in the hamules of the male, and by the brevity of the subgenital plate in the female. The caudal appendages have no teeth before the tip.

In the rather close consistency of the foregoing genital characters lies the justification for including *amnicola* and *scudderi* in the subgenus *Stylurus:* for in most other characters they more closely resemble species of *Gomphurus,* with which they have oftenest been associated hitherto.

The eleven species of this subgenus fall readily into two groups: a *Plagiatus-Notatus* group, which includes *olivaceus* and *spiniceps;* and a less homogeneous *Intricatus* group, which includes *ivae, laurae, potulentus, townesi, amnicola,* and *scudderi.* The two last named are among the more aberrant members.

Somewhat apart from all the other species stands *intricatus* by reason of its reduced venation and short triangles. The front side of the fore-wing triangle is only about as long as the inner side. Also, it has a longer gaff than any of the other species, a longer tornal cell, and a longer tenth abdominal segment; and there is scarcely any angulation of its vein A1.

Clearly there are two widely divergent trends in form of body in the genus: one, toward a long and narrow abdomen, reaches its culmination in *spiniceps;* the other, toward a broadly clubbed abdomen, in *scudderi.*

The nymph is of the elongate form shown in figure 142. Its structural characters are as stated in the table of species (p. 236). The lateral lobes of the labium have a long, strong, sharply incurved end hook with but few teeth on their opposed inner margin. The middle lobe is less prominent, its edge straight or very slightly convex. There is a middorsal groove on some of the middle abdominal segments.

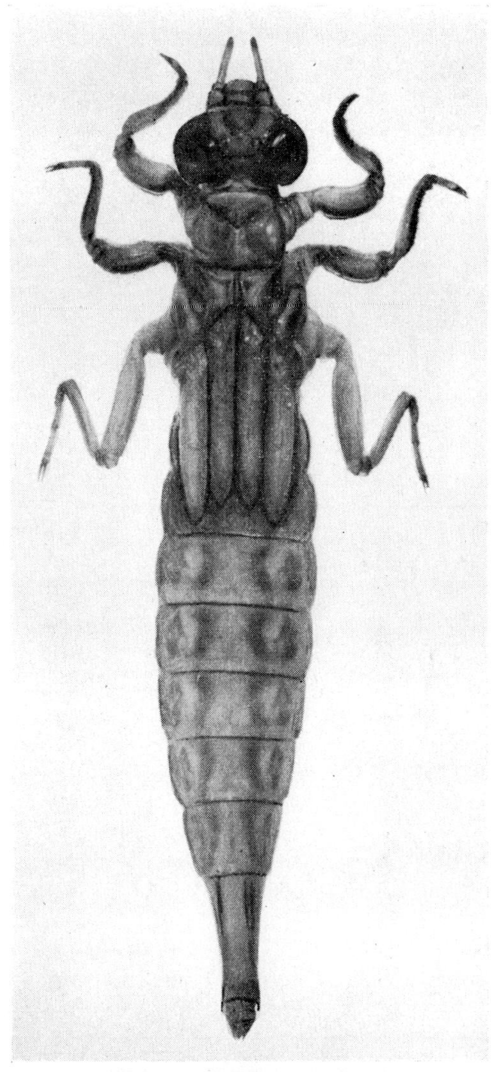

Fig. 142. *Gomphus spiniceps.*

There are small burrowing hooks on the front and middle tibiae; strong lateral spines on abdominal segments 6 to 9; no dorsal hooks at all save for a flat one on 9, and even that is lacking in several species. The lateral margins of 9 are not serrulate as in *Gomphurus.*

The form of the labium is more or less distinctive, the end hook of the lateral lobe being well developed and strongly inturned, and generally armed with a few strong teeth.

KEY TO THE SPECIES

ADULTS

1—Second hamule of male directed ventrally or slightly posteriorly, apical part of its anterior margin swollen before apex, then broadly excavated to form a shallow fork or hook with apex directed anteriorly; subgenital plate of female more than one-seventh as long as sternum of segment 9
INTRICATUS group 2

—Second hamule of male directed ventrally or more or less anteriorly, apical part of its anterior margin uniformly curved to apex, which is directed anteriorly; subgenital plate of female less than one-seventh as long as sternum of segment 9.................................PLAGIATUS group 8

TABLE OF SPECIES

ADULTS

Species	*Hind wing*	*Labr. blk.*[1]	*Collar*		*Color of*[4]		*Distr.*
			middle[2]	*ends*[3]	*carina*	*♂ ped.*	
amnicola	29-33	+	cont.	cont.	yellow	brown	E
intricatus	26-32	0	cont.	disc.	yellow	yellow	C, W
ivae	35-41	0	disc.	var.	brown	yellow	SE
laurae	36-42	+	disc.	disc.	brown	brown	SE
notatus	30-35	+	cont.	cont.	yellow	brown	E, C
olivaceus	35-36	0	cont.	cont.	yellow	brown	W
plagiatus	30-40	0	var.	disc.	var.	brown	E, C
potulentus	33	0	cont.	cont.	yellow	yellow	Miss.
scudderi	35-39	+	disc.	disc.	brown	brown	E
spiniceps	35-39	0	disc.	disc.	brown	brown	E
townesi	33	0	cont.	disc.	yellow	yellow	S. Car.

[1]Blackish border on front edge of labrum: present (+), or absent (0).

[2]Cross stripe of yellow or pale on collar of synthorax: continuous across middle (cont.), interrupted in middle (disc.), or variable (var.).

[3]Yellow of collar at its ends: continuous with yellow area between middorsal and antehumeral stripes (cont.), or separated from that area by joining of lower ends of these two dark stripes (disc.), or variable (var.).

[4]Color of edge of middorsal thoracic carina and of male peduncle, arbitrarily reduced to brown and yellow, disregarding intermediate shades and tints.

2—A small desert species; more yellow than black throughout . . . . **intricatus**
—Black or brown predominant on front of thorax and on middle abdominal segments 3

3—Blackish species; middle abdominal segments ringed with pale; end segments 7 to 9 very widely expanded . . . . **scudderi**
—Middle abdominal segments not conspicuously ringed with pale; 7 to 9 moderately expanded 4

4—Hind wing 29–33 mm.; yellow cross stripe on collar not interrupted in middle by blackish 5
—Hind wing 35–42 mm.; yellow cross stripe on collar interrupted in middle by blackish 7

5—Face blackish, labrum margined with black; second lateral suture sometimes with blackish stripe . . . . **amnicola**
—Labrum not margined with black; second lateral suture without black stripe . . 6

6—Front of thorax with pale areas not confluent with pale collar; occiput of male slightly concave . . . . **townsei**
—Front of thorax with pale areas confluent with pale collar; occiput of male slightly convex . . . . **potulentus**

7—Black or nearly black species, sharply patterned with yellow or green . . . . . . **laurae**
—Brown and yellow species, abdominal segments (3 to 6 at least) obscurely patterned . . . . **ivae**

8—Front of thorax with pale areas broadly confluent with pale collar; midlateral suture usually without black stripe; Western . . . . **olivaceus**
—Front of thorax with pale areas usually not confluent with pale collar; midlateral suture with blackish stripe 9

9—Abdomen blackish, marked with yellow or green 10
—Abdomen yellowish brown, marked with yellow or green; labrum largely green, with no distinct markings . . . . **plagiatus**

10—Abdominal segments 7 and 9 about equal in length; labrum black, with margins and a more or less distinct central spot yellowish . . . . **spiniceps**
—Abdominal segment 9 shorter than 7; labrum with margins and a longitudinal stripe (rarely reduced to central spot) black . . . . **notatus**

NYMPHS

1—Abdominal segment 9 twice as long as wide at base . . . . **spiniceps**
—Abdominal segment 9 less than twice as long as wide at base 2

2—End of lateral lobe of labium truncated and straight-edged to tip of end hook . . . . **intricatus**
—End of lateral lobe arched to tip of end hook 3

3—Vestigial dorsal hook, represented on segment 9 by small flattened triangular rearward projection on middorsal line, not hooklike in form 4
—Nothing representing a hook on 9 8

4—Small dorsal hook on 8; body very hairy . . . . **ivae**
—Nothing representing a hook on 8 5

5—Lateral spines of 9 less than half as long as 10 .... **olivaceus**
—Lateral spines of 9 more than half as long as 10 .... **6**

6—Segment 8 nearly as long as 9 .... **?laurae**
—Segment 8 about two-thirds as long as 9 .... **7**

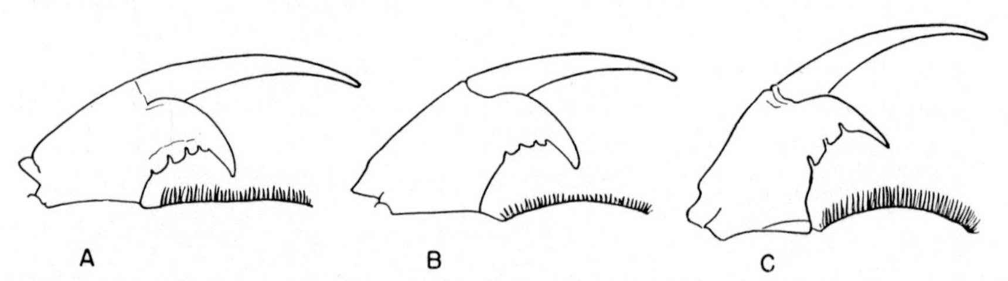

Fig. 143. Edge of median lobe of labium, with adjacent lateral lobe, to show three types in subgenus *Stylurus*. A, *Gomphus notatus;* B, *G. olivaceus;* C, *G. amnicola.* (Redrawn from Walker by Esther Coogle.)

TABLE OF SPECIES

NYMPHS

Species	*Total length*	*Labium*			*Abdomen*	
		Med. lobe[1]	*Lat. lobe*[2] *Teeth*	*Lat. lobe*[2] *Size*	*D. hook on 9*[3]	*Lat. sp. length*[4]
amnicola	27-29	C	2-3	large	0	5
intricatus	27-28	B	1-2	small	0	5
ivae	30-34	A	1-2	large	+	5
laurae	34-36	A	1-3	large	+	7
notatus	34-35	A	3-4	large	+	9
olivaceus	35-38	B	2-3	large	+	4
plagiatus	35-37	B	3-4	small	+	10
scudderi	35-39	B	2-4	large	0	7
spiniceps	43-45	A	2-3	large	+	8

[1]Edge of median lobe of labium reduced to three types: A, B, and C (see fig. 143).

[2]Number and size of teeth on inner edge of blade.

[3]Dorsal hook of segment 9: present (+), or absent (0).

[4]Lateral spines of segment 9: length expressed in tenths of middorsal length of segment 10.

7—Front border of median lobe straight; teeth on inner margin of lateral lobe large; lateral spines of 9 shorter than 10 .... **notatus**

—Front border of median lobe a little convex; median lobe very convex; lateral spines of 9 may be as long as segment 10 .... **plagiatus**

8—Length of body less than 34 mm. .... **amnicola**

—Length of body more than 34 mm.; median lobe slightly convex .... **scudderi**

Nymphs unknown: **potulentus** and **townesi.**

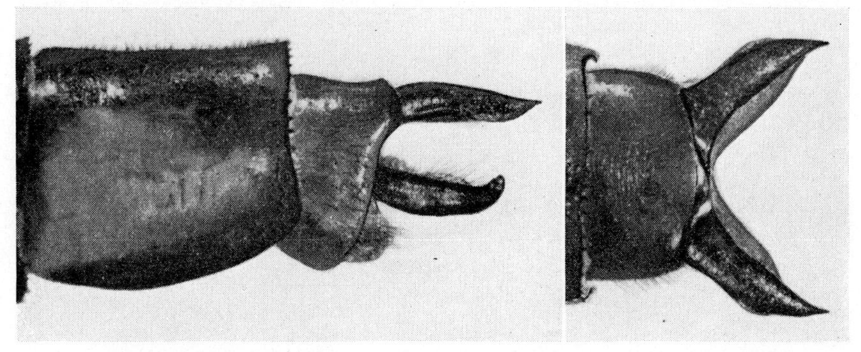

Fig. 144. *Gomphus amnicola.*

Gomphus amnicola Walsh

Syn.: abditus Butler

1862. Walsh, Proc. Acad. Phila., p. 396.
1900. Wmsn., Odon. of Ind., p. 294.
1901. Wmsn., Trans. Amer. Ent. Soc., 27:207 (figs.).
1914. Butler, Can. Ent., 46:347 (figs.) (as *abditus* n. sp.).
1917. Kndy., Bull. Univ. Kansas, 18:137 (figs.).
1928. Walk., Can. Ent., 60:87 (figs., nymph).
1929. N. & H., Handb., p. 95 (figs.).

Length 47–49 mm.; abdomen 34–36; hind wing 29–33.

A dainty species, with much black on front of thorax and yellow on sides of thorax. Face yellowish green, with lines of brown on all sutures, and with a black border on labrum and on margins of facial lobes of postclypeus. Vertex black in front of, yellow behind, postocellar ridge; occiput yellow, with black outer corners.

Middorsal thoracic stripe broadly widened downward, split lengthwise by yellow carina. Yellow collar stripe not divided by black on median line as in many related species, but continuous and greatly widened there, where broadly confluent with yellow of carina. Humeral and antehumeral black stripes almost entirely confluent and broadly conjoined with black of crest. Lateral stripes present and complete, or first may be obsolete at its upper end, and second at its lower end. Legs black.

Middorsal line of spots on middle segments of abdomen broadened on 8 to form a wide triangular spot, with a corresponding dot on 9. Large yellow side spots on 7, 8, and 9 do not entirely overspread black on lateral margin. Caudal appendages brown.

Species easily recognized by triradiate yellow that overspreads junction of collar and carina.

Distribution and dates.—CANADA: Que.; UNITED STATES: Ala., Ill., Ind., Iowa, Kans., Ky., Mass., Mich., Minn., Mo., Nebr., N. Y., N. C., Ohio, Pa., Wis.

May 5 (Ohio) to September 6.

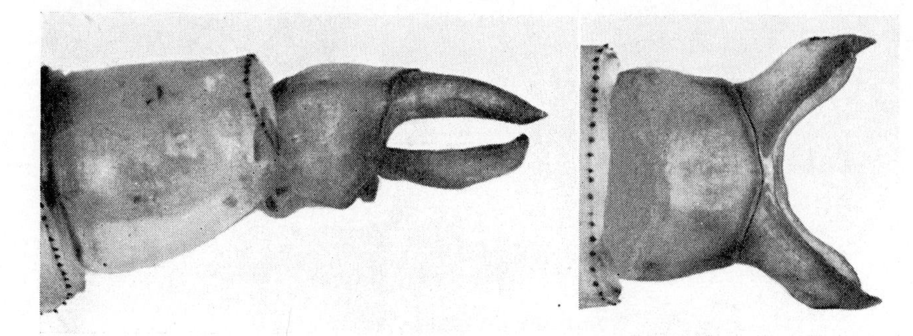

Fig. 145. *Gomphus intricatus.*

Gomphus intricatus Hagen

1857. Hagen, in Selys, Mon. Gomph., p. 418.
1861. Hagen, Syn. Neur. N. Amer., p. 108.
1917. Kndy., Bull. Univ. Kansas, 18:138 (figs.).
1917. Kndy., Proc. U. S. Nat. Mus., 52:550 (figs.).
1927. Seemann, J. Ent. Zool. Claremont, Calif., 19:22 (figs.).
1928. Walk., Can. Ent., 60:86 (figs., nymph).
1929. N. & H., Handb., p. 99 (figs.).

Length 41–55 mm.; abdomen 32–43; hind wing 26–32.

A delicate pale yellow desert species, nearly hairless. Head all yellow except eyes and two small spots on vertex, a crossband between ocelli, and a streak between antennae and compound eyes; these markings may be conjoined. Antennae brown, with pale rings on basal segments.

Prothorax pale yellow, with a dorsal wash of brown and ill-defined streaks on sutures at sides. Synthorax mostly yellow even in front, usual middorsal stripe being divided into two lengthwise, and abbreviated below. Pale brownish saddle mark at front below collar. Antehumeral stripe very narrow, fading out toward lower end; side stripes all vestigial or wanting. Legs pale, with black spines. Femora yellow, with a streak of pale brown near knees. Tibiae and tarsi black, with a wide yellow band covering outer face of each.

Abdomen yellow except for large triangular lateral spots of brown that overspread sides and are connected across back by black of joinings of segments. A smaller spot lies farther forward at same level on each of segments 3 to 7. Caudal appendages of male yellow, with blackish edgings at tips. Hamules also yellow.

Rests much of time on bushes and on driftwood, not on sand of shore (Kennedy). Flies low over water, and is difficult to catch there.

Distribution and dates—CANADA: Alta., Sask.; UNITED STATES: Ariz., Calif., Kans., Nebr., Nev., N. Mex., Tex.

June 8 (Tex.) to August 27.

Gomphus ivae Williamson

1924. Root, Ent. News, 35: 319 (no name).
1932. Wmsn., Occ. Pap. Mus. Zool. Univ. Mich., 247:12 (figs.).

Length 58–61 mm.; abdomen 42–46; hind wing 35–41.

A yellowish species, striped heavily with brown on thorax and suffused with brown on middle segments of abdomen. Face yellow or greenish yellow, including most of top of frons; vertex brown in front, with a very minute erect spine of same color between lateral ocellus and eye; occiput and rear of vertex yellowish; brighter yellow on straight-edged occiput.

Front of synthorax more brown than yellow, the broadly triangular middorsal stripe being fused at both ends with wide antehumeral, leaving only an isolated pair of bright yellow frontal stripes that are widely divergent downward. Humeral stripe incomplete above (not antehumeral this time; a difference worth noting), and separated from antehumeral by a hairline of yellow in bottom of humeral suture. In midst of black underneath outer end of crest is a large round spot of bright yellow. Sides of thorax yellow, with usual stripes of brown interrupted and inconstant.

Legs yellow at base and black beyond knees, with a diminishing streak of black running inward along front side of femora. Costa of wing yellow, touched with brownish toward base; stigma pale brown, bordered by black veins.

Abdomen moderately stout, well clubbed toward rear end, conspicuously yellow at both ends, and obscured with a wash of brown over middle segments. On either side of the two basal yellow segments is a lateral row of brown spots, one on segment 1 and three on 2. In obscure brown of middle segments is a suggestion of a paler cross ring covering about basal third of 3 to 7. These segments darken progressively to rearward; 7 to 10 patterned with light orange brown on dorsum, especially on 8 and 9. These end segments bright yellow beneath; appendages also yellowish.

"They rested usually on tree leaves, always in the sun, and were not wild or nervous. Coming out of the forest they appeared on the stream, a bright dash of brilliant yellow as conspicuous and lovely as a goldfinch." —Williamson (1932, p. 17).

Distribution and dates.—UNITED STATES: Fla., Ga., S. C.

September 2 (Fla.) to October 11 (S. C.).

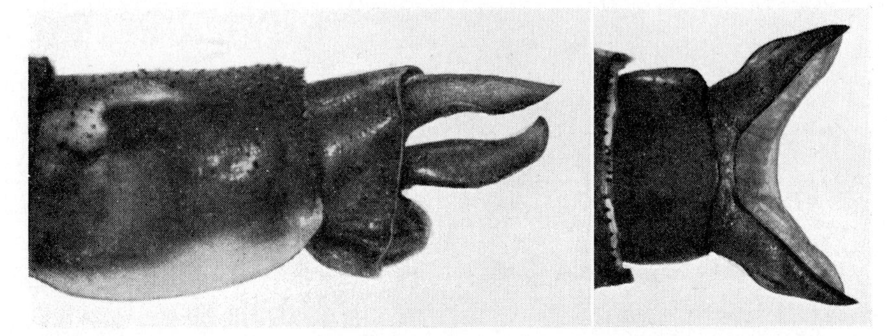

Fig. 146. *Gomphus ivae.*

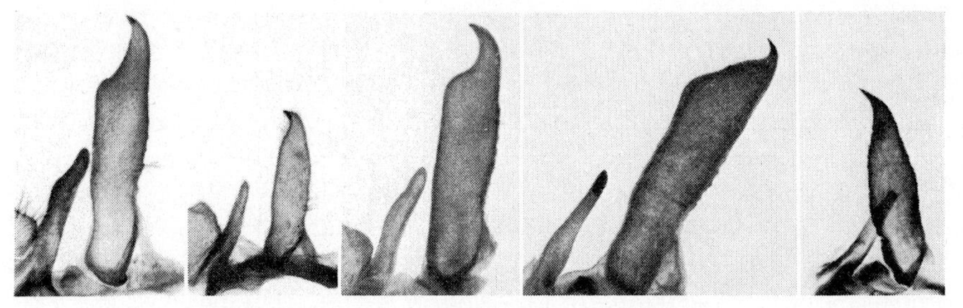

Fig. 147. *Gomphus amnicola; G. intricatus; G. ivae; G. laurae; G. notatus.*

Gomphus laurae Williamson

1932. Wmsn., Occ. Pap. Mus. Zool. Univ. Mich., 247:3 (figs.).
1940. Mtgm., J. Elisha Mitchell Sci. Soc., 56:289.
1942. Wstf., Ent. News, 53:98.

Length 60–64 mm.; abdomen 42–48; hind wing 36–42.

A fine big Gomphine of rather elongate form. Face greenish yellow, front margin of labrum edged with black. Entire clypeus blackish, green on all rounded prominences, darker on edges of facial lobes. Frons greenish yellow, black-edged to rearward above, where confluent with black vertex. Occiput relatively small, somewhat cross-wrinkled, yellowish or greenish brown.

Middorsal thoracic stripe widens downward into broad triangular form, conjoined both above and below with antehumeral; latter broad and closely conjoined with humeral, only a very narrow, irregular, generally interrupted yellow line between them. Side of thorax yellowish, with two lateral brown stripes complete and generally well defined; area between them grayish. Legs black beyond their pale bases, with some yellow on sides of both front and hind pairs.

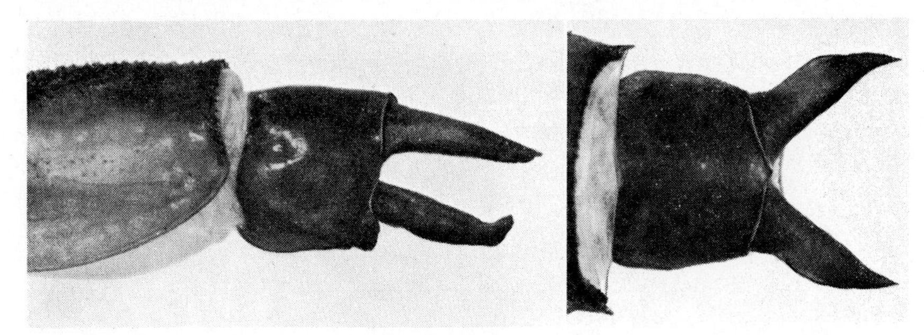

Fig. 148. *Gomphus laurae.*

Middorsal stripe of abdomen narrows as usual on segment 3 to a line of tapering dashes; dash on 7 very slender and nearly the full length of segment; on 8 wider, more tapering, only about half as long; on 9 reduced to a small triangular spot; 10 blackish above. Pale side spots on expanded end segments are a widening streak toward apex of 7; large half-height areas of bright yellow on 8 and 9 cover margin of segments; a basal inferior spot on 10. Caudal appendages and hamules of male brown.

Distribution and dates.—UNITED STATES: Ala., Fla., Ga., Ind., Miss., N. C., S. C., Va.

July 16 (N. C.) to October 6 (S. C.).

Gomphus notatus Rambur

Syn.: fluvialis Walsh, jucundus Needham

1842. Rbr., Ins. Neur., p. 162.
1861. Hagen, Syn. Neur. N. Amer., p. 110.
1900. Wmsn., Odon. of Ind., p. 294 (figs.).
1901. Wmsn., Trans. Amer. Ent. Soc., 27:210 (figs.).
1908. Mtk., Odon. of Wis., p. 91.
1927. Garm., Odon. of Conn., p. 161 (fig.).
1928. Walk., Can Ent., 60:83 (figs., nymph).
1929. N. & H., Handb., p. 100 (figs.).

Length 52–64 mm.; abdomen 37–42; hind wing 30–35.

A greenish yellow species, striped with brown. Face tawny; sides of mouth yellow and hairy. Labrum margined and traversed lengthwise with black; anteclypeus obscure; postclypeus with facial lobes edged with black. Frons yellowish; on its vertical part, in male, a diffuse cross-band of brown constricted in middle; this is wanting in female. Vertex black in front, brown on rear half. Occiput yellow, with a touch of brown on outer corners.

Middorsal thoracic stripe is a wide triangle of brown narrowly connected at both upper and lower corners with antehumeral; latter well

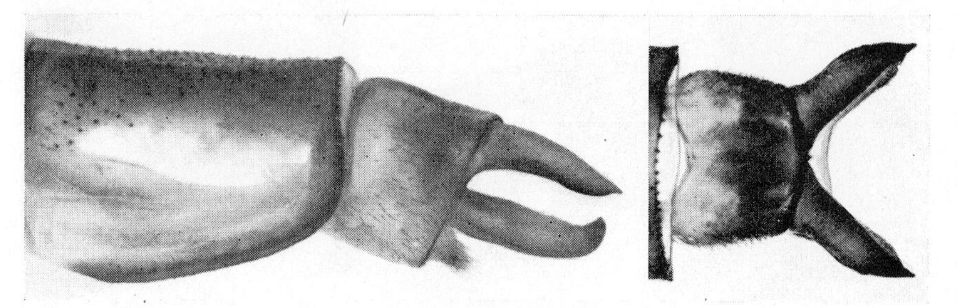

Fig. 149. *Gomphus notatus.*

separated from humeral by a yellow line widened at its rounded upper end. Lateral stripes entire but ill defined. Legs black; under side of front femora hardly paler.

Abdomen mostly black, its middorsal pale line reduced almost to a hairline on segments 4 to 6, widened to a spear-pointed spot on 7, a still wider but shorter one on 8, and reduced to a mere basal trace on 9 and 10. These pale markings wider in female; lower part of sides of middle segments edged with dull yellow. Brighter yellow on widened end segments in male covers less than half of sides of 8 and 9, much less on 7 and 10. Outside the yellow, a marginal border of brown at front end of 8 and 9. Caudal appendages and hamules of male blackish.

Distribution and dates.—CANADA: Man., Ont., Que.; UNITED STATES: Ala., Ill., Ind., Iowa, Ky., Md., Mich., N. Y., N. C., Ohio, Tenn., W. Va., Wis.

May 30 (Tenn.) to October 4 (Ky.).

Gomphus olivaceus Selys

Syn.: var. nevadensis Kennedy

1873. Selys, Bull. Acad. Belg., (2)35:749 (reprint, p. 21).
1903. Calv., Ent. News, 14:191 (fig.).

1917. Kndy., Proc. U. S. Nat. Mus., 52:557.
1927. Seemann, J. Ent. Zool. Claremont, Calif., 19:22 (figs.).
1928. Walk., Can. Ent., 60:85 (figs., nymph).
1929. N. & H., Handb., p. 99 (figs.).

Length 56–60 mm.; abdomen 41–45; hind wing 35–36.

A rather stout Western species, yellowish green or olive, striped with brown. Face olive green, with a black line on fronto-clypeal suture. Yellow frons with a narrow basal crossband of black next to vertex, notched on middle line. Vertex black in front, pale brown behind postocellar ridge, only a little darker than occiput.

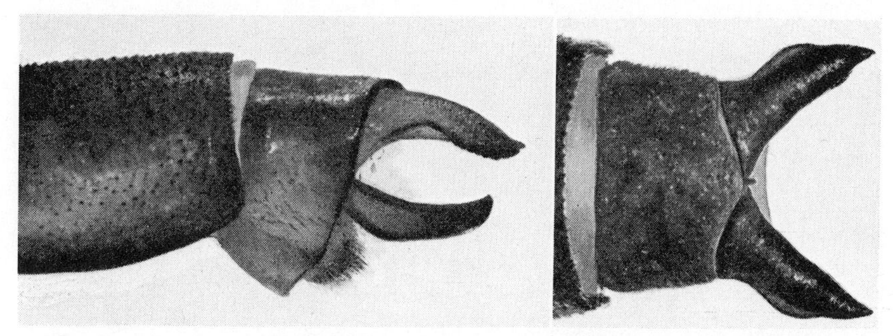

Fig. 150. *Gomphus olivaceus.*

Middorsal thoracic stripe widens downward, abbreviated at lower end, divided by a wholly yellow carina; lateral stripes wanting. Femora grayish, with a wide blackish stripe on upper side that darkens to knee. Tibiae and tarsi black.

Abdomen grayish at base, its lateral pale stripe becoming yellowish on slightly swollen segments 7 to 10. Dorsal and lateral pale lines meet and overspread 1, save for a small midlateral spot on each side. Middorsal pale line narrows to a point on 2, becomes an interrupted line of steeple-shaped spots on 3 to 7, a roundish spot on 8, a crossbar on 9; 10 black above. Lateral line of paler color continuous and wide. Sexes similar in pattern, with usual larger included yellow spots triangular on 8 and 9.

Female has less extensive brown coloration. Inhabits muddy, often alkaline, western rivers; perches in willows by streamside.

Distribution and dates—CANADA: B. C.; UNITED STATES: Calif., Nev., Oreg., Utah, Wash.

June (Calif.) to October 10 (Oreg.).

Gomphus plagiatus Selys

Syn.: elongatus Selys

1854. Selys, Bull. Acad. Belg., 21(2):57 (reprint, p. 38).
1861. Hagen, Syn. Neur. N. Amer., p. 109.
1893. Calv., Trans. Amer. Ent. Soc., 20:244.
1900. Wmsn., Odon. of Ind., p. 295 (figs.).
1901. Wmsn., Trans. Amer. Ent. Soc., 27:211 (figs.).
1928. Walk., Can. Ent., 60:83 (figs.).
1929. N. & H., Handb., p. 101 (figs.).
1930. Byers, Odon. of Fla., pp. 64, 252 (nymph).

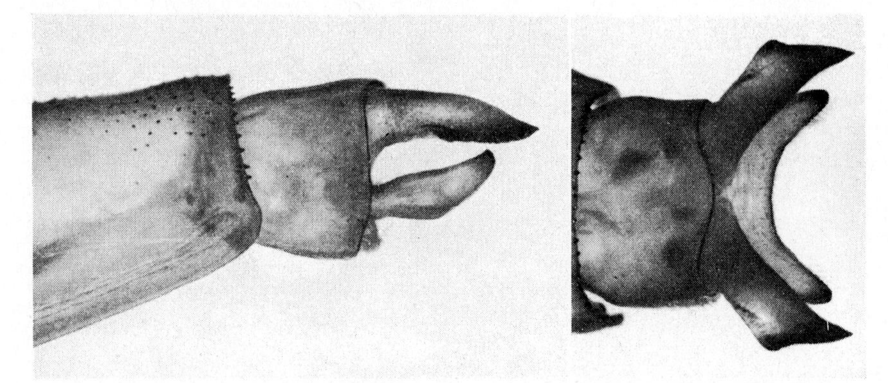

Fig. 151. *Gomphus plagiatus.*

Length 57–66 mm.; abdomen 40–50; hind wing 30–40.

A fine large species, handsomely striped with yellow and brown. Face brownish, with a greenish overcast on labrum, but no definite pattern. Anterior half of top of frons, posterior half of vertex, and occiput clear olive.

Dark brown stripes on front of thorax very broad, generally coalescent both above and below, thus leaving between the very broadly triangular middorsal and the antehumeral only a narrow included pair of oblique pale stripes that are strongly divergent downward; rarely widened at lower end to connect with yellow of collar. Another lesser isolated pale stripe generally persists between humeral and antehumeral. Sides of thorax olivaceous, with narrow midlateral black stripe generally interrupted in its upper half; similar second lateral more often interrupted in lower half.

Middorsal band of abdomen almost lacking; very narrow even on basal segments, and reduced to shortened spots or traces on middle segments. Segments 7 to 10 reddish brown or yellowish brown, with darker brown, patternless, expanded margins. Caudal appendages brown.

Female similar but much less brightly colored. A minute brown tubercle rises from ridge at each end of postocellar carina.

Distribution and dates.—CANADA: Ont.; UNITED STATES: Ala., Fla., Ga., Ill., Ind., Iowa, Kans., Ky., Md., Mich., Mo., N. J., N. Y., N. C., Ohio, Okla., Pa., S. C., Tenn., Tex.

May 21 (Ill.) to November 13 (Fla.).

Gomphus potulentus Needham

1942. Ndm., Can. Ent., 74:71 (figs.).

Length 52 mm.; abdomen 39; hind wing 33.

A blackish species, with sides of thorax and leg bases yellow. Face dull olivaceous, with edgings of brown on all clypeal sutures. Labrum greenish, with a brownish front border and paler margins at ends. Top of frons yellowish except for a narrow basal black band that merges into black of vertex; rear of vertex brown. Occiput dull yellow, its slightly convex border fringed with short stiff black hairs.

Double middorsal stripe of thorax widens downward, divided by yellow of carina, and isolated below by yellow of collar, which, conjoined with yellow stripes at sides, forms a pair of isolated and opposed 7-shaped marks on front. Middorsal and antehumeral brown stripes both join the black on crest. Antehumeral separated from humeral by a very narrow and irregular line of yellow, lower part of which is inclined upward toward depression in humeral suture, and detached and residual upper end of it is a conspicuous squarish spot beneath outer angle of crest, out of line with lower portion.

Sides of synthorax yellow except for an overwash of brown that spreads to rearward across humeral and midlateral sutures at their upper end; also a little touch of brown on latter suture below level of spiracle. No dark stripe on second lateral suture. Pale color of sides continues under thorax, envelops basal segments of legs and overspreads femora, increasingly from front to rear, leaving hind femora, mainly yellow, with only a dwindling line of black running toward body from knee. Tibiae, tarsi, and claws wholly black.

In venation of wing, triangles of unequal length, that of fore wing being three-fourths as long as hind wing. A single row of paranal cells in fore wing; in hind wing, three or four postanal cells (probably normally four); in first and second anal interspaces, two full-width cells and then two rows to wing margin. Gaff about three-fourths as long as inner side of triangle.

Abdomen black except on moderately dilated segments at ends, but middorsal pale line of yellow, present on segments 1 and 2, disappears near middle of 3. Sides of 1 and 2 and base of 3 yellow, including auricle. Hamules of male yellow. Sides of 7, 8, and 9 dull clay yellow, with a reddish tinge, this color extending to lower margin on 9 only; 10 wholly dull brown; caudal appendages brown.

Easily recognized by bicolored sides of synthorax, half black above, half yellow below. Much smaller than *plagiatus,* and wings less widened toward base.

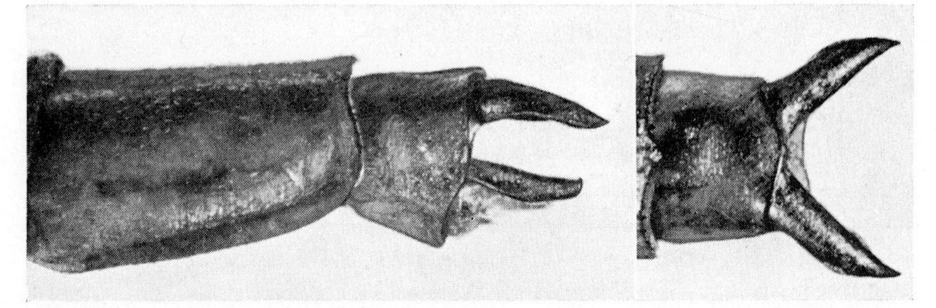

Fig. 152. *Gomphus potulentus.*

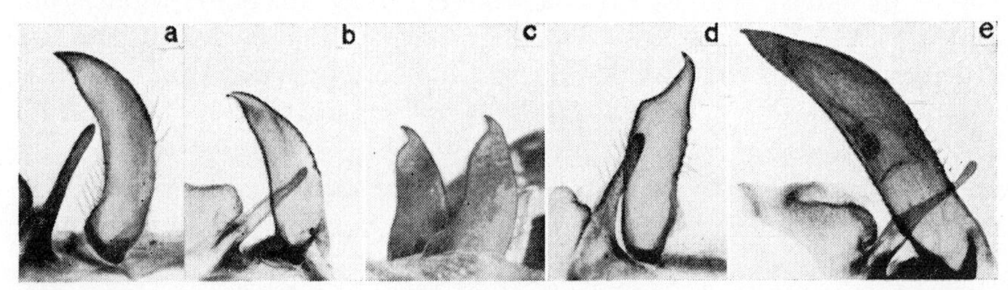

Fig. 153. A, *Gomphus olivaceus;* B, *G. plagiatus;* C, *G. potulentus;* D, *G. scudderi;* E, *G. spiniceps.* (Fig. 153, C, photographed from dry holotype specimen, showing both right and left posterior hamules.)

Known as yet from a single male specimen from Whisky Creek near Leaf, Mississippi, collected by Mrs. Alice L. Dietrich.

Distribution and dates.—UNITED STATES: Miss.

July 15.

Gomphus scudderi Selys

1873. Selys, Bull. Acad. Belg., (2)35:752 (reprint, p. 24).
1898. Harvey, Ent. News, 9:62 (figs.).
1901. Ndm., Bull. N. Y. State Mus., 47:456–457 (nymph).
1901. Wmsn., Trans. Amer. Ent. Soc., 27:208 (figs.).
1927. Garm., Odon. of Conn., p. 161 (figs.).
1928. Walk., Can. Ent., 60:85 (figs., nymph).
1929. N. & H., Handb., p. 95 (figs.).

Length 57–58 mm.; abdomen 40–44; hind wing 35–39.

A blackish species, with slender middle abdominal segments ringed with paler, and broadly dilated end segments brightly marked with yellow. Face dull yellow with three wide cross stripes of black: one on front margin of labrum; second at its rear, covering half of clypeus; and third on vertical face of frons. Front half of top of frons yellow; rear and all vertex blackish; occiput yellow, with triple swellings on its surface, and a marginal fringe of stiff black hairs.

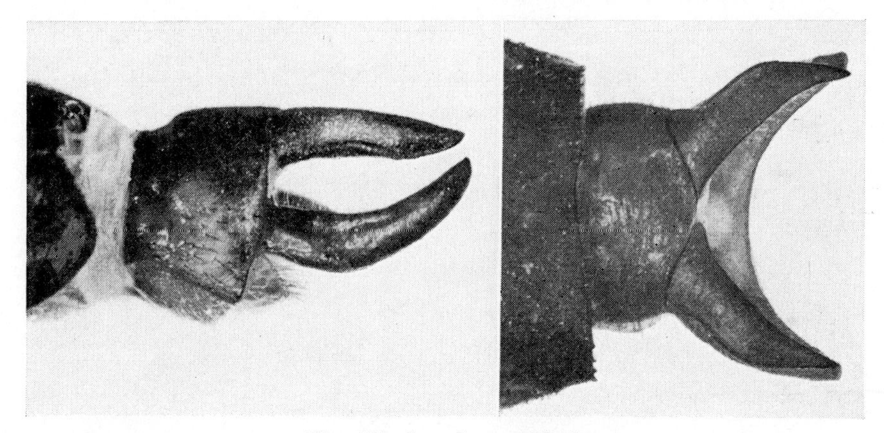

Fig. 154. *Gomphus scudderi.*

Very broadly triangular middorsal thoracic stripe, contracted below to a narrow point at crossing of long yellow cross stripe on collar. Humeral and antehumeral brown stripes fused into one broad shoulder band, with only a trace of yellow line that usually lies between them, that remnant near upper end of crest. Likewise, two side stripes fused to form oblique brown band, leaving only isolated greenish yellow stripes before and behind them; narrower stripe before this band parallel-sided, stripe behind it roughly semicircular in outline. Legs black, including femora, in mature specimens.

Abdomen blackish even on basal segments, paler middorsally and with spots of yellow on sides, one of which covers auricle. Slender middle segments black beyond their pale basal fourth. On 7 to 9 are yellow basal middorsal triangles abbreviated to rearward, and wide yellow spots on sides, basal and oblique in position, and largest on 9, there touching lateral margins of segments at base only; dorsum of 10 and appendages blackish.

Easily recognized by ringed abdomen and flare of leaflike expansion of segments of its club.

Distribution and dates.—CANADA: N. S., Ont., Que.; UNITED STATES: Conn., Ga., Maine, Mich., N. Y., N. C., S. C.

June 25 (Ont.) to October 6 (S. C.).

Gomphus spiniceps Walsh

Syn.: segregans Needham

1862. Walsh., Proc. Acad. Phila., p. 389 (in *Macrogomphus*).
1900. Wmsn., Odon. of Ind., p. 295 (as *segregans*).
1901. Ndm. & Hart, Bull. Ill. State Lab., 6:87.
1901. Wmsn., Trans. Amer. Ent. Soc., 27:209 (figs.).
1917. Howe, Odon. of N. Eng., p. 35 (figs.).
1928. Walk., Can. Ent., 60:81 (figs., nymph).
1929. N. & H., Handb., p. 101 (figs.).

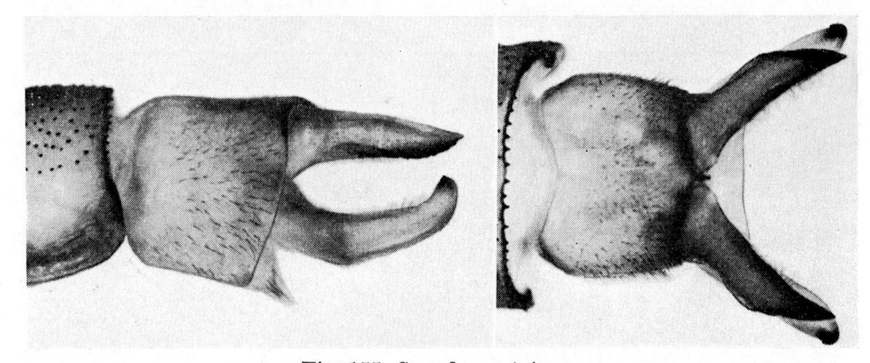

Fig. 155. *Gomphus spiniceps.*

Length 57–68 mm.; abdomen 44–51; hind wing 35–39.

A brownish species, with very elongate abdomen. Face blackish, with front margin of labrum shining brown. Front half of top of frons yellow and thickly beset with short stiff erect black hairs.

Stripes on front of thorax so broad and so coalescent by their ends that they leave between them only a pair of isolated, strongly divergent yellow lines; also two fragments of usual yellow line between humeral and antehumeral. Brown stripes of sides both broad and complete, but with ill-defined margins. Legs black, with paler bases. Wings lightly tinged with yellow in membrane toward base.

Abdomen black, with usual yellow touches toward ends greatly reduced. Middorsal stripe remains on segments 1 and 2, but on 3 to 7 is reduced to short basal greenish yellow spots. Sides of 2 black except for yellow on auricle. On sides of slightly expanded 8 and 9, remaining yellow amounts to hardly more than a streak on each; 10 wholly brown; caudal appendages blackish.

Easily recognized in both sexes by extremely long segment 9 of abdomen, one and a half times as long as 8; also, in male, by huge genital hamule, twice as large as in *plagiatus,* which turns strongly forward and reaches almost to level of hind end of thorax; in female, by presence of a pair of large horns on occiput.

Distribution and dates.—CANADA: Ont., Que.; UNITED STATES: Ill., Ind., Ky., Md., Mass., Mich., Mo., N. Y., N. C., Ohio, Pa., S. C., Tenn., W. Va., Wis.

June 23 (Ill.) to October 5 (S. C.).

Gomphus townesi Gloyd

1936. Gloyd, Occ. Pap. Mus. Zool. Univ. Mich., 226:5 (figs.).
1940. Mtgm., J. Elisha Mitchell Sci. Soc., 56:285, 289.

Length 52 mm.; abdomen 38; hind wing 33.

Greenish in ground color, with front of thorax heavily overspread by stripes of rich dark brown and with sides more yellow. Face pale, with only a hairline of brown on fronto-clypeal suture. Top of frons also pale, with only a narrow bilobed edging at its base next to brown vertex. Straight-edged occiput and adjacent strip on rear of vertex with prolongation on each side next to eye pale.

Very broadly triangular middorsal stripe of thorax cleft by pale carina with rounded angles at junction with yellow cross stripe of collar; confluent with broad antehumeral brown stripe at both upper and lower ends. Humeral and antehumeral stripes separated by a wide yellow stripe except at their upper ends under crest; humeral stripe itself longitudinally cleft by another narrower line that follows humeral suture and isolates posterior portion of dark humeral stripe. On more yellowish sides of thorax, narrower first lateral stripe is complete, second is wanting.

Abdomen long and slender, with moderate widening of end segments. Segments 1 and 2 mostly yellow; 3 to 6 dark brown with a narrowed median band of spots widened and shortened on 7 and 8; 9 and 10 dark brown dorsally. On side of 7, a full-length lateral yellow stripe half as high as segment; on 8, a wider rounded yellow area; on 9, a large yellow spot blends with reddish brown apically and along lateral margin; on 10, diffuse, blending with bright yellow ventrally; appendages golden brown, more yellowish ventrally.

Female unknown.

Distribution and dates.—UNITED STATES: S. C.

August 13, August 22.

Family AESCHNIDAE Selys

The Darners

Here we come upon a family that is dominant today, one that has made great advances over primitive conditions and has specialized in ways of its own. All the darners are above the average size of dragonflies generally, and some of them are the giants of the order Odonata.

These dragonflies are of rather brilliant coloration, mostly blues, greens, and brown. The huge eyes meet broadly on the top of a semiglobular head, crowding between and widely separating the visible part of vertex and occiput (fig. 3, p. 9).

The thorax is very robust. The wings are mostly hyaline. The abdomen is very long and for the most part slender, though the two basal segments are swollen, and the third is often markedly constricted. The caudal appendages are long and flattened.

The female is armed with an ovipositor; the lower rim of abdominal segment 10 is generally prolonged into a projecting plate that is variously armed with spines on its rear margin (fig. 13, p. 22).

In wing venation there are many striking family characters. The triangles are greatly elongated in the axis of the wing, the front one being a little longer than the hind. The moderate stigma has a brace vein at its inner end. The subtriangle is not well developed; the crossvein that might form its inner side is weak and inconstant in position. The portion of the bridge beyond the nodus is very short, the oblique vein being rather close to the subnodus. Vein M2 is strongly up-arched toward M1 near the stigma; it does not keep parallel to vein Rs as in other Odonata. Radial and median planates are strongly developed, and a trigonal planate, springing from the outer side of the triangle, is generally recognizable. There are two greatly thickened antenodal crossveins, the number of ordinary crossveins between them ranging from two to ten. The intermedian crossveins are likewise numerous, ranging from five to twelve. There is a row of large paranal cells in the fore wing, bordered by a single row of much smaller marginal cells. The anal loop is strongly developed. The number of cells in it is very inconstant, ranging from two to ten or more; in the broader wings these cells may be placed in two or three vertical or horizontal rows, according to genus and species. The anal triangle of the male is generally three-celled (often more in *Boyeria*), or two-celled and very narrow (a crossvein between first two paranal cells, *n*, and *o*, being eliminated). The last two or three of the five or six paranal cells of the hind wing lie within and form the front line of cells in the anal loop.

In the Aeschnidae there appear three new and peculiar features of venation that are of much evolutionary significance:

1. There is a tie-up of veins M3 and M4 far out toward the wing margin.

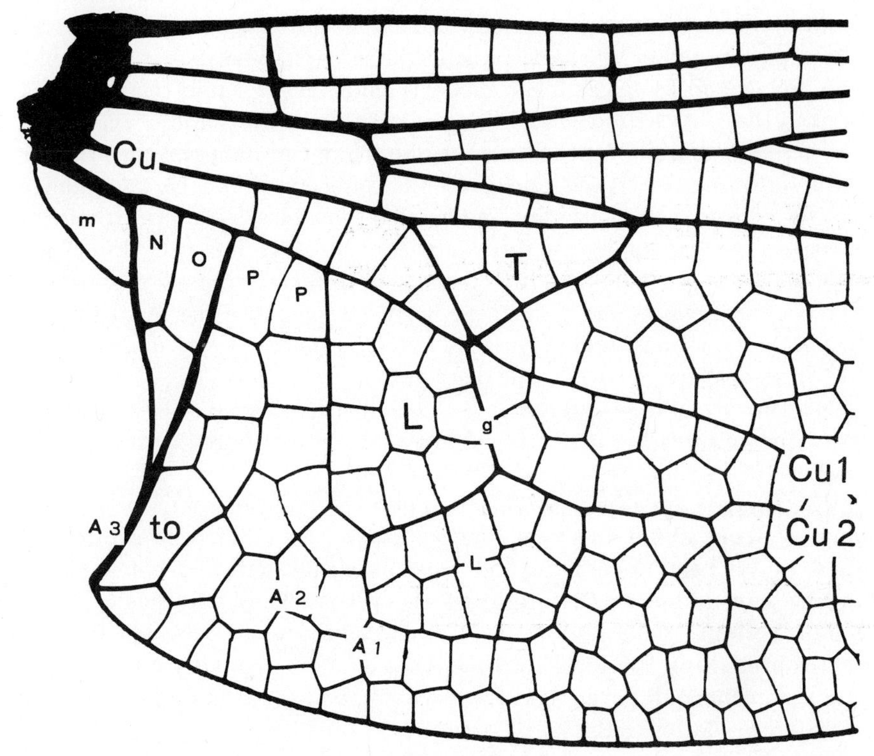

Fig. 156. Part of hind wing of *Oplonaeschna armata.* Note that first and seventh antenodal crossveins are thickened, supratriangular space is three-celled, triangle (T) itself is four-celled, and there is no definite subtriangle. Anal loop (L) is ten-celled, a newly appearing supplemental loop below it (L) also ten-celled, and there is a suggestion of two additional supplemental loops forming at right. Observe that there are seven paranal cells, four of them lettered *n, o, p, p,* the other three within anal loop. Note strength of vein A3. Large basal triangle of male occupies entire third anal interspace (*z*), and is composed of cells *n* and *o* and a third cell at their rear. Tornus (*to*) and other venational features as shown in introductory figure (p. 17).

2. An anterior branch of vein Rs is formed by the straightening out of the kinks between two rows of interlocking hexagonal cells, followed by strengthening of that line (fig. 172, p. 286).

3. Supplemental to the anal loop, the *plantar* and *patellar* loops emerge (fig. 166, p. 274).

These are the dragonflies that probably suggested the common name "darning needles," by which they were known of old. For the most part they are high flyers, with long, slim bodies, oftenest clearly seen against the background of the sky. They are remotely needle-like, to be sure, but simple folk have vivid imaginations. Perhaps these insects are more like arrows, with which some modern poets are wont to compare them. Both needles and arrows are familiar symbols of sharpness. At any rate, the common name "darners" recalls the superstitious belief that they can sew up folks' ears. That name is much more easily learned and spoken than "Aeschnids," and has a justifiable background of tradition.

The darners are a truly voracious lot. They can dart through the air with lightning-like speed and threaten collision. When captured alive and carelessly held with the fingers they can bite with jaws that are sharp and strong enough to draw a drop of blood from a thin skin. They can make a startling rustling sound with their wings. Perhaps the Devil's name was added to darning needles to enhance the fear of them.

They surpass most other groups of Odonata in size of eyes and strength of wings, and are of world-wide distribution.

Martin's *Aeschnines,* in the de Selys Collections series (1908–1910), is good for an introduction to the study of the world fauna in this family because of its many excellent illustrations.

The nymphs are climbers. They do not wallow in the mud. They do not burrow. They climb on stems, on stumps, and on submerged rough surfaces of any sort. Some live amid green aquatic vegetation, some among dead stems fallen in the edges of the water. They are mostly bare-skinned, thin-legged, and active. They are among the fiercest of fresh-water predators. Big ones eat little ones of their own species. They are found in nearly all permanent fresh waters, and are as abundant as their size and predatory habits will permit.

Their long and graceful bodies are smooth and clean, colored and patterned to suit their environment. The head is more or less flattened above. The eyes are lateral, rarely lifted to a high position on the fore corners. The labium is very long and its mentum flat, never spoon-shaped or masklike. Its middle lobe is convex and cleft, but never deeply cleft. The lateral lobes are relatively narrow, somewhat parallel-sided as far out as the base of the large movable hook, with the inner margin of the blade generally minutely denticulate. There are no raptorial setae (except some weak laterals in *Gynacantha* and *Triacanthagyna*).

The thorax is somewhat depressed, widened a little to rearward. The disc of the prothorax projects a little at the sides, and underneath its ends, on each side, is a pair of blunt supra-coxal processes. The sides of the synthorax slope steeply down to the leg bases. The legs are long and thin, and when pale are crossbanded with brown.

The abdomen is somewhat widened from the base to beyond the middle, then regularly narrowed to the end. It is without dorsal hooks (except for some low humps in *Nasiaeschna*), and has lateral spines on segments 5, or 6, or 7, or 8 to 9, according to genera and species.

KEY TO THE GENERA

Adults

1—Midbasal space with more than one crossvein........................ **Boyeria**
—Midbasal space with or without a single crossvein........................ 2

2—Sectors of arculus rising from its upper end; thorax uniform green...... **Anax**
—Sectors of arculus rising near its middle; thorax usually not uniform green.. 3

3—Vein Rs forked.. 4
—Vein Rs not forked.. 9

4—Stalk of Rs straight and fork symmetrical.............................. 5
—Stalk of Rs up-arched and fork askew forward........................... 6

5—Radial planate subtends one row of cells...................... **Nasiaeschna**
—Radial planate subtends more than one row of cells............. **Epiaeschna**

6—Vein Rs forked well beyond level of base of stigma; two rows of cells in fork of Rs .... **Coryphaeschna**
—Vein Rs forked at or before level of base of stigma; generally more than two rows of cells in fork of Rs.. 7

7—Supertriangle as long as, or shorter than, midbasal space............ **Aeschna**
—Supertriangle distinctly longer than midbasal space....................... 8

8—Two rows of cells between M1 and M2 beginning under stigma; ventral process of segment 10 of female three-pronged.............. **Triacanthagyna**
—Two rows of cells between M1 and M2 beginning at or before stigma in hind wing, and usually in fore wing; ventral process of segment 10 of female two-pronged .... **Gynacantha**

9—Two cubito-anal crossveins; one crossvein under stigma; supertriangle without crossveins .... **Gomphaeschna**
—Three or more cubito-anal crossveins; several crossveins under stigma; supertriangle with crossveins...10

10—Triangle of male two-celled; base of wings with large brown spot; one row of cells between Cu1 and Cu2............................ **Basiaeschna**
—Triangle of male three-celled; base of wings hyaline; two rows of cells between Cu1 and Cu2.................................. **Oplonaeschna**

Nymphs

1—Hind angles of head angulate.. 2
—Hind angles of head rounded or in **Aeschna eremita** sometimes slightly angulate 6

2—Head flattened and elongate-rectangular seen from above; low, broadly rounded eyes directed laterally........................... **Coryphaeschna**
—Head roughly trapezoidal, strongly narrowed to rearward.................. 3

TABLE OF GENERA

ADULTS

Genera	*Hind wing*	*Fork*[1] *of Rs*	*Rpl*[2] *rows*	*Anal loop* *Rows*[3]	*Anal loop* *Cells*[4]	*Cells in ♂ triangle*	*♀No. of spines*[5]	*Distr.*
Aeschna	34-52	2,*3*,4	3-4	2-3	7-14	2-3	∞	G
Anax	45-67	2	6-7	3	8-17	...	∞	G
Basiaeschna	32-42	0	2-3	2	5-8	2	∞	E
Boyeria	39-46	2-3±	1-2	3	6-14	3-5	∞	E
Coryphaeschna	42-59	2	4-6	3	7-13	2	∞	S
Epiaeschna	52-59	4	2-2+	2	5-8	3	10±	E
Gomphaeschna	29-36	0	1	2	2-5	2	∞	E
Gynacantha	42-56	3-4	4-6	3	10-19	3	2	S
Nasiaeschna	45-50	3-4	1	3	7-9	3	6±	E
Oplonaeschna	45-55	0	3	2+	8-10	3	∞	SW
Triacanthagyna	34-47	3-4	4-5	3	6-10	3	3	S

[1]Number of included rows of cells. Prevailing number italicized.

[2]Number of rows of cells subtended by radial planate.

[3]Number of vertical rows of cells.

[4]Number of included cells.

[5]Number of spines or prickles on swollen edge of ventral side of segment 10 — ♀ (∞ = more than 10 prickles).

3—Blade of lateral lobe of labium wide and squarely truncated on outer end **Boyeria**
—Blade of lateral lobe narrowed toward tip and with stronger end hook...... 4

4—Dorsum of abdomen broadly rounded.......................... **Basiaeschna**
—Dorsum of abdomen with low median ridge............................ 5

5—Blunt dorsal hooks on median ridge.......................... **Nasiaeschna**
—No dorsal hooks on median ridge.............................. **Epiaeschna**

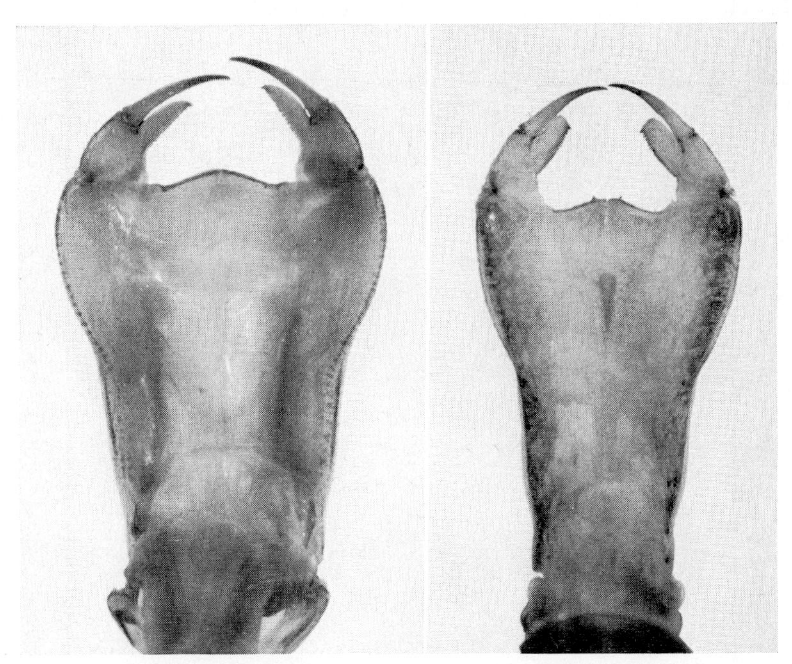

Fig. 157. Labia of nymphs of *Basiaeschna janata* and *Boyeria vinosa*.

6—Lateral spines on segments 7 to 9 only (also on 7 to 9 in **Aeschna sitchensis**).. 7
—Lateral spines on 5 or 6 to 9 (on 6 to 9 in most species of **Aeschna**, where spine on 6 may be minute; on 5 to 9 in **A. eremita**).................. 8

7—Antenna longer than distance from its base to rear of head........ **Gomphaeschna**
—Antenna about half as long as this distance.......................... **Anax**

8—Raptorial setae on lateral lobes of labium............................ 9
—No raptorial setae 10

9—Lateral setae of labium nearly uniform in length; total length of nymph less than 40 mm.................................... **Triacanthagyna**
—Lateral setae more numerous, less robust, very unequal in length, diminishing to very small ones at proximal end of row; total length of nymph more than 40 mm. .... **Gynacantha**

10—Lateral spines well developed on 5 to 9...................... **Oplonaeschna**
—Lateral spines on 6 or 7 to 9 (on 5 to 9 in **A. eremita**)................ **Aeschna**

TABLE OF GENERA—NYMPHS

Genera	Total length	Lateral lobe		Lat. spines on abd. segs.	Caudal append.[3]			
		Blade[1]	Teeth[2]		Lat.	Sup.	Inf.	Tip[4]
Aeschna	30-48	variable	small	5, 6, 7, 8, 9	5-7	7-8	10	cleft
Anax	40-62	variable	small	7, 8, 9	5-7	9	10	cleft
Basiaeschna	29-35	narrow, pointed	10± low	5, 6, 7, 8, 9	4	6	10	cleft
Boyeria	34-39	truncate	small	5, 6, 7, 8, 9	5	10	10	var.
Coryphaeschna	41-65	roundly pointed	very low	7, 8, 9	9	10	10	blunt
Epiaeschna	47-60	narrow, truncate	large	5, 6, 7, 8, 9	7	9+	10	blunt
Gomphaeschna	30	truncate	large	7, 8, 9	9	10	10	blunt
Gynacantha	42	squarely truncate	20± small	±6, 7, 8, 9	9	10	10	blunt
Nasiaeschna	54	narrow	15± small	5, 6, 7, 8, 9	4	10	10	blunt
Oplonaeschna	35	broadly truncate	15-18 small	5, 6, 7, 8, 9	5	9	10	var.
Triacanthagyna	37	squarely truncate	35± small	6, 7, 8, 9	10	10	10	blunt

[1]The part of lateral lobe that extends beyond base of its movable hook.

[2]Teeth on inner edge of blade.

[3]Relative length of caudal appendages, lateral, superior, and inferior, expressed in tenths of length of inferiors taken as 10.

[4]Tip of superior appendage.

Genus GOMPHAESCHNA Selys 1871

These clear-winged dragonflies have slender reddish legs and wide-meshed wing venation. The T spot on the frons is mostly stem and base, the top bar being short and diffuse. Before the T the frons is transversely wrinkled, and the face greenish or brownish, faintly streaked with paler. The pattern in dried specimens is usually obscure.

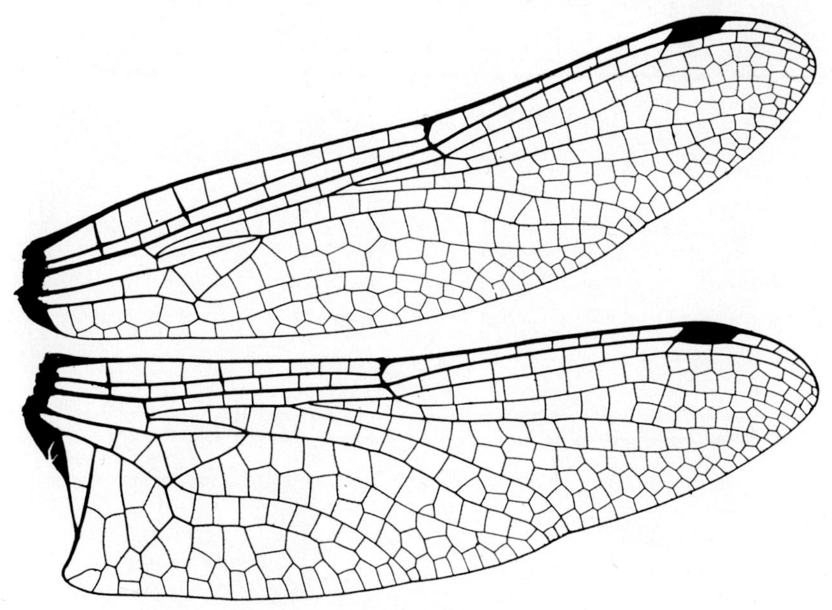

Fig. 158. *Gomphaeschna furcillata.*

The abdomen of the male has a slight waistlike constriction of segment 3, beyond which the long segments widen slightly to rearward. The abdomen in the female is shorter and stouter, and regularly tapers to rearward beyond its swollen basal segments. The caudal appendages in the male are long and thin, about as long as segments 9 and 10 together; in the female, short and blunt, hardly as long as 9 alone; in both, flattened and widened from the base outward. The inferior appendage of the male is deeply forked. The ovipositor of the female is very short, ending in a truncate tip under the middle of 10; it bears a pair of rather long and slender bristle-tipped palps.

The venation is greatly reduced as compared with our other representatives of the Aeschnidae. There are no crossveins in the supertriangle, and often there is but one in the triangle and but one beneath the stigma beyond the brace vein. The triangle is nearly twice as long as wide in the axis of the wing. The gaff is nearly or quite as long as the inner side of the hind-wing triangle. The large paranal cells of the fore wing are five or six, accompanied by a shorter row of smaller marginals. The tallest marginal is only about a third as high as the paranal cell above it. In the hind wing there are generally two paranal cells before

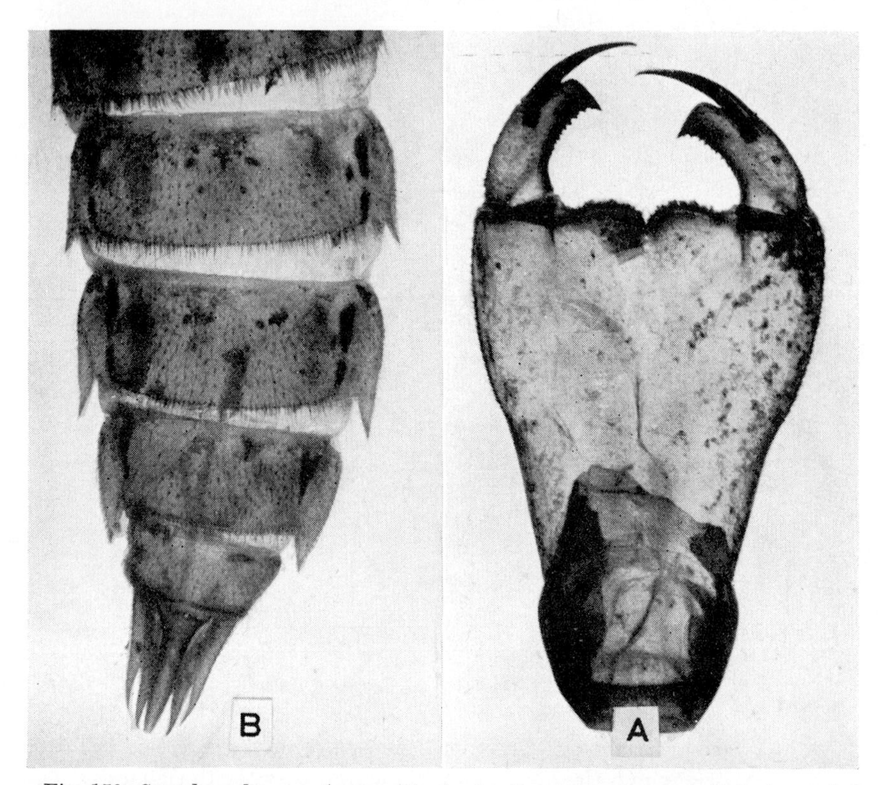

Fig. 159. *Gomphaeschna* sp. (supposition), showing parts of a cast-off last-nymphal skin. A, labial mentum (with a bit of submentum attached); B, end of abdomen, showing lateral spines and caudal appendages. (Photographs by Dr. Robert S. Hodges.)

the anal loop and two within it, sometimes three before the loop in the female.

The nymph appears not to have been reared beyond the third instar (Kennedy, 1936). However, a cast skin of a female specimen, obtained by Mr. Ernest Martin near New Orleans and photographed by Dr. R. S. Hodges of the University of Alabama, seems to be safely referable by exclusion to *Gomphaeschna,* since the nymphs of all other regional genera of Aeschnidae are known, and none of them is like it.

It is distinguished by the following characters. The mentum of the labium is widened forward to be more than twice as wide at the distal end as at the basal hinge. Its middle lobe is deeply notched, the end roundly bilobed at the sides of the open V-shaped notch, and fringed with short scurfy hairs. Each lateral lobe is rather slender, its blade truncate on the end, with the truncation rounded a little externally and prolonged internally to form a short sharp straight end hook; its inner margin is beset by about ten denticles. The movable hook is long and strong, tapered to a slowly incurving tip. The hind angles of the head are rounded. The legs are pale, with faint rings of darker color near the knees. The length of the nymph is about 30 mm.

The abdomen is widest on segment 6, with strong outstanding lateral spines on 7 to 9 only. These spines are of nearly equal size.

The proportionate length of the three terminal abdominal segments is about as 12:10:7, with the caudal appendages 12 on the same scale. Other characters are as in the table of genera for nymphs.

KEY TO THE SPECIES

Adults

1—Hind wing at level of nodus not wider than distance on costa from nodus to stigma in fore wing; two rows of cells for a distance in each of paired-vein interspaces, M2–Rs and M3–M4; more than one bridge crossvein present; distance between bases of superior caudal appendages of male, viewed from above, about twice as great as breadth of one of them....**furcillata**

—Hind wing at level of nodus as wide as, or wider than, distance on costa from nodus to stigma in fore wing; a single row of cells between paired veins, with sometimes a few cells doubled near wing margin; a single bridge crossvein generally present; distance between bases of superior appendages much less, about as great as width of one of them......**antilope**

Gomphaeschna antilope Hagen

1874. Hagen, Proc. Boston Soc. Nat. Hist., 16:354 (in *Aeschna*).
1940. Gloyd, Occ. Pap. Mus. Zool. Univ. Mich., 415:2 (figs.).

Length 52–60 mm.; abdomen 38–45; hind wing 30–36.

Similar to following species, lighter in color, with broader wings that have fewer crossveins. Blackish areas less extended; T spot on frons covers less than half its upper surface. Occiput and costal margin of wings yellowish. Generally but a single bridge crossvein. Thickened antenodals often first and third or fourth.

Sides of middle abdominal segments less extensively infuscated; pale areas more extensive than dark ones. Superior appendage of male, when viewed from side, shows no tooth projecting downward at base of terminal enlargement. Inferior appendage less deeply and widely forked.

Distribution and dates.—UNITED STATES: D. C., Fla., Ga., La., Md., Miss., N. J., N. C., Ohio, Pa., S. C., Va.

March 3 (Fla.) to July 10 (Pa.).

Gomphaeschna furcillata Say

Syn.: quadrifida Rambur

1839. Say, J. Acad. Phila., 8:14 (in *Aeschna*).
1861. Hagen, Syn. Neur. N. Amer., p. 131 (in *Aeschna*).
1929. N. & H., Handb., p. 125.
1936. Kndy., Proc. Ind. Acad. Sci., 45:315 (figs., eggs and nymph).
1940. Gloyd, Occ. Pap. Mus. Zool. Univ. Mich., 415:1–13 (fig.).

Length 53–60 mm.; abdomen 39–45; hind wing 29–36.

Face obscure dull greenish brown, paler on fronto-clypeal suture and across labrum. A black crossband envelops vertex, and extends forward to cover most of frons in an ill-defined T spot with broadly spreading base. Tip of vertex paler. Occiput olivaceous brown.

Thorax greenish under thin cover of grayish hairs. On its front is a pair of parallel stripes of cream color; higher up on front above them, a pair of short dashes extends in line crosswise. These paler markings sometimes wholly darkened in dried specimens. Sides of thorax greenish above, blackish below in streaks that extend upward from leg bases into paler areas. Membrane of wings in female sometimes diffusely tinged with brown.

Abdomen mainly blackish, with paler streakings; more blackish than paler on sides of middle segments, where pattern is best preserved. Caudal appendages blackish; superiors, when viewed from side, show an inferior projecting angle at a third of their length. Inferior appendage deeply and widely forked.

Distribution and dates.—CANADA: N. S.; UNITED STATES: Ala., Ark., Conn., Fla., Ga., Ky., Maine, Md., Mass., Mich., Miss., N. H., N. J., N. Y., N. C., Pa., Vt.

January 8 (Fla.) to August (Mich.).

Genus BASIAESCHNA Selys 1883

This is a brownish darner of rather slender stature, with two conspicuous parallel yellowish stripes on each side of the thorax. There is a brown twin-spot at the base of each of the narrow wings, beyond which the membrane is generally hyaline in the male and tinged more or less deeply with amber brown in the female. The eyes are smaller than in most members of this family, and the eye-seam in which they fuse on top of the head is shorter. There is a slight constriction of segment 3 of the abdomen in the male; none in the stouter abdomen of the female.

The venation is as shown in the table of genera, with triangles shorter than in other genera, except *Gomphaeschna*. The stigma is narrow. The triangle in the male is generally two-celled. The abdomen is thinly hairy on the two basal segments, both above and below; elsewhere smooth.

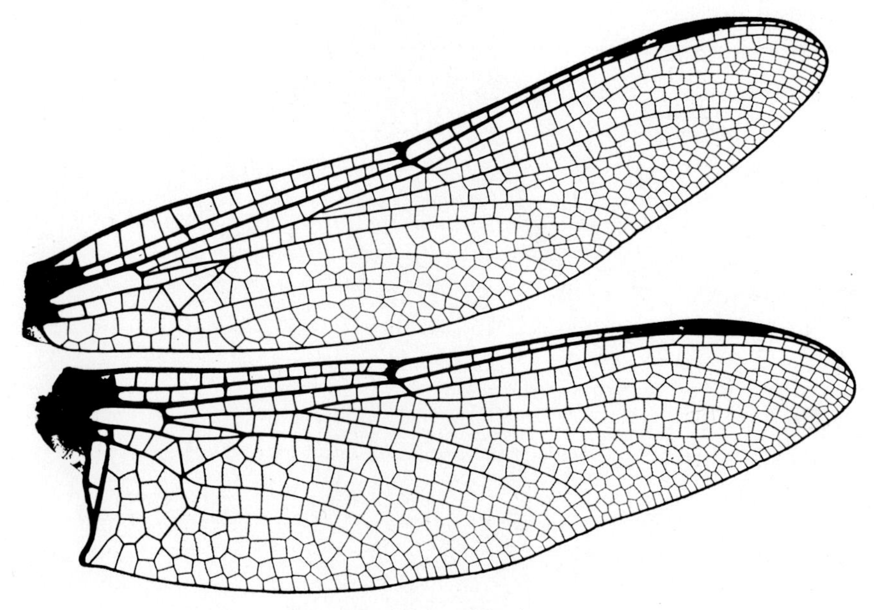

Fig. 160. *Basiaeschna janata.*

The large, half-conic, appressed yellow auricles of the male are armed each with about four rather large, strong, sharply incurved denticles. Segment 10 in the female is not longer on the ventral side than on the dorsal; it is not produced to rearward on its lower apical rim, but only thickened there and beset with minute prickles. The caudal appendages are about as long as 9 and 10 together in both sexes, but they are much more slender in the female.

The adults haunt the riffles in small woodland streams. The females fly low over the surface of the water, and settle on floating dead leaves and trash to deposit their eggs, placing them in regular punctures in soft decaying tissues. Dr. Helen E. Murphy reports capturing with her bare hands a female that she found busily ovipositing in a floating dead cattail (*Typha*) stem.

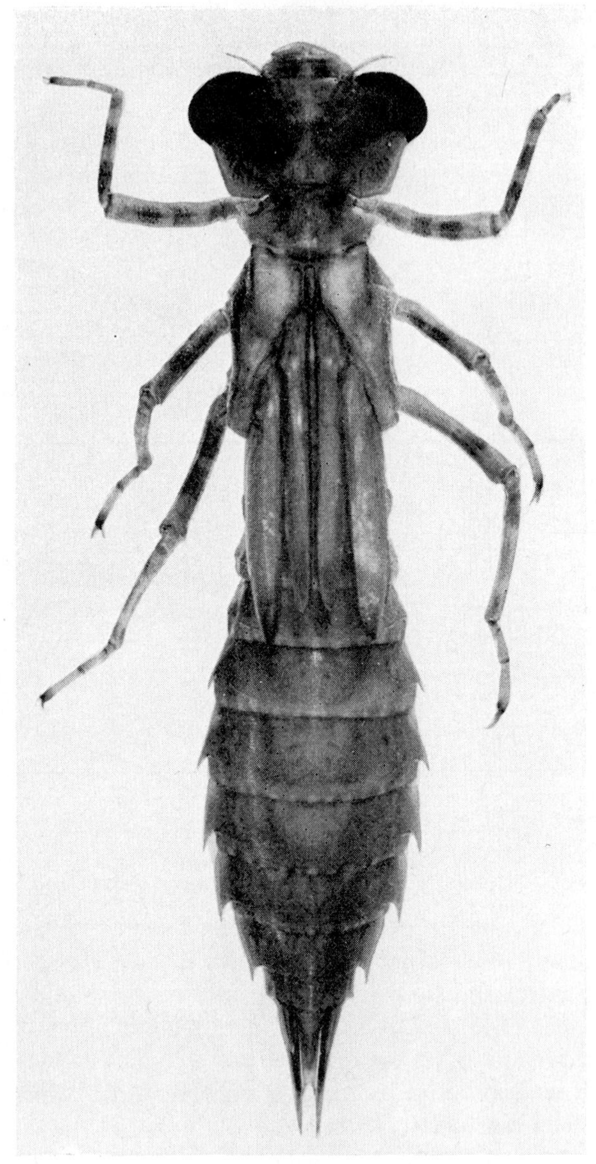

Fig. 161. *Basiaeschna janata.*

The senior author made the following observations on the oviposition of *B. janata* at Saranac Inn, New York:

The eggs were deposited in leaves of a species of bur-reed, *Sparganium*, which, where it grew in the deeper water of the creek, trailed its long leaves on the surface of the stream. The female flitted from plant to plant, making a few thrusts with her ovipositor into each at the water line, and then settled and balanced herself carefully on a long, floating leaf; this was doubtless a favorable place for the eggs, and she settled down to more extensive operations. Backing down into the water till the abdomen was wholly submerged, she began thrusting with her ovipositor, first to right and then to left, moving forward a little between thrusts, leaving behind a double row of egg punctures, as regular as the neatest double stitching that might be done with a needle. Several such double rows of eggs were placed in the tissues of this leaf before she left it.

The nymph is slender, widest across segment 7 of the lanceolate abdomen. The inferior appendages are as long as the last three abdominal segments. The body is neatly patterned in green and brown, dotted with darker brown, and ringed with paler across the joinings of the abdominal segments. The shortness of the superior caudal appendage and the slenderness of the inferior beyond its deeply notched tip are distinctive.

The nymphs are commonly found in the shelter of submerged dead sticks and rubbish stranded in the edges of slow streams.

Basiaeschna janata Say

Syn.: minor Rambur

1839. Say, J. Acad. Phila., p. 13 (in *Aeshna*).
1895. Klct., Odon. of Ohio, p. 81.
1899. Wmsn., Odon. of Ind., p. 301.
1901. Ndm., Bull. N. Y. State Mus., 47:466.
1929. N. & H., Handb., p. 126.

Length of male 50–60 mm.; abdomen 38–41; hind wing 32–36.
Length of female 55–67 mm.; abdomen 43–51; hind wing 36–42.

A brownish species, daintily marked with green, and with conspicuous yellowish or creamy white stripes on sides of thorax. Face greenish, becoming obscured with brown in old specimens, thinly besprinkled with short black hairs. Rear half of frons greenish or yellowish, divided on middorsal line by stem of a diffuse brown T spot. Eyes above narrowly bordered with yellow within their shining black rims. Occiput brown, becoming yellow toward crest.

Two faint pale stripes on front of thorax, often wholly obscured in old specimens. Two on each side conspicuous, being margined with black. Legs brown, paler basally. Brown spots at wing bases extend outward to first thickened antenodal crossvein.

Abdomen brown, fenestrate with usual lines of paired pale spots, which are bluish in life, but subject to darkening and more or less complete disappearance in museum specimens. All carinae black.

Distribution and dates.—CANADA: N. B., N. S., Ont., Que.; UNITED STATES: Ala., Conn., Fla., Ill., Ind., Iowa, Ky., Maine, Md., Mass., Mich., N. H., N. J., N. Y., N. C., Ohio, Okla., Pa., S. C., Tenn., Tex., Vt., Va.

March 25 (Fla.) to September.

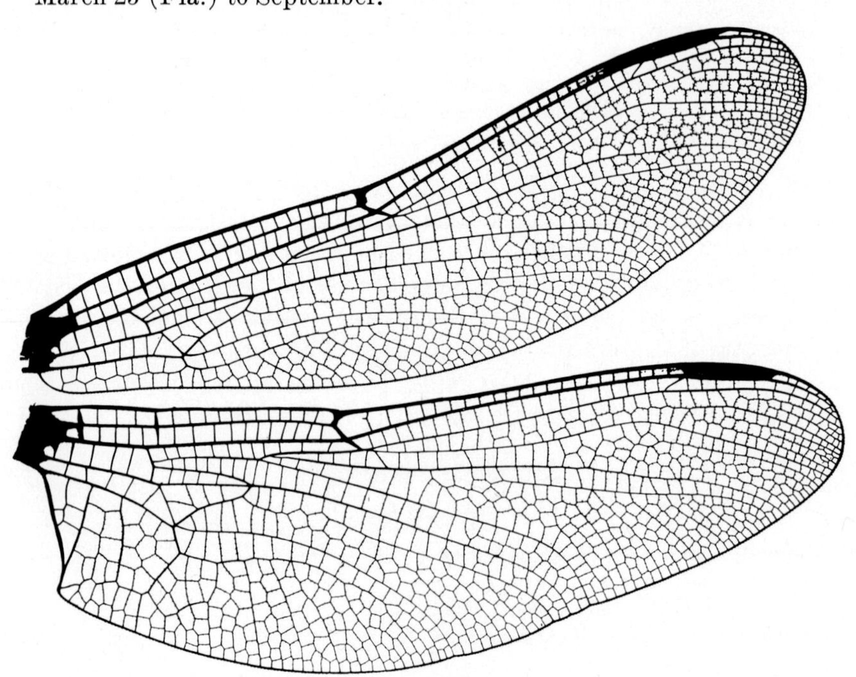

Fig. 162. *Boyeria vinosa.*

Genus BOYERIA McLachlan 1896

Syn.: Fonscolombia Selys

These are large brownish darners, with two big pale spots on each side of the thorax by which they may be recognized even in flight. They are inhabitants of woodland streams, where the adults fly near the water, especially on sunny afternoons. They glide along a little above the shining surface of the water on well-poised, transparent, well-nigh invisible wings. They haunt the shadows in the edges of the woods, and when discovered they are not hard to capture with a net.

A structural character in which this genus differs from all other American genera is the presence of crossveins in the midbasal space of

Fig. 163. *Boyeria grafiana.*

the wings. In the hind wing, anal and plantar loops are well developed, and the median planate runs out straight and strong (not zigzagged) to the margin of the wing.

The nymphs are to be found under stones and heavier trash in the edges of slow-flowing woodland streams. They are blackish, and the legs are triple-ringed with dull yellow and have pale edgings and claws. A middorsal pale line on the abdomen widens into a rather conspicuous spot on segment 8. The tip of the superior caudal appendage is usually, but not always, divided by a narrow median cleft. Transformation occurs about a foot above the edge of the water.

There are four known species: two occur in the eastern United States, one in Europe, and one in Japan.

For a full account of structure and habits of our species, see the paper by E. B. Williamson, first cited below.

KEY TO THE SPECIES

Adults

1—Wings hyaline; side spots of thorax bluish; abdominal segments 9 and 10 similar in coloration, dark .... **grafiana**

—Wings with brown spots at base of each; side spots of thorax yellowish; 10 much more yellow than 9 .... **vinosa**

Nymphs

1—Length when grown 37–39 mm.; lateral spines on abdominal segments 4 to 9, on 4 minute; mentum of labium more than twice as long as its median width; inferior caudal appendages stouter, scarcely incurved at tips; apex of superior anal appendage uncleft and sharply pointed; superior as long as inferiors .... **grafiana**

—Length when grown 34–37 mm.; lateral spines on 5 to 9; mentum less than twice as long as its median width; inferior appendages more slender, distinctly incurved at tips; apex of superior anal appendage deeply emarginate (cleft); superior distinctly shorter than inferiors .... **vinosa**

Boyeria grafiana Williamson

1907. Wmsn., Ent. News, 18:1 (figs.).
1913. Walk., Can. Ent., 45:163 (figs., nymph).
1927. Garm., Odon. of Conn., p. 172.
1929. N. & H., Handb., p. 128.

Length 63–65 mm.; abdomen 46–51; hind wing 40–43.

A more robust species, blackish, marked with bluish; face green; only a trace of brown in wing roots, and sometimes a diffuse tinge of it in membrane beyond stigma; costa greenish, stigma pale; inferior appendage of male brown. Other distinctive characters as given in key.

Distribution and dates.—CANADA: N. B., N. S., Ont., Que.; UNITED STATES: Ky., Mass., N. Y., N. C., Ohio, Pa., Tenn., Va.

June 29 (Ont.) to October 8 (Tenn.).

Boyeria vinosa Say

1839. Say, J. Acad. Phila., 8:13 (in *Aeshna*).
1893. Calv., Odon. of Phila., p. 247.
1901. Ndm., Bull. N. Y. State Mus., 47:465 (nymph).
1901. Ndm. & Hart, Bull. Ill. State Lab., 6:36 (fig., nymph).
1913. Walk., Can. Ent., 45:163 (figs., nymph).
1929. N. & H., Handb., p. 127.

Length 60–71 mm.; abdomen 45–56; hind wing 39–46.

A more common species, brownish, with a tinge of fulvous in both body and wings; legs pale; face green, washed with brownish; costa pale, stigma yellowish; inferior appendage of male yellowish. See also the key.

Distribution and dates.—CANADA: N. S., Ont., Que.; UNITED STATES: Ala., Ark., Conn., D. C., Fla., Ga., Ill., Ind., Iowa, Ky., La., Maine, Md., Mass., Mich., Miss., Mo., N. H., N. J., N. Y., N. C., Ohio, Okla., Pa., S. C., Tenn., Tex., Vt., Va., W. Va.

May 9 (New England) to October 31 (Ala.).

Genus ANAX Leach 1815

This is a genus of large, strong-flying dragonflies. The males are unique among our Aeschnines in lacking auricles on abdominal segment 2, and the adjacent wing margin is rounded off, not notched or stiffened with an anal triangle. Some abdominal segments, at least 7 to 10, bear supplementary lateral carinae. The inferior abdominal appendage of the male is quadrangular. The distal border of segment 10 of the female is rounded below and armed with prickles, but does not form a strong projecting shelf as in some other genera.

The venation is rather primitive. The triangles are much elongated in the axis of the wings, as in most genera of Aeschnidae. The stigma is long and narrow. M2 makes a *sharp* upward bend toward M1 at the level of the distal end of the stigma. The several supplementary sectors arising from the posterior side of Rs are strongly developed. The cells subtended by Rpl are not well settled into definite rows. The lower end of the anal loop is often rather indefinite, and sometimes the loop appears not to be closed there.

The nymphs climb actively in submerged vegetation. Their long smooth bodies are decorated with a protective pattern of greens and browns. The eyes cover most of the sides of the head, but are not strongly outstanding, giving the head a distinctive appearance easily recognized when one becomes familiar with it. There are lateral spines only on segments 7 to 9; however, Calvert (1934) reports very small ones occasionally on segment 6 in *A. junius*. There are no dorsal hooks. The basal tubercle of the superior appendage of the male is truncate or concave at the apex. The inferior appendages are longer than segments 9 and 10 together. The ovipositor of the female nymph does not reach the hind margin of 9.

This genus is cosmopolitan, containing some twenty-five species, of which four are found within our limits. One of these, *A. junius,* is very widely distributed; the other three are more restricted in their distribution.

Calvert (1934) published the results of a comprehensive study of the nymphs of the genus *Anax,* discussing in particular the rate of growth and larval development.

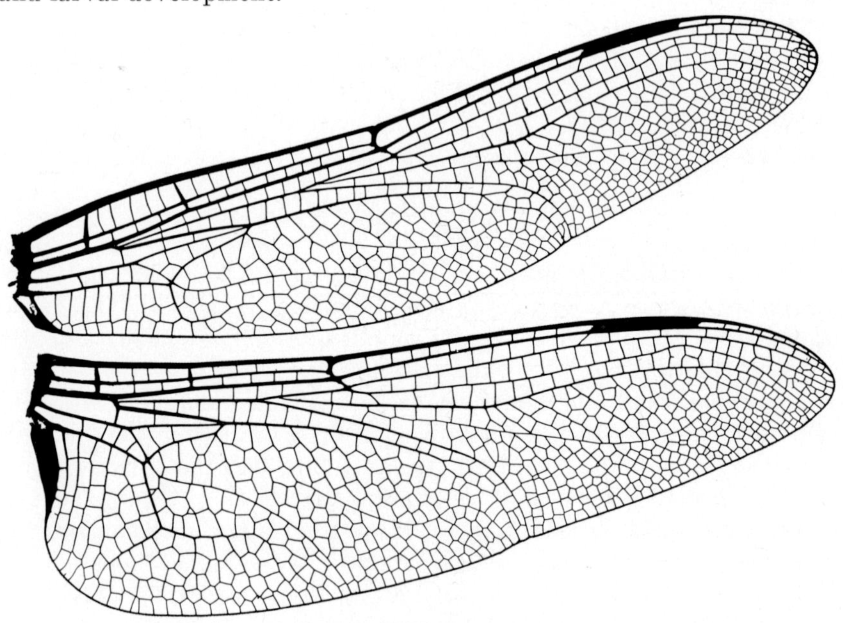

Fig. 164. *Anax junius.*

KEY TO THE SPECIES

Adults

1—Frons with no dark markings; occiput black . . . . **longipes**
—Frons with dark markings above; occiput yellowish 2

2—Frons above with a more or less triangular central spot of black or brown, usually with triangular spot of blue on each side . . . . **amazili**
—Frons above with circular central spot surrounded anteriorly at a short distance by semicircle of blue 3

3—Smaller species (abdomen 47–58 mm.; hind wing 45–56 mm.); superior appendage of male not bifid in profile view; occiput of female with two blunt teeth on hind margin . . . . **junius**
—Larger species (abdomen 67–90 mm.; hind wing 56–67 mm.); superior appendage of male bifid in profile view; occiput of female without blunt teeth on hind margin; Southwestern . . . . **walsinghami**

Nymphs

1—Lateral lobes of labium tapering to hooked point . . . . **junius**
—Lateral lobes squarely truncate, a little rounded on superior angle 2

Fig. 165. *Anax junius.*

2—Labium exceedingly long, four times as long as width at base, its middle hinge when appressed to under side of body reaching abdominal segment 1; hind femora long, one and a half times as long as width of head. . . **longipes**

—Labium shorter, its length not more than three times width at base; hind femora as long as width of head or less. 3

3—Labium longer, reaching posterior border of hind coxae when appressed; movable hooks with row of heavy setae; median cleft of labium deep, reaching beyond level of chord of arc formed by mentum. . . . . **walsinghami**

—Labium shorter and wider, scarcely reaching anterior border of hind coxae when appressed; movable hooks with row of very fine almost invisible setae; median cleft of labium not so deep, not reaching level of chord of arc formed by mentum .... **amazili**

Anax amazili Burmeister

Syn.: maculatus Rambur

1839. Burm., Handb., p. 841 (in *Aeschna*).
1861. Hagen, Syn. Neur. N. Amer., p. 119.
1908. Mrtn., Coll. Selys Aeschnines, p. 13 (figs.).
1927. Byers, J. N. Y. Ent. Soc., 35:67 (fig., nymph).
1929. N. & H., Handb., p. 129.

Length 70–75 mm.; abdomen 48–54; hind wing 49–52.

A greenish species with black legs. Face greenish yellow. Labrum margined with black. Frons above with a triangular black or brown spot bounded by yellow and usually with a triangular blue spot on each side. This blue spot may be almost brown, and central spot more circular, especially in females. Ocular ridge brown; occiput brown. Thorax bright green. Legs dark brown to black, anterior femora paler beneath. Costa greenish; stigma brown.

Abdomen much inflated at base. Segments 1 and 2 greenish; 3 to 10 brownish, with a wide dorsal band of black, narrower on middle of segments, 3 to 7 with two spots of green or blue on each side; 8 to 9 with a single light spot or almost entirely black; 10 obscure. Supplementary lateral carinae on sides of 4 to 10. Abdominal appendages of male brown, superiors narrowed at base, somewhat roundly widened on inner margin at half their length, notched, and tapering to a long thin tip.

Distribution and dates.—UNITED STATES: Fla., La.; ANTILLES: Cuba, P. R.; also south to Brazil.

July 2–16 (Fla.).

Anax junius Drury

Syn.: spiniferus Rambur

1770. Drury, Ill. Exot. Ins., 1:112 (pl. 47, fig. 5) (in *Libellula*).
1861. Hagen, Syn. Neur. N. Amer., p. 118.
1903. Ndm., Proc. U. S. Nat. Mus., 26:709 (figs.).
1908. Mrtn., Coll. Selys Aeschnines, p. 11 (figs.).
1927. Byers, J. N. Y. Ent. Soc., 35:6 (figs., nymph).
1927. Garm., Odon. of Conn., p. 178.
1929. N. & H., Handb., p. 129.
1948. Zimmerman, Aquatic Insects of Hawaii, 2:327 (figs.).

Length 68–80 mm.; abdomen 47–58; hind wing 45–56.

A large species, with green thorax and bluish abdomen. Face yellowish green. Labrum brown on distal border. Frons above with a rounded black or brown spot surrounded anteriorly by a yellow and then a blue semicircle. Ocular ridge yellow or greenish yellow dorsally. Occiput yellow. Prothorax brown to black; mesothorax and metathorax mostly green, with sutures slightly brownish; katepisterna and area near coxae brownish. Spiracle margined with black; a brown spot just above it. Legs black beyond reddish brown femora; first femora yellowish below. Wings hyaline, often tinged with amber; costa yellow, stigma slightly darker. Abdominal segments 1 and 2 greatly swollen, their diameter being two or three times that of following segments. Segment 1 mostly green, with a brown spot on each side and very hairy above; 2 mostly blue, green above as far back as first transverse carina, with a brown spot on each side; remaining segments mostly brown, sometimes reddish in females, with black spots laterally; strong supplementary carinae on 4 to 8; weaker carinae on 9 and 10, those on first segments being short; apex of 10 yellowish. Inferior appendage of male short, broad, with short tubercles on apex and dorsal surface; superiors brown, about as long as 9 plus 10, with a sharp pointed tooth at outer apical angle and a rounded protuberance at inner apical angle; also a dense brush of hairs on inner margin near apex.

Distribution and dates.—Alaska; CANADA: B. C., Man., N. S., Ont., P. E. I., Que., Sask.; UNITED STATES: Ala., Ariz., Calif., Colo., Conn., Fla., Ga., Ill., Ind., Iowa, Kans., Ky., La., Maine, Md., Mass., Mich., Minn., Miss., Mo., Nebr., Nev., N. H., N. J., N. Y., N. C., Ohio, Okla., Oreg., Pa., R. I., S. C., Tenn., Tex., Utah, Vt., Va., Wash., Wis.; MEXICO: Baja Calif., Chihuahua, Coahuila, Tamaulipas; ANTILLES: Cuba, Dom. Rep., Haiti, Jamaica, P. R.; also from Hawaiian Islands and western coast of Asia.

Probably found in every state of the United States.

The idea that it is widespread in Canada is far from being the case. It is essentially Austral and in Canada its numbers rapidly fall off in the Transition. In the Canadian Zone it occurs only as an occasional straggler. Its habit of migrating northward in the spring is perhaps responsible for the occasional occurrence of individuals far north of their usual territory.—Walker (1927, p. 10).

Year-round in South, taken every month in Florida. In Canada, Whitehouse reports migrants in April, with normal flight from first week in August to second week in October.

The paper by Calvert (Proc. Amer. Phil. Soc., 73:1–70, 1934) contains an excellent account of the rate of growth and development of the nymph of this species which he reared from egg to adult.

Anax longipes Hagen

Var.: concolor Brauer

1861. Hagen, Syn. Neur. N. Amer., p. 118.
1905. Calv., B. C. A., p. 176 (figs.).
1908. Mrtn., Coll. Selys Aeschnines, p. 12 (figs.).
1929. N. & H., Handb., p. 129.

Length 75–87 mm.; abdomen 50–60; hind wing 47–56.

A large species, with green thorax and reddish abdomen.* Face and rear of head green in life, fading to yellowish in some dried specimens. Ocular ridge brown. Occiput dark brown to black. Thorax almost entirely greenish in life except for brownish katepisterna and leg bases. Legs very long, hind femora measuring as much as 18 mm. in some specimens. Legs black beyond reddish femora; under side of first femora lighter. Wings hyaline, occasionally slightly tinged with amber; costa greenish, stigma tawny; wings quite narrow. Abdominal segments 1 and 2 greenish; remainder reddish, including appendages. Supplementary lateral carinae on sides of 4 to 10, with a slight ridge on distal end of 3 in male. Inferior appendage of male short and broad, about one-third as long as superiors, which are armed on outer apical angle with a very minute tooth. A dense brush of hairs on inner margin of superiors.

Distribution and dates.—UNITED STATES: Ala., Fla., Ga., Ind., Md., Mass., Miss., N. J., N. Y., N. C., Ohio, Pa., S. C.; ANTILLES: Haiti, Jamaica; also from Bahamas, and from Mexico south to Brazil.

February (Fla.) to September 24 (Fla.).

Anax walsinghami McLachlan

1882. McL., Ent. Mon. Mag., 20:127.
1895. Calv., Proc. Calf. Acad. Sci., (2)4:511 (figs.).
1908. Mrtn., Coll. Selys Aeschnines, p. 14 (figs.).
1927. Byers, J. N. Y. Ent. Soc., 35:66 (figs., nymph).
1929. N. & H., Handb., p. 130.

Length of male 100–116 mm.; abdomen 74–90; hind wing 56–67.

Length of female 88–98 mm.; abdomen 67–75; hind wing 56–60.

Largest dragonfly found in our limits. Face greenish yellow; labrum dark-margined distally. Top of frons with a brown spot surrounded anteriorly by a yellow and then a blue semicircle. Ocellar ridge yellowish to dark brown. Occiput brown. Prothorax dark. Thorax greenish, except

* "And it was *Anax longipes* . . . Heavenly Day, isn't he a beautiful thing on the wing! With that emerald green of the thorax and blood red of the abdomen, and that striking flash of white from the base of the abdomen."—Extract from a letter from Elsie Broughton Klots.

at bases of legs, where it is brown. Legs mostly black; femora brown; under side of front femora yellow. Costa yellow; stigma brown.

Abdomen with segment 1 reddish brown above, blue on sides; 2 chiefly blue; 3 to 8 mostly brown, with blue markings, usually an elongate lateral spot near base and a small distal one; 9 and 10 with a blue spot on each side, sometimes absent on 9; 10 with a median and two lateral keels on dorsal surface. Supplementary lateral carinae on 7 to 10. Robust appendages of male brown, inferior paler and about half as long as superiors. In female, blue markings of abdomen are replaced by green.

A. walsinghami, though largest species, has most sparse venation. Number of crossveins in triangles fewer, loop cells and cell rows subtended by Rpl not so numerous, and venation more open throughout wings. In this species, cells in anal loop often have only two rows with one interpolated central cell.

Distribution and dates.—UNITED STATES: Ariz., Calif., Tex., Utah; MEXICO: Baja Calif.; also south to Guatemala.

April (Baja Calif.) to October (Baja Calif.).

Genus OPLONAESCHNA Selys 1883

These are large clear-winged dragonflies, brown varied with blue and black. The face is rather broadly rounded, the frons with neither cross ridge nor frontal furrow, but with a low area in front of the very large middle ocellus.

The wings are broad, with wide-meshed network of veins. Vein Rs is unbranched. Vein M4 is separated by a single row of cells from M3 for most of its length; then it sags, widening that interspace to two rows thereafter all the way to the wing margin. Both radial and median planates subtend three rows of cells. The intercubital space includes two rows of cells for a short distance and is thereafter narrowed in the hind wing by an upbend of Cu2 to include but a single cell row to the wing margin. The anal loop is strongly widened to rearward, with a supplemental roundish plantar loop behind it and a narrow patellar loop of larger cells along its inner side. The anal triangle in the male is three-celled.

The abdomen is stout at the base, constricted on segment 3, and nearly parallel-sided thereafter. The rather large auricles of the male are straight-edged and armed with a row of about six denticles. On the dorsum of 10 of the male is a large and conspicuous prominence shaped like a finger tip, blunt, polished, and pointed straight to rearward, continuing the skyline of the abdomen; the tip of 10 behind it slopes downward.

The nymphs are dark-colored and hairless, with roughly granulate skin. The body is mottled with dark green and dull brown; the head a little paler, with narrow incomplete rings of black partly surrounding each of the ocelli. The legs are ringed with light and dark color bands about half and half. The largest of the nymphs that are available to us for description (apparently in the antepenultimate stage) are 32 mm.

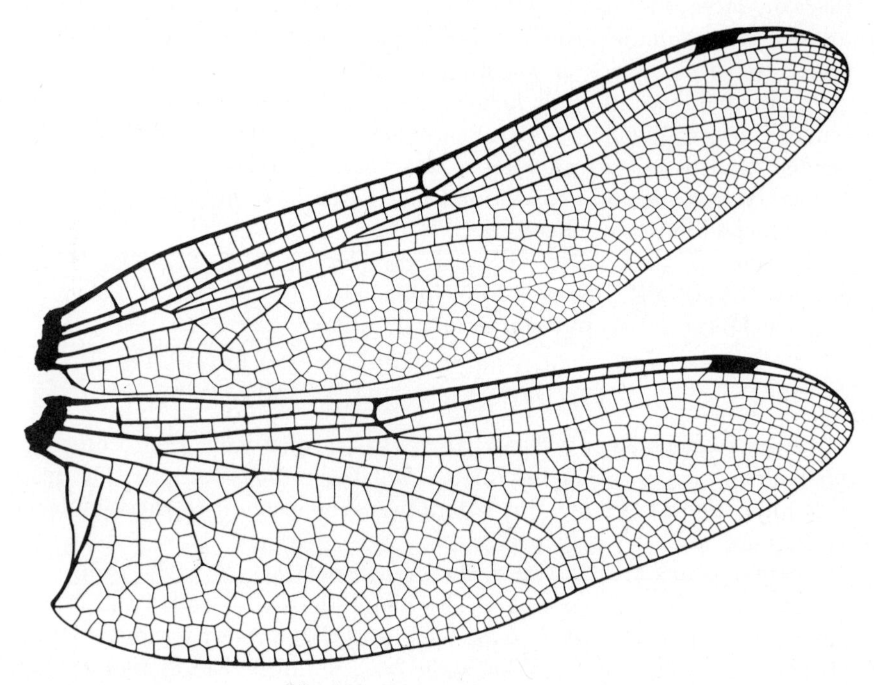

Fig. 166. *Oplonaeschna armata.*

long, with wings reaching backward only to the level of the middle of abdominal segment 2.

The head is very wide across the front of the eyes and suddenly narrowed behind them to broadly rounded and rough-edged hind angles. The mentum of the labium is widened rather regularly forward from the middle hinge almost to the base of the lateral lobes. The blade of the lateral lobe is broadly and squarely truncate at its end, with a low blunt tooth at the inner angle. This tooth is preceded on the inner margin by a row of about eighteen minute flat-topped denticles. The median lobe is slightly produced forward, with a narrow closed cleft in the middle, on each side of which the straight front edge is fringed with a line of short scurfy hairs.

Fig. 167. *Oplonaeschna armata,* young nymph.

The abdomen is lanceolate in outline, widest just behind the middle, where it is scarcely wider than the head. The dorsum is smoothly rounded. There are lateral spines on segments 5 to 9: on 5 small, on 6 to 9 subequal in length but markedly increasing in stoutness to rearward. The abdominal appendages are very sharp-edged. The superior is a little shorter than the inferiors; its tip seems to vary from an acute point to a very narrow cleft with sharp points beside the cleft. The inferiors are strongly incurved at the tips, their upper (supero-internal) edge roughly serrulate by a marginal line of close-set blackish denticles.

The nymphs are found in residual pools of rocky canyon streams that are subject to flash floods. At lowest water stages they are found clinging to the under sides of stones or stranded slabs of bark, often at the edges of pools exposed to air, but where the air is always saturated with moisture.*

These nymphs bear all the usual marks of a lotic environment—adaptations to withstanding the impact of swift waters: thick skin with rough surface, compact form, depressed and more or less wedge-shaped front of head, thickened front edge of labrum, and dark coloration.

Oplonaeschna armata Hagen

1861. Hagen, Syn. Neur. N. Amer., p. 124 (in *Aeschna*).
1883. Selys, Bull. Acad. Belg., (3)5:735 (reprint, p. 27).
1903. Ndm., Proc. U. S. Nat. Mus., 26:735 (fig.).
1905. Calv., B. C. A., p. 195 (figs.).
1929. N. & H., Handb., p. 130.
1949. Tinkham, Ent. News, 60:40.

Length 66–75 mm.; abdomen 49–59; hind wing 45–55.

A handsome brownish species, marked with porcelain blue, varied with tan and yellow. Face blue; in fully adult male, tinged with gold above and below, and marked with black as follows: a line on fronto-clypeal suture; two deeply impressed dots on postclypeus, at both front and rear margins of labrum; and a triangular patch at top of face. This patch spreads upward over a wrinkled frontal carina and forms head of a black **T** spot that covers about half of upper surface of frons. Sides of frons of a brighter blue. Sutures of top of head broadly black, fringed with erect black hairs; top of vertex and most of occiput blue

Thorax brown, with more or less interrupted blue stripes: two on front and two on each side. Hairy beneath and low down at front and sides, with a notably long and dense fringe of erect tawny hairs on hind lobe of prothorax. High middorsal carina thin and pale. Blue stripes on front when complete form a pair of opposed **L**-shaped marks; when interrupted, inturned upper ends of these marks become isolated triangular spots. Anterior side stripe sinuous and irregular, posterior one straight-edged. All these paler markings may disappear in old and faded specimens.

* We are under deep obligation to Dr. E. R. Tinkham, who made a special trip to Ramsay Canyon in the Huachuca Mountains of Arizona to get for us the nymphs from which the preceding description is drawn. Dr. Tinkham had previously (1940) obtained two nymphs and reared them; but the reared specimens were not available at our time of need.

Legs blackish, paler on sides of femora and on middle of tarsi and claws. Wings have yellow costa, a tawny stigma, and brown veins. Radial sector (Rs) and radial planate (Rpl) strongly convergent at both ends and separated by three rows of cells. End of vein M4 strongly undulate. Gaff about as long as inner side of hind-wing triangle. Anal loop strongly widened to rearward; plantar loop weakly developed. Anal triangle of male three-celled.

Abdomen moderately enlarged at base, constricted sharply on segment 3, and slowly widened thereafter. Segments 1 and 2 thinly hairy above; auricles on 2 stout and edged with about nine serrately arranged denticles. Segment 2 mainly blackish above, with two black-edged streaks of paler on each side converging toward hind border of segment; 3 mostly pale out to girdle, where it is narrowest, then brown with a medially interrupted crossband of paler bordering girdle, two pairs of pale spots farther to rearward, and a long pale streak low down on lateral margin; 4 to 9 similar, increasingly overspread with brown, the repeated pale spots smaller, pale streak of side margin often interrupted. In male, obscurely colored segment 10 markedly longer on dorsal side than on ventral; prolonged in a rounded lobe between superior appendages; just before base of this lobe it bears a thumblike median process about as long as lobe and inclined to rearward above it. Superior appendages of male black, inferior paler. Female unknown to us.

Distribution and dates.—UNITED STATES: Ariz.; also from Mexico south to Guatemala.

June 23 (Ariz.).

Genus CORYPHAESCHNA Williamson 1903

This is a genus of large, in some species very large, Neotropical dragonflies. They are swift-flying rovers of the upper air. The eye-seam is very long; the occiput very short as seen from above, and deeply recessed between the greatly swollen eyes. The abdomen is very long and slender beyond the swollen basal segments.

The wings are hyaline, with open-meshed venation. The triangles are very long; the front side is more than twice as long as the inner side, and there is only one cell next to the inner side, the several crossveins in it being parallel. There is a well-developed trigonal planate. The high-arching fork of the radial sector is far out in the wing, with its base at or near the level of the stigma. The fork is widest before the middle and narrows slightly beyond; in it are generally two rows of cells. The paranal cells of the fore wing are unusually large and regular, with the small marginals behind them grading upward distally in a regular series to hardly half of their height. Vein M4 in the hind wing is so switched on thickened crossveins as to appear to end in vein R3. In the female, abdominal segment 10 is a little longer on its ventral margin, where it is thickly beset with short coarse prickles.

The nymphs are distinguished by the great flattening of the head to rearward, and by having its hind margin cut off squarely; also by the slenderness of the lateral spines on abdominal segments 7 to 9, the trun-

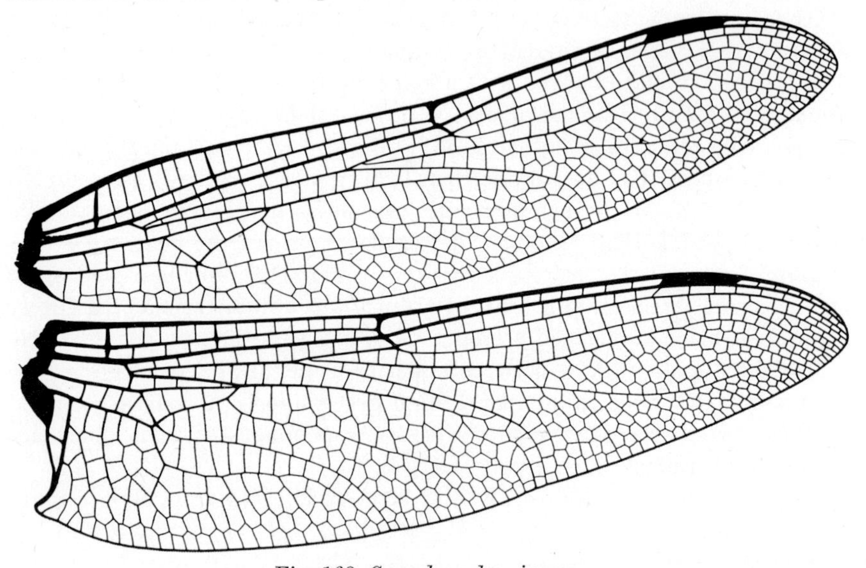

Fig. 168. *Coryphaeschna ingens.*

cate tip of the superior appendage, and the length of the lateral appendages, almost as long as the inferiors. So far as is known, they are found living in tangles of waterweeds. Four of the ten or more known species occur within our limits.

A most useful summary of the present knowledge of our species, both nymphs and adults, is that of Geijskes (1943), cited below.

KEY TO THE SPECIES

Adults

1—Caudal appendages of male simple in form; those of female longer than in male 2

—Caudal appendages of male with quadrangular notch on inner margin; those of female shorter than in male, not more than 2 mm.; Mexico, Baja California . . . . **luteipennis**

2—Face blue; hind wing shorter than 45 mm. . . . . **adnexa**

—Face green; hind wing longer than 50 mm.; radial planate subtends four or more rows of cells 3

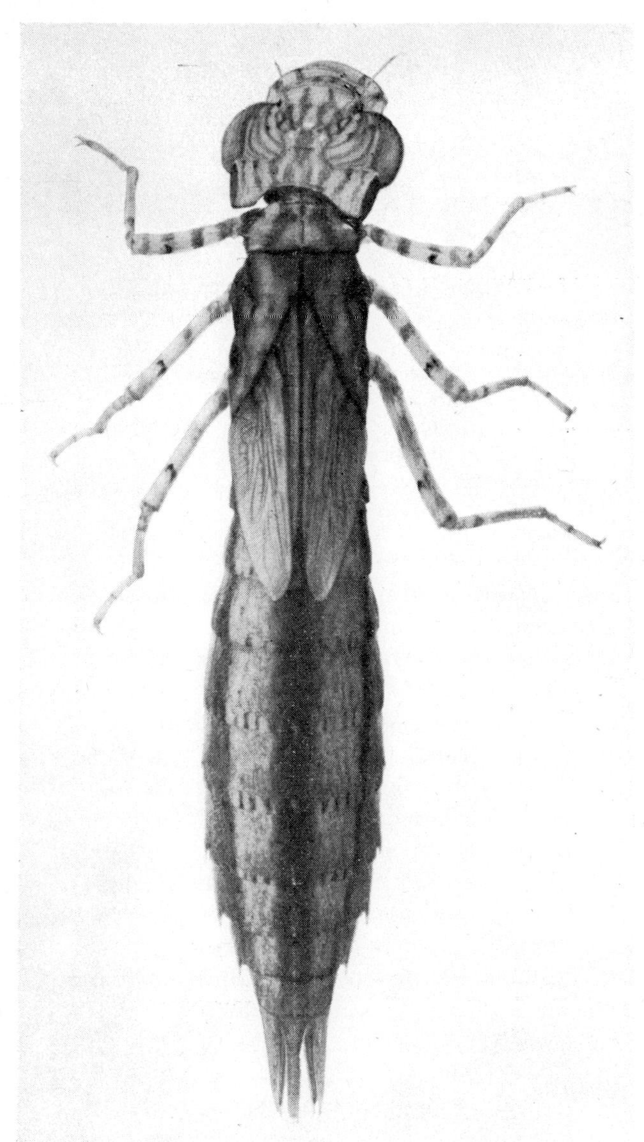

Fig. 169. *Coryphaeschna ingens.*

3—Thoracic brown stripes on green thorax wide; radial planate subtends four or five rows of cells; superior appendages of male 8 mm. long; inferior half as long, and all black . . . . **ingens**

—Thoracic brown stripes narrow; radial planate subtends six or more rows of cells; superior appendages of male 6 mm. long; inferior three-fifths as long, its basal half yellow . . . . **virens**

NYMPHS

1—Median lobe of labium with a pair of long slender spines beside its median cleft 2
—Median lobe with only a pair of short blunt tubercles beside its median cleft **adnexa**
2—Length of body 47–49 mm. .... **virens**
—Length of body 62–65 mm. .... **ingens**

Nymph unknown: **luteipennis.**

Coryphaeschna adnexa Hagen

Syn.: macromia Brauer

1861. Hagen, Syn. Neur. N. Amer., p. 127 (in *Aeshna*).
1905. Calv., B. C. A., p. 188 (figs.).
1932. Klots, Odon. of P. R., p. 20.
1938. Garcia, J. Agr. Univ. Puerto Rico, 22:75 (figs., nymph).
1943. Geijskes, Ent. News, 54:61 (figs.).
1943. Whts., Bull. Inst. Jamaica, Sci. Ser., 3:9.

Length 66–69 mm.; abdomen 48–52; hind wing 42–45.

A wide-ranging Neotropical species that enters our southernmost limits. Face bright porcelain blue, sometimes with a pale streak down middle line of anteclypeus. Fronto-clypeal suture narrowly black. Heavy T spot above overspreads ridge of frons halfway down to postclypeus. Front margin of labrum fawn color. Vertical tubercle bright green; occiput yellow and brown, half and half.

Thorax green, with brown sutures, and a black carina on front; venter brown, clothed with brown hair. Legs black. Wings hyaline, with more or less tinting of amber in membrane, especially in females.

Abdomen blackish, narrowly marked with green except on segments 1, 2, and 7, where green prevails. Caudal appendages black. Auricles of male on segment 2 armed each with two denticles.

Distribution and dates—MEXICO: Tamaulipas; ANTILLES: Cuba, Dom. Rep., Haiti, Jamaica, P. R.; also south to Brazil.

August (Cuba) to April 13 (Jamaica).

Coryphaeschna ingens Rambur

Syn.: abboti Hagen

The Sky Pilot

1842. Rbr., Ins. Neur., p. 192 (in *Aeshna*).
1861. Hagen, Syn. Neur. N. Amer., p. 127.
1905. Calv., B. C. A., p. 187 (figs.).
1929. N. & H., Handb., p. 131.
1943. Geijskes, Ent. News, 54:61.

Length 86–99 mm.; abdomen 65–77 (incl. app.) ♂ 8, ♀ 13; hind wing 54–59.

A very large species; a high flyer, very difficult to capture. Face green, with heavy black T spot on frons. Thorax green, with wide brown stripes: two wide stripes of brown in front, and others on lateral sutures.

Abdomen streaked about equally with green and brown on swollen basal segments, beyond which it is blackish, with narrower green markings, ranged in series of three above and three below on sides of middle segments. To rearward, second and third markings tend to be connected along lower margin. Segment 10 blackish, with two dull greenish spots on dorsum. Auricles on segment 2 of male armed each with three denticles.

Distribution and dates.—UNITED STATES: Ala., Fla., Ga., La., Miss., N. C., S. C.; ANTILLES: Cuba; also south to Panama.

February 12 (Fla.) to October 3 (Fla.).

Coryphaeschna luteipennis Burmeister

Syn.: excisa Brauer; var. florida Hagen, peninsularis Calvert

1839. Burm., Handb., p. 837 (in *Aeshna*).
1908. Mrtn., Coll. Selys Aeschnines, p. 73 (figs.).
1941. Calv., Ann. Ent. Soc. Amer., 34:389 (figs.).
1943. Geijskes, Ent. News, 54:63.

Length 78–80 mm.; abdomen 59–61; hind wing 45–49.

A little-known Mexican species that may occur across Mexican border in Arizona.* Face pale blue, with a blackish line on fronto-clypeal suture. T spot variable. Eye-seam long, twice as long as occiput.

Thorax bluish green, with black carinae, and has in front a pair of very wide blackish stripes that are divergent downward. Crest green, its bounding carinae black. On the green sides, blackish stripes in first and third lateral sutures. Legs black, lightening to reddish brown on femora. Wings tinged with clay yellow in membrane.

Distribution and dates.—UNITED STATES: Ariz. (?); MEXICO: Baja Calif., Nuevo León; also south to Brazil.

June 22 (Nuevo León) to October 7 (Baja Calif.)

* Three nymphs of a species of *Coryphaeschna* that differ from the three hereinbefore described were collected by the senior author at Arivaca, Arizona, in 1937. They may well be one of the variants (*florida* Hagen or *peninsularis* Calvert) of *luteipennis*. These appear to be distinguished by scarcely discernible differences in the sex characters and in these alone; coloration is the same in all. They are discussed in the paper by Calvert cited above.

Coryphaeschna virens Rambur

1842. Rbr., Ins. Neur., p. 193 (in *Aeshna*).
1861. Hagen, Syn. Neur. N. Amer., p. 127.
1905. Calv., B. C. A., p. 187 (figs.).
1929. N. & H., Handb., p. 132.
1943. Geijskes, Ent. News, 54:61 (figs., nymph).

Length 76–85 mm.; abdomen 57–60; hind wing 50–59.

Another very large species, similar to *C. ingens* but smaller. Face green, also top of vertex. Black T spot on frons more slender and less heavily black. Green areas of thorax more extensive, brown stripes narrower on abdomen; black rings on joinings of segments more conspicuous, but general color pattern similar. Auricles on abdominal segment 2 of male each armed with three denticles; basal half of inferior caudal appendage yellowish.

In both *virens* and *ingens*, girdle groove on abdominal segments 2 to 7 nearer base of segments than in *adnexa*.

Distribution and dates.—UNITED STATES: Fla.: MEXICO: Tamaulipas; ANTILLES: Cuba, Dom. Rep., Haiti, Jamaica; also south to Brazil and Bolivia.

Apparently year-round; Whitehouse gives November to March for Jamaica; other dates are June and July from Tamaulipas, also August 2.

Genus NASIAESCHNA Selys 1900

These long, slender, narrow-winged darners are well marked by the prominence of the frons, which doubtless suggested the generic name. The wings are of rather open-meshed venation, the interspaces between veins filled with cells in notably regular rows. The anal triangle in the male is wide; the tornus, low; the second anal interspace (x) widens greatly to the rear margin. The trigonal planate is well developed. It runs outward one cell row below the median planate. The paranal cells of the fore wing are but little larger than their accompanying marginals.

The abdomen is long and tapers regularly from segment 2 to the tip, without the usual constriction on 3. The caudal appendages of the male are about twice as long as those of the female.

The nymphs inhabit woodland streams, and are to be found clinging to dead wood and sticks and among fallen leaves and trash along the edges of the current. They are thick-skinned, bare, and brown or blackish. The head is wide in front and tapers abruptly to rearward, behind the eyes, with a single low tubercle above each hind angle. The mentum of the labium is long-stalked, rather narrow and parallel-sided in its basal half, and suddenly and widely expanded beyond. The sharp lateral spines on abdominal segments 5 to 9 are minute on 5 and increase regularly in size to rearward. Other characters are shown in our table for genera of nymphs of Aeschnidae (p. 256).

The nymphs are distinguishable from all our other darner nymphs by the low dorsal hooks on several abdominal segments, and by two pairs of rather prominent rounded tubercles on top of the head.

The single known species of *Nasiaeschna* is North American.

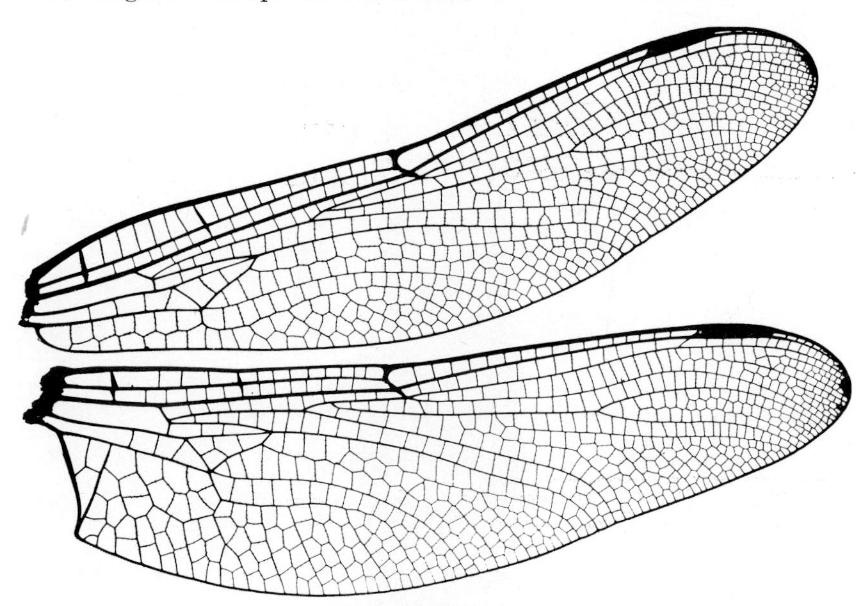

Fig. 170. *Nasiaeschna pentacantha.*

Nasiaeschna pentacantha Rambur

1842. Rbr., Ins. Neur., p. 208 (in *Aeshna*).
1861. Hagen, Syn. Neur. N. Amer., p. 129 (in *Aeschna*).
1901. Ndm. & Hart, Bull. Ill. State Lab., 6:33 (fig., nymph).
1929. N. & H., Handb., p. 132.

Length of males 70–73 mm.; abdomen 51–55; hind wing 45–48.

Length of females 62–67 mm.; abdomen 47–49; hind wing 46–50.

Face of this darner flat and bare below cross ridge at top of high frons. A black band across front of labrum, and narrow black lines in all sutures of face. A short cross streak of brown on postclypeus close to interclypeal suture. Sides of face sloping back to eyes are bluish or tan, thinly besprinkled with black hairs. On outer side of mandible are

Fig. 171. *Nasiaeschna pentacantha.*

several serrately arranged prickles. Top of frons blue, with no black T spot, but with a diffuse blackish band at its confluence with vertex. Occiput brownish, with a pair of little spots of blue in female.

Rather small thorax, brown, brightly marked with blue. Two blue stripes of its front abbreviated below and widened above at crest on inner side. Top of crest blue within its surrounding black carinae; other lesser spots of blue about wing roots. Three more or less confluent ill-defined streaks of blue on convexities of sides of thorax. Venation of wings given in table on page 254.

Abdomen long and thin, tapering rather regularly from segment 2 to end; blue and brown, with a middorsal black stripe interrupted only on joinings of segments. Widely interrupted girdle and accompanying cross ring of brown near base of segments. Long stripes of blue of dorsum with two pairs of blackish dots on each of middle segments. Paler colors prevail on basal segments, and brown on apical ones.

Auricles on segment 2 of male yellow, with blackish edgings. Each margined with about a dozen minute denticles. Caudal appendages of male rather simple, subequal. Inferior notched at tip between upturned ends of its inrolled edges. In female, a slight scooplike prolongation of ventral margin of segment 10, armed on edge with six or more sharp thornlike spines. Caudal appendages of female about as long as segment 10; of male, about twice as long as 10.

Distribution and dates.—CANADA: Ont., Que; UNITED STATES: Ala., Conn., Fla., Ga., Ill., Ind., Iowa, La., Mass., Miss., N. H., N. Y., N. C., Ohio, Okla., S. C., Tenn., Tex.

March 2 (Fla.) to August 28 (Ind.).

Genus EPIAESCHNA Hagen 1877

These are among the largest of our darners, attaining a wing expanse of 116 mm. The head is very broad, nearly covered by the huge bulging eyes. The occiput is small. The strongly braced synthorax bulges with powerful wing muscles. The abdomen is very long. The caudal appendages are complicated and hairy on the inner side in the male; smooth and leaflike in the female. They are a little shorter in the female than in the male and also wider, and flatter; they are paddle-shaped. Segment 10 in the female is twice as long on the ventral as on the dorsal side.

Venational characters additional to those shown in our table of genera are: outer side of the triangles sinuate, with a well-developed trigonal supplement springing from near its upper end and extending obliquely downward one cell row below the median planate; anal loop relatively small, generally of five cells.

The nymph is smooth, blackish, and thick-skinned. The ample labium has a narrow-based parallel-sided mentum, which suddenly widens at two-thirds of its length to a broad dinner-plate form. The lateral lobe has a narrow blade ending in a strong, squarely inturned end hook. The abdomen has a middorsal ridge on segments 4 to 10, weakly developed on

4 to 7, conspicuously on 8 to 10 (sometimes showing a free tip to the ridge on 9, suggesting a dorsal hook).

The one known species is North American.

Epiaeschna heros Fabricius

Syn.: multicincta Say

1798. Fabr., Ent. Syst., Suppl., p. 285 (in *Aeshna*).
1861. Hagen, Syn. Neur. N. Amer., p. 128.
1881. Cabot, Mem. M. C. Z., 8:39 (figs., nymph).
1901. Ndm., Bull. N. Y. State Mus., 47:469 (fig., nymph).
1927. Garm., Odon. of Conn., p. 196 (fig.).
1929. N. & H., Handb., p. 134.

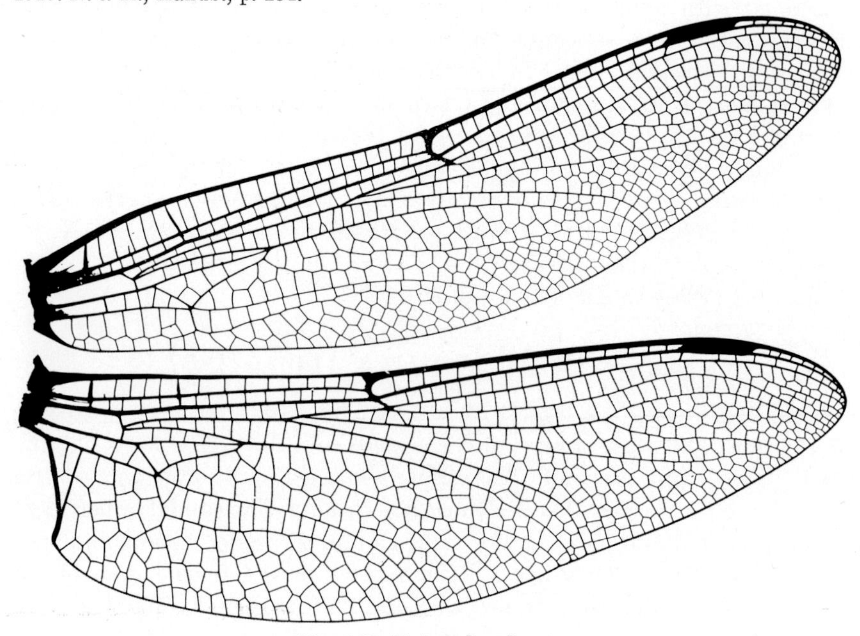

Fig. 172. *Epiaeschna heros.*

Length 82–91 mm.; abdomen 62–70 (incl. app. 8–9); hind wing 52–59.

A very large, strong-flying darner, with brown body and more or less smoky wings. Face brown, smooth, shining and somewhat bulging below well-developed transverse frontal carina. Frons black above in male; paler in female, which has a pair of included spots of green. Pale occiput enclosed at rear by raised crestlike margin of occiput; raised much higher in female than in male, and divided on median line by a deeper notch.

Thorax brown, with a pair of green or greenish stripes on front, widening upward to end just below crest. On each side, two broader oblique

Fig. 173. *Epiaeschna heros.*

green stripes; between them, an isolated triangular spot of green high up next to subalar carina. Legs black, with reddish femora. Wings usually with pronounced tinge of brown in membrane. Narrow yellowish line on costa widens beyond nodus. Stigma brown.

Abdomen much swollen at base, slightly constricted on segment 3, and nearly parallel-sided thereafter; dark brown or blackish, with long middle segments narrow and generally with obscure rings of paler at base, at girdle, and at distal end; all rings interrupted on middorsal line. Basal segments have greater, apical ones smaller, pale areas. Segments 8 and 9 nearly all black; 10 paler, with a single large triangular middorsal spine strongly flattened laterally.

Auricles on segment 2 thin and flat, with seven or eight denticles serrately arranged on rear margin. Inferior caudal appendage of male broad to apex; more than half as broad as at base, concavely truncate on end, with an upturned tooth on each outer angle. Subanal plates end in a pair of strong, converging teeth. In female, segment 10 is produced downward in a scooplike spine-margined plate that extends to rearward over tip of ovipositor.

Distribution and dates.—CANADA: Ont., Que.; UNITED STATES: Ala., Conn., Del., Fla., Ga., Ill., Ind., Ky., La., Maine, Md., Mass., Mich., Miss., Mo., N. H., N. J., N. Y., N. C., Ohio, Okla., Pa., R. I., S. C., Tenn., Tex., Va., Wis.; also from Mexico.

March 7 (Fla.) to October 1 (Ind.).

Genus AESCHNA Fabricius 1775

These are the dominant blue darners of the Northern Hemisphere. Their colors are blue and brown: brown striped with blue, or sometimes green, on the thorax, and spotted with blue on the abdomen, with here and there small touches of yellow. The light colors are generally very bright in life, but unfortunately they fade and are usually badly dulled in preserved specimens. The main features of the color pattern are a pale face, a black T spot on the blue upper surface of the frons, two pale stripes on the front and two on each side of the synthorax, and two rows of blue spots on each side of the abdomen. When a full count of these abdominal spots is present on the middle segments, there are in each row basal, middle, and apical spots, the middle one being alongside the oblique girdle groove. In the darker species some of these spots may be reduced or wanting; in the paler species the spots may become elongated streaks, tending to run together, especially low down on the sides, along the carina of the lateral margin. The joinings of the segments are black.

The wing venation is as shown in the table of species. Its main features are the arched and unsymmetrical fork of the radial sector (Rs), its base well before the level of the stigma; the long sag in the radial planate; and the short sag in vein M4 toward its outer end, where it moves away from M3 and then back into line again.

In the female the lower margin of abdominal segment 10 is only a little produced and thickened, and is beset with minute prickles.

This is the dominant genus of Aeschnidae in North America. Its species outnumber those of all our other Aeschnid genera put together. In structure it represents the flowering of the Aeschnid type. Though a rather homogeneous group, it appears to be still in process of increase and differentiation if the existence of varieties be considered the sign of that condition. Among the species, however, are two groups that may be rec-

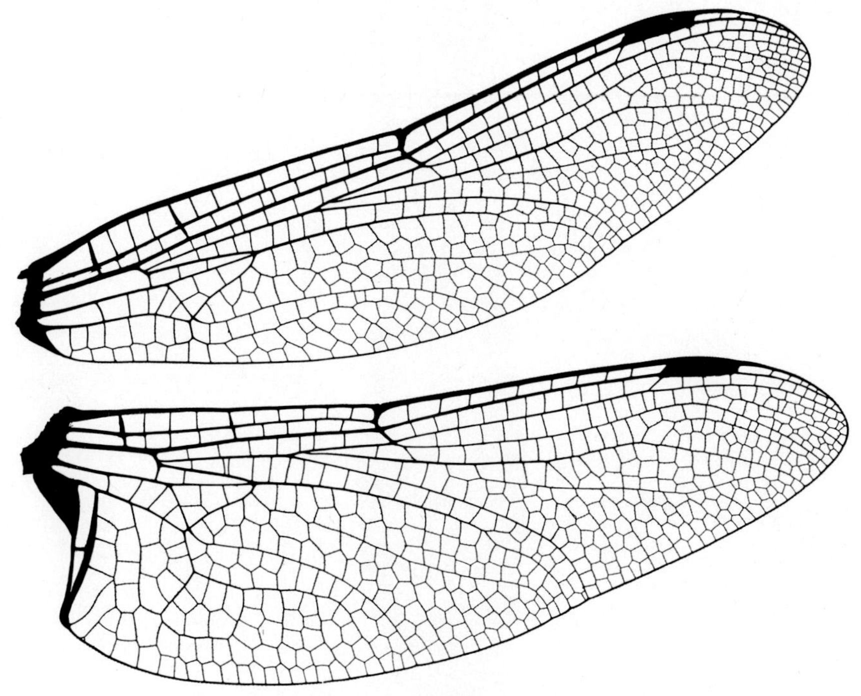

Fig. 174. *Aeschna juncea.*

ognized most easily by the conformation of the caudal appendages of the male: the *Constricta* group and the *Multicolor* group.

The *Constricta* group includes five species: *arida, constricta, palmata, umbrosa,* and *walkeri,* recognizable at a glance by the peculiar in-bent tips of the broad paddle-like superior appendages. "Paddle-tails" is a convenient name for them. Unfortunately, no easy recognition mark has yet been found for the females of the group. In the male the anal triangle is three-celled, and the sternum of abdominal segment 1 and the dorsum of 10 are smooth.

The *Multicolor* group contains three species within our limits: *dugesi, multicolor,* and *mutata,* recognizable in the males by a peculiar beaklike formation of the tip of the caudal appendages. These have a two-celled

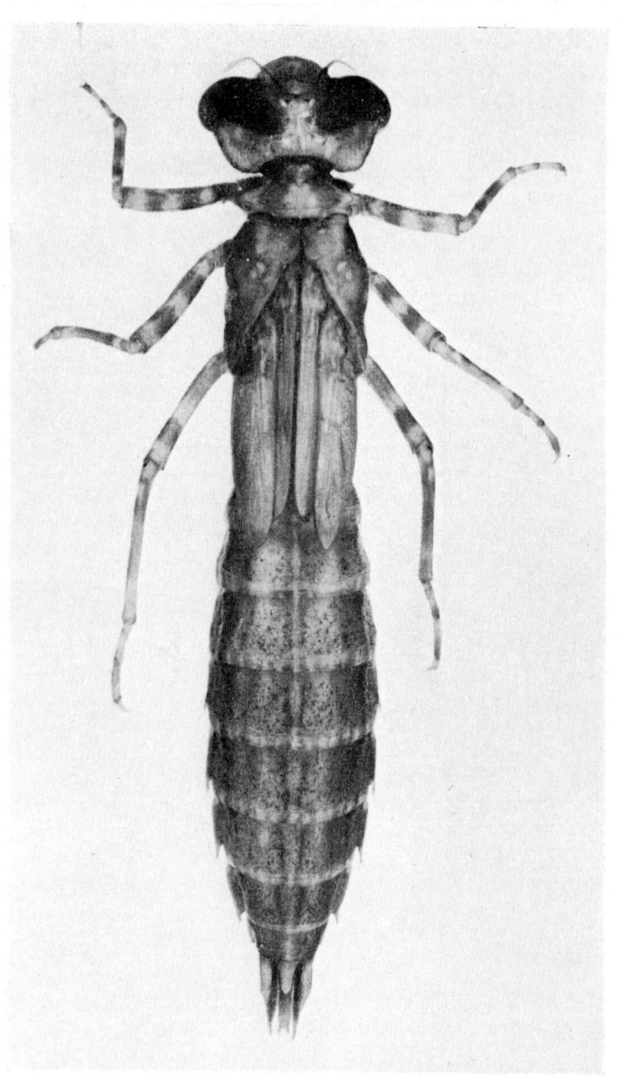

Fig. 175. *Aeschna umbrosa.*

anal triangle, a little hairy tubercle on the sternum of abdominal segment 1, and a hard black middle tubercle on the dorsum of 10.

Two small Far Southern species, *psilus* and *manni,* are distinguished by having the caudal appendages of the female longer than those of the male. They are of slender build and rather open venation, with the hairy ventral tubercle on segment 1 and the dorsal tooth on 10. Only one other of our species, *californica,* has both of these marks, and it is of a slightly different type, with its nearest related species ranging southward the length of the South American continent.

Then there are two pairs of Far Northern species: *juncea* and *subarctica,* allied in having a shorter junction of the eyes on the top of the head—a primitive character; the other two, *sitchensis* and *septentrionalis,* are allied by having the first stripe on the side of the thorax very narrow and nearly Z-shaped, also by having the plate that covers the base of the ovipositor in the female distinctly and broadly bilobed, not cut off bluntly as in all our other species of the genus. The remaining species show fewer marks of affinity.

The nymphs of *Aeschna* are among the most graceful of odonate nymphs, streamlined of body and neatly patterned in markings of green and brown that tend to run in longitudinal bands when among the green stems of water plants, in camouflage. The head is a little flattened. The legs are slender and pale, usually ornamented with rings of brown or of lighter and darker greens. The abdomen is widest in the middle and tapers gracefully to its slender tip. All are much alike in general appearance; but they differ in minor details, such as are shown in the table of nymphs which follows.

The one indispensable work for further study of this genus is Dr. E. M. Walker's fine monograph, *The North American Dragonflies of the Genus Aeshna* (pp. 1–213, 28 pls., in part colored), published by the University of Toronto in 1912.

KEY TO THE SPECIES*

Adults

MALES

1—Anal triangle of male three-celled 2
—Anal triangle of male two-celled 12

2—Bare middorsal tubercle on abdominal segment 10; hairy low midventral tubercle on 1 3
—Neither of these tubercles present 8

3—Hind wing 40 mm. or less; black cross stripe on fronto-clypeal suture . . . . **californica**
—Hind wing generally more than 40 mm.; no black cross stripe on fronto-clypeal suture 4

4—Superior caudal appendages simply carinate 5
—Superior caudal appendages with inferior carina prolonged downward into anteapical tooth 6

5—Stripes on thoracic dorsum long, reaching almost up to crest . . . . **psilus**
—These stripes short, less than 3 mm. long . . . . **manni**

6—Eastern . . . . **mutata**
—Western 7

* Key to the genera of Aeschnidae is on p. 253.

7—Anal loop with three paranal cells .... **dugesi**
—Anal loop with two paranal cells .... **multicolor**

8—Rear of head marked with yellow or brown; row of pale spots on each side of venter of segments 4, 5, and 6 .... **umbrosa**
—Rear of head black; no such spots on under side of 4, 5, and 6 9

9—Lateral thoracic pale stripes conspicuously bordered with black .... **arida**
—No such borders present 10

10—Anal loop generally with two paranal cells .... **constricta**
—Anal loop generally with three paranal cells 11

11—Face with two black cross stripes: one on fronto-clypeal suture and one on edge of labrum .... **palmata**
—Face with no black cross stripes present .... **walkeri**

12—Eye-seam short, hardly as long as occiput; first pale stripe on side of thorax narrow (less than 1 mm. wide), and bent twice almost at right angles 13
—Eye-seam longer than occiput; first pale stripe more than 1 mm. wide, or, if narrower, never so bent 14

13—Caudal appendages acute at tip; upper edge of dorsal carinae near their tips denticulate .... **sitchensis**
—Superior caudal appendages rounded at apex; upper edge of dorsal carinae smooth or nearly so .... **septentrionalis**

14—Dorsal thoracic pale stripes reduced to elongate spots, sometimes nearly wanting .... **interrupta**
—These stripes wider, generally full length 15

15—Face with black stripe on fronto-clypeal suture 16
—Face with no black stripe on fronto-clypeal suture 19

16—Large pale spots present between lateral stripes of thorax, sometimes confluent with first lateral stripe .... **clepsydra**
—These spots small or absent 17

17—Paranal cells three in anal loop and two before it .... **eremita**
—Paranal cells two in anal loop and one before it 18

18—Front edge of first lateral stripe of thorax emarginate; upper end produced in a spur to rearward .... **subarctica**
—Front edge convex; no such spur to rearward .... **juncea**

19—Segment 10 of abdomen black .... **tuberculifera**
—Segment 10 of abdomen with pale spots on dorsum 20

20—First lateral stripe of thorax blue above, green below; dorsal carina of superior abdominal appendage denticulate on its upper edge .... **canadensis**
—First lateral thoracic stripe all green; dorsal carina of appendage smooth on its upper edge .... **verticalis**

FEMALES

1—Distinct ventral tubercle on abdominal segment 1 2
—No distinct ventral tubercle on 1 7

2—Hind wing less than 40 mm.; black line generally present on fronto-clypeal suture .... **californica**
—Hind wing more than 40 mm.; no black line on fronto-clypeal suture 3

3—Stem of T spot on frons parallel-sided; abdominal appendages as long as segments 8 + 9 + 10 4
—Stem of T spot on frons distinctly widened to its base; abdominal appendages scarcely longer than 9 + 10 5

4—Top of frons with yellow continuous down its sides; stripes on thoracic dorsum long, reaching almost up to crest .... **psilus**
—Top of frons with yellow isolated, forming two spots; stripes on thoracic dorsum short, less than 3 mm. long .... **manni**

5—Eastern .... **mutata**
—Western 6

6—Anal loop with two paranal cells .... **multicolor**
—Anal loop with three paranal cells .... **dugesi**

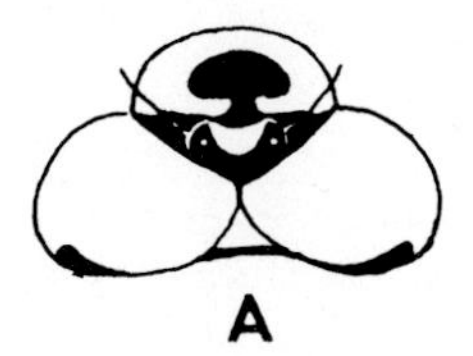

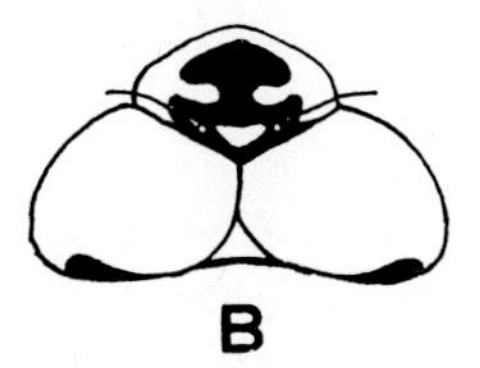

Fig. 176. Top of heads showing form of T spots. A, *Aeschna septentrionalis;* B, *A. sitchensis.* (After Walker.)

7—Eye-seam little if any longer than occiput 8
—Eye-seam distinctly longer than occiput 9

8—Top of frons as in figure 176, A .... **septentrionalis**
—Top of frons as in figure 176, B .... **sitchensis**

9—Basal plate of ovipositor distinctly bilobed 10
—Basal plate of ovipositor with edge straight or nearly so 11

10—First lateral thoracic stripe wide and straight, narrowed at top .... **juncea**
—First lateral thoracic stripe narrow and angulated, with long projection to rearward at top .... **subarctica**

11—Palp of ovipositor much shorter than dorsum of abdominal segment 10 12
—Palp of ovipositor as long as dorsum of 10 20

12—Face with black line on fronto-clypeal suture 13
—This line not present 17

13—Genital valves 2–2.5 mm. long, with a very minute terminal pencil of hairs .. 15
—Genital valves 3–3.5 mm. long, without a terminal pencil of hairs 14

14—Stigma black; caudal appendages spatulate; front margin of pale stripe behind humeral suture straight or regularly convex .... **walkeri**
—Stigma brown; caudal appendages narrowly oval; front margin of pale stripe behind humeral suture undulate .... **palmata**

15—Stripes on thoracic dorsum represented by narrow spots or wanting .. **interrupta**
—Stripes on thoracic dorsum well developed 16

16—Front of humeral suture with large triangular pale spot below .... **clepsydra**
—No such spot present .... **eremita**

TABLE OF SPECIES—ADULTS

Species	Hind wing	Paranal cells			Stripes		♂ Aur. teeth [6]	Distr.
		In z [1]	In y [2]	In x [3]	fr-cl [4]	Labr. [5]		
arida	47-52	2	*1*,2	2,3	+	0	3	SW
californica	34-40	2	1	2	+	+	2	W
canadensis	43-47	1	1,*2*	3	0	+	3	N
clepsydra	40-47	1	2	2,3	+	0±	3	NE
constricta	43-47	2	2	*2*,3	0	0	3,*4*,5	N, C
dugesi	48-52	2	2	3	0	0	2,3	SW
eremita	45-52	1	1,*2*	3	+	+	3+	N
interrupta	43-49	1	1,2	2,*3*	+	+	*3*,4	NE
juncea	39-46	1	1	2	+	+	4	N
manni	43-45	2	2	3?	0	0		SW
multicolor	42-45	2	*1*,2	2	0	+	2	W
mutata	44-51	2	*1*,2	2,3	0	+	2	NE, C
palmata	41-47	2	1	3	+	+	3,4,5	W
psilus	36-43	2	1,*2*	3	0	+	2	S
septentrionalis	34-40	1	*1*,2	3	+	+	*4*, 5	N
sitchensis	36-41	1	1,2		+	+	2,*3*	N
subarctica	39-46	1	1		+	+	3,4,5	N
tuberculifera	44-51	1	1,*2*	2,*3*	0	+	3,*4*	NE, C
umbrosa	42-47	2	2	3	0	+	4	N, E, W
verticalis	43-50	1	2	3	0	+	3	E, C
walkeri	43-48	2	2	3	0	0	3	W

[1] Number of paranal cells falling within third anal interspace (*z*); when only one, anal triangle of ♂ is two-celled; when two, is three-celled. Italics, in this and succeeding columns, indicate prevailing numbers present.

[2] Number of paranal cells falling within second anal interspace (*y*).

[3] Number of paranal cells falling within first anal interspace (*x*). These form front row of anal loop.

[4] Black cross stripe on fronto-clypeal suture: present (+) or absent (0).

[5] Black cross stripe on front border of labrum: present (+) or absent (0).

[6] Number of teeth on auricles of abdominal segment of male.

17—First lateral thoracic stripe distinctly sinuate and not margined with blackish 18
—This stripe straight and margined with blackish 19

18—Upper end of second lateral stripe with a projection to rearward .... **canadensis**
—This stripe without such projection to rearward .... **verticalis**

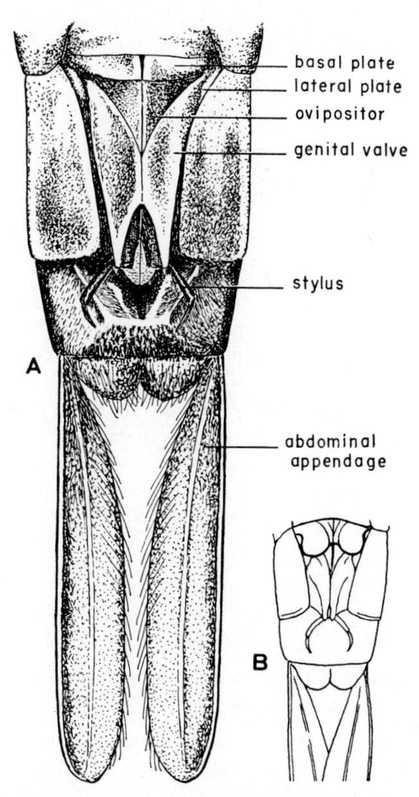

Fig. 177. Ventral view of end segments of female in *Aeschna*. A, *Aeschna clepsydra;* B, same parts in *A. subarctica* to show a bilobed basal plate that is a specific character. (Drawings by Esther Coogle; modified from Walker.)

19—Rear of head with some yellow or brown .... **umbrosa**
—Rear of head entirely black .... **arida**

20—First lateral thoracic stripe straight or nearly so .... **tuberculifera**
—This stripe distinctly sinuate .... **constricta**

Nymphs

1—Lateral spines on abdominal segments 5 to 9 (on 5 minute); hind angles of head obtusely angulate .... **eremita**
—Lateral spines well developed on 6 to 9; hind angles of head rounded 2
—Lateral spines on 7 to 9 (if present on 6, minute) 13

2—Blade of lateral lobe of labium shaped as shown in A in fig. 178........**constricta**
—Blade wider than A..3

3—Blade shaped about like B..4
—Blade wider than B..5

4—Width to length of mentum of labium about as 10:13; lateral spines of 8 about as long as those of 9....................................**canadensis**
—Width to length of mentum about as 10:14; lateral spines of 8 longer than those of 9...**clepsydra**

5—Blade about like C...6
—Blade wider than C..8

6—Blade with minute triangular tooth on innermost angle..............**verticalis**
—Blade merely sharply angulate there.......................................7

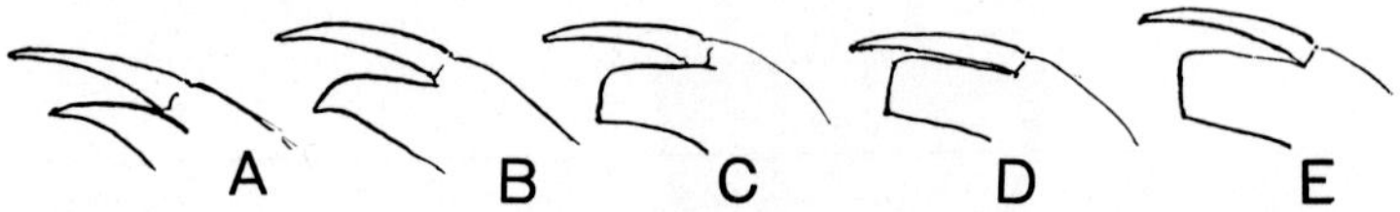

A B C D E

Fig. 178. Five types of lateral labial lobes of nymphs in genus *Aeschna*.

7—Lateral spines of 8 longer than those of 9; relative length of caudal appendages as 6:8:10..**californica**
—Lateral spines of 8 shorter than those of 9; relative length of caudal appendages as 7:8:10..**multicolor**

8—Blade about like D...9
—Blade about like E...12

9—Lateral spines of 8 longer than those of 9; width to length of mentum of labium as 10:17..**tuberculifera**
—Lateral spines of 8 shorter than those of 9; mentum much shorter.........10

10—Blade of lateral lobe of labium with minute tooth on innermost angle.......11
—Blade merely sharply angulate there..................................**walkeri**

11—Length of body 30 mm.; spine of segment 6 very small.................**psilus**
—Length of body 39–42 mm.; spine of 6 larger.......................**interrupta**

12—Width to length of mentum of labium about as 10:13; ovipositor of female about one and one-third times as long as segment 9; Western........**palmata**
—Width to length of mentum about as 10:16; ovipositor of female about one and one-tenth times as long as 9....................................**umbrosa**

13—Length of body more than 36 mm.; lateral spines of 8 shorter than those of 9; caudal appendages about as 6:8:10......................................14
—Length of body less than 35 mm.; lateral spines of 8 as long as or longer than those of 9; caudal appendages about as 5:7:10.........................15

14—Width to length of mentum of labium about as 10:13; its front border broadly rounded on each side of median cleft; blade of type C; ovipositor of female reaches tip of ninth sternite.............................**juncea**
—Width to length of mentum about as 10:12; sides of its median cleft straight, suddenly incurved at cleft; blade of type B; ovipositor one and a fifth times as long as sternum of segment 9....................**subarctica**

15—Blade of type E; lateral spines of 8 as long as those of 9; ovipositor of female about as long as ninth sternite .... **septentrionalis**

—Blade of type D; lateral spines of 8 longer than those of 9; ovipositor about one and a fifth times as long as ninth sternite .... **sitchensis**

Nymphs unknown: **arida, dugesi, manni,** and **mutata.**

TABLE OF SPECIES

NYMPHS

Species	*Total length*	*Labium*		*Lat. spines on abd. segs.*[3]
		W. x L.[1]	*Shape*[2]	
californica	33-37	10:12	C	6<7<8>9
canadensis	35-39	10:13–	B	6<7<8=9
clepsydra	36-40	10:14	B	6<7<8<9
constricta	36-38	10:12	A	6<7<8<9
eremita	41-48	10:13	C	5<6<7<8<9
interrupta	39-42	10:13	D	6<7<8<9
juncea	37-41	10:13	C	7<8<9
multicolor	35-40	10:12	C	6<7<8<9
palmata	40-41	10:13	E	6<7<8<9
psilus	30	10:12	D	6<7<8<9
septentrionalis	34	10:15+	E	6<7<8=9
sitchensis	31-33	10:13+	D	7<8<9
subarctica	40-42	10:12	B	6<7<8<9
tuberculifera	41-45	10:17	D	6<7<8>9
umbrosa	38-44	10:16	E	6<7<8<9
verticalis	36-40	10:13	C	6<7<8<9
walkeri	34-37	10:14	D	5?6<7<8<9

[1]Width (W.) to length (L.) of mentum of labium.

[2]Shape of blade of lateral lobe as viewed directly from above most nearly resembling A, B, C, D, or E of figure 178.

[3]Lateral spines present on segments indicated: (>) stands for decrease in size, (<) stands for increase in size of hooks on successive segments, (=) indicates that lateral spines of adjacent segments are equal in length.

Aeschna arida Kennedy

1918. Kndy., Can. Ent., 50:298 (figs.).
1929. N. & H., Handb., p. 144 (fig.).

Length 73 mm.; abdomen 54; hind wing 47–52.

A southwestern desert species, with black-bordered side stripes on thorax. Face greenish, with a black line on base of labrum and a narrower paler line in fronto-clypeal suture. T-shaped black spot on frons widens to its junction with black of vertex; tip of vertex yellowish; occiput pale; rear of head black.

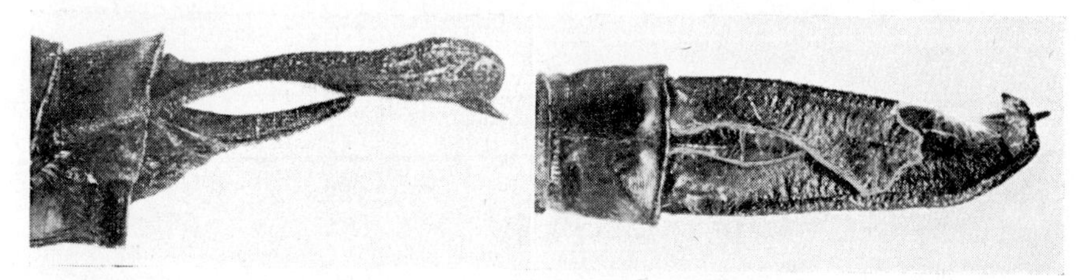

Fig. 179. *Aeschna arida*. (Photograph of type.)

Thorax brown, with its pair of dorsal dark stripes widening upward to crest. Two yellowish side stripes well developed and made more prominent by black edgings; anterior stripe slightly emarginate on front border, and narrowed above emargination; second stripe one-half wider than anterior one. Legs black, with femora dark brown. Wings hyaline, with stigma yellowish brown above and yellow beneath.

Abdomen brown, intricately spotted with blue, paler on basal segments, and a little constricted on segment 3. Female similar to male, with all paler areas more extended; caudal appendages heavy, with nearly straight upper margin. Caudal appendages of male (see fig. 179) show close affinity with others of *Constricta* group.

Distribution and dates.—UNITED STATES: Ariz., N. Mex.

August (Ariz.) to September 5 (N. Mex.).

Aeschna californica Calvert

1895. Calv., Proc. Calif. Acad. Sci., (2)4:504.
1901. Ndm., Proc. U. S. Nat. Mus., 26:736 (fig.).
1908. Mrtn., Coll. Selys Aeschnines, p. 47 (figs.).
1912. Walk., N. Amer. Aeshna, p. 184 (figs.).
1929. N. & H., Handb., p. 140 (fig.).

Length 57–61 mm.; abdomen 41–49; hind wing 34–40.

A rather small early-season Western species. Face bluish, merging into yellow on top of frons at sides of a well-defined black T spot. Stalk of

T greatly widened to its basal confluence with black of vertex. A bilobed spot of yellow covers tip of vertex. Occiput yellowish, bordered with black.

Dorsum of prothorax black, its sides paler; its hind lobe bears a dense fringe of long erect tawny hairs. Front of synthorax olivaceous, with two short darker stripes, almost lost among the hairs. Of two oblique pale stripes on each side of thorax, front one is narrowed upward toward wings and other downward toward leg bases.

Wings hyaline, with yellowish costa and antenodal crossveins. Venation scantier than in any of our other species of genus: range in nodal crossveins 12–15:7–10/8–10:9–12; in supra-triangular crossveins 0–1/0–1.

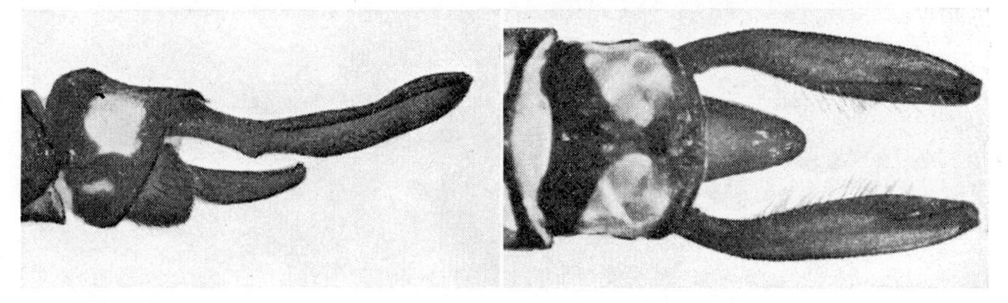

Fig. 180. *Aeschna californica.*

Four paranal cells in hind wing in male, five in female, two of these in each being within anal loop, when they are front cells of two vertical rows within loop. Anal triangle of male three-celled. Membranule black and white, about half and half.

Abdomen brown, conspicuously and copiously spotted with sky blue; all carinae black. Postero-lateral spots continue large and prominent to end segments. Basal segments moderately swollen; 3 markedly constricted. Remaining segments parallel-sided, dorsally depressed; 4 to 7 bear on each side a short curved longitudinal extra carina, line of which presents a sinuate appearance. On dorsum of 10, a low eroded median tubercle. On sternum of 1, a very low-conic median tubercle, thickly beset with minute prickles. In male, auricles on 2 rather slender; each armed with two terminal inturned teeth, inner one slightly larger. Segment 2 mostly pale, with a longitudinal black stripe through auricle in male, from which a long-stalked, Y-shaped black mark extends upward on dorsum.

Easily recognized in male by eroded tubercle on dorsum of 10 and low sagging of superior appendages.

Distribution and dates.—CANADA: B. C.; UNITED STATES: Ariz., Calif., Idaho, Nev., Oreg., Utah, Wash.

April 1 (Calif.) to July 26 (Utah).

Aeschna canadensis Walker

1908. Walk., Can. Ent., 40:384, 389.
1912. Walk., N. Amer. Aeshna, p. 135 (figs.).
1927. Garm., Odon. of Conn., p. 183 (figs.).
1929. N. & H., Handb., p. 148 (fig.).

Length 68–74 mm.; abdomen 49–57; hind wing 43–47.

A species with a greenish blue face and a wasp-waisted abdomen. A black band across front of labrum, but none on fronto-clypeal suture. Top of frons has a heavy black T spot with a thick stem to the T. Occiput yellow in middle, blackish at sides.

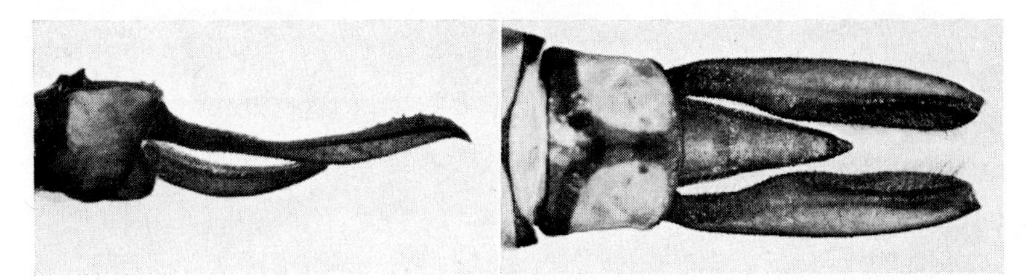

Fig. 181. *Aeschna canadensis.*

Thorax brown, striped with blue. Two stripes on front obliquely truncated below, their points strongly divergent; their upper ends widened laterally beneath crest. Two broad lateral stripes: first narrowed by a deep quadrangular excavation of its front margin and widened at top by extension to rearward; second still more widened above; between the two is often a small yellowish spot near subalar carina.

Wings hyaline, with tawny costa and stigma; cubito-anal crossveins usually 6/5. Cell rows in fork of vein Rs 3/3; in radial planate 4/4; in median planate 4/4 (sometimes 4/3). Anal triangle of male two-celled; in second anal interspace of hind wings, generally a single cell in male; two in female. Paranal cells five to seven, with three of them (rarely two) in anal loop; three vertical rows of cells in loop.

Abdomen long and slender, strongly constricted on segment 3. Abdomen brown, spotted and streaked with blue. Sides of swollen basal segments mostly brownish, with three large blue side spots in a row, middle one just above auricle in male; edges of ventral genital pocket blue, beset with minute black prickles. Long middle segments with usual rows of blue spots, becoming smaller to rearward. Segment 10 has a black basal ring that widens into a middorsal triangle, apex of which extends back to yellowish transverse apical carina of 10, separating two rather large spots of blue. Middorsal spine on 10 rises in an acute angle.

Distribution and dates.—CANADA: Alta., B. C., Man., N. B., Nfld., N. S., Ont., P. E. I., Que., Sask.; UNITED STATES: Conn., Ill., Ind., Maine, Mass., Mich., Minn., Mo., N. H., N. J., N. Y., Vt., Wash., W. Va., Wis.

June 14 (Wis.) to October 2 (N. S.).

Aeschna clepsydra Say

Syn.: arundinacea Selys, maxima Hisinger

The Spotted Blue Darner

1839. Say, J. Acad. Phila., 8:12.
1861. Hagen, Syn. Neur. N. Amer., p. 122.
1900. Wmsn., Odon. of Ind., p. 304 (figs.).
1912. Walk., N. Amer. Aeshna, p. 129 (figs.).
1927. Garm., Odon. of Conn., p. 184 (figs.).
1929. N. & H., Handb., p. 146 (fig.).

Fig. 182. *Aeschna clepsydra.*

Length 65–70 mm.; abdomen 51–54; hind wing 40–47.

Another wasp-waisted species, as the specific name indicates (*clepsydra,* an hourglass), with a cross-striped face and an exceedingly complicated pattern of stripes on thorax. Face greenish, with a black band on fronto-clypeal suture that narrows to a hairline at each end, but with no marginal black band on front of labrum; clypeus brownish; occiput yellow, with blackish lateral angles. Black hairs of face longer at sides. Vertex blackish. Heavy black T spot on top of frons has a thick stem that widens a little toward its base.

Thorax complexly patterned in brown and blue, clothed with white hair about leg bases below and at junction with abdomen above. Front of thorax brownish, with usual two greenish blue bands divergent downward and dilated laterally beneath crest. Sides brown, fenestrate with irregular bluish stripes and spots, among which may be recognized usual pair of oblique lateral stripes, both very large and irregular, and two other smaller pale areas between them. Legs brown, paler externally. Wings hyaline, with tawny costa and stigma. Generally six cubito-anal cross-veins in fore wing, four of them before subtriangle and one in it. In hind wing, five or six paranal cells: one in triangle of male, two in second anal interspace (*y*), and two or three in anal loop. Loop widens to rearward, and is composed of two or three vertical rows of cells.

Abdomen patterned in brown and blue, with black joinings to all segments. Rather slender, strongly constricted on segment 3, swollen as usual on 1 and 2. Sides of 2 largely brownish, with two large pale areas above line of auricles in male. Auricles generally three-toothed. Area below auricles black. Paler areas and spots more or less confluent on middle segments. Postero-lateral spots become better defined on end segments, becoming confluent on 10 around a black middorsal triangle, in which tubercle is situated. Female similar to male, with pale areas more extensive. Black middorsal triangle on 10 almost eliminated.

Distribution and dates.—CANADA: N. S., Ont.; UNITED STATES: Conn., Ind., Iowa, Maine, Mass., Mich., N. H., N. J., N. Y., Ohio, Wis.

June (N. Y.) to October 12 (New England).

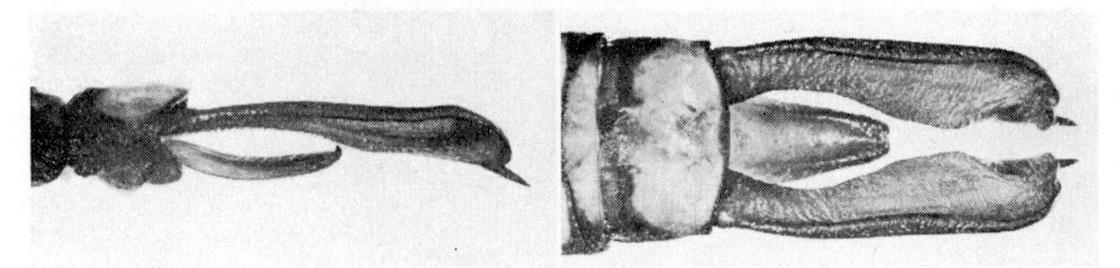

Fig. 183. *Aeschna constricta.*

Aeschna constricta Say

1839. Say, J. Acad. Phila., 8:11.
1912. Walk., N. Amer. Aeshna, p. 176 (figs.).
1927. Garm., Odon. of Conn., p. 189 (figs.).
1929. N. & H., Handb., p. 145 (fig.).

Length 68–72 mm.; abdomen 50–57; hind wing 43–47.

A wide-ranging and rather common Eastern species. Face greenish, with no black lines crossing it. Black T spot on frons has its stalk scarcely widened back to its junction with black of vertex. Occiput yellow, with black lateral angles.

Thorax brownish, with a not very conspicuous pair of pale stripes on front that widen upward to end just under crest. Two lateral yellowish stripes also widen upward to subalar carina. Anterior stripe has a sinuate front edge and gives off to rearward a short superior spur next to subalar carina. Second stripe much broader than first. Paranal cells of hind wing generally six, with three before and three within anal loop, but there may be two or three before and two or three within loop. Generally, loop narrower than in closely related species, having but two vertical rows of cells.

Abdomen brown, spotted with blue, in usual *Aeschna* pattern. Female similar to male but paler.

Distribution and dates.—CANADA: B. C., Man., N. S., Ont., Que., Sask.; UNITED STATES: Calif., Colo., Conn., Ill., Ind., Iowa, Kans., Maine, Md., Mass., Mich., Minn., Nebr., Nev., N. H., N. J., N. Y., N. Dak., Ohio, Pa., S. Dak., Utah, Vt., Wash., Wis.; MEXICO: Baja Calif.

May 30 (Mich.) to October 19 (Maine).

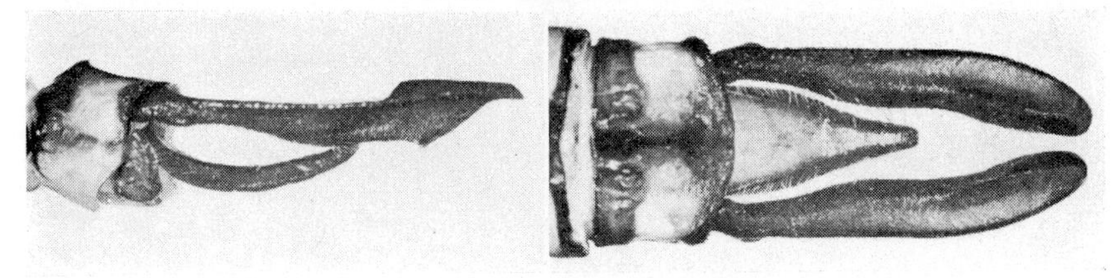

Fig. 184. *Aeschna dugesi.*

Aeschna dugesi Calvert

1905. Calv., B. C. A., p. 184 (figs.).
1932. Gloyd, Ent. News, 43:189.
1934. Tinkham, Can. Ent., 66:217.

Length 70–74 mm.; abdomen 50–55; hind wing 48–52.

A large blue-faced Mexican species; has been found in United States only on southwest border of Texas. Face light blue, becoming white about mouth. Pale brown hairline in fronto-clypeal suture, and a narrow edging of same color in concave portion of front margin of labrum. Black T spot on frons has a very thick stem and a small diffusely outlined head. On either side of stem a squarish spot of white lies between stem and blue of side of frons. On top of black vertex, a little stripe of olivaceous. Occiput yellow, with black corners.

Thorax robust, thinly clothed with whitish hairs. Carina black. Front brown, with usual two pale stripes. Of two side stripes, first widens downward and second upward; one or two smaller pale spots may lie between, or larger upper spot may be joined to upper end of first stripe. Wings broad, hyaline, with short stigma; under stigma are but two crossveins. Membrane of wings may be tinged with brown, sometimes rather heavily in female. As to venation of wing, we may note, in addition to what is shown in table, that median planate (Mpl) sags heavily to rearward, and vein M4 near wing margin is switched forward on crossveins so as to appear to terminate in M3.

Abdomen stout, blackish, with the usual lines of spots of blue. Segment 1 mostly pale; 2 black in posterior half both above and below, including genital lobes, and is clothed above with long tawny hairs, as is also low ventral tubercle on 1. Segment 10 black, with two yellow side spots on dorsum and a rather prominent tubercle on middorsal line.

Distribution and dates—UNITED STATES: Tex.; MEXICO: Baja Calif.; also south to Guanajuato, Mexico.

June 25 (Baja Calif.) to October 7 (Baja Calif).

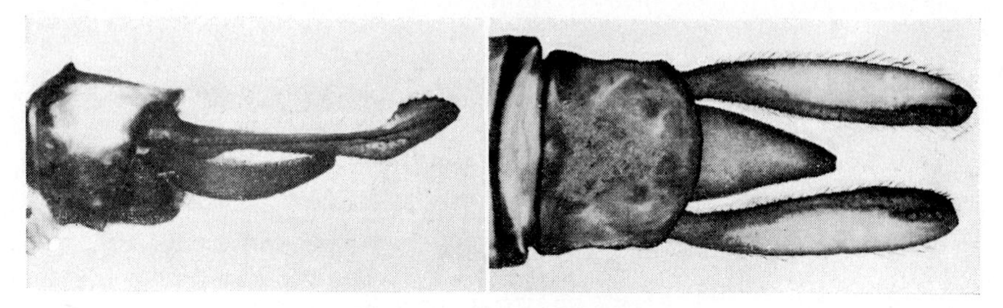

Fig. 185. *Aeschna eremita.*

Aeschna eremita Scudder

Syn.: hudsonica Selys

1866. Scudder, Proc. Boston Soc. Nat. Hist., 10:213.
1908. Mrtn., Coll. Selys Aeschnines, p. 35 (figs.).
1912. Walk., N. Amer. Aeshna, p. 119 (figs.).
1929. N. & H., Handb., p. 146 (fig.).

Length 74–76 mm.; abdomen 53–65; hind wing 45–52.

A Northern species of wide distribution. Face yellowish green, cross-lined with black on both front and rear margins of labrum and on fronto-clypeal suture, diffusely washed with brown. Black T spot on frons has a thick stem and its head runs over and down a little on face.

Thorax robust and thinly hairy. Black carina very prominent. Stripes on front fairly well defined. Two broad ones on each side both more or less deeply emarginate on front edges; second stripe widens to almost triangular form. Two or three pale spots between these stripes, a large one at top and a small one below spiracle. Legs reddish brown, with black spines and claw tips. Wings hyaline, with pale costa and narrow brown stigma. Venation as shown in our table of species, with added feature that subtriangle is generally a little lacking in development: cubito-anal cross-vein that forms its inner side does not quite reach triangle.

Abdomen brown, with usual lines of blue spots. Segments 1 and 2, swollen and hairy above, mostly pale on sides from auricles upward, except for a narrow half ring of black that connects auricles across dorsum. Three or four teeth on each auricle; area below them black. Segment 3 much constricted; 10 armed with a moderate middorsal tooth on a middorsal black stripe, with a single large blue spot on each side of it.

Distribution and dates.—Alaska; Labrador; CANADA: Alta., B. C., Man., N. B., Nfld., N. S., NW. Terr., Ont., Que., Sask., Yukon; UNITED STATES: Maine, Mass., Mich., N. H., N. Y., Utah, Vt., Wis., Wyo.

June 11 (NW. Terr.) to October 1 (New England).

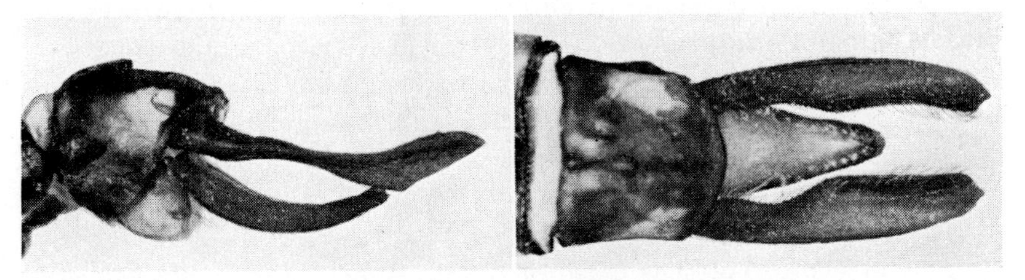

Fig. 186. *Aeschna interna.*

Aeschna interna Walker

1908. Walk., Can. Ent., 40:381, 388, 450.
1912. Walk., N. Amer. Aeshna, p. 116 (figs.).

Length 67–73 mm.; abdomen 47–53; hind wing 43–45.

A mountain variety of *Aeschna interrupta;* its range overlaps a little that of *A. lineata* (see p. 307). First lateral thoracic stripe not more than 1 mm. wide in lower half, then narrows to half that width in upper half, but variable. Distinguished from other varieties of *interrupta* by presence of a basal swelling on under side of superior caudal appendage of male. This is best seen when viewed obliquely from above. Viewed in profile, a slight undulation of nearly straight upper margin, and a slight angulation of inferior margin at base of uptilted and swollen terminal portion. In other varieties, only a smooth curve.

Distribution and dates.—CANADA: Alta., B. C.; UNITED STATES: Colo., N. Mex., Oreg., Utah, Wash.

July 9 (Oreg.) to September 10 (Alta.).

Aeschna interrupta Walker

1908. Walk., Can. Ent., 40:381, 387 (figs.).
1912. Walk., N. Amer. Aeshna, p. 103 (figs.).
1915. Kndy., Proc. U. S. Nat. Mus., 52:581 (figs.).
1927. Garm., Odon. of Conn., p. 187 (figs.).
1929. N. & H., Handb., p. 145 (fig.).

Length 72–77 mm.; abdomen 50–59; hind wing 43–49.

A wide-ranging Northern species, distinguishable from associated Aeschnas even in flight by its generally darker coloration. Thinly clad with blackish hairs. Face olive green, with black fronto-clypeal stripe narrowed to a hair streak at each end. Black T spot on frons heavy, its short stem generally much widened to its junction with vertex band. Both tip of vertex and occiput dull yellow.

Thorax rather thinly hairy, with usual stripes greatly reduced, being limited to elongated spots or entirely wanting. The two on each side broken in middle into spots, with the two smaller spots in area between stripes. Legs black beyond knees, with femora reddish brown. Wings hyaline, with brown veins and stigma; membranule wholly brown.

Abdomen brown, becoming black toward end, with usual bands of blue spots, among which posterior dorsal pair on segments 8, 9, and 10 stand out prominently.

"The race interrupta is a characteristic inhabitant of the Canadian Zone east of the Great Plains. It occurs only occasionally in the Transition Zone, except in its most northern parts, and is apparently absent from the Upper Austral Zone."—Walker (1912).

Distribution and dates.—CANADA: N. B., Nfld., N. S., Ont., Que.; UNITED STATES: Maine, Mass., Mich., N. H., N. Y., Vt.

June 30 (Que.) to September 18 (N. S.).

See also, in their proper alphabetic order, *A. interna, A. lineata,* and *A. nevadensis.*

Aeschna juncea Linnaeus

1758. Linn., Syst. Nat., 1:544 (in *Libellula*).
1861. Hagen, Syn. Neur. N. Amer., p. 120.
1908. Mrtn., Coll. Selys Aeschnines, p. 83 (figs.).
1912. Walk., N. Amer. Aeshna, p. 83 (figs.).
1929. N. & H., Handb., p. 147 (fig.).
1934. Walk., Can. Ent., 66:267 (fig., nymph).

Length 66–75 mm.; abdomen 47–55; hind wing 39–46.

A species of Far North, circumpolar; a pale species, bluish areas being ampler than in most other Aeschnas. Face greenish blue, more or less overspread with brownish except on sides of the frons and facial lobes of postclypeus. Black crossbands on fronto-clypeal suture and on both front and rear margins of labrum. Black T spot above has an ill-defined front margin; its stalk is widened to its confluence with black of vertex. Top of vertical tubercle broadly yellow, occiput obscurely so.

All pale stripes of thorax broad and all carinae narrowly black. Two stripes on front broadly widened laterally under crest. Between the two on each side, a short intervening half stripe terminates wide at top and tapers to a point halfway down toward spiracle. Legs brown, paler basally. Wings dull hyaline, with tawny costa and stigma. Their venation as shown in table for species. Cells in fork of radial sector and on both radial and median planates rather more numerous and irregular than is usual in *Aeschna.*

Abdomen brown, broadly marked with blue; black on all carinae and on joinings of middle segments. Two swollen basal segments have a middorsal yellow line; sides of 2 streaked with brown and yellow, and all yellow below auricle in male. Each auricle armed with four minute teeth. Segment 3 moderately constricted. Darkening segments beyond 3 have usual spots larger than in other species, postero-dorsal one increasing markedly to rearward, covering most of depressed dorsum of 10. Middorsal tubercle of 10 low and erect.

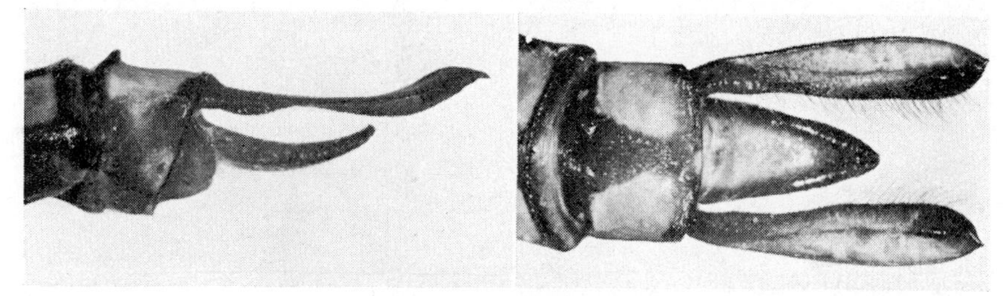

Fig. 187. *Aeschna juncea.*

Distribution and dates.—Alaska; Labrador; CANADA: Alta., B. C., Man., Nfld., NW. Terr., Ont., Que., Yukon; UNITED STATES: Colo., N. H., Wyo.; also from boreal regions of Europe and Asia.

June 16 (B. C.) to September 1 (Labrador).

Aeschna lineata Walker

1908. Walk., Can. Ent., 40:382.
1912. Walk., N. Amer. Aeshna, p. 112 (figs.).

Length 66–72 mm.; abdomen 49–53; hind wing 44–46.

A variety of *Aeschna interrupta* (see p. 305); differs in having pale side stripes of thorax continuous but very narrow (not more than 0.75 mm. wide), and blue spots on abdomen somewhat larger. Inner margin of superior caudal appendages of male gently sinuate. Their dorsal margin when viewed in profile runs in a uniformly concave curve.

"This race of *Ae. interrupta* is probably the most characteristic dragonfly of the Canadian Prairies, ranging over the whole of this region from

Manitoba to the foothills of the Rocky Mountains and northward into the wooded country as far as Great Slave Lake."—Walker (1912).

Distribution and dates.—Alaska; CANADA: Alta., B. C., Man., NW. Terr., Ont., Sask., Yukon; UNITED STATES: Minn., N. Dak.

First week in June to first week in October (Canada, Whitehouse).

Aeschna manni Williamson

1930. Wmsn., Occ. Pap. Mus. Zool. Univ. Mich., 216:27 (figs.).

Length 63–66 mm.; abdomen 47–54; hind wing 43–45.

A recently discovered Mexican species, closely allied to *Aeschna psilus*. Face blue, labrum tinged with greenish or yellowish. Black T spot on frons well defined. Stem of T parallel-sided, half surrounds and separates two spots of yellow on top of frons. Top of vertex grayish green. Occiput yellowish.

Thorax light brown, with a black median carina. Usual pale stripes on front reduced to elongated blue spots about 2 mm. long below, and a minute dot of the same color above it. Side stripes roughly wedge-shaped, narrowed downward to blunt lower ends, and greatly widened at top. Wings hyaline, with black veins.

Abdomen light brown, the extent of the blue rapidly diminishing all the way to rearward.

Distribution and dates.—MEXICO: Baja Calif.

April to October 8.

Aeschna multicolor Hagen

Syn.: furcifera Karsch

1861. Hagen, Syn. Neur. N. Amer., p. 121.
1895. Calv., Proc. Calif. Acad. Sci., (2)4:508 (figs.).
1905. Calv., B. C. A., p. 183.
1908. Mrtn., Coll. Selys Aeschnines, p. 48 (fig.).
1908. Wmsn., Ent. News, 9:265, 301 (figs.).
1912. Walk., N. Amer. Aeshna, p. 190 (figs.).
1915. Kndy., Proc. U. S. Nat. Mus., 49:344.
1929. N. & H., Handb., p. 141 (fig.).

Length 68–72 mm.; abdomen 48–52; hind wing 42–45.

A widely distributed Western species, similar to *dugesi*, but less robust. Face blue, paler on fronto-clypeal suture and about mouth, and thinly beset with short black hairs. Stalk of black T spot on frons widens strongly to rearward to its junction with black of vertex. On each side of frons is a rectangular spot of light blue, beyond which sides of frons are a darker blue. Tip of vertex and occiput also blue.

Thorax, half hidden among whitish hairs, with usual two pale stripes on a dark ground. Two lateral dark stripes fairly regular and of moderate width; between their upper ends is a smaller tract of blue. Legs black; wings hyaline, with venation as shown in our venation table. Special features described under *dugesi* all present, only less markedly developed.

Abdomen long, and much contracted on segment 3, quite hairy above on much swollen basal segments. Brownish auricles of male each armed with two black teeth; area above and below auricles is blue. Beyond segment 3, joinings of segments black, with usual rows of spots on intervening areas. Segment 10 has a broad, tapering middorsal black band,

Fig. 188. *Aeschna multicolor.*

with a large blue spot on each side of it, and a low ridgelike tubercle in midst of black dorsum.

Distribution and dates.—CANADA: B. C.; UNITED STATES: Ariz., Calif., Colo., Idaho, Kans., Mo., Nebr., Nev., N. Mex., Okla., Oreg., Tex., Utah, Wash.; MEXICO: Baja Calif.; also south to Panama.

May 18 (B. C.) to October 9 (Calif.).

Aeschna mutata Hagen

1861. Hagen, Syn. Neur. N. Amer., p. 124.
1908. Wmsn., Ent. News, 19:264, 302.
1912. Walk., N. Amer. Aeshna, p. 198 (figs.).
1927. Garm., Odon. of Conn., p. 192 (figs.).
1929. N. & H., Handb., p. 142 (fig.).

Length 74–76 mm.; abdomen 51–59; hind wing 44–51.

A Northern representative of *Multicolor* group of species. Body brown, with markings that in life are of deepest and most intense blue. Eyes of a darker shade, and face of a lighter tint of blue. No black crossbands on face. Stem of T spot on frons nearly parallel-sided. Sides of frons, tip of blackish vertex, and middle of occiput also blue.

Thorax brown, with two stripes on its front wide and prominent; two on sides greatly widened to rearward at upper ends. Legs black, except under side of front femora, which is blue. Costal margin of wings yellowish, stigma short, with only two crossveins under it beyond brace vein. Venation as shown in our table of genera.

Abdomen brown, with segment 2 swollen and 3 constricted as usual. In male, a wide area of blue above and another below black auricles, from each of which a blackish side stripe extends to hind margin of segment.

Distribution and dates.—CANADA: Ont.; UNITED STATES: Ind., Ky., Mass., Mich., Ohio, Pa.

May 17 (Mich.) to July 20 (Ind.).

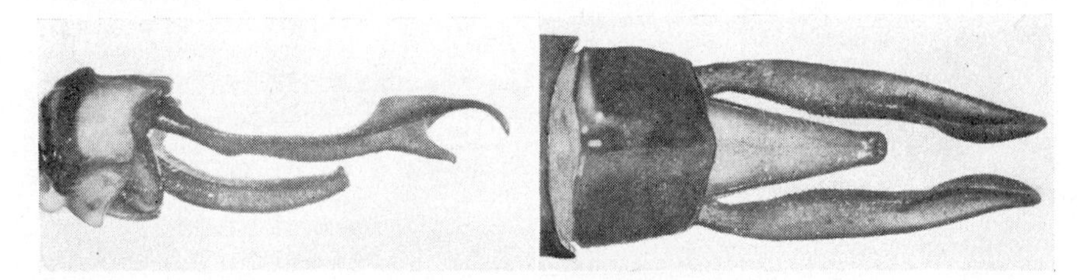

Fig. 189. *Aeschna mutata.*

Aeschna nevadensis Walker

1908. Walk., Can. Ent., 40:382.
1912. Walk., N. Amer. Aeshna, p. 111 (figs.).
1917. Kndy., Proc. U. S. Nat. Mus., 52:581 (figs.).

Length 69–74 mm.; abdomen 51–53; hind wing 45–47.

A variety of *A. interrupta* occurring in arid Southwest (see p. 305); differs in having lateral pale stripes of thorax entire and narrow (about 1 mm. wide). Superior caudal appendage of male markedly sinuate on inner margin, and bears a minute tooth at tip. Female and nymph unknown.

Distribution and dates.—CANADA: B. C.; UNITED STATES: Calif., Colo., Nev., Utah.

First week in July (B. C.) to second week in October (B. C.).

Aeschna occidentalis Walker

1912. Walk., N. Amer. Aeshna, p. 174.
1915. Kndy., Proc. U. S. Nat. Mus., 49:344.
1941. Whts., Odon. of B. C., p. 523.

A West Coast variety of *Aeschna umbrosa* (see p. 316), like that species except in minor particulars, most important of which seems to be a shorter and thicker abdominal segment 3. Postero-dorsal pale spots on 5 to 10 green in *umbrosa;* blue, except on 10, in *occidentalis.*

Distribution and dates.—CANADA: B. C.; UNITED STATES: Calif., Nev., Oreg., Utah, Wash.; probably grades eastward into the typical *umbrosa.*

First week in July (B. C.) to first week in November (B. C.).

Aeschna palmata Hagen

1856. Hagen, Stettin. Ent. Ztg., 17:369.
1912. Walk., N. Amer. Aeshna, p. 157 (figs.).
1917. Kndy., Proc. U. S. Nat. Mus., 52:586 (figs.).
1929. N. & H., Handb., p. 144 (fig.).

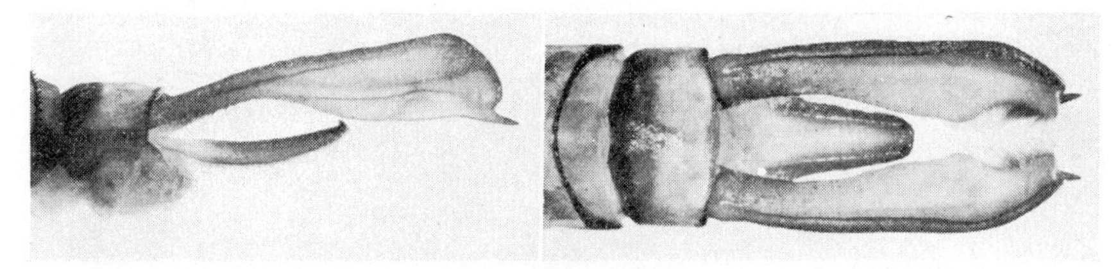

Fig. 190. *Aeschna palmata.*

Length 65–75 mm.; abdomen 47–58 mm.; hind wing 41–47.

A Western paddle-tail darner, often locally abundant. Face greenish or darker, with a black stripe across fronto-clypeal suture, another across front border of labrum, and another shorter one on base of labrum that sends a spur forward to end in a median dot. Black stalk of black T spot on frons parallel-sided. Occiput brownish, broadly bordered with black.

Two stripes on dorsum of thorax and two on each side parallel-sided. Between the two laterals there may be one or two small pale spots. Legs black. Wings hyaline, with blackish veins and brown costa. Usually six paranal cells in hind wing of male, three before and three within anal loop, but occasionally four before loop. Generally three vertical rows of cells within loop.

Abdomen brown, spotted with blue after pattern of closely related species, but darker than others in general coloration. Ventral side of middle abdominal segments black. Black auricle on segment 2 of male armed with four (sometimes three or five) teeth.

Distribution and dates.—Alaska; CANADA: Alta., B. C., Sask.; UNITED STATES: Calif., Colo., Idaho, Nev., Oreg., Utah, Wash.; also from Kamchatka (type locality).

July 1 (B. C.) to November 1 (B. C.).

Aeschna psilus Calvert

1918. Ris, Arch. Naturgesch., 82 Abt. A, (9):157 (in part) (as *cornigera*).
1932. Klots, Odon. of P. R., p. 18 (as *cornigera*).
1938. Garcia, J. Agr. Univ. Puerto Rico, 22(1):55 (as *cornigera*).
1947. Calv., Notulae Naturae, Acad. Nat. Sci. Phila., 194:4 (figs.).

Length 58–60 mm.; abdomen 41–51; hind wing 36–43.

A Neotropical species, not as yet known from the United States. A trim and elegant species, light and airy in aspect, with a very slender abdomen. Face blue, paler toward mouth. Labrum has a black front border. No black stripe on fronto-clypeal suture. T spot on frons dark

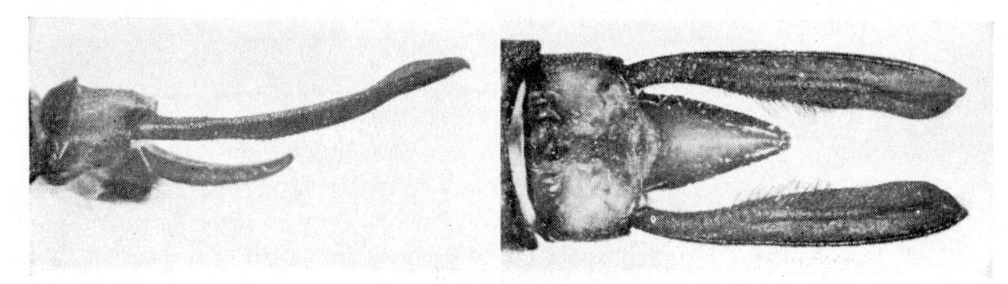

Fig. 191. *Aeschna psilus.*

brown, with head of T overspreading prominence of frons and traversed there by a sharp black frontal carina; stem of well-defined T nearly parallel-sided. Two bands of brown extend from tips of T rearward to black vertex, leaving two rectangular yellow spots on top of frons with stem of T between them. Top of hairy vertex and small occiput yellow. Eye-seam unusually long.

Thorax hairy and brownish, with usual stripes: two in front and two on each side, ill defined. Legs black. Wings shining hyaline, with brown veins, tawny stigma, and rather open venation. Even in wider spaces between veins, cells run in rather regular rows. Paranal cells of front wing are twice the height of marginals beside them. Paranals of hind wing six or seven: three within anal loop, and one or two between it and three-celled triangle of male. A plantar loop bordering inner side of anal loop well defined and definitely curved. Between veins Cu1 and Cu2, two rows of cells for a space; thereafter but a single row out to their converging tips at wing margin. Stigma small and tapers outward from a brace vein that is strongly aslant, with two or three other cross-veins behind it.

Abdomen brown, marked with usual blue spots, most conspicuous of which are two pale half rings, both of which are interrupted on middorsal line: one apical, other following girdle groove, with a prolongation to rearward along middorsal carina. Dorsum of segments 8 and 9 nearly covered by two large pale spots. Abdomen decidedly swollen at base in both sexes, constricted on 3 (less so in female), very slender and slowly widened all the way to rearward, with last two segments dorsally depressed. Black carina that traverses 2 at its highest point runs obliquely down each side from a short, straight, prominent middle portion. On slopes of dorsum of segment beyond this carina lie a pair of large impressed transverse scars in both male and female. Thin auricles on side of 2 in male bear each a pair of sharp incurved teeth. Segment 3 strongly constricted in male; less so in female. Farther to rearward, joinings of segments all black, and along middorsal line runs a sharp carina. Appendages of male a millimeter shorter than those of female.

Distribution and dates.—ANTILLES: Cuba, Dom. Rep., Jamaica, P. R.; also from Mexico south to Ecuador and Peru.

Apparently year-round; dates include June 8–13 (P. R.), August 24, August 26, January 10–11 (Jamaica), and March 30 (Cuba).

Aeschna septentrionalis Burmeister

1839. Burm., Handb., p. 839.
1861. Hagen, Syn. Neur. N. Amer., p. 120.
1912. Walk., N. Amer. Aeshna, p. 72 (figs.).
1941. Whts., Odon. of B. C., p. 513 (figs., nymph).

Length 53–61 mm.; abdomen 38–46; hind wing 34–40.

A moderately robust species, rather dull green and brown. Face dark, with a heavy black line on fronto-clypeal suture. Labrum paler, banded with black across both front and rear margins. Paler frons has a well-developed blackish **T** spot, with a well-rounded top bar above.

Thorax dull brownish, streaked narrowly with paler. Usual pair of pale stripes on front reduced to small isolated spots low down near collar. Two pale stripes on sides correspondingly reduced in width, more or less **Z**-shaped, interrupted and strongly sinuous but obtusely angled at their two bends. Between the two lateral stripes a little fleck of paler color is generally present. Second lateral stripe much more **T**-shaped than **Z**-shaped when present and well developed, and often almost wanting. All pale markings more extended in female than in male.

Legs pale reddish brown. Wings hyaline, with brown costa and stigma.

Abdomen blackish, with large spots of blue in usual segmental arrangement.

Distribution and dates.—Alaska (Hagen, 1856); Labrador; CANADA: Alta., B. C., Man., Nfld., NW. Terr., N. S. (Hagen, 1861), Ont., Que., Sask.; UNITED STATES: N. H.

June 16 (Man.) to September 10 (Que.).

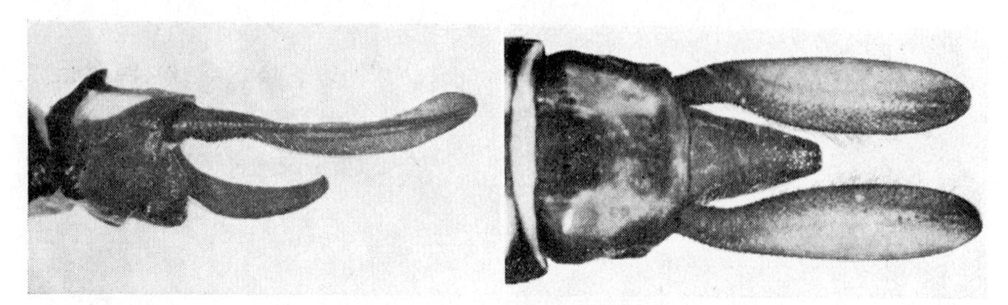

Fig. 192. *Aeschna septentrionalis.*

Aeschna sitchensis Hagen

1861. Hagen, Syn. Neur. N. Amer., p. 119.
1908. Mrtn., Coll. Selys Aeschnines, 18:41 (figs.).
1909. Adams, Geol. Surv. of Mich., p. 263.
1912. Walk., N. Amer. Aeshna, p. 77 (figs.).
1917. Whts., Can. Ent., 49:99.
1921. Walk., Can. Ent., 53:221 (figs., nymph).
1929. N. & H., Handb., p. 139 (fig.).

Length 54–64 mm.; abdomen 42–49; hind wing 36–41.

A Far Northern species, small and dully colored, with very narrow **Z**-shaped lateral pale stripes on sides of thorax. Face greenish, with black cross stripes on sutures and on both front and rear margins of labrum; anteclypeus black. From basal stripe on labrum a half-length black stripe projects forward on median line. **T** spot on top of frons short-stalked and thick. Vertex black save for a pale streak across its low summit. Occiput pale, with brown edgings.

Hairy thorax olive brown in front, with black-edged crest and carina. First lateral pale stripe more or less interrupted, as well as being narrow and **Z**-shaped. Wings hyaline, with dull yellow costa and brown veins and stigma.

Abdomen black, with large paired blue spots on dorsum of middle segments; these spots become confluent on segments 9 and 10. Caudal appendages pale brown, with darker tips; inferior darker than superiors.

Distribution and dates.—Alaska; Labrador; CANADA: Alta., B. C., Man., Nfld., NW. Terr., Ont., Que., Sask.; UNITED STATES: Maine, Mich., Minn.

June 2 (B. C.) to September 14 (Ont.).

Aeschna subarctica Walker

1908. Walk., Can. Ent., 40:385, 390, 451 (figs.).
1912. Walk., N. Amer. Aeshna, p. 93 (figs.).
1929. N. & H., Handb., p. 148 (fig.).
1941. Whts., Odon. of B. C., p. 518.

Length 66–76 mm.; abdomen 49–57; hind wing 39–46.

A Canadian species, with a bright yellow face and a heavy black T spot on frons. Stem of T spot widens to its base. Top of vertex and occiput lemon yellow.

Robust thorax brown, with stripes of bright green. Stripes on its front strongly divergent downward and much widened laterally above. First lateral stripe broad below, narrowed and angulated near middle, and extended at top to rearward in a beaklike prong under subalar carina. Two yellowish spots between stripes. Legs brown. Wings hyaline, with stigma ochre yellow.

Abdomen brown, broadly marked with green on segment 1 and on basal half of 2; blue on distal half of 2 and on more slender segments beyond. Spots of lower row on each side large and conspicuous.

Distribution and dates.—CANADA: B. C., Man., NW. Terr., N. S., Ont., Que.; UNITED STATES: Mich.

July 7 (Ont.) to September 11 (Ont.).

Aeschna tuberculifera Walker

1908. Walk., Can. Ent., 40:385.
1912. Walk., N. Amer. Aeshna, p. 152 (figs.).
1927. Garm., Odon. of Conn., p. 187 (figs.).
1929. N. & H., Handb., p. 148 (fig.).
1940. Lincoln, Proc. Amer. Phil. Soc., 83:589 (growth).
1941. Whts., Odon. of B. C., p. 521.

Length 71–80 mm.; abdomen 52–63; hind wing 44–51.

A rather slender Northeastern species of rather dark coloration. Face olivaceous, with pale labrum. Blackish crossline on fronto-clypeal suture rather narrow. Stem of blackish T spot on frons widens to its base. Vertex green at summit; occiput yellowish.

Thorax reddish brown, with complete and nearly regular stripes on its front pea green. Side stripes greenish at upper end and blue below. These stripes full length and regular, with first one a little wider than second, and bent to rearward at its tapering top end. Legs reddish brown, with under side of front femora pale. Wings hyaline or lightly tinged with brown in membrane, especially in teneral specimens. Costa and stigma dark reddish brown.

Abdomen slender, with usual basal swelling and segment 3 very long. Succeeding segments diminish regularly in length to end. Blue spots of dorsal row markedly larger than those of row low down on sides.

Distribution and dates.—CANADA: B.C., N. S., Ont., Que.; UNITED STATES: Conn., Ind., Maine, Mass., Mich., N. H., N. Y., Pa., R. I., Wis.

July 3 (Ont.) to September 30 (Ont.).

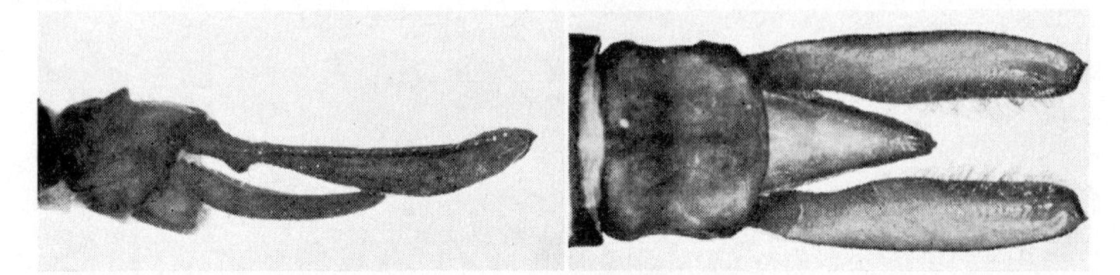

Fig. 193. *Aeschna tuberculifera.*

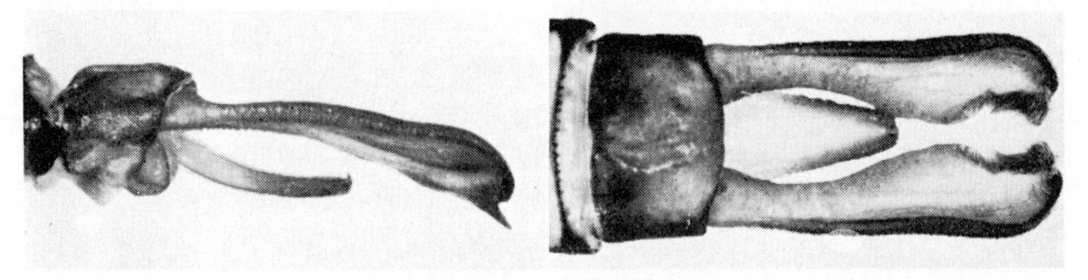

Fig. 194. *Aeschna umbrosa.*

Aeschna umbrosa Walker

1908. Walk., Can. Ent., 40:380, 390.
1912. Walk., N. Amer. Aeshna, p. 165 (figs.).
1927. Garm., Odon. of Conn., p. 191 (figs.).
1929. N. & H., Handb., p. 142 (fig.).

Length 68–78 mm.; abdomen 49–60; hind wing 42–47.

A very common and widespread Eastern midsummer species. Face greenish or brownish, without black cross stripes; at most, a darkening in sutures of face. Stem of black T spot on frons narrowest in middle, widening fore and aft.

Thorax brownish, its two stripes on front parallel-sided, also the two wider stripes on side. Front ones may widen downward; they are bordered with black. Legs brown. Wings hyaline, with brown costa and tawny stigma. Paranal cells of hind wing generally seven, with four before anal loop and three in it. Loop generally composed of three vertical rows of cells.

Abdomen, spotted above after usual pattern, blue on brown, has a row of blue spots on ventral side of segments 4 to 6 or 7 inside lateral carina, which is very markedly sinuate on 7. This is distinctive of the species.

Distribution and dates.—Labrador; CANADA: Alta., B. C., Man., N. B., Nfld., NW. Terr., N. S., Ont., P. E. I., Que., Sask., Yukon; UNITED STATES: Ala., Conn., D. C., Ind., Ky., Maine, Md., Mass., Mich., Minn., Miss., Mo., Nebr., N. H., N. J., N. Y., N. C., Ohio, Okla., Pa., R. I., S. C., Tenn., Vt., Va., Wis.

May 6 (New England) to November 8 (Miss.).

(See also, in its proper order, the variety *A. occidentalis.*)

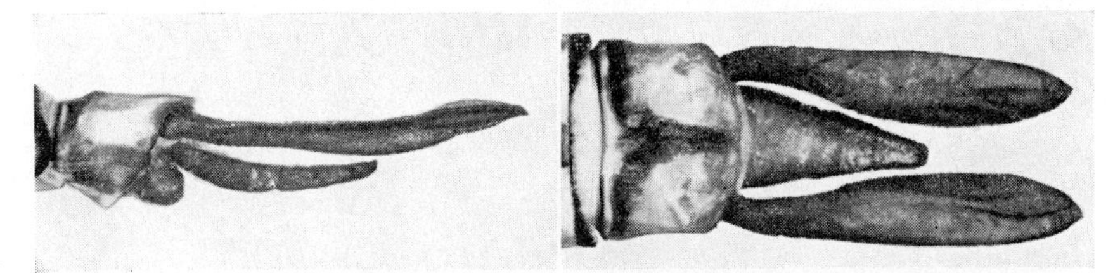

Fig. 195. *Aeschna verticalis.*

Aeschna verticalis Hagen

Syn.: propinqua Scudder

1861. Hagen, Syn. Neur. N. Amer., p. 122.
1912. Walk., N. Amer. Aeshna, p. 145 (figs.).
1927. Garm., Odon. of Conn., p. 188 (figs.).
1929. N. & H., Handb., p. 149 (fig.).
1941. Walk., Can. Ent., 73:229 (figs., nymph).

Length 74–79 mm.; abdomen 49–60; hind wing 43–50.

A brownish species, darkening to blackish on abdomen marked with green and blue and a few touches of yellow. Face green and bare in front, scantily hairy at sides, with only a hairline of brown in fronto-clypeal suture, and a brown edging to notched front border of labrum. Stalk of black triangle on frons regularly or slightly widened to its junction with black of vertex. From its base a narrow yellow streak runs laterally down base and sides of frons. Tip of vertex and middle of occiput dull yellow.

Thorax brown, thinly clothed with white hairs both above and below. Stripes of yellowish green irregular, with pair on front pointed at both ends in opposite directions and constricted toward upper end. Side stripes more deformed: first one ajog near upper end, then prolonged to rearward along subalar carina; rear one with a marked crescentic widening above middle. Legs black, reddish brown on femora. Wings hyaline, with tawny costa and stigma. Venation as shown in table of species.

Swollen base of long abdomen blue on sides above black auricle in male, and in a streak across rear of dorsum of both segments 1 and 2, and on genital lobe below auricle. Beyond constricted segment 3, abdomen brown, spotted as usual with blue, darkening to rearward. Segment 10 flattened, about half covered dorsally by two large pale bluish spots. Apical margin of segment narrowly edged with yellow between bases of superior appendages. Middorsal tubercle on 10 rather sharp-pointed.

Distribution and dates.—CANADA: N. B., N. S., Ont., Que.; UNITED STATES: Conn., Ill., Ind., Iowa, Maine, Md., Mass., Minn., N. Y., N. C., Ohio, Pa., R. I., Vt., Wis.

June 13 (Ill.) to October 18 (Pa.).

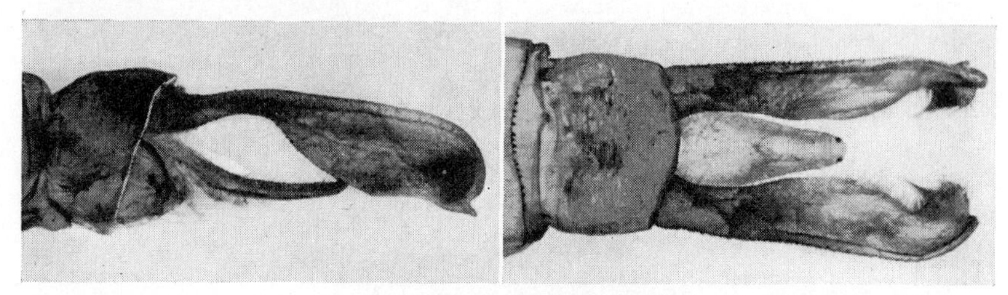

Fig. 196. *Aeschna walkeri.*

Aeschna walkeri Kennedy

1917. Kndy., Proc. U. S. Nat. Mus., 52:587 (figs.).

Length 64–66 mm.; abdomen 51–57; hind wing 43–48.

Face bluish gray, with labrum paler; no black crosslines, only a brown hairline in fronto-clypeal suture. Tip of vertex and middle of occiput yellowish. T spot on frons short and wide, its stem narrowest in middle and widened to both ends.

Thorax brown, with two stripes on front pale blue, lateral stripes whitish. Edges of latter nearly parallel.

Abdomen of usual form, swollen on segment 2, contracted moderately on 3. Hairiness of 2 mainly on a short middorsal ridge. Auricle on 2 armed with three teeth, often with a vestige of a fourth. Inferior appendage of male rather broad, truncate at tip, with a minute upturned tooth on each corner.

Inhabits frost-free streams of coastal mountains of California south of San Francisco Bay and also in Baja California, Mexico.

An excellent account of the habits of this species accompanies the original description.

Distribution and dates.—UNITED STATES: Calif.; MEXICO: Baja Calif.

June 28 to November.

Genus GYNACANTHA Rambur 1842

This large genus includes more than forty species, some of which are found on almost every major land mass. They are especially numerous in the Tropics, but only two species are known from our range. A third species, *G. bifida* Rambur, has been reported from Florida, but the records are questionable and probably may be due to misdeterminations.

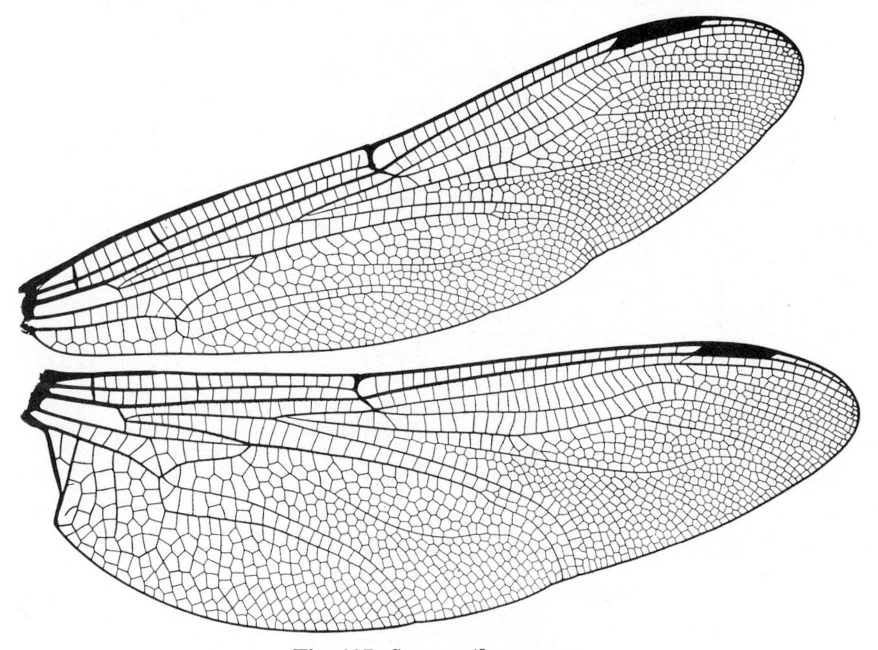

Fig. 197. *Gynacantha nervosa.*

The wings are broad and the venation dense. The triangles are very long and of about five to ten cells. The basal triangle in the male contains three cells. The midbasal space is free and the supertriangle contains several crossveins. The fork of Rs is distinctly basal to the stigma in the hind wing and usually in the front wing.

The two-pronged ventral process on segment 10 of the female is distinctive of this genus. The eyes are very large and the habit often crepuscular.

The nymph of *G. nervosa* has been reared in Guatemala by F. X. Williams, who has published an account of it together with a discussion of the habits of oviposition. The nymph of this species and that of *Triacanthagyna trifida* are outstanding in our fauna because of the presence of raptorial setae on the lateral lobes of the labium. The nymph of *nervosa* is

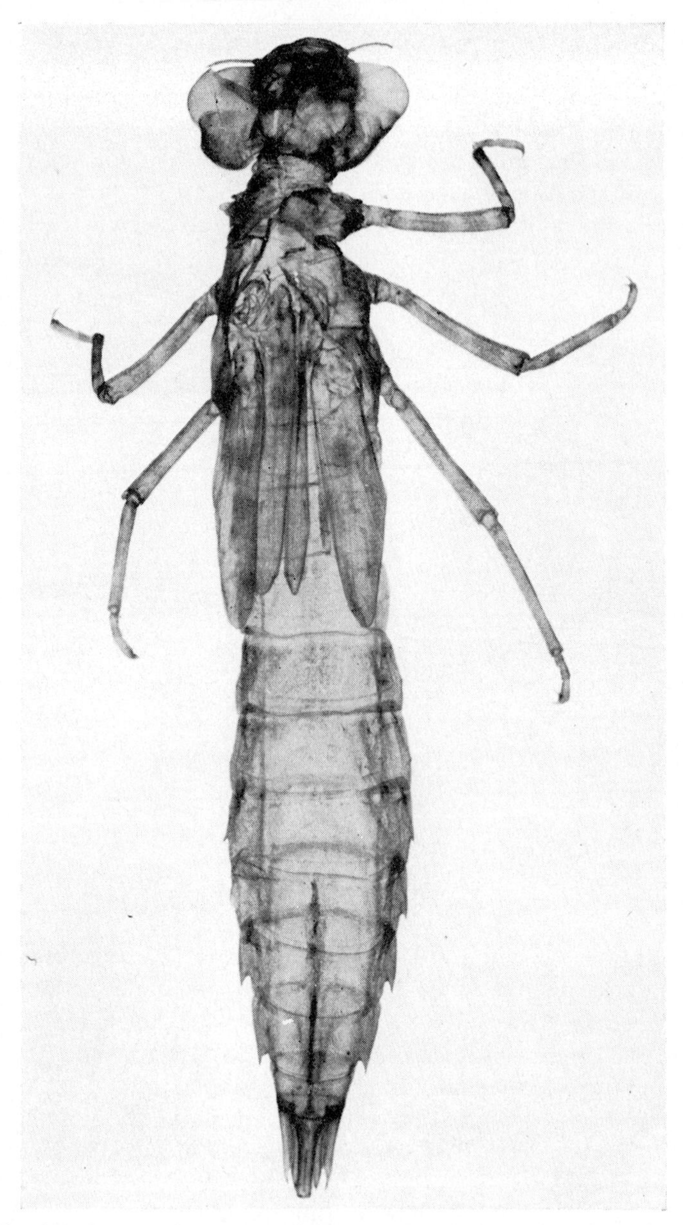

Fig. 198. *Gynacantha nervosa*, empty nymphal skin of a reared specimen.

larger than that of *T. trifida,* the median lobe of the labium on each side of the median cleft is more convex, and the end hook is much larger. The lateral setae are not so strong as in *T. trifida;* otherwise the nymphs are quite similar.

An excellent account of the American species of this genus was published by E. B. Williamson in 1923.

KEY TO THE SPECIES

ADULTS

1—Hind wing 42–46 mm.; postnodal crossveins in front wing fewer than eighteen .... **ereagris**
—Hind wing 47–56 mm.; postnodals in front wing more than eighteen..... **nervosa**

Gynacantha ereagris Gundlach

1888. Gundlach, Contribucion a la Entomología Cubana, 2:243.
1919. Calv., Trans. Amer. Ent. Soc., 45:359, 386.
1923. Wmsn., Misc. Pub. Mus. Zool. Univ. Mich., 9:38.

Length of male 64–65 mm.; abdomen 49–52; hind wing 42–43.
Length of female 59 mm.; abdomen 44–48; hind wing 43–46.

A brownish species, similar to *nervosa,* but smaller and with hyaline wings. Labrum and face pale olive straw color. Frons of both sexes with a well-defined black T spot. Thorax greenish brown, with about six small black spots on each side: one surrounding stigma, another just above and one just below stigma, another at upper end of second lateral suture, one at lower anterior corner of metepimeron, and one posteriorly on latero-ventral carina. Legs yellowish.

Wings hyaline and venation less dense than in *nervosa;* front-wing antenodals usually 19–22, postnodals 13–15; hind-wing antenodals 14–16, postnodals 15–18; cells in triangle 5–6; cubito-anal crossveins 5–6; supratriangle with 3–5 crossveins.

Abdomen darker brown, greenish in life, with transverse carinae black; abdomen of male strongly constricted at segment 3, appendages brown.

Distribution and dates.—ANTILLES: Cuba.
August 18.

Gynacantha nervosa Rambur

Syn.: gracilis Burmeister

1842. Rbr., Ins. Neur., p. 213.
1909. Mrtn., Coll. Selys Aeschnines, p. 167 (figs.).
1923. Wmsn., Misc. Pub. Mus. Zool. Univ. Mich., 9:40.

1929. N. & H., Handb., p. 150.
1937. Williams, Pan-Pacific Ent., 13:1 (figs., nymph).
1945. Ndm., Bull. Brooklyn Ent. Soc., 40:106.

Length 75–80 mm.; abdomen 50–57; hind wing 47–56.

A large brownish species with little differentiation in color pattern. Face tawny; frons above diffusely blackish around margin and may show a distinct T spot; small occiput yellow. Thorax greenish to brownish above, with four small well-defined black spots or stripes on sides as described for *G. ereagris*. Legs pale.

Venation in general as follows: front-wing antenodal crossveins 26–32, postnodals 22–23; hind-wing antenodals 19–21, postnodals 24–28; cells in triangle 6–10; cubito-anal crossveins 6–8; supratriangle with 6–10 crossveins. Wings tinged with yellowish brown, more intense between costa and subcosta.

Abdomen brown, with narrow black lines on sutures encircling segments, not constricted at segment 3, or only slightly so.

Chiefly crepuscular in flight; often encountered in large numbers in Florida and southward.

Distribution and dates.—UNITED STATES: Calif., Fla.; ANTILLES: Cuba, Dom. Rep., Haiti, Jamaica, P. R.; also south to Bolivia and Brazil.

Year-round.

Genus TRIACANTHAGYNA Selys 1883

This is a genus of some six species, which are almost wholly Neotropical in distribution. Only two valid species have been reported from our range.

The wings are quite broad and the venation somewhat dense. The triangles are usually of five to six cells. The basal triangle in the male contains three cells. The midbasal space is free, and the supertriangle contains several crossveins. The fork of Rs is near the proximal end of the stigma in the front wing (except in *trifida*), and more basal in the hind wing than in the front.

The three-pronged ventral process on segment 10 of the female is characteristic of the genus. The eyes are very large, the size perhaps being correlated with the crepuscular habit.

The nymph of *T. trifida* only is known to us. It was reared in Santo Domingo by the senior author. Like *Gynacantha nervosa,* it has raptorial setae on the lateral lobes of the labium, this apparently being characteristic of only these two genera of Aeschnines in our fauna. Except for the smaller size and differences mentioned under *Gynacantha*, it is very much like *G. nervosa*.

Williamson (1923) has given a good account of the American species in this genus.

KEY TO THE SPECIES

ADULTS

1—Legs entirely pale; thorax without definite markings; abdomen pale; anterior edge of frons seen from above convex; abdomen of male not constricted at segment 3 . . . . **septima**

—Legs more or less dark; thorax with definite dark markings; abdomen dark; anterior edge of frons seen from above more or less angled; abdomen of male constricted at segment 3 . . . . **trifida**

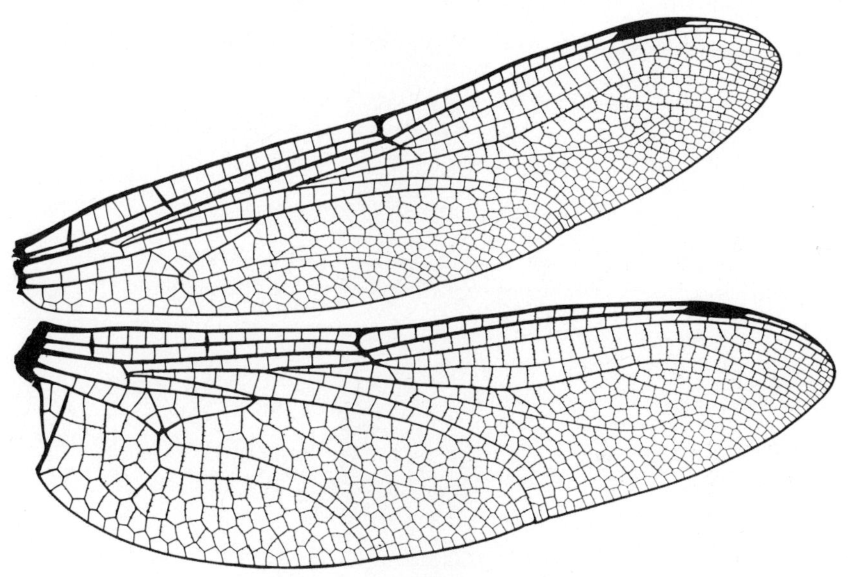

Fig. 199. *Triacanthagyna trifida.*

Triacanthagyna septima Selys

1857. Selys, in Sagra, Hist. Cuba, Ins., p. 460 (in *Gynacantha*).
1861. Hagen, Syn. Neur. N. Amer., p. 132 (in *Gynacantha*).
1906. Calv., B. C. A., p. 191 (in *Gynacantha*).
1909. Mrtn., Coll. Selys Aeschnines, p. 150.
1923. Wmsn., Misc. Pub. Mus. Zool. Univ. Mich., 9:16.

Length of male 59–63 mm.; abdomen 40–50; hind wing 34–41.

Length of female 61–66 mm.; abdomen 47–53; hind wing 37–43.

A rather small species, more or less uniformly light brown, without dark markings on thorax. Face olivaceous; dorsum of frons with an-

Fig. 200. *Triacanthagyna trifida*, young nymph.

terior margin blackish; occiput small and yellow; thorax uniformly greenish brown, light green in life, with few light brown markings. Legs yellowish, without dark markings. Wings usually hyaline.

Abdomen light brown, with green markings; carinae black; opposing hairs on blades of superior appendages of male reduced in length from base to middle of blade, apical half bearing ordinary short hairs.

Distribution and dates.—ANTILLES: Cuba, Jamaica, P. R.; also from Mexico and Bolivia.

January 9 (P. R.).

Triacanthagyna trifida Rambur

Syn.: needhami Martin

1842. Rbr., Ins. Neur., p. 210 (in *Gynacantha*).
1905. Calv., B. C. A., p. 191 (figs.) (in *Gynacantha*).
1909. Mrtn., Coll. Selys Aeschnines, p. 148 (figs.).
1923. Wmsn., Misc. Pub. Mus. Zool. Univ. Mich., 9:24.
1929. N. & H., Handb., p. 150 (in *Gynacantha*).

Length 62–75 mm.; abdomen 42–50; hind wing 40–47.

A slender species, with greenish thorax and brownish abdomen. Face obscure yellowish; apex of frons above blackish, a T spot sometimes evident; vertex blackish and narrow; occiput brown. Thorax brown and green, thinly clad with short, pale pubescence; front brown with two broad green stripes narrowed below; sides predominantly green, with a brown stripe on first lateral suture and touches of brown on other sutures. Legs brown, with tarsi and usually tips of femora black. Wings hyaline, with brown veins and tawny stigma, generally suffused with yellowish at extreme base and along costal border.

Abdomen brown, with blackish carinae and obscure median and subapical pale rings; median rings dilated on segments 3 to 7. Swollen basal segments greenish at sides and brownish above along sutures. Appendages brown, those of female very long, 9–10.6 mm. Opposing hairs on blades of superior appendages of male about equally long and numerous entire length of blade.

Distribution and dates.—UNITED STATES: Calif., Fla., Ga.; ANTILLES: Cuba, Dom. Rep., Haiti, Jamaica, P. R.; also from the Bahamas and south to Bolivia.

May 27 (Cuba) to January 8 (Fla.).

Family LIBELLULIDAE Rambur

The principal characters of this great group have already been stated briefly in our table of families (p. 64). The labium has no median notch in its front margin. The eyes meet on the top of the head. The triangles of the fore and hind wings are unlike in form and in the direction of their more acute angle. The triangle in the fore wing is farther out from the arculus than in the hind wing. There is no special brace vein to the stigma. There are no specially thickened antenodal crossveins.

The most characteristic venational feature of the family is the outward slant and extension of the anal loop. This loop begins in the Macrominae with very little change of form from that of Petaluridae, with only a skewness, an outward slant, to its longest axis (in certain exotic genera longer, and with its constituent cells arranged strictly in two rows, but never foot-shaped). In the Cordulinae the loop begins to have a toe, and in most Libellulinae the foot-shaped outline is fully attained. This unmistakable vein pattern will distinguish adult Libellulinae at a glance.

The nymphs are characterized by a wide head with a more or less vertical face, covered below the antennae by a wide spoon-shaped labium, which is armed with definite rows of mental and lateral raptorial setae. It is like no other except that of the Cordulegasteridae. In that family it seems to have had an independent and parallel development.

The nymphs are squatters or climbers, but never burrowers.

Macromian nymphs almost equal those of Hagenius in length of legs, flatness of abdomen, and prominence and size of middorsal hooks along the abdomen. The blunt head, long antennae, and spoon-shaped labium are more like *Cordulegaster*.

The main differences between the three subfamilies Macrominae, Cordulinae, and Libellulinae are stated in our key (p. 66). Beyond the key characters there are very many divergencies, to the consideration of which we shall now proceed.

Subfamily MACROMINAE Needham

These are long-legged dragonflies of large size. They are brown or blackish, marked with yellow in various tints, never with green. Characteristic features of color pattern are a yellow band across the dark face, a pair of yellow cross streaks in the top of the crest on the thorax, and a yellow stripe engirdling the thorax, passing between fore and hind wings and forming a conspicuous oblique midlateral stripe down each side. The frontal furrow is very wide, its sloping sides very flat and high at the outer side. A black streak occupies the bottom of the furrow, overspreads the frontal ridge, and then extends laterally to the eyes and rearward to join the black of the vertex, thus enclosing one or two yellow spots of variable size on each side of the frons. The eyes are contiguous on the dorsum for a little way. There is a large, rounded tubercle (vestige of the nymphal eye) on their hind border at the outer side. The legs are long, spined, and strong, with tibial keels generally present on the legs of the males. The claws of the tarsi are two-parted at the tip.

The venation of the wings has a number of distinctive characters. The triangle is about twice as far out beyond the arculus in the fore wing as it is in the hind wing, and its long axis is pointed rearward. The sectors of the arculus are united for some distance beyond it. Veins

Fig. 201. *Macromia taeniolata* standing above cast-off skin from which it emerged an hour earlier. Colors black and bright yellow. (Photo from life by Mr. Richard Archbold.)

M3 and M4 are distinctly undulate. The paranal cells in the fore wing are much larger than the accompanying single row of marginals, and the last two or three of them are turned out of line to border on the subtriangle. In the hind wing the paranals are only four or five, two of them before the anal loop (three in the female) and two or three within it, forming its front row. The loop itself is compact in form, about as broad as long, somewhat askew outward with a slant that foreshadows the foot-shaped anal loop of the subfamilies, which are to follow this one. The anal triangle of the male is two-celled. There are no planates, no thickened antenodal crossveins, and there is no brace to the stigma.

The abdomen is stout, and only moderately swollen on its end segments. It is black, often with half rings of bright yellow laid across the rather sharply ridged middle segments. The subgenital plate in the female is short and bilobed.

These dragonflies are often very difficult to capture. They generally fly in long sweeping curves on a regular beat; and the collector, to get within reach of them with his net, may seek a vantage point to which they return frequently, station himself there, and, with a net in readiness, await their passing. It will not pay to chase them.

The nymphs are very long-legged, and their tarsi bear very long and simple claws. They may be recognized at a glance by the presence of a large upcurving pyramidal horn on the head before the eyes, and a low tubercle on each hind angle to rearward. The labium is short, very wide at the front, and markedly spoon-shaped. The teeth on its lateral lobes are deeply and obliquely incised, and armed well with short stout spinules.

The whole body is depressed, the abdomen very flat, with a median row of high, laterally flattened dorsal hooks down the back. The caudal appendages are short and stout.

The nymphs sprawl on bottom sand and silt in shallow lakes and in the settling basins and wide beds of larger streams. They are of strictly sedentary habits. They wait for their prey, well concealed by a coloring of closely mottled grays and browns when living on clean sand, and by an accumulated covering of silt when on a mud bottom.

This subfamily is represented in our fauna by only two genera, *Didymops* and *Macromia.* There are more in the Tropics, especially the Old World Tropics. For an introduction to the world fauna, Martin's *Cordulines,* in the de Selys Collections series (1906), is available, with excellent illustrations. It covers Macrominae and Cordulinae. The best general account of a single species is that of Kennedy, cited hereinafter, for *Macromia magnifica.*

KEY TO THE GENERA

Adults

1—Nodus of fore wing about midway between base and apex of wing; at top of head, vertex smaller than occiput; latter has bulbous swelling on posterior side .... **Didymops**

—Nodus of fore wing distinctly beyond middle of wing; vertex larger than occiput; latter not bulging prominently to rearward; top of vertex bears pair of small cones that are rather acutely pointed .... **Macromia**

NYMPHS

1—Lateral spines of abdominal segment 9 reach to rearward to level of tips of inferior appendages; bulging sides of head hardly narrowed behind eyes; lateral setae of labium three or five; no dorsal hook on 10.......... **Didymops**

—Lateral spines of 9 do not reach to rearward to level of tips of inferior appendages; sides of head somewhat convergent behind eyes to pair of low tubercles on hind angles; lateral setae of labium (so far as known) six; small dorsal hook on 10.................................. **Macromia**

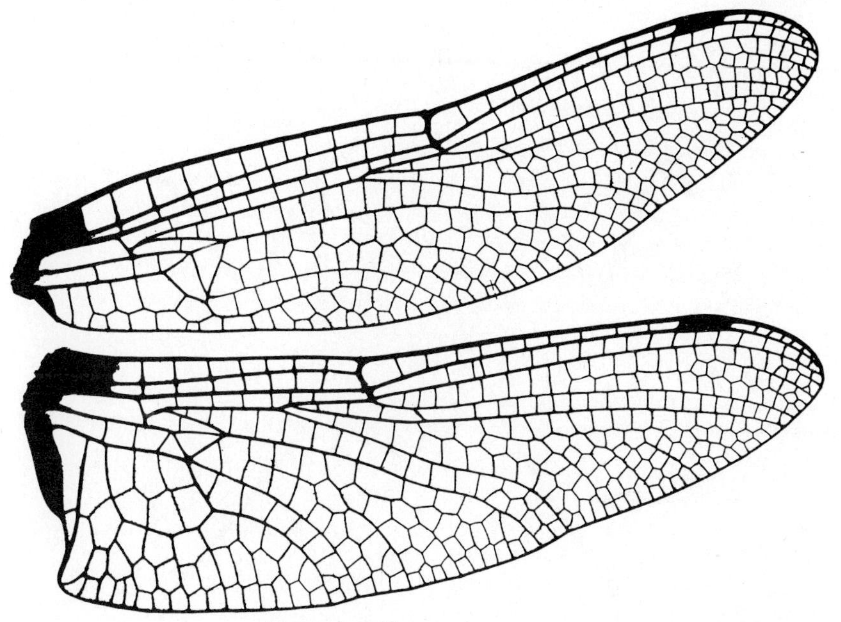

Fig. 202. *Didymops transversa.*

Genus DIDYMOPS Rambur 1842

These are large brownish dragonflies without metallic reflections. The face is conspicuously yellow, with two wide cross stripes of black: one on the anteclypeus, the other on the face of the frons. The frontal furrow is not so deep as in *Macromia*. The stem of a diffuse T spot lies in the median sulcus, its top far overspreading the front and sides. The eye-seam is very short, the eyes meeting in hardly more than a single point. The thorax is without the usual antehumeral pale stripes. The legs are chiefly brown; the tarsi, black. The wings with costa are paler than the remaining veins; triangles and subtriangles are free from crossveins. There are generally three cubito-anals in both front and hind wings.

The nymphs are long-legged, flat sprawlers upon the bottom sand and mud in the borders of lakes and slow streams. The sexes are distinguishable in exuviae by impressions of the developing genitalia on the mid-ventral line of the abdomen, on segment 2 in the male, and at the apex of 8 in the female.

This is a genus of only two species that are restricted in distribution to the eastern United States.

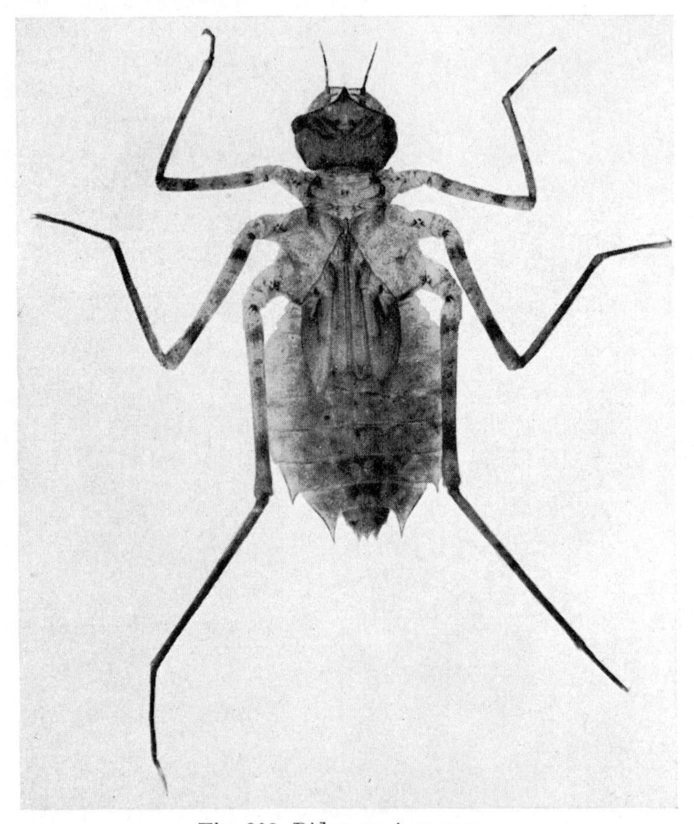

Fig. 203. *Didymops transversa.*

KEY TO THE SPECIES

Adults

1—Length of body less than 63 mm.; basal antenodal cells of both wings tinged with brown .... **transversa**

—Length of body more than 63 mm.; basal antenodal cells of both wings clear **floridensis**

Nymphs

1—Length of body about 35 mm.; teeth on opposed margins of lateral lobes of labium sharply pointed and bare; lateral setae three.............. **floridensis**

—Length of body about 30 mm.; teeth on lateral margins of lateral lobes crenate, blunt, and edged with short spinules; lateral setae five.... **transversa**

Didymops floridensis Davis

1921. Davis, Bull. Brooklyn Ent. Soc., 16:110 (fig.).
1929. N. & H., Handb., p. 170.
1930. Byers, Odon. of Fla., p. 94.

Length 65–68 mm.; abdomen 44–48; hind wing 37–41.

A stocky, clear-winged brownish species. Thorax thickly clad with white hair; blackish abdomen, bare and shining. (See fig. 1, p. 5.)

Labrum yellowish olive, entirely bordered with black. Postclypeus and top of frons yellowish olive; remainder of face black, including a mid-dorsal line in bottom of wide frontal furrow. Ends of thoracic crest and encircling middle band of synthorax bright yellow. Legs shining brown, becoming black distally. Venation of wings black, with costa entirely bright yellow, and stigma tawny.

Stout abdomen about equally brown and dull yellow on its moderately swollen basal segments, with large quadrangular yellow saddle marks on 3 to 7, bright yellow on 3 and 7. It may blacken with age in old males; in general, female paler than male. Caudal appendages almost entirely pale, darkening slightly with age. Segment 2 in male has edge of genital pocket, genital lobes, and posterior hamules yellow.

Distribution and dates.—UNITED STATES: Fla.

February 28 to May 3.

Didymops transversa Say

Syn.: cinnamomea Burmeister, servillei Rambur

1839. Say, J. Acad. Phila., 8:19 (in *Libellula*).
1861. Hagen, Syn. Neur. N. Amer., p. 135.
1890. Cabot, Mem. M. C. Z., 17:14 (fig., nymph).
1901. Ndm., Bull. N. Y. State Mus., 47:481 (nymph).
1907. Mrtn., Coll. Selys Aeschnines, 17:75.
1929. N. & H., Handb., p. 170.

Length 56–60 mm.; abdomen 34–43; hind wing 34–38.

Head marked as in preceding species, save that labrum is wholly brown. Frons and stripes dark brown, not black; yellow of dorsum not so extensive or so bright as in *floridensis*. Occiput greenish yellow, very much swollen and broadly triangular.

Legs have outer face of tibiae pale yellowish. Wings washed with brown across extreme base at front. Costa, media, and basal part of cubitus brown or brownish, becoming yellowish farther out. Stigma tawny.

Abdominal segments 2 to 8 have basal yellow spots that become quite obscure in old males and in museum specimens, and abdomen appears almost uniformly brown. Segment 2 and edge of genital pocket, genital lobes, and posterior hamules brown. Caudal appendages almost entirely pale.

Nymphs of species often travel far from water's edge, and may climb high in trees to find a place for transformation.

Distribution and dates.—CANADA: N. B., N. S., Ont., Que.; UNITED STATES: Ala., Conn., D. C., Fla., Ga., Ind., Ky., Maine, Mass., Mich., Minn., Miss., N. H., N. J., N. Y., N. C., Ohio, Okla., Pa., S. C., Tenn., Tex., Vt.

March 5 (Miss.) to August (New England).

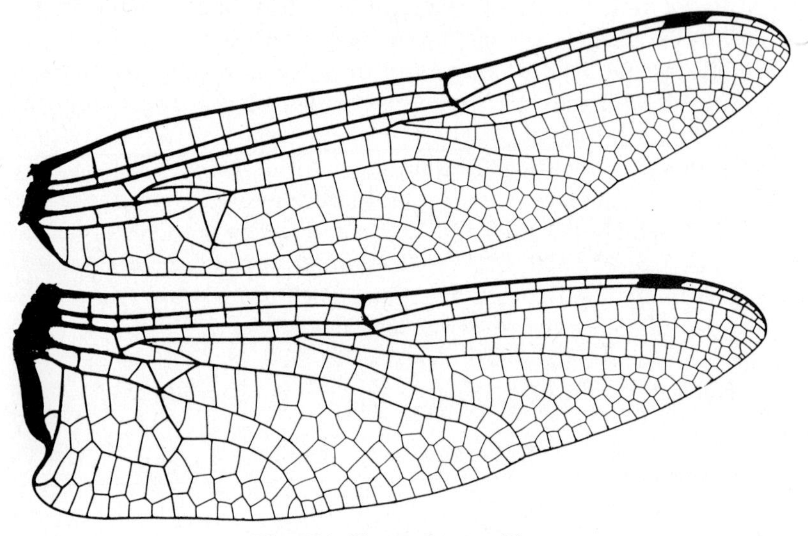

Fig. 204. *Macromia magnifica.*

Genus MACROMIA Rambur 1842

These are fine large dragonflies of robust form and swift meandering flight. The coloration is metallic. The top of the vertex is a double cone, with the two apices rather sharply pointed. The small occiput is brown or blackish, its top shining, its rear hairy. The legs are black. The pale stripes on the front of the thorax vary greatly, according to species, from conspicuous yellow bands of even width to half length or shorter streaks tapering upward and lost in a clothing of hairs, or they may even be absent; their depth of color varies from light yellow to tawny and darker. Likewise, the yellow markings of the abdomen vary in extent and in depth of color. The spots on the two sides, generally separate, may be united into rings, especially where largest, on segments 2 and 7.

The venation of the wings is as stated for the subfamily (p. 327), with a stronger tendency than in *Didymops* for the formation of a panel of about six tall cells in an even row between the anal loop and the marginal row of the hind-wing border. In the fore wing, two or three paranal cells border on the subtriangle. The usually clear wing membrane is sometimes suffused with amber brown, especially in the female.

Genitalia are as shown for several species in figure 206.

The genus is of world-wide distribution. Two important papers on the North American species of *Macromia* are that of Williamson (1909), in which were published the results of an extensive study made of the adults; and that of Walker (1937), which presents a key to the adults of the known species of North America. Both papers are repeatedly cited below.

KEY TO THE SPECIES

Adults

1—Vertex marked with yellow.......... 2
—Vertex wholly blackish.......... 5

2—Yellow of vertex broadly covers its upper surface.......... 3
—Yellow of vertex nearly or quite restricted to its double-cone-like summit.... 4

3—Two yellow spots on each side of abdominal segment 3 confluent; sides of segment 1 marked with yellow.......... **annulata**
—These spots separate; no yellow on sides of 1.......... **magnifica**

4—Yellow on labrum a complete crossband.......... **pacifica**
—Yellow reduced to pair of large spots.......... **rickeri**

5—Yellow stripe on front of thorax present.......... 6
—Yellow stripe wanting or merely vestigial.......... 8

6—Yellow ring on abdominal segment 2 narrowly or not at all interrupted on middorsal line.......... **georgina**
—Yellow ring on 2 widely interrupted on middorsal line.......... 7

7—Costa dark.......... **taeniolata**
—Costa yellow.......... **wabashensis**

8—Auricles of male black.......... **illinoiensis**
—Auricles of male yellow.......... 9

9—Fore-wing triangle generally with crossvein; two cells of paranal row border on subtriangle.......... **margarita**
—Fore-wing triangle without crossvein; three paranal cells border on subtriangle.......... **alleghaniensis**

Nymphs

1—Tubercles on rear of head erect, conical, each as high as its width at base; dorsal hook on abdominal segment 10 projecting prominently from middorsal ridge to rearward and decurved at tip.......... **caderita**
—Tubercles on rear of head low, moundlike or wartlike; dorsal hook on 10 mere middorsal ridge prolonged more or less to rearward, not decurved at tip.......... 2

TABLE OF SPECIES

ADULTS

Species	Hind wing		Cells in wing[2]			Fore w. Cu-A cvs.[3]	Yellow spots[4]	Distr.
	Length	An. cvs.	Triangles	Super T	Sub T			
alleghaniensis	45-50	10-*11*[1]	1/1	*3*-4/*2*-3	1-*2*/1	*4*-5	separ.	E
annulata	44-50	*9*-10	1/1	2-3/2	1/1	*3*-4	confl.	E, SW, C
georgina	43-53	9-*11*-12	*1*-2/*1*-2	3-*4*/*2*-3	*1*-2/1	4-*5*	separ.	S
illinoiensis	40-49	10-*11*-12	*1*-2/*1*-2	2-*4*/*2*-3	1/1	4-5-6	separ.	E, C
magnifica	43-46	8	1/1	*2*-3/2	1/1	3	confl.	W
margarita	46-52	10-*11*-12	1-*2*/1-2	3-*4*-5/*2*-3	1-*2*/1	4-5-6	separ.	N. Car.
pacifica	41-50	8-9	*1*-2/*1*-2	2-*4*/*2*-3	*1*-2/1	*4*-5	confl.	C
rickeri	40-44	10-*11*-13	1/1	2/2	1/1	*3*-4	separ.	B. C.
taeniolata	46-59	10-*12*-14	1-*2*/1-*2*	3-*4*-5/3	1-*2*/1	*4*-6	separ.	E, C, S
wabashensis	46-48	11-*12*-14	*1*-2/*1*-2	3-*4*/2-*3*	1-*2*/1	4-5-6	separ.	C

[1]Prevailing numbers italicized.

[2]Numerals indicate range in number of cells present in triangle, supertriangle, and subtriangle; oblique line (solidus-/) used as a shorthand symbol for words "in fore and hind wing, respectively" in all three columns.

[3]Number of crossveins in cubito-anal space of fore wing, not counting the one that forms front side of subtriangle.

[4]Pair of yellow spots on top of frons: separate spots (separ.) or confluent into single spot across middle line of frons (confl.).

Key to genera of Macromians on page 328.

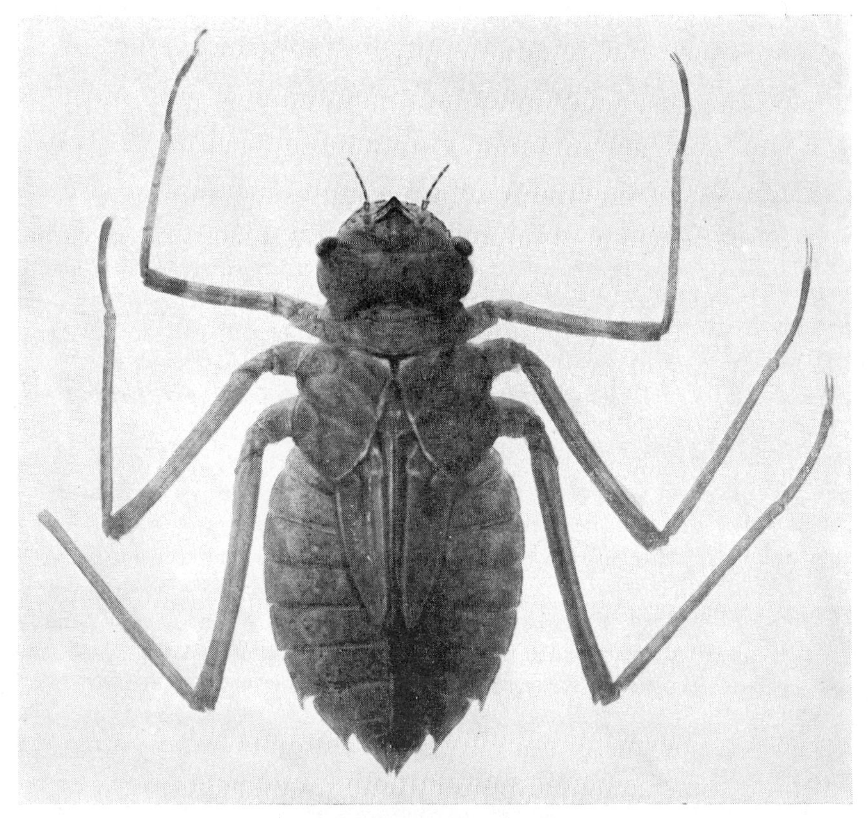

Fig. 205. *Macromia taeniolata.*

2—Horn on front of head directed forward; dorsal hook on 9 a mere ridge **magnifica**

—Horn on front of head turns upward at angle of 45° or more from horizontal.. 3

3—Lateral spines of 8 and 9 with axes directed straight to rearward; dorsal hooks on 6 and 7 with bird-head curvature viewed from side...... **illinoiensis**

—Lateral spines of 8 and 9 incurved, especially 8........................... 4

4—Tip of dorsal hook on 9 projects only to level of tip of dorsal hook on 10 **alleghaniensis**

—Dorsal hook on 9 projects beyond level of tip of dorsal hook on 10........... 5

5—Lateral spines of 8 about equal in size to those of 9................. **taeniolata**

—Lateral spines of 9 nearly twice as large as those of 8, extending almost to level of tips of caudal appendages............................ **georgina**

Nymphs unknown: **annulata, pacifica,** and **wabashensis** (**rickeri** not included herein; see Walker, 1937).

Macromia alleghaniensis Williamson

1909. Wmsn., Proc. U. S. Nat. Mus., 37:376.
1929. N. & H., Handb., p. 167.
1947. Wstf., J. Elisha Mitchell Sci. Soc., 63:32–36 (fig.).

Length 65–72 mm.; abdomen 51–56; hind wing 45–50.

Face heavily clothed with short dark hair. Labrum yellowish brown, with front border black; postclypeus yellow, often divided by brown into a central and two lateral pale areas; frons dark brown, blackish on dorsum, with two small superior yellow spots (sometimes wanting) and a lateral spot on each side, laterals slightly larger and elongated.

Thorax without trace of antehumeral pale stripes; latero-ventral metathoracic carina very narrowly yellow posteriorly. Wings hyaline; costa and stigma dark. First tibia in male 8.3–9 mm. long, hind tibia 12–13 mm., keel on middle tibia one-seventh to one-fifth as long as tibia.

Abdominal segment 2 with ring of yellow interrupted in middorsal line, but not at auricles; 3 with a lateral pale stripe not connected with dorsal spot; 3 to 6 with a pair of dorsal spots anterior to transverse carina, confluent or narrowly separated middorsally on 3 and 4, more widely separated on 5 and 6 by a median dark stripe; 7 with basal pale spot completely encircling segment basally and extending posteriorly in middorsal line a little beyond transverse carina; 8 with a large basal middorsal spot and a pale spot on anterior front corner; 9 with basal inferior margin pale. Appendages as seen from side slightly constricted just beyond base and slightly widened just before apex; seen from above there is a small lateral external tooth midway of appendage. Genital lobes black; posterior hamules yellow basally and black at tips, tips long and slender.

Female similar to male, but with dorsal spots of frons larger. Pale spot on segment 7 interrupted laterally. Basal lateral inferior spots of 7 to 9 may not be conspicuous; 8 may or may not have a pale area basally. Wings hyaline or slightly tinged with brown; costal and subcostal interspaces show a trace of brown. Legs slightly longer in female, front tibia about 9 mm., hind tibia 13 mm.

Male may at once be separated from *illinoiensis* and *margarita* by short keel of middle tibia and by abdominal markings. Female can be separated from *illinoiensis* by longer tibiae.

Distribution and dates.—UNITED STATES: Ala., Ga., Ky., N. J., N. C., Pa., Va.

June 23 (Ky.) to September 7 (Ga.).

Macromia annulata Hagen

1861. Hagen, Syn. Neur. N. Amer., p. 133.
1907. Mrtn., Coll. Selys Cordulines, p. 66.
1909. Wmsn., Proc. U. S. Nat. Mus., 37:387.
1929. N. & H., Handb., p. 168.
1937. Walk., Can. Ent., 69: 13.

Length 68–73 mm.; abdomen 52–55; hind wing 44–50.

Face very light yellow, with frons in front and anteclypeus somewhat darker; rear of eyes also pale except above. A very narrow brown line in sulcus of frons; usual pale lateral areas of frons extensive, reaching the postclypeus.

Thorax with antehumeral stripe long, separated from crest above by a distance equal to its own width or slightly more. Latero-ventral metathoracic carina broadly yellow. Wings hyaline, costa distinctly yellow to wing tip; stigma yellowish brown. Femora pale basally.

A lateral yellow streak on abdominal segment 1. Only species of North American *Macromia* known in which this segment is not concolorous laterally. Segment 2 with yellow ring very wide and uninterrupted either above or at level of auricles; dorsal spots on 3 to 8 not separated in mid-dorsal line as in *pacifica,* dorsal spot on 3 broadly connected posteriorly with a pale stripe on inferior margin of segment; spots on 7 and 8 encircle segments basally; 7 to 10 largely pale-colored beneath. Superior caudal appendages, when viewed from above, are seen to be straighter and much wider beyond median external tooth than in any other North American species of *Macromia;* they widen from tooth to before broadly rounded apex, whereas in *pacifica* they taper from tooth to apex; there is an indistinct, pale dorso-basal area.

Female similar to male, but with dorsal light spots on segments 4 to 6 connected just in front of transverse carina, with extensive and suffused stripes on inferior margin of segments; 9 and 10 with traces of dorso-basal spots. Wings hyaline, with faint traces of basal brown spots in costal spaces.

Distribution and dates.—UNITED STATES: Ill., Okla., Tex.

May 30 (Tex.) to August 27 (Tex.).

Macromia caderita Needham

1950. Ndm., Trans. Amer. Ent. Soc., 76:11.

Adult of species not known. Nymph (fig. 18, p. 29) 25–27 mm. long.

Familiar macromian form: long-legged and sprawling, with a skyline of high dorsal hooks along middle ridge of flat abdomen. Eye-tips elevated at fore corners of head. Two tubercles at rear corners erect and cone-shaped, with rounded tips. Pyramidal horn on front of head sharply up-curved to tip. Huge labium has usual six deeply cut spinule-fringed teeth on terminal margin of lateral lobe of labium; six lateral setae, five close-set mental setae, and several additional smaller setae placed nearer middle line of mentum.

Broad longitudinal band of darker brown markings covers middle third of abdomen. Outer third on each side paler, more thinly speckled with brown. Abdomen very flat; hooks along its middorsal line on segments 2 to 9 very prominent and of nearly equal height. Lateral spines sharp and slightly curved, nearly equal on the two segments. Their outer edge bare save for a thin line of long soft hairs. Caudal appendages triangular pyramidal, somewhat attenuate to their paler tips; superior and inferiors about of equal length, laterals a little shorter.

Distribution.—UNITED STATES: Texas; MEXICO: Nuevo León.

Macromia georgina Selys

Syn.: australensis Williamson

1878. Selys, Bull. Acad. Belg., (2)45:197 (reprint, p. 19) (in *Epophthalmia*).
1907. Mrtn., Coll. Selys Cordulines, p. 64 (figs.).
1909. Wmsn., Proc. U. S. Nat. Mus., 37:383, also 381 (as *australensis* n. sp.).
1929. N. & H., Handb., p. 169.
1937. Walk., Can. Ent., 69:12.
1947. Wstf., J. Elisha Mitchell Sci. Soc., 63:32–36 (fig.).

Length 73–79 mm.; abdomen 50–56; hind wing 43–53.

Labrum yellowish brown, with a black margin; postclypeus bright yellow, with entire front border brown; frons brown to black, with the two dorsal spots of yellow larger than lateral spots.

Thorax with antehumeral stripes varying from 1.5 mm. to 3 mm. long; latero-ventral metathoracic carina yellow. Wings hyaline, without trace of color; costa brown; stigma yellowish brown to black. First tibia 7.25–8 mm. long, hind tibia 11 mm., keel of middle tibia one-fourth to one-half as long as tibia.

Abdomen black, with following yellow: ring on segment 2 continuous over dorsum and on sides; 3 to 6 with dorsal spots which may be united over dorsum or separated by a very thin line on 3 to 5 or by a wider line of black on 6, 3 usually with dorsal spot prolonged ventrally in male and female to meet longitudinal stripe on inferior margin of segment; 7 with yellow band usually slightly interrupted ventro-laterally; 8 with a shorter spot which may be almost confluent on anterior margin with inferior spot; 9 with a basal inferior spot; 10 may show a light area on each side. Appendages viewed in profile with upper and lower margins of superiors about parallel, slightly upturned at tip; from above, a prominent external tooth is seen on superiors at mid-length.

Female similar to male, but with abdominal spots more uniform in size. Wings with basal brown streaks in costal or costal and subcostal areas and slightly fumose at tips. Female of this species, like male, has dorsal spot of segment 3 joined below with longitudinal stripe of inferior margin—a good characteristic for species.

Distribution and dates.—UNITED STATES: Ala., Fla., Ga., Ill., Ind., Kans., Md., Miss., N. C., Okla., S. C., Tex.

June 15 (Miss.) to October 11 (Ga.).

Macromia illinoiensis Walsh

1862. Walsh, Proc. Acad. Phila., p. 397.
1890. Cabot, Men. M. C. Z., 17:3, 16 (nymph).
1907. Mrtn., Coll. Selys Cordulines, p. 67.
1909. Wmsn., Proc. U. S. Nat. Mus., 37:377 (figs.).
1927. Garm., Odon. of Conn., p. 204.
1929. N. & H., Handb., p. 166.
1947. Wstf., J. Elisha Mitchell Sci. Soc., 63:32–36 (fig.).

Length 65–76 mm.; abdomen 47–51; hind wing 40–49.

Apparently our most common and widespread *Macromia.* Labrum yellowish brown, with a black band on front border; postclypeus brighter yellow, with a dark stripe on anterior border widening slightly at sides; frons dark brown to metallic black, with the two pale spots above smaller than those on lateral face of frons.

Thorax without trace of antehumeral light stripes. Latero-ventral metathoracic carina brown. Wings hyaline, tips sometimes slightly fumose; costal and subcostal areas may be marked with brown basally to first antenodal; venation black, stigma yellowish brown in tenerals to black in fully mature specimens. First tibia 7 mm. long, hind tibia 11 mm., keel on middle tibia about one-half as long as tibia—only very rarely shorter.

Abdominal segment 2 with usual yellow band interrupted middorsally and at auricles, forming four yellow spots; 3 to 6 with a pair of dorsal spots confluent over dorsum, spots becoming smaller posteriorly and often entirely absent on 4 or 5 to 6; 3 with a ventro-lateral stripe not connected with dorsal spot; 7 with a basal spot of yellow reaching transverse carina laterally and slightly produced beyond transverse carina in middorsal line, not encircling segment but limited beneath on sides by black; 7 sometimes showing a trace of yellow on basal inferior margin; 8 variable, but usually with a pair of yellow spots, one on each side of middorsal carina, these sometimes confluent over dorsum and separated only by darker carina; 8 and 9 with a prominent yellow spot on basal inferior margin; 9 and 10 black above. Caudal appendages viewed from side: upper edge of superiors almost straight, slightly elevated at apex, lower edge beyond swollen base almost parallel with upper, edges converging at tip; seen from above, superiors lyre-shaped, usually bearing a small lateral external tooth which may be absent. Inferior equals or slightly exceeds superiors in length. Genital lobes black; posterior hamules very slender, with long narrow blackish tips.

Female similar to male, but with dorsal spots on frons as well as abdominal spots larger. Segment 2 with dorsal spot usually prolonged ventrally along transverse carina, but not meeting inferior spot; 3 to 6 with paired spots not confluent over dorsum, spot on each side of 3 not meeting pale stripe on ventral border of that segment; 7 with a large basal spot similar to one in male; 8 with a larger spot than in male; 8 and 9 with only small spots on basal inferior angle, or these spots missing entirely. Wings sometimes yellowish brown, especially in tenerals, with basal brown spots more extensive, occasionally reaching beyond second antenodal.

Species may be separated from other two which lack antehumeral stripes as follows: long tibial keels of male and shorter tibiae of both sexes distinguish it from *alleghaniensis*, and latter character separates both sexes from *margarita*. *M. illinoiensis* only species which has yellow band of segment 2 interrupted in males at level of auricles.

This fine species is very common about the Chara beds on the bottom of the lake (Walnut Lake, Michigan), especially where the growth is sparse, exposing bare bottom areas. There are few finer examples of protective coloration than these larvae present; nevertheless they are much eaten by the fishes. If overturned by the fishes in fanning the bottom they would be readily seen with the light yellow color of the under surface of the body exposed . . . They have been taken in many places in the lake, always in connection with a scanty growth of Chara.—Needham (*Rep. Geol. Surv. Mich. for 1907*, p. 263).

Distribution and dates.—CANADA: N. B., N. S., Ont., Que.; UNITED STATES: Ala., Ga., Ill., Ind., Iowa, Kans., Ky., Maine, Md., Mass., Mich., Miss., Mo., N. H., N. J., N. Y., N. C., Ohio, Pa., R. I., S. C., Tenn., Va.,W. Va., Wis.

May 3 (Ohio) to September 30 (Tenn.).

Macromia magnifica McLachlan

1874. McL., in Selys, Bull. Acad. Belg., (2)37:22 (reprint, p. 11).
1909. Wmsn., Proc. U. S. Nat. Mus., 37:389 (fig.).
1915. Kndy., Proc. U. S. Nat. Mus., 49:313 (figs., adult and nymph).
1929. N. & H., Handb., p. 168.
1937. Walk., Can. Ent., 69:13.
1938. LaRivers, J. Ent. Zool. Claremont, Calif., 30:79.

Length 69–74 mm.; abdomen 46–52; hind wing 43–46.

A very brightly colored Far Western species. Old specimens show a marked pruinosity of sides of thorax. Face largely yellow, with two black crossbands: the longer, larger one covers vertical face of frons; shorter band covers anteclypeus and basal third of labrum. A cap of yellow covers most of vertex.

Thorax brown, with half-length antehumeral stripes and a wide mid-lateral girdle of clay yellow; top of crest brighter yellow. Inferior carina at junction with abdomen bright yellow. Wings hyaline, with bright yellow costa and brown or black stigma.

Abdomen broadly half-ringed with dull clay yellow. Girdle band of segment 2 almost or quite divided on middorsal line by a basal triangle of black; it may or may not be interrupted laterally at level of auricles. Saddle marks broad on basal half of middle segments, diminishing in breadth on 3 to 6, increasing on 7 and 8, becoming small again on 9; 10 largely commingled yellow and brown, blackish about middorsal carina. Yellow spots low down on sides decrease in proportionate area on 8, 9, and 10.

From LaRivers' (1938) observations on this species in southern Nevada we take these excerpts:

A large yellow and black "belted skimmer," a conspicuous element in any libelluloid assemblage . . . at well formed ponds, a marginal beat is most common, with occasional sorties over the open pond . . . At creeks, a short beat up or down stream with return trip out over the surrounding country . . . period of greatest activity, morning, after other species had been on the wing for some time . . . almost continuously on the wing.

Distribution and dates.—CANADA: B. C.; UNITED STATES: Ariz., Calif., Nev., Wash.

June 6–9 (Calif.) to September (Sask.).

Macromia margarita Westfall

1947. Wstf., J. Elisha Mitchell Sci. Soc., 63:32–36 (figs.).

Length 72–78 mm.; abdomen 49–57; hind wing 46–52.

Face much less hairy than in *alleghaniensis,* with lateral faces of frons almost devoid of hair. Labrum yellowish brown, with a black anterior border. Postclypeus bright yellow except for a very narrow edging of brown on anterior margin. Frons metallic blue, with a large yellow spot on each side and a pair of spots above, only about half as large as, or at least smaller than, lateral spots.

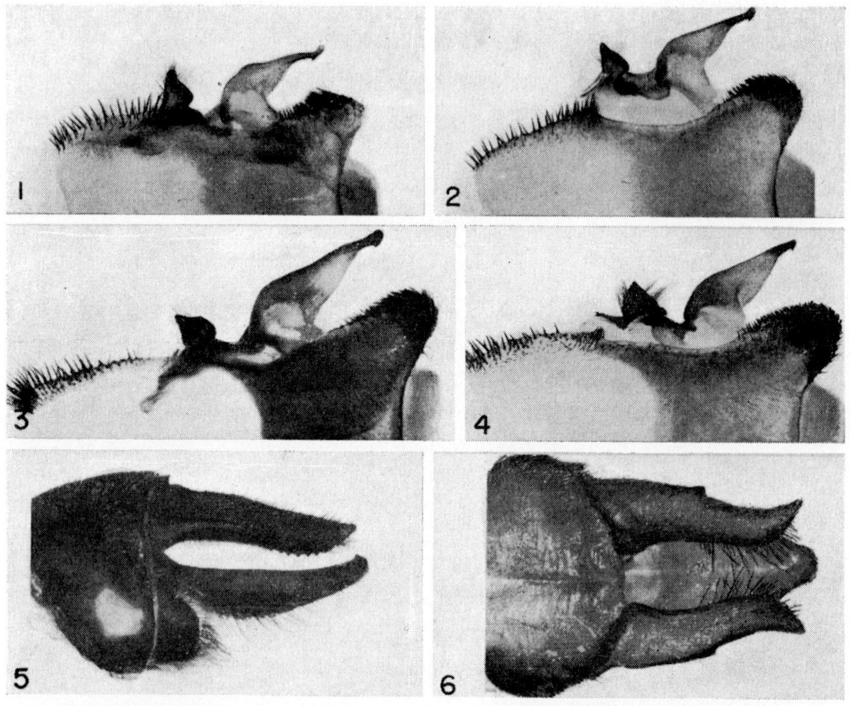

Fig. 206. Genitalia of several closely related species of *Macromia*. Four lateral views of abdominal segment 2 of male, showing hamules, genital lobe, and edge of genital pocket: 1, *Macromia georgina;* 2, *M. illinoiensis;* 3, *M. margarita;* 4, *M. alleghaniensis.* Nos. 5 and 6 *M. margarita,* are lateral and dorsal views of segment 10 and appendages (from paper by the junior author, *J. Elisha Mitchell Sci. Soc.,* 63:32, 1947). (Figs. 1–4 photographed to same scale.)

Thorax without trace of antehumeral stripes; latero-ventral metathoracic carina narrowly yellow posteriorly. Wings hyaline, with only faintest suggestion of brown at base in costal and subcostal spaces; extreme tips faintly tinged with brown; venation and stigma black. First tibia 9.5 mm. long, hind tibia 14 mm., keel of middle tibia one-half to seven-twelfths as long as tibia.

Abdomen black, with following yellow: segment 2 with usual band generally interrupted middorsally and narrowed but not interrupted at level of auricles; 3 to 6 with a pair of dorsal spots not confluent over dorsum, decreasing slightly in size to rearward but always present, 3 with dorsal spot not joining stripe on basal inferior edge; 7 with a large basal spot which encircles segment or is slightly interrupted laterally; 8 with a smallér spot dorsally not reaching large spot on basal inferior margin of 8; 9 with a basal inferior spot, sometimes also with a distal inferior spot and a small lateral spot toward apex; 10 with a small lateral spot. Caudal appendages viewed in profile: dorsal line of superiors almost straight, slightly elevated at apex; inferior line almost parallel with dorsal, not so noticeably constricted beyond base and widened near apex as in *alleghaniensis;* seen from above: a small midlateral tooth externally; genital lobes black; posterior hamules broad and stout, with a short comparatively broad tip, black, paler basally.

Female similar to male, but with yellow spots of frons larger; band on 2 interrupted latero-ventrally as well as dorsally; dorsal spots on 3 to 6 larger than in male, basal inferior spot on 3 almost absent; yellow of 7 reduced to a large basal spot; 8 with a much smaller dorsal spot; 8 and 9 with small basal inferior spots; 10 all black. Front tibia 10 mm. long, hind tibia 15 mm.

Separated from *illinoiensis* by longer tibiae, yellow ring on segment 2 of male not interrupted at auricles, much stouter posterior hamules, and other differences shown in table and description. From *alleghaniensis* it is distinguished by longer keels of middle tibiae in males, longer tibiae, stouter posterior hamules with short tips, superior appendages of male, and venational characters shown in table (p. 334). From *georgina* it is distinct because of absence of antehumeral stripes, stouter posterior hamules with short tips, shorter anterior hamules, longer keels on middle tibiae, and several venational characters as shown in table.

Distribution and dates.—UNITED STATES: N. C.

June 20.

Macromia pacifica Hagen

Syn.: flavipennis Walsh

1861. Hagen, Syn. Neur. N. Amer., p. 134.
1862. Walsh, Proc. Acad. Phila., p. 398 (as *flavipennis* n. sp.).
1909. Wmsn., Proc. U. S. Nat. Mus., 37: 389 (fig.).
1917. Kndy., Bull. Dept. Entomol., Univ. Kansas, 18:139 (figs.).
1929. N. & H., Handb., p. 168.
1937. Walk., Can. Ent., 69:12.

Length 74 mm.; abdomen 48–54; hind wing 41–50.

A brightly colored species, with full-length antehumeral yellow stripes. Labrum dull yellow, with a very narrow black stripe on anterior margin. Postclypeus greenish yellow. Frons with entire dorsum yellow except median sulcus, yellow of dorsum confluent down sides with lateral spots.

Thorax with antehumeral stripes separated from crest above by about their own width. Latero-ventral metathoracic carina brown or occasionally very narrowly yellow. Wings hyaline to strongly tinged with yellow; a very small basal brown spot sometimes present in costal space; costa yellow to wing tip; stigma black.

Abdomen dark brown to black, with following yellow: wide ring on segment 2 usually very narrowly interrupted above, but entire on sides; 3 to 6 with pair of large spots narrowly separated in middorsal line, and extending from transverse carina forward almost to anterior end of segment; spots on 3 to 7 usually not connected with inferior spots; 7 and 8 with spots confluent over dorsum and prolonged to rearward in a triangle on each side of middorsal line, with yellow on 8 encircling segment; 9 with a small dorsal spot; inferior spots most noticeable on 8 and 9. Superior appendage with external tooth on lateral surface and a small external yellow spot at base.

Female similar to male, but with yellow ring on 2 interrupted dorsally; dorsal spots on 8 not connected with inferior spots, which are prominent only on 8.

Distribution and dates.—UNITED STATES: Calif. (?), Ill., Ind., Kans., Nev., Ohio, Okla., Tex., Wis.

April 13 (Ind.). to August 30 (Ind.).

Macromia rickeri Walker

1937. Walk., Can. Ent., 69:7 (figs., adult and nymph).

Length 64–70 mm.; abdomen 46–50; hind wing 40–44.

Head very dark brown, almost black, with metallic reflections and yellow markings. A small bilobate spot on frontal vesicle yellow, also two rather large widely separated spots on dorsum of frons which are narrowly separated from spots on lateral face of frons, dorsal and lateral spots of about equal size; postclypeus yellow with an impressed area on each side of base and a median distal spot dark; labrum entirely dark except for a pair of sharply delimited pale spots on lower half; pile of face not very dense, and blackish.

Thorax dark brown, with greenish reflections. Antehumeral pale stripe very short, separated from crest above by a distance equal to one and one-half to more than three times its own length, and not confluent with yellow spot on mesinfraepisternum below. Band of yellow on side of thorax narrow. Latero-ventral metathoracic carina narrowly yellow. Wings hyaline, with black venation and stigma; faintly tinged with yellow in adults, more so in tenerals. Middle tibiae without trace of a keel.

Abdomen dark greenish brown, with yellow markings: segment 2 with a pair of widely separated dorso-lateral spots extending over auricles but not reaching inferior margin of segment; 3 to 6 with dorsal yellow spots diminishing in size to rearward and separated on middorsal line; 9 and 10 entirely dark.

Female similar to male but with reduced yellow spots on 2 not extending so far toward inferior margin; spots on 3 to 6 more uniform in size than in male; those on 7 and 8 about as large as those of male, and uninterrupted in middorsal line.

Distribution and dates.—CANADA: B. C.

June 16 to September 10.

Macromia taeniolata Rambur

1842. Rbr., Ins. Neur., p. 139.
1909. Wmsn., Proc. U. S. Nat. Mus., 37:372 (fig.).
1929. N. & H., Handb., p. 167.
1937. Walk., Can. Ent., 69:13.

Length 75–91 mm.; abdomen 51–67; hind wing 46–59.

Our largest *Macromia,* a broad-winged blackish species. Face densely clothed with short black pubescence. Labrum brown, with a wide black front border. A yellow band covers top half of postclypeus, widening toward its ends. Hollow of top of frons metallic blue except for two yellow spots, one before each lateral ocellus; lateral spots on frons indistinct.

Thorax brown, its lateral convexities faintly metallic blue. Low down on front it is densely clothed with whitish hair. Seen dimly through the hair, antehumeral yellow stripes fade out at half the height of front. Wings hyaline, tinged with yellow in teneral specimens, sometimes heavily tinted with amber brown except that a basal area about triangles may be clear. Costa dark.

Abdomen black, with usual spots of yellow not very prominent, slightly diminishing in size to segment 7: inferior abdominal spots almost lacking, most pronounced on 2; ring on 2 narrow and slightly interrupted above, entire laterally; 8 with a subbasal spot which may or may not be interrupted in mid-line; 9, 10, and appendages blackish.

Female similar to male, but with lateral spots on frons more distinct, though still smaller than dorsal spots. Basal spots on 8 may be reduced or entirely wanting.

Distribution and dates.—UNITED STATES: Ala., D. C., Fla., Ga., Ill., Ind., Iowa, Ky., La., Md., Miss., Ohio, Pa., S. C., Tenn., Tex., Va.

April 26 (Fla.) to November 11 (Fla.).

Macromia wabashensis Williamson

1909. Wmsn., Proc. U. S. Nat. Mus., 37: 374 (fig.).
1929. N. & H., Handb., p. 167.

Length 70–79 mm.; abdomen of male 51–57; hind wing of male 46–48.

Very similar to *taeniolata*, but may be recognized by the following characters: costa and antenodals and postnodals yellow or yellowish as far as stigma or tip of wings; labrum less obscured, face paler and brighter colored; lateral spots on frons distinct. Wings tinged with pale yellowish, or hyaline, the extreme apex frequently slightly fumose. Abdominal appendages similar to those of *taeniolata*, but seen in profile the apices of the superiors are less curved and upturned, the apical portion between the median external tooth and apex less inflated, approaching the form of *illinoiensis;* median lateral external tooth present in every case.—Williamson (1909).

Distribution and dates.—UNITED STATES: Ind., Okla.(?).
June 11 (Ind.) to August 11 (Ind.).

Subfamily CORDULINAE Selys

These are strong-flying dragonflies, some of large size and many of brilliant metallic coloration. The low tubercle on the rear margin of each eye is less prominent than in the Macrominae. The hind-wing triangle is retracted to the arculus. There is a long anal loop, somewhat foot-shaped, a little widened at the distal end, with small development of the toe. The basal triangle in the male generally contains but one dividing crossvein, which is obliquely placed far down in the posterior angle. Adjacent to this triangle is a long tornal cell (*to*) somewhat crescentic in form. Outside the loop at its upper end a single large cell rests in the place for a kneecap (whence this cell is called the *patella* and is labeled *p*).

In the formation of the triangle of the hind-wing base in the male, the first two paranal cells ($n + o$) have been consolidated, so that there appear to be but five paranals in all, three of them in the base of the anal loop; but in the female there are three before the loop as usual. Tornal cell and patella are both extremely elongated, with a long spliced contact between them. There is no subtriangle, since the second cubito-anal crossvein does not tip over far enough to meet the triangle at its upper end. Anal crossing (Ac), middle fork (Mf), and long veins are as shown in the introductory figure of wing venation (fig. 7, p. 17).

Note the fine mechanical adjustment of the cells of the anal area, and the graded row of six "panel cells" (p-a-n-e-l-s) filling the space between vein A2 and the marginal row.

In most genera there is a single bridge crossvein. On the under side of fore and middle legs is a longitudinal carina in males.

The principal papers dealing with this subfamily as a whole are those of Martin (*Coll. Selys Cordulines,* 1906), especially valuable for its fine illustrations; Williamson (*Ent. News,* 18:428–434, 1908); and Needham (*Ann. Ent. Soc. Amer.,* 1:273–280, 1908). All these are based on adults only.

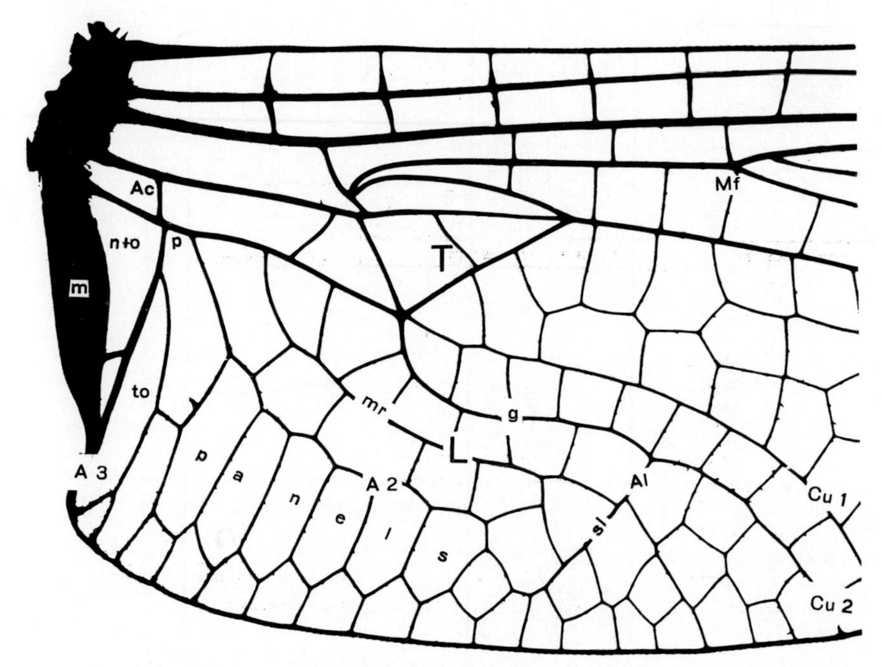

Fig. 207. Basal part of hind wing of *Somatochlora tenebrosa,* showing venational features of Cordulinae.

The nymphs are all sprawlers on the bottom or climbers over bottom trash and waterweeds. The ones that lie in the silted bottom (*Cordulia,* etc.) are generally hairy; the climbers (*Tetragoneuria,* etc.) are often rather prettily patterned in green and brown. Our genera have a predominantly northward distribution.

KEY TO THE GENERA

Adults

1—Veins M4 and Cu1 in fore wing divergent to wing margin 2
—These veins convergent to wing margin 3

2—Triangles and subtriangles each of a single cell .... **Williamsonia**
—Fore-wing triangle divided into two or three cells (includes **Platycordulia** and **Rostrocordulia**) .... **Neurocordulia**

TABLE OF GENERA

ADULTS

Genera	Hind wing	Cells in triangles[1]					Crossveins				Distr.
		Fore wing			Hind wing		Hind wing		Both wings		
		Sub T	T	Super T	T	Super T	Ante-nodals	Cubito-anals	Over bridge	Under stigma	
Cordulia	29-32	2-*3*	1-*2*	1	1	1	5-6	1	1-2	1	N
Dorocordulia	26-31	2-3	1	1	1	1	5	1	1	1	NE, C
Epicordulia	38-52	3	*2*-3	1	1-*2*	1	5-6	1	1-2	1	E, C
Helocordulia	26-29	2-*3*	2	1	1	*1*-2	5	1-*2*	1-3	1-2	E
Neurocordulia	29-40	1-*3*	2-*3*-4	1-2	2	1-2	5	*2*-3-4	2-5	1-*2*-3	E, C, S
Somatochlora	25-47	3	2	1	1-2	1	5-6	2	1-2	*1*-2	N, C, E
Tetragoneuria	24-34	3	2	1	*1*-2	1	*4*-6	1	1-3	1	General
Williamsonia	22-23	1	1	1	1	1	5	1	1	1	N

[1]Number of cells in triangles, subtriangle, and supertriangles. Italics indicate prevailing numbers.

3—Wings with large spots of brown at nodus and stigma; hind-wing triangle two-celled; large species, hind wing 38 mm. or more .... **Epicordulia**
—Wings with no spots at nodus or stigma 4

4—Hind wing generally with two cubito-anal crossveins 5
—Hind wing generally with one cubito-anal crossvein 6

5—Hind wing with spots at base and on some antenodal crossveins . . **Helocordulia**
—Hind wings clear .... **Somatochlora**

6—Fore-wing triangle one-celled .... **Dorocordulia**
—Fore-wing triangle generally two-celled 7

7—Outer side of fore-wing triangle straight or nearly so .... **Cordulia**
—Outer side of fore-wing triangle distinctly convex .... **Tetragoneuria**

NYMPHS

1—Pair of small tubercles on top of head; lateral lobe of labium with four or five setae (sometimes six or seven in **Neurocordulia**) 2
—No tubercles on top of head; lateral lobe with six or seven setae 3

2—Strong lateral spines of abdominal segment 8 often very divergent and as strong as parallel spines of 9 .... **Neurocordulia**
—Lateral spines of 8, if divergent at all, much weaker than those of 9; small setae on middle lobe of mentum numerous .... **Epicordulia**

3—Dorsal hooks large, laterally flattened, generally cultriform 4
—No dorsal hooks, or rudimentary; not flattened laterally or cultriform, but small obtuse or pointed prominences 6

4—Lateral spines of 9 more than half its middorsal length; dorsal hooks on 2 to 9, highest on 6 or 7, often cultriform and sharp .... **Tetragoneuria**
—Lateral spines of 9 less than half its middorsal length; dorsal hooks less developed 5

5—Dorsal hooks on segments 2 to 9 laterally flattened, but obtuse at apices **Somatochlora**
—Dorsal hooks on segments 6 to 9, longest on 8 and cultriform ...... **Helocordulia**

6—Hind angles of head rounded; lateral spines of 9 one-fifth as long as segment 7
—Hind angles of head angulate superiorly; spines of 9 about one-third as long as segment .... **Dorocordulia**

7—Teeth on lateral lobes of labium deeply cut and separated by rather wide notches .... **Somatochlora**
—Teeth low, separated only by shallow crenulations .... **Cordulia**

Nymph unknown: **Williamsonia.**

Genus NEUROCORDULIA Selys 1871

These dragonflies are of moderate size and soft brown nonmetallic coloration. The face is yellowish olive, clothed with blackish hairs. The frons is low and rounded, with little development of longitudinal furrow. The top of the vertex is a narrow transverse shelflike ridge that overshadows the middle ocellus. The yellow occiput is nearly bare.

TABLE OF GENERA

NYMPHS

Genera	*Length*	*Setae*[1]		*Dorsal hooks*[2]			*Lat. spines*[3]		*Caudal append.*[4]		
		Lat.	*Ment.*	*On segs.*	*High-est*	*Form*	*8*	*9*	*Lat.*	*Sup.*	*Inf.*
Cordulia	20-21	7	14	7-9	8	vest.	3	3	7	9	10
Dorocordulia	20-21	7	12-13	4-9	5	stubby	3	4	7	9	10
Epicordulia	27-29	4(5)	4-5	2-9	6	cultr.	4	5	8	9	10
Helocordulia	20-21	6-7	13-14	6-9	8	cultr.	2	3	6	10	10
Neurocordulia	18-25	5(6)	7-9	2-9	6	blunt	3	4	7	10	10
Somatochlora	17-25	5-10	9-15	2-9[5]	7 or 8	spinous	0-3	0-4	9	9	10
Tetragoneuria	19-23	6-7	9-12	2-9	6 or 7	cultr.	2	3-4	4-8	8-9	10

[1]Number of raptorial setae on labium; lateral lobe and one side of mentum respectively.

[2]Middorsal abdominal hooks: (1) on which segments; (2) on which highest; (3) form as viewed from side, expressed as cultriform (like a hawk's beak), blunt, spinous, stubby, or vestigial (vest.).

[3]Length of lateral spines on segments 8 and 9 expressed in tenths of entire length of lateral margin of which they are a part.

[4]Relative length of caudal appendages, lateral, superior, and inferior, expressed in tenths of length of inferior.

[5]On segments 2 to 9 when present; in some species, entirely wanting.

The thorax is rather short and compactly built. It is thickly clothed with hair in front. In coloration it is almost patternless. Its darker front is sometimes divided by a yellow middorsal carina. On each side of the synthorax is a faint midlateral spot or streak of pale yellow that is best developed about the spiracle. The legs are slender and pale, becoming blackish on the edgings of the tarsi, and narrowly lined with yellow on the outer edges of the tibiae. The wings are subhyaline; their membrane toward the front is more or less tinged with saffron yellow. There is generally some sort of pattern in brown at the extreme wing base, and there are roundish spots of paler brown on the rearward half of some of the antenodal crossveins, the number and intensity of the spots varying with the species.

The venation of the wings is as shown in our key and table for Corduline genera. The most distinctive feature is the divergence of veins M4 and Cu1 in the fore wing. Coupled with that is a copious cross-venation; bridge crossveins 3–6; under the stigma, generally 2; the first of these and several more proximal in the same interspace have a brace-vein slant. The intermedian crossveins are 5–7/3–4; paranals 5–2/2–3. The anal loop has little development of the toe, and within its borders there is often an extra heel cell or ankle cell or both.

The abdomen is moderately swollen on its basal segments, contracted at the apex of 3, depressed beyond, and widens slowly thereafter to 9. The basal segments are pale and a little hairy. Beyond 3 the color is brown. The surface is bare and shining, with all carinae black; 10 and the appendages are mostly yellow. The superior appendages of the male are slender and arched in their basal half, then thickened to more or less elliptical form in their outer half, with acutely pointed tips. The subgenital plate of the female is very short, with a more or less quadrangular notch in the hind margin.

Two species stand apart from the other five, and from each other:

N. xanthosoma differs in having a blunt spine or tooth under the narrow basal fourth of the superior appendages of the male, two rows of cells between the toe of the anal loop and the wing margin, deeper saffroning in both sexes, and more copious cross-venation.

N. molesta differs in having in the male an internal process on the trochanter of the middle leg—a conspicuous truncated process, as long as the joint is wide. In both sexes the caudal appendages are longer than in the other species.

The dainty adults fly on silent wings in the twilight. Garbed in soft tints of brown, they are difficult to see, even when flying in the open, and are impossible to see in the dimness of the shadow. Therefore they are not so common in insect collections as their abundance in nature warrants.

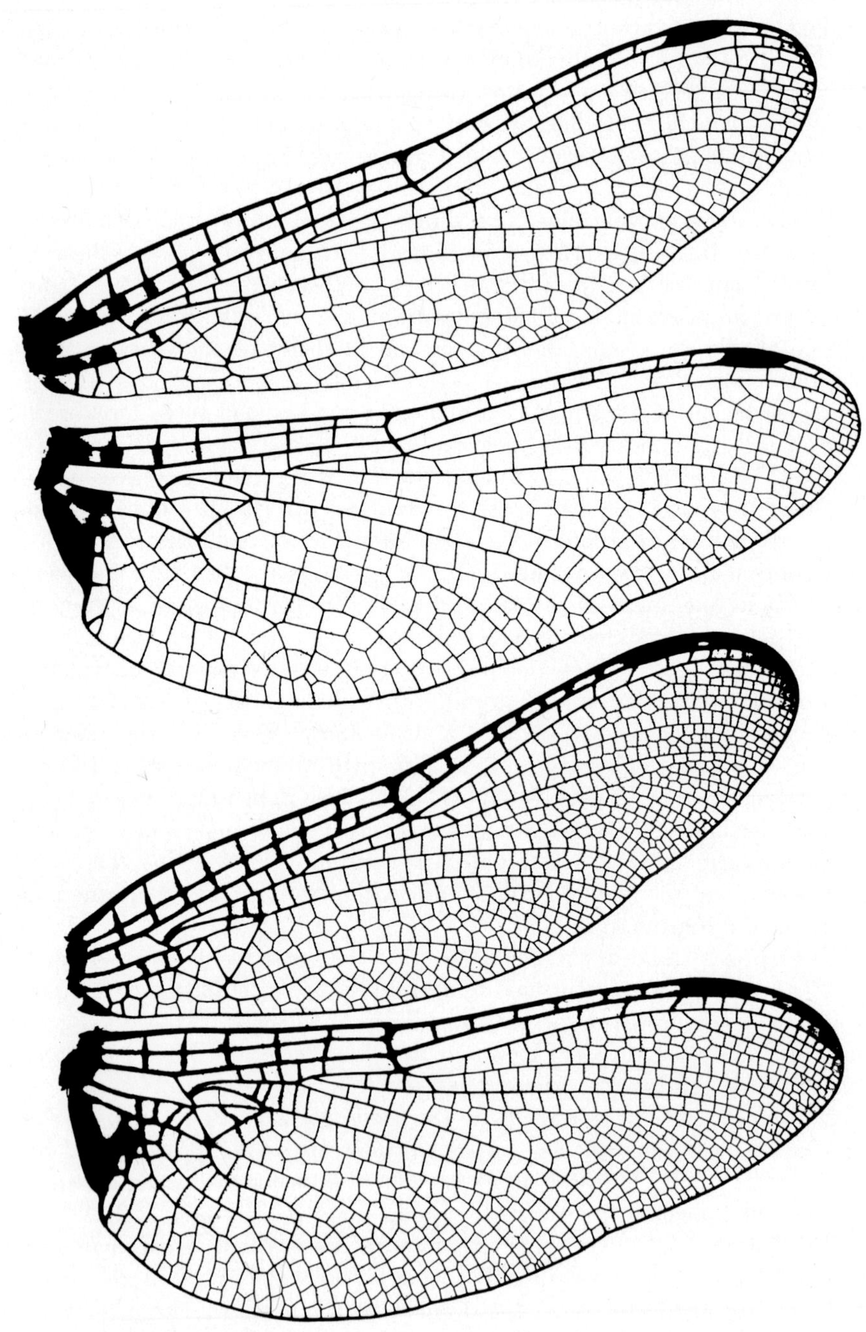

Fig. 208. Above, *Neurocordulia virginiensis;* below, *N. xanthosoma.*

The nymph in this genus is stockily built, with short legs and a thick brown skin. It is the color of the rocks or logs to which it clings. The head has a prominent transverse frontal ridge (produced forward in the

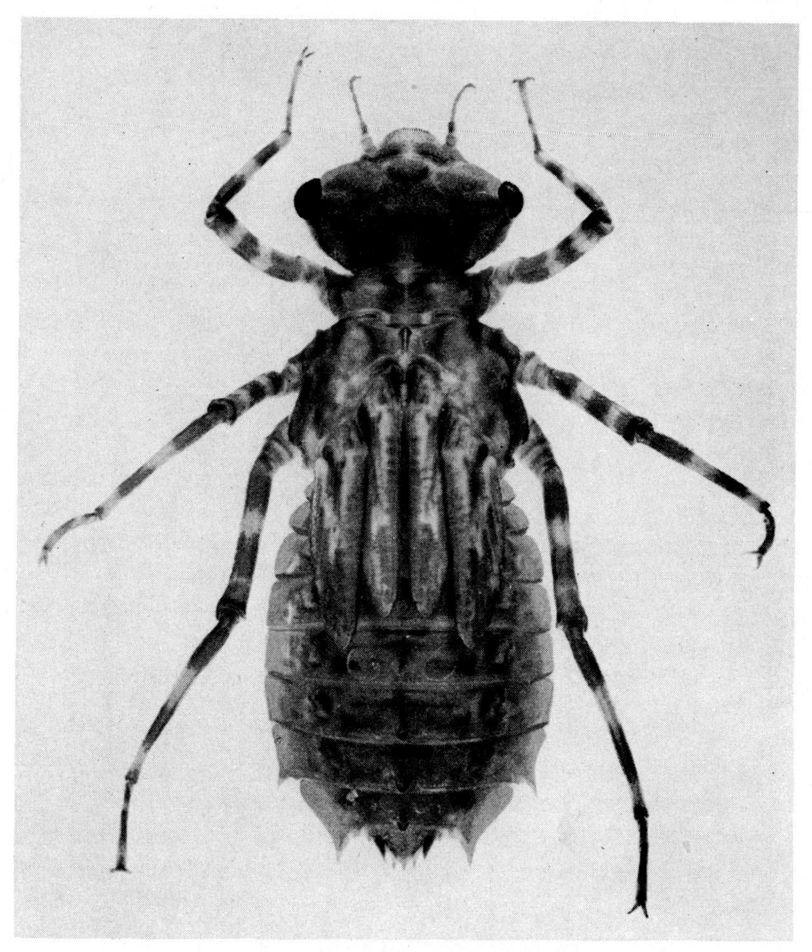

Fig. 209. *Neurocordulia virginiensis.*

middle in pyramidal horn in *N. molesta*) and a pair of little tubercles on the vertex. The labium is short and wide, with seven to nine mental setae and five or six laterals. The teeth on the inner margin of the lateral lobe are semielliptic and separated by incisions almost as deep as they are wide apart. There is a row of middorsal hooks on the abdomen. They are slender, erect, and spinelike on the basal segments between the wing pads; thick, blunt-tipped, and somewhat cultriform on 4 to 9; and generally

TABLE OF SPECIES

ADULTS

Species	Hind wing	Crossveins				Cells		Distr.
		Ante-nodals[1]	Under stigma[2]	Midb. space	Cubito-anal[3]	Triangles f.w./h.w.[4]	♂ Basal triangle	
alabamensis	29-33	5,*6*	1,*2*	0	2/2	3/2	2	SE
clara	38	5	2	*0*,1	2/	3/2	?	Ala.
molesta	33-38	*5*,6	*1*,2	0	2/2	3/2	2	SE, C
obsoleta	30-33	5,6	3	0,*1*	2-3/2-4	3/2-3	3	E, C
virginiensis	32-35	5,*6*,7	1,*2*	0	2/2	3/2	2	SE
xanthosoma	36-40	5	2	*0*,1	3-4/3-4	3-4/3	3	SW
yamaskanensis	33-35	5,*6*,7	2	0	2/2	3/2	2	NE, C

[1]Number of antenodal crossveins in hind wing. Strongly prevalent numbers italicized.

[2]Number of crossveins under stigma and in contact with it; first one strongly aslant.

[3]Number of cubito-anal crossveins in fore and hind wing (f.w./h.w.).

[4]Normal number of cells in triangle of fore and hind wing (f.w./h.w.).

highest on the middle segments. There are short triangular lateral spines on 8 and 9. The caudal appendages are short and stubby.*

KEY TO THE SPECIES

Adults†

1—Midbasal space of all wings generally traversed by crossvein.......... **obsoleta**
—Midbasal space without a crossvein 2

2—Vein network dense; two rows of cells between toe of anal loop and wing margin 3
—Venation more open; one row of cells between toe of anal loop and wing margin 3

3—Three rows of cells between vein Cu2 and hind-wing margin; middle and hind tibiae of male keeled for a fourth of their length or more... **yamaskanensis**
—Only two rows of cells between vein Cu2 and hind-wing margin; middle and hind tibiae of male keeled for a fifth of their length or less.............. 4

4—Males 5
—Females 7

5—Trochanter of middle legs with conspicuous truncate process on inner side, as long as segment is wide .... **molesta**
—Trochanter of middle legs with no such process 6

6—Front tibia with vestigial keel, scarcely discoverable............. **virginiensis**
—Front tibia with keel a little longer than width of end of tibia...... **alabamensis**

7—Caudal appendages about 1.6 mm. long.......................... **alabamensis**
—Caudal appendages about 2.0 mm. long.......................... **virginiensis**
—Caudal appendages about 2.4 mm. long.......................... **molesta**

Nymphs

1—Pyramidal horn on front of head.................................. **molesta**
—No such horn present 2

2—Lateral spines of segment 8 strongly divergent.......................... 3
—Lateral spines of 8 point directly rearward, parallel...................... 4

---

* We are greatly indebted to Dr. R. S. Hodges of Tuscaloosa, Alabama, for aid in the treatment of this genus. With kind intent to make our volume more complete, he has given us for advance publication the description of a new species, *Neurocordulia alabamensis*. He also has permitted us to examine the Neurocordulias in a large collection that he and Dr. Septima Smith together made during seven years of active field work, mainly in Alabama; and he has allowed us to read an important unpublished manuscript on *Neurocordulia* in which he described a new genus, *Rostrocordulia*, for Walsh's *Cordulia molesta*. Further research on the Neurocordulias revealed agreement in so many characters that he came to doubt the desirability of splitting the genus. Following his latest advice, we have treated both *Rostrocordulia* and Williamson's *Platycordulia* as subgenera of *Neurocordulia*. The story of the rediscovery of Walsh's *molesta* and identification of its nymph will have to await the publication of his manuscript.

The two names *Rostrocordulia* and *alabamensis*, if and when used again, are to be credited to Dr. Hodges and not to us.

† The data for this key were kindly supplied by Dr. Hodges.

3—Lateral spines of 9 surpass tips of appendages; dorsal hooks high, blunt-tipped, erect .... **obsoleta**

—Lateral spines of 9 end on level with tips of appendages; dorsal hooks lower and aslant to rearward .... **yamaskanensis**

4—Dorsal hooks on 5 to 8 high, subequal, obtusely cultriform (fig. 210) **alabamensis**

—Dorsal hooks scarcely hooked backward, on 8 lower than on 6 and 7 **virginiensis**

Nymphs unknown: **clara** and **xanthosoma.**

Fig. 210. *Neurocordulia molesta,* head, showing frontal horn; *N. alabamensis,* lateral view of abdomen showing dorsal hooks.

Neurocordulia alabamensis Hodges

Description supplied by Dr. Robert S. Hodges from his unpublished MS

Length 42–46 mm.; abdomen 30; hind wing 29–33.

A rather pale and slender species, with faintly and rather uniformly saffronated wings. Head wholly pale save for dark eyes. Thorax without any pattern of darker markings; yellow spot on metathoracic spiracle very small and ill defined. Legs wholly pale, with tibial spines only a little darkened. Wings saffron yellow and diffusely marked with brown on both antenodal and postnodal crossveins; wings darker at subnodus.

Usual dark basal markings greatly reduced, with only one small spot of black on rear side of first and second paranal cells in hind wing. First crossvein under pale stigma very strongly aslant; second very variable in position, sometimes out beyond, not under, stigma. Membranule brown and white, half and half.

Abdomen, beyond its pale and little-swollen basal segments, clouded with blackish, more heavily to rearward, and black forms rings on joinings of middle segments. Caudal appendages wholly pale.

Distribution and dates.—UNITED STATES: Ala., Fla., Ga., S. C.

March 21 (reared, Fla.) to July 18 (S. C.).

Neurocordulia clara Muttkowski

1910. Mtk., Bull. Wis. Nat. Hist. Soc., 8:170.
1929. Davis, J. N. Y. Ent. Soc., 37:449 (fig.).
1937. Byers, Misc. Pub. Mus. Zool. Univ. Mich., 36:17 (fig.).

Length 48 mm.; abdomen 38; hind wing 38.

This dragonfly, known as yet from a single female specimen from Alabama, was described as a subspecies of *obsoleta;* perhaps it is a mere variety, though Davis (1929) considered it a distinct species and published a photograph of the type to prove it. It is known to us only from the description and brief note in the first two papers cited above.

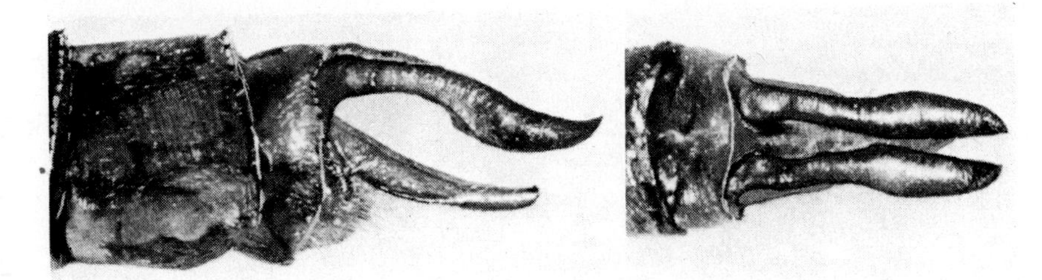

Fig. 211. *Neurocordulia alabamensis.*

Face described as olive brown. Thorax usual general brownish color, darker in front, with yellow middorsal carina; brown sides paler. A yellow streak covers spiracle and points obliquely upward toward a small yellow spot at base of front wing; a quadrate yellow spot at base of hind wing. On dorsum, yellow spots at bases of all four wings. Legs olivaceous, with carinae yellowish, spines and tips of claws black. Brown spots on antenodal crossveins of wings similar to those of *obsoleta,* only less extensive, but brown basal quadrangle on base of hind wings of that species altogether lacking; entire anal area uncolored.

Abdomen brown, about two and a half segments at its base olivaceous. Segment 2 with a basal yellow spot low down on each side; on 3, a narrow basal ring of yellow. Caudal appendages brown beyond their pale basal third.

Distribution.—UNITED STATES: Ala.

Neurocordulia molesta Walsh

1863. Walsh, Proc. Ent. Soc. Phila., 2:254 (in *Cordulia*).

Length 45–53 mm.; abdomen 36–37; hind wing 33–38.

This long-lost species, described by Walsh ninety-one years ago and thereafter mistakenly relegated to synonymy, was rediscovered by

Smith and Hodges in 1938. They also have its nymph. It is distinct from all others in both nymphal and adult stages.

General color dull brownish olive, with little pattern. Face wholly pale save for a very narrow blackish ocellar crossband that runs down at each end in groove along front margin of eye. Behind tubercle at rear of eyes are some yellow markings. Thorax brown in front, shading off paler halfway to humeral suture, with middorsal yellow carina. A diffuse half-length midlateral yellow stripe, with an ill-defined yellowish spot above its upper end. Legs pale.

Wings faintly yellowish, with brown spots on antenodal crossveins of second series suddenly darker on inner and outer margins, "so that they

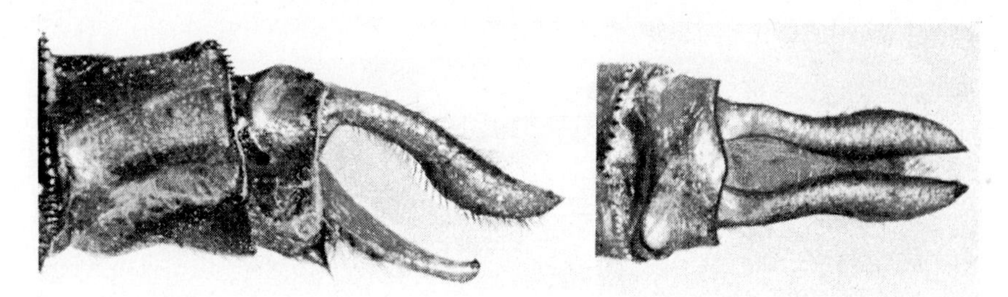

Fig. 212. *Neurocordulia molesta.*

appear to be banded by crossveins" (quoted from original description). These spots fade out at last antenodal crossvein in fore wing but not in hind. A larger transverse brownish spot on both nodus and subnodus. No brown spots on postnodal crossveins. Spots all better developed in female than in male. This species has first crossvein under stigma strongly aslant as usual; second one generally moved so far out toward wing tip that it is not under stigma at all (this occurs occasionally in other species).

Abdomen brown, darker beneath, with hind margin of segments 2 to 5 faintly edged with yellowish. Appendages very pale.

Distribution and dates.—UNITED STATES: Ala., Ga., Ill., Iowa, La., Okla., S. C., Tenn., Wis.

June 10 (Ala.).

Neurocordulia obsoleta Say

Syn.: polysticta Burmeister

1839. Say, Proc. Ent. Soc. Phila., 8:29 (in *Libellula*).
1861. Hagen, Psyche, 5:367–373 (figs.) (in *Cordulia*).
1929. Davis, J. N. Y. Ent. Soc., 37:449 (fig.).
1929. N. & H., Handb., p. 174 (figs.).
1937. Byers, Misc. Pub. Mus. Zool. Univ. Mich., 36:11 (figs.).

Length 43–48 mm.; abdomen 33–35; hind wing 30–33.

Ground color olivaceous, heavily suffused with brown. Face olivaceous, with a yellow labrum; top of vertex and most of occiput dull yellow. A thin fringe of white hairs borders occiput in rear.

Thorax olive brown, darker on a wide middorsal band that is divided by a yellowish carina. Wedge-shaped streaks of darker brown above leg bases on each side. Legs pale, darkening toward tarsi, and with black spines. Wings yellowish hyaline, with fawn-colored veins and tawny stigma. A large basal spot of brown on hind wings covers paranal cells. A diffuse brown spot on each antenodal crossvein, a larger spot or streak

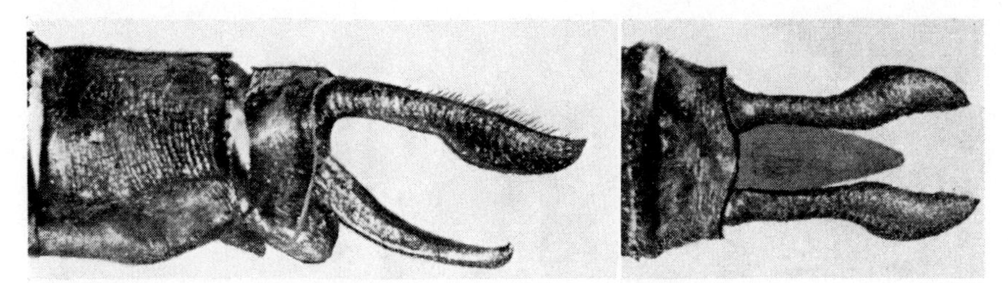

Fig. 213. *Neurocordulia obsoleta.*

covering nodus and subnodus and fading on postnodal crossveins. A larger basal spot in hind wings, trapezoidal in form, covering area between midbasal space and hind margin. Some cells in this spot have clear centers.

Abdomen bare and shining beyond swollen basal segments. Caudal appendages pale brown, thinly besprinkled with black hairs.

Distribution and dates.—UNITED STATES: Ala., Ill., Ind., Ky.(?), La., Maine, Md., Mass., Mich., N. H., N. J., N. Y., N. C., Ohio, Okla., Pa., Tenn., Va.

May 26 (Tenn.) to October 23 (Tenn.).

Neurocordulia virginiensis Davis

1927. Davis, Bull. Brooklyn Ent. Soc., 22:155.
1929. Davis, J. N. Y. Ent. Soc., 37:449 (fig.).
1929. N. & H., Handb., p. 175 (fig.).
1937. Byers, Misc. Pub. Mus. Zool. Univ. Mich., 36:21 (figs.).

Length 42–49 mm.; abdomen 33–35; hind wing 32–35.

Face pale olivaceous yellow, thickly beset with black hairs of reddish cast. Low down on face, hairs are shorter. Top of frons becomes ruddy with age.

Thorax dull brown in front, with usual yellow carina, and with black edgings on crest. Sides paler brown, with only a small roundish spot of yellow overspreading spiracle. Hair on thorax tawny above, thinner and whitish underneath. Legs dull yellow with outer faces of tibiae of a lighter tint and tarsi mostly black. Spots restricted to crossveins that are before level of triangles in both wings. A black spot covers lower half of membranule; faint ones cover the two cubito-anal crossveins. A filling of orange in basal triangle of male.

Abdomen brown, with a wash of black along the very black transverse carinae. Swollen basal segments more olivaceous in female, more

Fig. 214. *Neurocordulia virginiensis.*

yellowish in male. Genital lobe of male conspicuously black-margined, but hamules and caudal appendages yellow.

Distribution and dates.—UNITED STATES: Ala., Fla., Ga., Miss., Okla., Tenn., Va.

April 13 (Fla.) to June 24 (Ala.).

Neurocordulia xanthosoma Williamson

1908. Wmsn., Ent. News, 19:432 (figs.) (in *Platycordulia* n. gen.).
1917. Kndy., Bull. Univ. Kansas, 18:132 (figs.).
1929. N. & H., Handb., p. 173.
1937. Byers, Misc. Pub. Mus. Zool. Univ. Mich., 36:25 (figs.).

Length 48–50 mm.; abdomen 37–40; hind wing 36–40.

A handsome species, largest of genus. Face and occiput yellow, becoming deeply saffronated with age. Pale reddish or yellowish brown thorax, clad in tawny hairs. Almost without color pattern save for a yellow middorsal carina and on each side a faint pale spot that surrounds spiracle. Legs pale, markedly yellow on outer face of tibiae, and armed with brown spines.

Wings subhyaline, with tints of gold and saffron in membrane. Along costal strip of both wings are large roundish brown spots on all antenodal and postnodal crossveins. Tinted area broadens toward wing base to cover all triangles, including basal triangle in male. Stronger veins in this basal area yellow, bordered by dark brown. Membranule is half white; rear half, dark brown.

Abdomen darker than thorax, except on basal segments and on segment 10 and its caudal appendages. All carinae dark brown. Strong lateral carina begins at transverse groove on segment 4 and ends on 9. Superior appendages of male black at extreme tip. Under middle of each, a small tooth is visible from side.

Distribution and dates.—UNITED STATES: Kans., Okla., Tex.

June 4 (Okla.) to August 2 (Okla.).

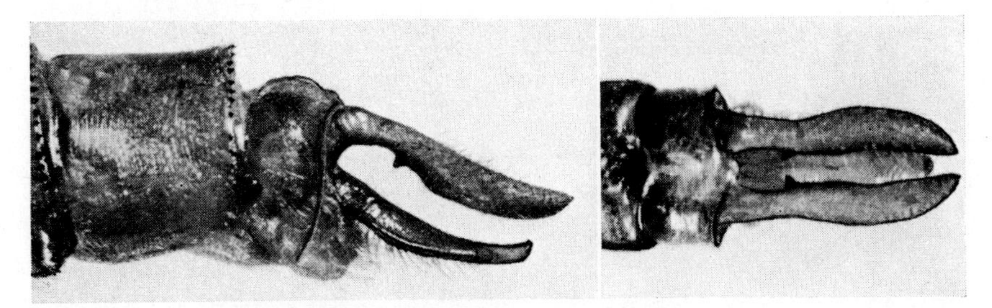

Fig. 215. *Neurocordulia xanthosoma.*

Neurocordulia yamaskanensis Provancher

1875. Provancher, Nat. Can., 7:248 (in *Aeshna*).
1913. Walk., Can. Ent., 45:161–168 (nymph).
1929. Davis, J. N. Y. Ent. Soc., 27:449.
1929. N. & H., Handb., p. 175 (figs.).
1937. Byers, Misc. Pub. Mus. Zool. Univ. Mich., 36:17 (figs.).

Length 45–55 mm.; abdomen 39–41; hind wing 33–35.

A handsome species, partly clear-winged. Sides of body dark olive brown, polished and shining on all bulges. Inner side of tibia, also tibial spines, very black.

Round spots on antenodal crossveins largely restricted to subcostal side, fading out before reaching nodus. Few spots beyond level of arculus in either fore or hind wing. In hind wing, clear area between midbasal space and wing border has diffuse splashes of brown on several crossveins. Some cells bordered with brown in hind wing on a golden background. No spot on subnodus.

Abdomen darker in any other species, with age becoming wholly black beyond swollen basal segments; superior appendage of male black beyond its olivaceous base.

Distribution and dates.—CANADA: Ont., Que.; UNITED STATES: Conn., Ky., Maine, Mich., Mo., Ohio, Pa., W. Va.

June 7 (Ont.) to August 10 (Ky.).

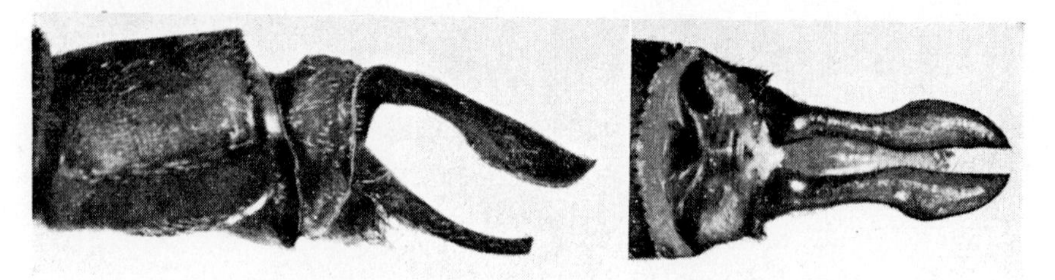

Fig. 216. *Neurocordulia yamaskanensis.*

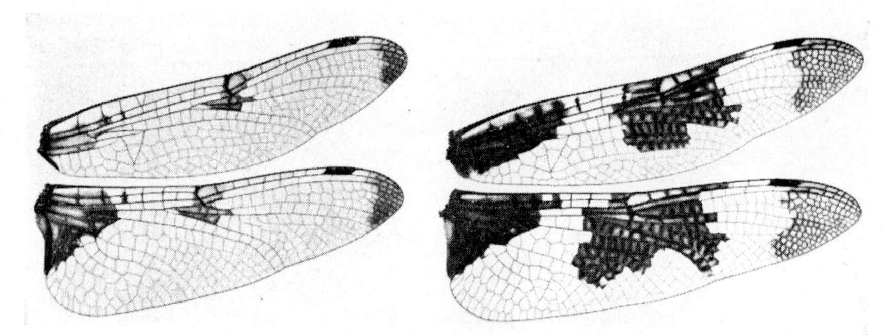

Fig. 217. *Epicordulia princeps; E. regina.*

Genus EPICORDULIA Selys 1871

These are large, strong-flying dragonflies. The wings are heavily marked with brown at base, nodus, and apex. The spots vary in extent, and the nodal spot may sometimes be wanting in *princeps.* The head and thorax are olive or yellowish brown.

The thorax is clothed with long shaggy gray hair, under which metallic reflections shine in a good light. There is usually a blackish streak along the ventral border of the mesepisternum and along the first lateral suture up to the spiracle. Sometimes there are also dark areas above on the humeral and at both ends of the second lateral sutures. The legs are brown. All tibiae of the males possess a ventral keel.

As to wing venation, in addition to the characters shown in our table of genera, half-antenodal crossveins are of frequent and scattering occurrence, and the number of extra bridge crossveins is very variable.

The abdomen becomes quite black on the dorsum, with obscure pale markings on the sides.

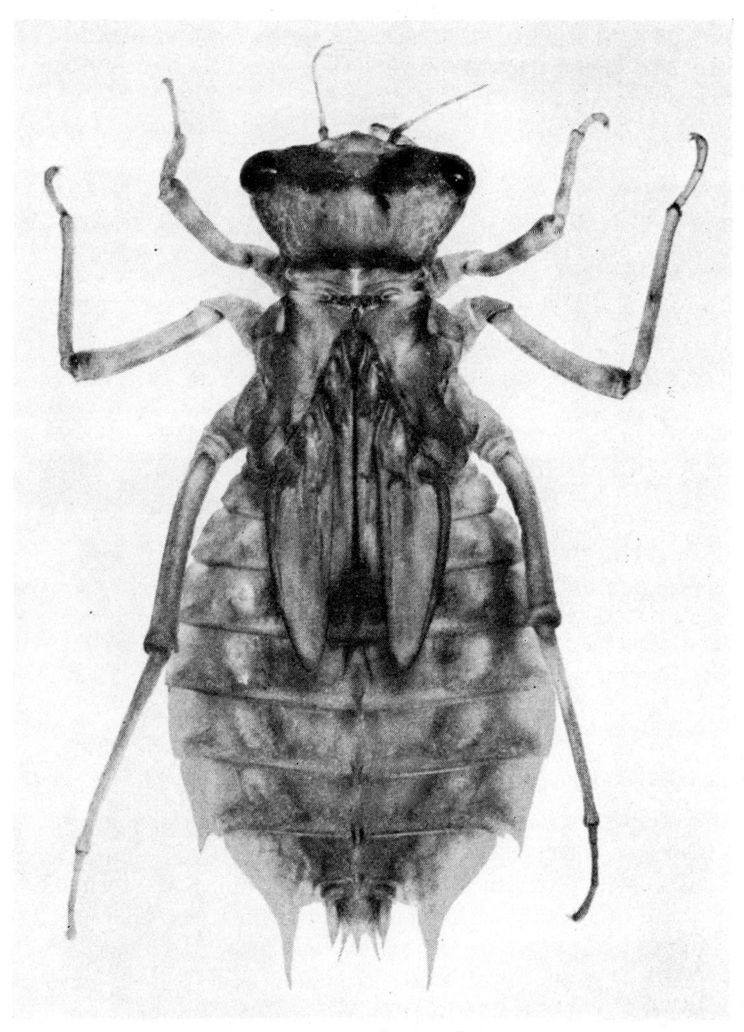

Fig. 218. *Epicordulia princeps.*

The nymphs are marked by a pair of nipple-shaped tubercles on the top of the head, and by a middorsal row of high cultriform dorsal hooks on the abdomen.

The nymphs of the two species are very similar. That of *regina* is usually larger and possesses a wider head, but individuals of the two species overlap in total length. The lateral spines on segment 9 are long in both species, but in a long series of specimens their length and curvature are quite variable. There are generally five lateral setae in *princeps* and only four in *regina,* but this is not quite constant. The two known species in this genus are restricted to the eastern United States.

KEY TO THE SPECIES

Adults

1—Hind wings less than 45 mm.; spots light brown, covering one-fourth or less of wing area .... **princeps**

—Hind wings more than 45 mm.; spots dark brown, covering one-third or more of wing area .... **regina**

Nymphs

1—Two tubercles on top of head roundish or nipple-shaped; lateral spines of abdominal segment 8 generally comprise about one-third of its side margin **princeps**

—Two tubercles less circular and not nipple-shaped, extended a little fore and aft to form a very short ridge; lateral spines of 8 comprise about half of its side margin.. **regina**

Epicordulia princeps Hagen

1861. Hagen, Syn. Neur. N. Amer., p. 134 (in *Epitheca*).
1883. Garm., Bull. Ill. State Lab., 3:179 (nymph) (in *Libellula*).
1890. Cabot, Mem. M. C. Z., 17:25 (figs., nymph).
1893. Calv., Trans. Amer. Ent. Soc., 20:251.
1901. Ndm., Bull. N. Y. State Mus., 47:488 (nymph).
1929. N. & H., Handb., p. 176 (fig.).

Length 58–68 mm.; abdomen 42–49; hind wing 38–43.

Brown spots of wings may be much reduced in size, nodal spot sometimes almost or quite absent. Size generally smaller.

Genitalia of males distinct from those of *regina:* posterior hamules not so thick, and other slight differences exist which are difficult to define. Superior caudal appendages of males dark brown to black, equal to or shorter than segments 9 plus 10; expanded and strongly divergent in apical two-thirds when viewed from above. Inferior about four-fifths as long as superiors and generally paler.

A persistent flyer.*

Distribution and dates.—CANADA: Ont., Que.; UNITED STATES: Ala., Conn., Ga., Ill., Ind., Iowa, Ky., La., Maine, Md., Mass., Mich., Minn., Miss., Mo., Nebr., N. H., N. Y., N. C., Ohio, Okla., Pa., R. I., Tenn., Tex., Vt., Va.

May 12 (Tenn.) to September 14 (Ohio).

---

* In the gray twilight just before sunrise . . . *princeps,* misty and indistinct, floated by. After sunset . . . there he was again out over the water, hurrying along in the gathering dusk, as though his day were not completed."—Williamson.

Epicordulia regina Hagen

1871. Hagen, in Selys, Bull. Acad. Belg., (2)31:277 (reprint, p. 43) (in *Cordulia*).
1922. Davis, Bull. Brooklyn Ent. Soc., 17:111 (figs.).
1928. Broughton, Can. Ent., 60:33 (nymph).
1929. N. & H., Handb., p. 177 (fig.).

Length 69–78 mm.; abdomen 52–60; hind wing 46–52.

Brown spots in wings very large. Superior caudal appendages of male brown, darker apically, and distinctly longer than segments 9 plus 10; expanded in apical three-fourths, but not strongly divergent as in *princeps*. Inferior shorter in proportion to length of superiors than in *princeps*, being only about two-thirds as long as superiors; generally of same color as bases of superiors.

Distribution and dates.—UNITED STATES: Fla., Ga., Miss., S. C.

March 23 (Fla.) to August 25 (Fla.).

Genus TETRAGONEURIA Hagen 1861

These brownish dragonflies are of moderate size, brown or olive, hairy, and with mostly dull yellowish markings. They conform so closely to one type of coloration that a general statement will save much repetition in the descriptions which follow.

The face is suffused with yellow on labrum and frons, and thinly beset with short erect black hairs. The synthorax is olivaceous or brownish, sharply rimmed with black on carina and crest and along the ridges beside the wing roots; diffusely lined with black in the lateral sutures; on the midlateral one, only in its lower half. The hair on the thorax is tawny or whitish, longer toward each end, especially long and dense next to the prothorax; thinner on the sides, but dense enough everywhere to more or less obscure the color pattern. A paler ground color, sometimes bright yellow, prevails to rearward and below. The legs are blackish, with the front femora paler. The wings have color pattern varied from species to species, as will be noted in the following key and descriptions, and their membrane in its transparency varies from hyaline to deep amber in individuals of some of the species.

The abdomen is predominantly black above, very scantily hairy except toward the ends, more or less constricted on abdominal segment 3, especially in males, and often depressed on the middle segments. These segments are broadly marked with yellow low down on the sides and beneath, with the very black lateral carina sharply dividing the yellow areas lengthwise; the sides of segments 2 and 3 are more yellow. The caudal appendages are black. The small auricles on segment 2 of the male are not denticulated.

The venational characters that collectively characterize the genus are: antenodal crossveins of the hind wing reduced to four or five, generally four, rarely six; fore-wing triangle wide and usually very convex on the outer side; one or two crossveins behind the stigma. The radial planate is well developed, the median planate, less so. In the hind

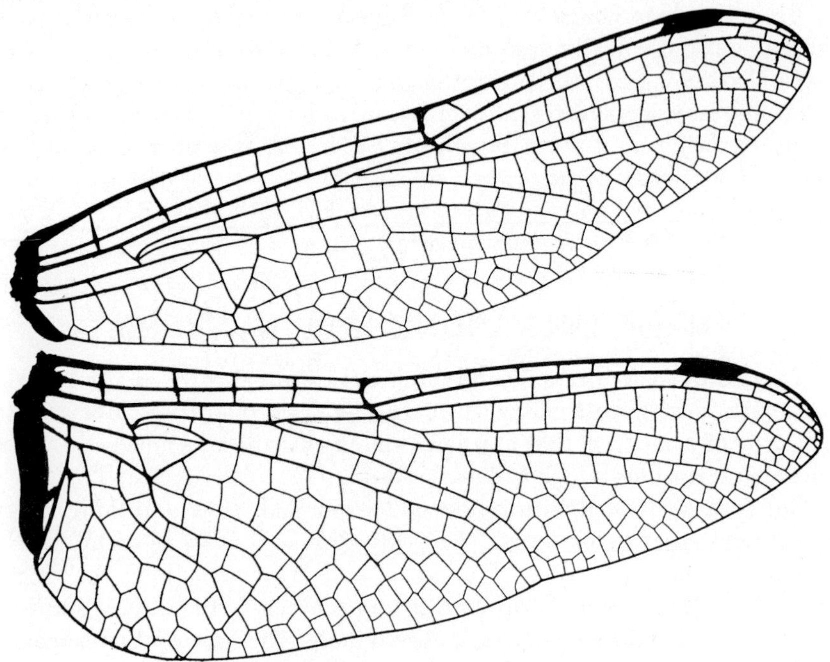

Fig. 219. *Tetragoneuria sepia.*

wing: anal loop well developed, containing about a dozen cells in two rows, with three cells on the sole and scant development of the toe; four paranal cells, followed by a very long semilunar tornal cell that reaches down from the second paranal to the prominent hind angle of the wing.

This is a difficult genus. The species are all of about the same size, and similar in form. Chief reliance must be placed on the shape of the caudal appendages of the male, and these we show in photographs. The subgenital plate of the females is divided deeply into a pair of long strap-like flaps that in drying may suffer shrinkage and pressure distortions; they may thus be made to look unlike in different specimens of the same species. We are unable to offer a satisfactory key for females. They may have to be named by their association with the males.

The two most important papers dealing with adults of this genus are those of Muttkowski (1911) and Davis (1933); for nymphs, that of Needham (1901).

The nymphs are similar in form to those of *Epicordulia,* having

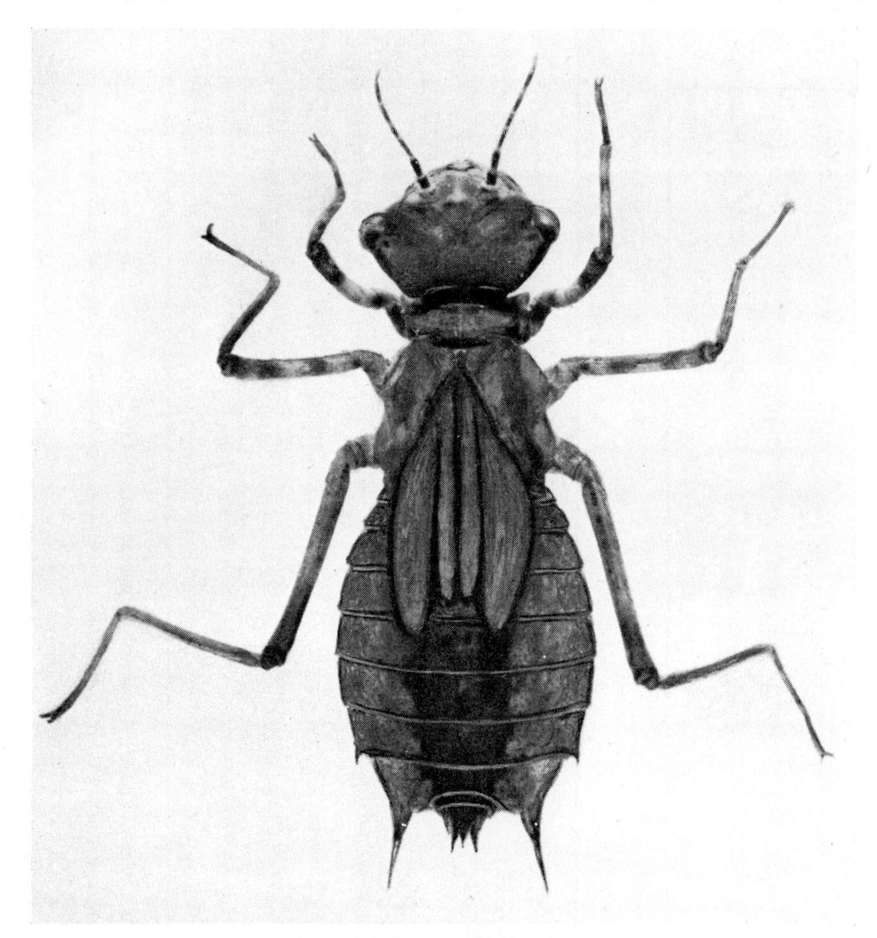

Fig. 220. *Tetragoneuria sepia.*

broadly depressed abdomen, armed with large laterally flattened dorsal hooks and prominent lateral spines. The top of the head is smooth, lacking tubercles and having only a low convexity of the vertex, and there are six or seven raptorial setae on the lateral lobe of the labium. The head when seen from above is trapezoidal, widest in front across the forwardly directed eyes. It is narrowed straight back to rounded hind angles, between which the hind border is straight. The nymphs of this genus, so far as known, seem to be distinguishable from those of

TABLE OF SPECIES

ADULTS

Species	*Hind wing*	*Abdomen*[1]	T *spot*[2]	*Nodal crossveins*[3]		*Distr.*
				Ante.	*Post.*	
canis	30-31	31-36	0	7-8-9/5-6	5-6-7/6-7-8	N, W
cynosura	26-31	27-31	±	6-7-8/4-5	5-6-7/5-6-7	E, C
morio	30-33	32-33	±	6-7-8/4-5	5-6-7/5-6-7-8	NE
petechialis	27-31	27-31	0	7/4	5-6/5-6	S
semiaquea	24-31	22-27	0	5-6-7/4	4-5-6/5-6-7	SE
sepia	28-33	25-29	+	6-7/4	5-6/5-6	Fla.
spinigera	29-34	30-34	+	6-7-8/4-5	6-7-8/5-6-7-8	N, W
spinosa	30-33	30-32	0	7-8/4-5	5-6/5-6	SE
stella	27-32	30-33	0	6-7-8/4-5	4-5-6/5-6-7	S
williamsoni	26-28	27-28	±	6-7/4	5-6-7/5-6	S

[1]Length of abdomen without appendages (in part from Muttkowski, 1911).

[2]Blackish T spot on top of frons: present (+), absent (0), variable, either present or absent (±).

[3]Number of antenodal and postnodal crossveins in fore and hind wing: f.w./h.w.; prevailing numbers italicized.

Epicordulia by the smooth inner surface of the middle labial lobe (that surface bearing a dense patch of short setae in *Epicordulia*) as well as by their smaller size.

The thorax is wider than the head, and moderately depressed. The legs are slender and thinly fringed with hairs; the body is for the most part smooth and clear of silt, and neatly patterned in green and brown.

The abdomen is wider than the thorax and rounded at the sides, as seen from above, with short lateral spines on segment 8 and much longer ones on 9. There are dorsal hooks on 2 to 9, spinelike and erect on 2 to 4, larger and laterally flattened on 5 to 7, slowly diminishing on 8 and 9, with increasing slope to rearward all the way back.

KEY TO THE SPECIES

Adults (Males)

1—Superior caudal appendages of male each with a sharp spine projecting downward at about two-fifths of its length; swollen end of appendage upcurved and very hairy.. **spinigera**

—These appendages with not more than a low tubercle or angle on under side projecting downward; tips not upcurved or so hairy................ 2

2—These appendages with outer end bent downward, and with a tubercle on upper inner side where slope begins.................................. 3

—These appendages without tubercle on upper side, and with no downward slope to end 4

3—These appendages with tubercle on lower side......................... **canis**

—These appendages with no tubercle on lower side..................... **spinosa**

4—Hind wing with brown extending outward to nodus or beyond...... **semiaquea***

—Hind wing with brown much less extended outward....................... 5

5—Both fore and hind wings generally with small transverse brown spot on nodus; hind wings also with smaller spots on antenodal crossveins. . **petechialis**

—No spot on nodus; fore wing hyaline.................................... 6

6—Hind wing hyaline or with a mere trace of brown in extreme base; inferior caudal appendage seven-eighths as long as superiors.................. **sepia**

—Hind wing with more brown at base; inferior appendage not more than three-fourths as long as superiors 7

7—Hind-wing base with brown generally extending outward to cover and enclose first antenodal crossvein................................ **cynosura**

—Wing with brown not extending so far outward........................... 8

8—Top of frons with no T spot; swollen part of caudal appendage smoothly rounded on lower side; stigma 3 mm. long......................... **stella**

—Top of frons generally with a blackish T spot; swollen terminal part of superior caudal appendage with projecting angle on lower side; stigma about 2 mm. long.. 9

9—Abdomen slender; hind wing 26–28 mm....................... **williamsoni**

—Abdomen stouter; hind wing 30–33 mm. .... **morio**

* Specimens of **semiaquea** from Georgia and Florida do not have the brown of the wings reaching the nodus. When in doubt follow both paths in the key and refer to photographs, table, and descriptions.

NYMPHS

1—Lateral spines of segment 9 pointing directly backward, their axes parallel.. 2
—Lateral spines of 9 distinctly divergent.. 5

2—Lateral spines of 9 short, tips not reaching level of tips of inferior abdominal appendages; dorsal hooks large and rough-clad................**canis**
—Lateral spines long, surpassing level of tips of caudal appendages........... 3

3—Spines of 9 not quite so long as distance across 9 from their bases to 8; dorsal hooks low, rather sharp-pointed..........................**petechialis**
—Spines of 9 equal to or longer than shortest distance from their bases to 8.... 4

4—Length of grown nymph about 22 mm.; spines of 9 as long as distance to 8; dorsal hooks large, rough-clad, and scurfy........................**cynosura**
—Length of grown nymph about 19 mm.; spines of 9 distinctly longer than distance to 8; dorsal hooks lower and less scurfy..................**?semiaquea**

5—Side margins of 8 distinctly concave before base of their thornlike lateral spines; dorsal hooks large....................................**spinigera**
—Side margins of 8 straight to base of lateral spines........................ 6

6—Lateral setae of labium 6, mental setae 6+3...........................**sepia**
—Lateral setae of labium 7, mental setae 7+3...........................**?stella**

Nymphs unknown: **morio, spinosa,** and **williamsoni.**

TABLE OF SPECIES

NYMPHS

Species	*Total length*	*Setae*[1]		*Spine*[2] *of 9*	*Caudal append.*[3]		
		L.lobe	*M.lobe*		*Lat.*	*Sup.*	*Inf.*
canis	23	7	8+3	6	6	9	10
cynosura	22	6	6-7+4-5	10	8	9	10
petechialis	22	6	6-7+3	9	7	8	10
semiaquea ?	19	6	6+3	11	4	8	10
sepia	21	6	6+3	12	7	9	10
spinigera	21	7	8+4	12	8	9	10
stella ?	19	7	7+3	14	7	9	10

[1]Number of raptorial setae on labium.

[2]Length of lateral spines on abdominal segment 9 in terms of tenths of length of body of 9.

[3]Length of lateral and superior caudal appendages in terms of tenths of length of inferiors.

Tetragoneuria canis McLachlan

Dog's Head

1886. McL., Ent. Mon. Mag., 23:104.
1901. Ndm., Bull. N. Y. State Mus., 47:493 (nymph) (as *spinosa*).
1906. Mrtn., Coll. Selys Cordulines, p. 43 (figs.).
1911. Mtk., Bull. Wis. Nat. Hist. Soc., 9:130 (figs.).
1927. Garm., Odon. of Conn., p. 210.
1929. N. & H., Handb., p. 181 (fig.).
1933. Davis, Bull. Brooklyn Ent. Soc., 28:101 (figs.).

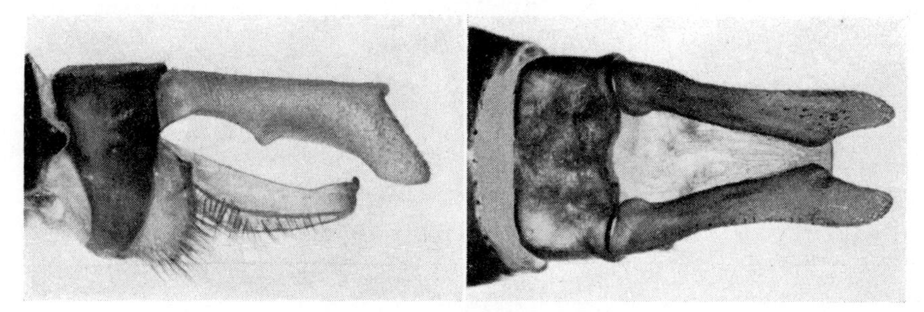

Fig. 221. *Tetragoneuria canis.*

Length 43–48 mm.; abdomen 34–39; hind wing 30–31.

A Northern species of transcontinental range, stout and hairy, with hair of thorax white on front and tawny on sides. Membrane of wings tinted to a very variable degree, ranging from glassy clearness in some males to smoky brown (as far out as stigma at least) in some females; more opaque brown spots sometimes cover antenodal crossveins in dark specimens. Stigma black in fully mature specimens. A help in determining females of species is presence of five antenodal crossveins in hind wing, with six only occasionally.

Auricles of male shining black on a paler ground color. Abdomen of usual pattern, with segments 9 and 10 generally wholly black. Subgenital plate of female, deeply cleft as usual, has inner edge of lance-oval flaps straight, outer edge rounded, but tips not convergent as in some other species. Appendages of female 2.3–2.7 mm. long.

Distribution and dates.—CANADA: B. C., N. B., N. S., Ont., P. E. I., Que.; UNITED STATES: Calif., Conn., Maine, Mich., N. H., N. Y., Okla., Oreg., Wash., Wis.

May 7 (Que.) to August 2 (N. Y.).

Tetragoneuria cynosura Say

Syn.: basiguttata Selys, diffinis Selys, lateralis Burmeister; var. simulans Muttkowski

1839. Say, J. Acad. Phila., 8:30 (in *Libellula*).
1871. Selys, Bull. Acad. Belg., (2)31:270 (reprint, p. 36) (in *Cordulia*).
1890. Cabot, Mem. M. C. Z., 17:28 (nymph) (in *Epitheca*).
1901. Ndm., Bull. N. Y. State Mus., 47:493.
1906. Mrtn., Coll. Selys Cordulines, p. 41 (figs.).
1911. Mtk., Bull. Wis. Nat. Hist. Soc., 9:104 (figs.).
1929. N. & H., Handb., p. 180 (fig.).
1933. Davis, Bull. Brooklyn Ent. Soc., 28:91 (figs.).

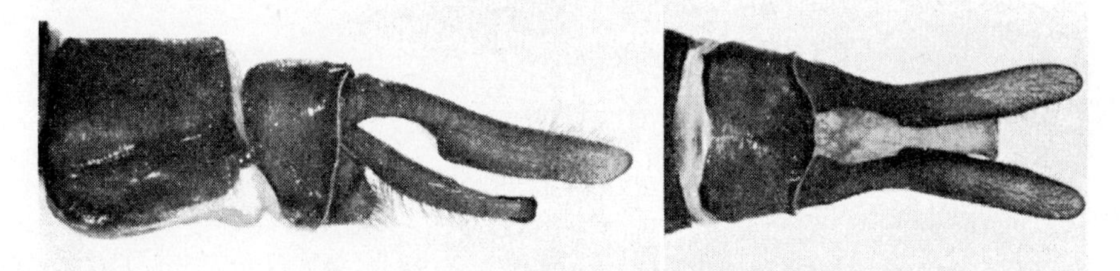

Fig. 222. *Tetragoneuria cynosura.*

Length 38–43 mm.; abdomen 25–34; hind wing 26–31.

Commonest species in northeastern United States, a brownish species becoming blackish with full maturity. Frons and labrum may become very bright yellow. T spot on frons very variable, always with a wide stem, often with nothing more—no crossbar at top; sometimes stalk is reduced to a mere prominence of black in frontal furrow next to black vertex. A little less hairy than preceding species; thoracic pile often conspicuously gray.

Wings hyaline, with variable brown markings at base of hinder pair. Antenodal crossveins of hind wing four, rarely five. Legs brown, becoming black with age.

Abdomen of usual pattern; spots of yellow on sides a little wider in female than in male, with small pale spots on segment 10. Superior appendages of male have a prominent angle near mid-length on under, inner side and a low and conspicuous longitudinal lateral carina just above it. Halves of split subgenital plate of female divergent at tips. Appendages of female very short, slightly more than 1.5 mm. long.

Distribution and dates.—CANADA: N. B., N. S., Ont., Que.; UNITED STATES: Ala., Conn., D. C., Fla., Ga., Ill., Ind., Iowa, Kans., Ky., La., Mich., Maine, Md., Mass., Minn., Miss., Nebr., N. H., N. J., N. Y., N. C., Ohio, Okla., Pa., S. C., Tenn., Tex., Va., Wis.

March (N. C.) to August 12 (N. Y.).

Tetragoneuria morio Muttkowski

1911. Mtk., Bull. Wis. Nat. Hist. Soc., 9:125 (figs.).
1919. Howe, Odon. of N. Eng., p. 63 (figs.).
1933. Davis, Bull. Brooklyn Ent. Soc., 28:99 (figs.).

Length 47 mm.; abdomen 30–34; hind wing 30–33.

A somewhat more slender and less hairy species, much less common than *cynosura.* Front and rear brown streaks in base of hind wing about equally well developed. Latter extends outward over anal crossing, then runs around borders of first and second paranal cells, leaving centers of these cells clear. A dull brown color covers inner half of membranule in a diffuse lengthwise band; marginal part gray. Costa yellow or tawny externally toward wing base. Hind wing with four or five antenodal crossveins.

Abdomen a little more slender than in *cynosura,* the lighter specimens of which *morio* much resembles. Segment 10 yellow on sides and beneath. Superior caudal appendages of male similar to those of *cynosura,* but more hairy toward end. Viewed from above, their tips less divergent, and inferior angulation before swollen end portion more prominent than in that species, and more acute. Female easily separated from *cynosura* by appendages, which are quite long, 3–3.3 mm., and stout. From *spinigera* it is not so easily separated, but appendages of *morio* are somewhat stouter and longer than in former species.

Distribution and dates.—UNITED STATES: Maine, Mass., Mich, N. H., Wis.

May 18 (New England) to June 29 (New England).

Tetragoneuria petechialis Muttkowski

Spot Wing

1911. Mtk., Bull. Wis. Nat. Hist. Soc., 9:101 (figs.).
1933. Davis, Bull. Brooklyn Ent. Soc., 28:97 (figs.).

Length 41–43 mm.; abdomen 30–34; hind wing 27–31.

A rather elegant Southern species, easily recognizable by little spots of brown on nodus and antenodal crossveins. A pale-faced species. Top of vertex pale under a dense covering of erect black hairs. Darkened sutures of thorax show dimly under tawny hairs, but three yellow spots in a row on each side generally stand out clearly: two above and one below spiracle.

Slender abdomen well-nigh covered by a dorsal band of black, surface of which is thinly besprinkled with very minute white hairs. Segment 2 mainly yellow, but traversed on sides by an encircling oblique band of black in which lies polished black auricle. Yellow of line of spots along each side very bright on 4 to 8. Abdomen slightly narrowed on 3 and again on 8, with intervening segments considerably depressed; 10 black. Superior appendages of male thinly hairy, without tooth or sudden angulation and slightly divergent. Divisions of subgenital plate of female more slender, recurved, with a longer taper to rearward than in other species of genus. Appendages of female about 2 mm. long.

Distribution and dates.—UNITED STATES: Fla. (?), Kans., N. Mex., Tex. June 16 (Kans.).

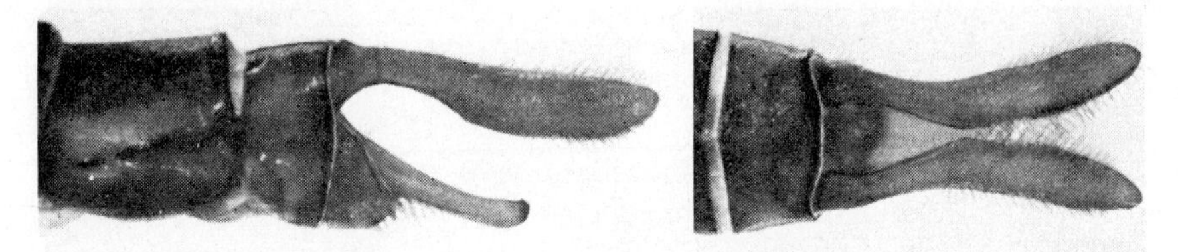

Fig. 223. *Tetragoneuria petechialis.*

Tetragoneuria semiaquea Burmeister

Syn.: complanata Rambur; var. calverti Muttkowski

1839. Burm., Handb., 2:858 (in *Libellula*).
1911. Mtk., Bull. Wis. Nat. Hist. Soc., 9:118 (fig.).
1915. Mtk., Bull. Wis. Nat. Hist. Soc., 13:49 (relations with *cynosura*).
1933. Davis, Bull. Brooklyn Ent. Soc., 28:94 (fig.).

Length 34–38 mm.; abdomen 24–30; hind wing 24–31.

An Eastern species, smallest of genus. Has scantiest venation and broadest areas of brown color pattern on hind wing. Front of frons overspread by a broad crescent of bright yellow; in furrow on top of frons lies black stump of an incomplete black T spot. Thoracic pile usually brown instead of gray, as is more common in *cynosura.* Wings hyaline beyond the brown. Venation as in *cynosura;* somewhat fewer crossveins, but otherwise same. Extent of dark color pattern exceedingly varied, being least in specimens of southward distribution, where overlap of patterns with *cynosura* is considerable.

Caudal appendages of male much alike in the two species. When seen in profile, top and bottom lines of superior appendages of male beyond ventral angle more parallel than in *cynosura,* and rounded tip more suddenly attained. Subgenital plate of female very much like that of *cynosura;* tips perhaps less divergent. Appendages short, like those of *cynosura,* about 1.2–1.5 mm. long. In larger Florida specimens of form *calverti,* appendages about 1.8 mm. long.

Distribution and dates.—UNITED STATES: Fla., Ga., Mass., N. J., N. C., S. C.

January 8 (Fla.) to August 30 (S. C.).

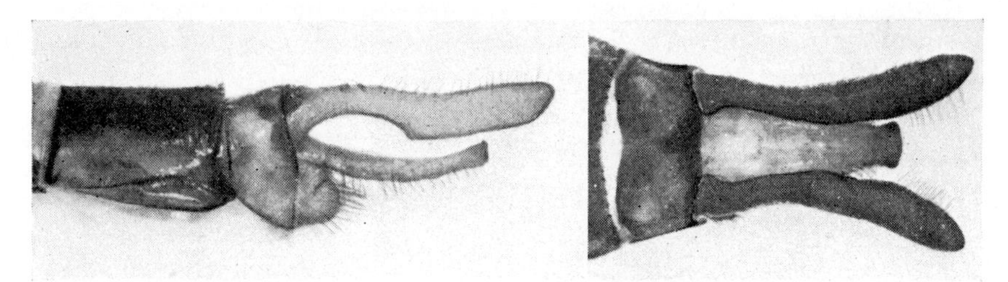

Fig. 224. *Tetragoneuria semiaquea.*

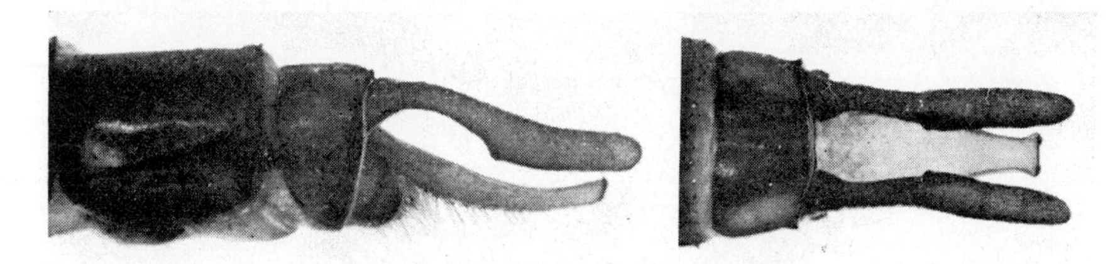

Fig. 225. *Tetragoneuria sepia.*

Tetragoneuria sepia Gloyd

1933. Gloyd, Occ. Pap. Mus. Zool. Univ. Mich., 274:1 (figs.).
1941. Wstf., Ent. News, 52:15.
1951. Wstf., Fla. Ent., 34:9 (figs., nymph and female).

Length 40–45 mm.; abdomen 25–29; hind wing 28–33.

A very pretty species, more slender in form and lighter in coloration than others of genus. Top of vertex yellowish; top of occiput olivaceous, with only its margin narrowly black. Long silky hair, very white on rear of head and low down on sides of body. Front of thorax sepia brown beneath blackish crest; sides olivaceous, with spots above and below spiracle often bright yellow. Only lower end of humeral suture black. Wings very clear, with only a light touch of brown on crossvein in basal triangle of male sometimes present. As in other species, wing membrane may have a light tinge of amber. Claws reddish in part.

Abdomen long and slender, widest on segment 2, and regularly tapered all the way to rearward. Yellow spots along each side wider than usual, being narrowly separated by black at joinings of segments; they dwindle on 9 and disappear on 10. Segment 2 mostly yellow, with shining black auricle rising in an oblique and encircling streak of black. Side of genital lobe yellow. Superior appendages of male not longer than 9 plus 10; inferior about seven-eighths as long as superiors, about three-fourths as long in other species.

Resembles *stella* in clearness of wings, generally lighter coloration, and more slender form, but thorax, especially on sides, much less hairy, and yellow spots beside spiracle more pronounced. Halves of subgenital plate slightly less divergent than those of *cynosura,* with tips slightly convergent. Appendages about 1.8 mm. long in female.

Distribution and dates.—UNITED STATES: Fla.

March 28 to November 15.

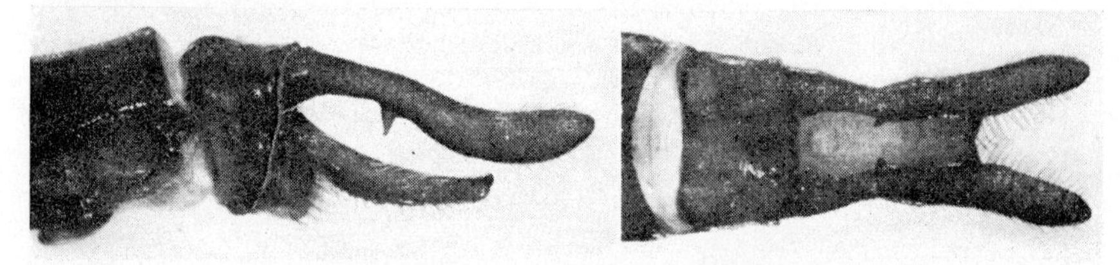

Fig. 226. *Tetragoneuria spinigera.*

Tetragoneuria spinigera Selys

Syn.: indistincta Morse; var. suffusa Davis

1871. Selys, Bull. Acad. Belg., (2)31:269 (reprint, p. 35).
1874. Selys, Bull. Acad. Belg., (2)37:20 (reprint, p. 9).
1901. Ndm., Bull. N. Y. State Mus., 47:493 (nymph).
1911. Mtk., Bull. Wis. Nat. Hist. Soc., 9:127 (fig.).
1913. Walk., Can. Ent., 45:161 (nymph).
1929. N. & H., Handb., p. 181 (fig.).
1933. Davis, Bull. Brooklyn Ent. Soc., 28:99 (figs.).

Length 43–47 mm.; abdomen 29–37; hind wing 29–34.

A rather stout, densely hairy blackish Northern species. Hair whitish on a dark body, with sutures of thorax showing through it very broad. Wings lightly tinted in membrane. In typical specimens, brown coloration at base restricted generally to three short and separate streaks: one runs out far enough to cover first antenodal crossvein; next one, far enough to cover anal crossing; third lies across junction of second paranal and tornal cells. When two latter are confluent, a clear spot remains in first paranal cell.*

* In 1933 Davis recognized a variety to which he gave the name *Tetragoneuria spinigera* var. *suffusa.* This variety he designated as differing from typical *spinigera* in the same way that *simulans* differs from typical *cynosura.* The dark markings of the wings reach the triangle or beyond.

Abdomen broadly black above, with lateral yellow spots correspondingly reduced in size. Constricted on segment 3 and depressed on widened following segments. Male easily recognized by large sharp tooth it bears near middle on under side. Branches of subgenital plate of female concave on inner face, forming a U, with tips parallel. Appendages of female about 3.5 mm. long.

Distribution and dates.—CANADA: Alta., B. C., Man., N. B., N. S., Ont., Que., Sask.; UNITED STATES: Calif., Conn., Ind., La., Maine, Mass., Mich., Minn., N. H., N. J., N. Y., Wash., W. Va., Wis.

May 17 (Mich.) to September (N. Y.).

Tetragoneuria spinosa Hagen

1878. Hagen, in Selys, Bull. Acad. Belg., (2)45:188 (reprint, p. 10) (in *Cordulia*).
1906. Mrtn., Coll. Selys Cordulines, p. 44 (figs.).
1929. N. & H., Handb., p. 181.
1933. Davis, Bull. Brooklyn Ent. Soc., 28:101 (figs.).

Length 45–46 mm.; abdomen 33–35; hind wing 30–33.

A little-known Southern coastal species, said to be most like *spinigera*, with more hair about front of thorax and base of abdomen, and hair more nearly white. Only four antenodal crossveins in hind wing. In specimen figured by Davis (1933), narrow basal brown strip at base of hind wing extends its three prominences a little farther outward to cover basal antenodal crossvein, anal crossing, and hind margin of second paranal cell, respectively. Keel on hind tibia of male beneath is unusually long, projecting like a spur to extreme level of tip of that joint.

Male easily distinguished from *canis* by form of superior caudal appendages. A sharp anteapical tooth at top of terminal declivity where *canis* has a very blunt one, and no inferior tooth. In dorsal view, rather broadly truncated inferior appendage appears between widely outspread ends of superiors. Subgenital plate of female reaches only to end of segment 9, and its rather broad and very blunt divisions are outspread in a broad U, with a V-shaped median notch in middle of base of U. Appendages of female about 2 mm. long.

Distribution and dates.—UNITED STATES: Ga., N. J., N. C., S. C.

April 2 (S. C.) to June 6 (N. J.).

Tetragoneuria stella Williamson

1911. Wmsn., Bull. Wis. Nat. Hist. Soc., 9:96 (figs.).
1933. Davis, Bull. Brooklyn Ent. Soc., 28:96.

Length 44–47 mm.; abdomen 32–36; hind wing 27–32.

Another rather large, almost clear-winged Southern species, in which abdomen is long and slender beyond tapered segment 3. A light-colored species. Short black hair on yellow frons; hair lower down on face whitish. On thorax, dark sutural stripes show faintly, most plainly at lower ends where first and third stripes are conjoined by upturned ends that meet at and enclose spiracle. Thorax clothed with a moderately thick layer of white hairs, abdomen is mostly bare and shiny.

Wings clear except for very scanty brown markings at base that do not reach outward quite far enough to cover basal crossveins. Brown markings continuous across extreme wing base; hindmost one may extend outward to cover crossvein of second paranal cell. Stigma of both

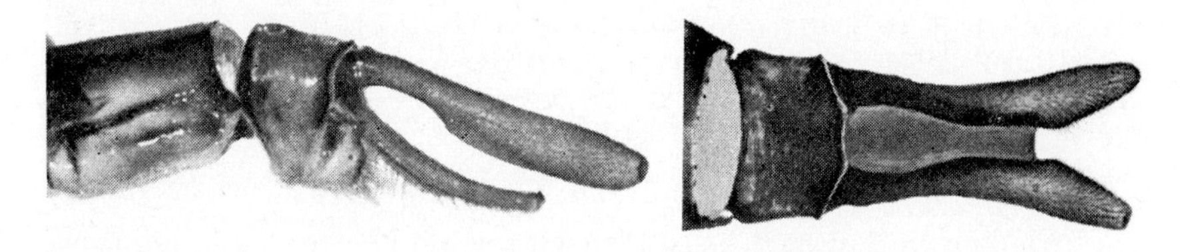

Fig. 227. *Tetragoneuria stella.*

fore and hind wings longer than in *cynosura,* about 3 mm. in *stella* and 2 mm. in *cynosura.*

Abdomen quite bare beyond its swollen basal segments. Largely yellowish on segments 1 to 3; yellow spots along side margin diminish in width. Segment 10 and appendages blackish. Superior appendages of male similar to those of *cynosura;* but their basal third is nearly straight, not highly arched as in *cynosura,* and angle at beginning of larger swollen part underneath, when viewed from side, is much less prominent than in that species. Subgenital plate of female similar to that of *cynosura,* but with lobes less divergent, tips more parallel or slightly convergent. Appendages of female slightly longer than those of *cynosura,* about 2 mm. long.

Distribution and dates.—UNITED STATES: Fla., Ga., La.

February 3 (Fla.) to April 25 (Fla.).

Tetragoneuria williamsoni Muttkowski

1911. Mtk., Bull. Wis. Nat. Hist. Soc., 9:122 (figs.).
1933. Davis, Bull. Brooklyn Ent. Soc., 28:98.

Length 43–45 mm.; abdomen 29–32; hind wing 26–28.

Another almost clear-winged species, very much like *cynosura* but less hairy, and with a slender abdomen in which middle segments are less depressed. Dark stripes on lateral sutures of thorax very diffuse, but plainer and stronger low down on sides where first and third are conjoined by an irregular longitudinal streak that includes spiracle. A spot of yellow lies above spiracle and another below it. Pattern of hind-wing base quite as in *stella* except that brown is a little more extended in basal subcostal interspace.

Abdomen contracted on segment 3, and slowly widened but not depressed thereafter to rearward toward, but not quite to, end. Genital lobe on 2 black; 10 and appendages black. Superior caudal appendages of male like those of *cynosura,* but not quite so highly arched in basal third, inferior angle more prominent, and tips of pair as viewed from

Fig. 228. *Tetragoneuria williamsoni.*

above much less divergent. In that view they appear slightly undulate, two slight constrictions dividing them into three parts of about equal length. Lobes of subgenital plate of female smoothly curved and subparallel at tips, forming somewhat of a U, not broadly divaricate and V-shaped as in *cynosura.* Appendages about 2 mm. long.

Distribution and dates.—UNITED STATES: Ala., Fla., Miss., Okla.

April 1 (Ala.) to June 17 (Ala.).

Genus HELOCORDULIA Needham 1901

These are Cordulines of small size a few of which have some of the antenodal crossveins spotted with brown, and a brown spot at the base of each wing. The thorax is heavily clothed with a shaggy pubescence. The wings generally show the following characters of venation in addition to those indicated in the generic key and table: the stigma narrow and very oblique at its ends; two crossveins under the stigma, with a normal space at each side of them; antenodals of the hind wing six; the anal loop almost squarely truncate at its apex, with three cells on the distal end along the sole; two cubito-anal crossveins in the hind wing and one in the fore wing; sectors of the arculus separate at origin. The male has tibial keels on all the legs.

The nymphs are rather stocky, the head nearly twice as wide as long. The eyes are small, not very prominent, the contour of their front margins being continuous with the contour of the front of the head. Each occipital prominence bears a lateral and a dorsal row of long spinelike hairs. The labium is about as wide as long, and the median lobe forms almost a right angle on its front border. There are 13 to 14 mental setae on each side, the outer 8 being very long, the inner ones much shorter; the lateral setae are 6 to 7. The movable hook is slender and not much longer than the lateral setae. The abdomen is elliptical, widest at about

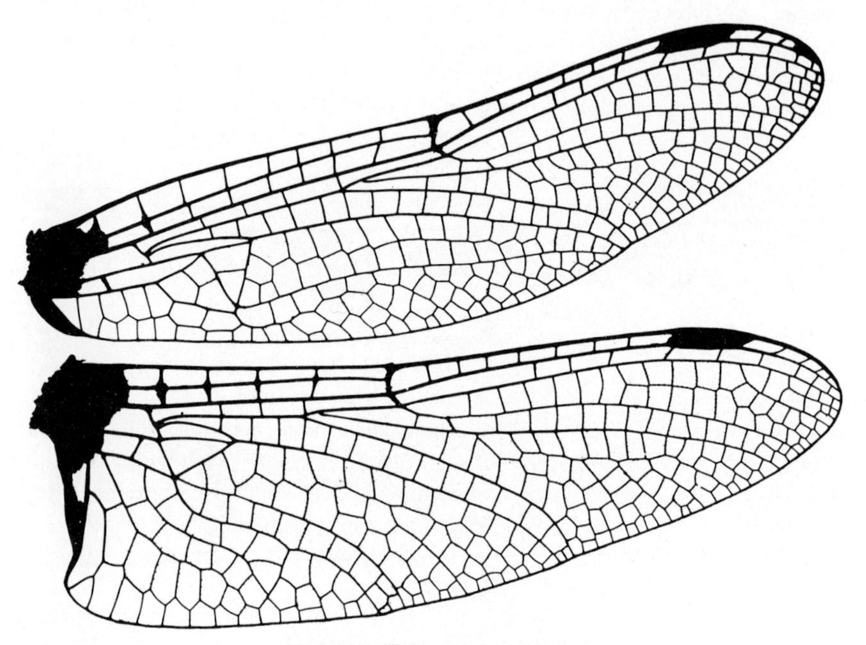

Fig. 229. *Helocordulia uhleri.*

segment 6, tapering regularly to the apex. Segment 10 is reduced to a mere ring inserted in the apex of 9. Segment 9 is only about half as long middorsally as it is midventrally.

This is a small genus of only two known species; found only in the eastern United States.

KEY TO THE SPECIES

ADULTS

1—Hind wing without golden yellow spot in midst of basal brown spot; subgenital plate of female less than one-third as long as segment 9, emarginate, tips widely divaricate and lateral margins subparallel........... **selysii**

—Hind wing with golden yellow spot in midst of basal brown spot; subgenital plate of female from one-half as long to almost as long as segment 9, deeply bifid, its divisions divaricate................................. **uhleri**

NYMPHS

1—Dorsal hooks on 7 to 9; lateral setae seven; lateral spine on 8 half as long as on 9...selysii

—Dorsal hooks on 6 to 9; lateral setae usually six; lateral spine on 8 about as long as on 9...uhleri

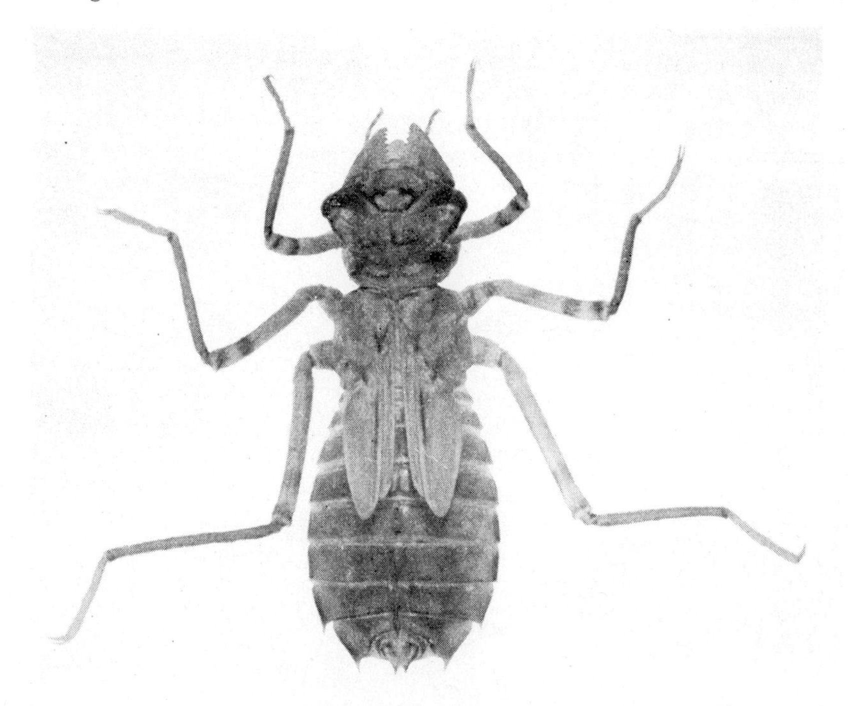

Fig. 230. *Helocordulia uhleri.*

Helocordulia selysii Hagen

1878. Hagen, in Selys, Bull. Acad. Belg., (2)45:189 (reprint, p. 11) (in *Cordulia?*).
1901. Ndm., Bull. N. Y. State Mus., 47:496 (figs.).
1906. Mrtn., Coll. Selys Cordulines, p. 40 (figs.).
1924. Kndy., Proc. U. S. Nat. Mus., 64(12):1 (figs., nymph).
1929. N. & H., Handb., p. 183.

Length 38–41 mm.; abdomen 29–31; hind wing 26–28.

A Southern species, very similar to more Northern *uhleri,* but much less common in collections. Adult seems to prefer open sunny glades in woods.

Kennedy found the cast skin of this species on the side of a boathouse on the shore of a mud-bottomed artificial pond at Raleigh, N. C.

Distribution and dates.—UNITED STATES: Ala., Ga., Miss., N. C., S. C. March 19 (Miss.) to April 29 (Miss.).

Helocordulia uhleri Selys

1871. Selys, Bull. Acad. Belg., (2)31:274 (reprint, p. 40) (in *Cordulia*).
1901. Ndm., Bull. N. Y. State Mus., 47:496 (figs., adult and nymph).
1906. Mrtn., Coll. Selys Cordulines, p. 40 (figs.).
1924. Kndy., Proc. U. S. Nat. Mus., 64(12):1 (figs., nymph).
1927. Garm., Odon. of Conn., p. 208.
1929. N. & H., Handb., p. 182.

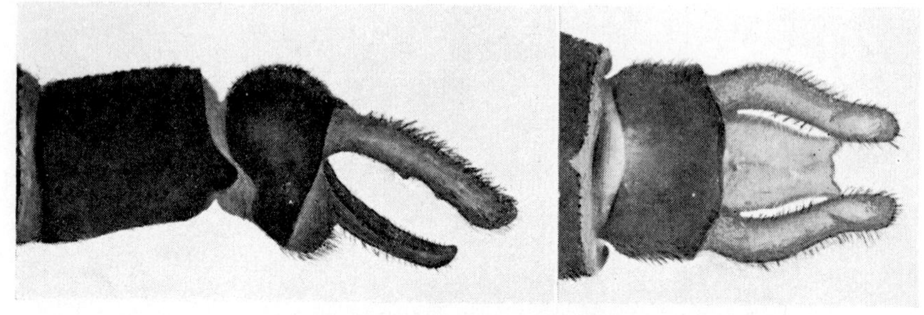

Fig. 231. *Helocordulia selysii.*

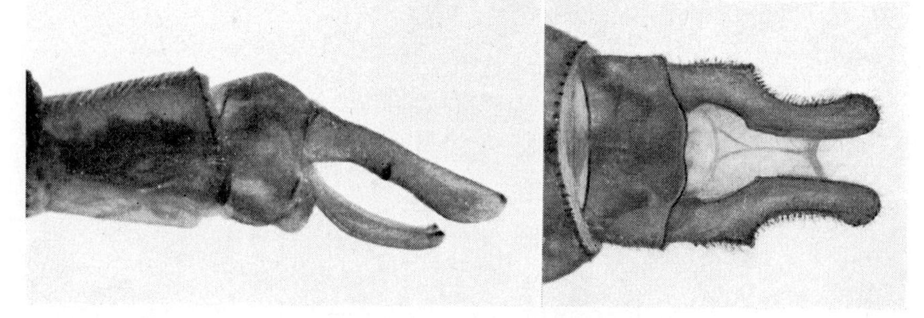

Fig. 232. *Helocordulia uhleri.*

Length 41–46 mm.; abdomen 28–32; hind wing 25–29.

Face mostly yellowish; labium, labrum, and front surface of frons yellowish orange; clypeus and lateral and dorsal surfaces of frons olivaceous; frons with a rather deep median depression on its dorsal surface, floor of depression often dark brown or black; occiput brown, with its rear border as well as rear border of eyes clothed with long whitish hairs.

Thorax clothed with thick, shaggy, silky pubescence under which may be seen olivaceous-to-brown color of dorsum and sides; middorsal carina black, humeral and lateral sutures rather heavily streaked with black. Legs black, anterior femora brownish basally.

Abdomen mostly black; sides and dorsum of segment 1 and sometimes dorsum of 2 brown. Following areas pale: a large lateral spot on 2 below auricles on anterior half of segment, basal fourth to third of 3, and basal lateral spots on 4 to 8. Segments 9, 10, and appendages mostly black. Lateral carinae present on 4 to 9. In male, 3 rather strongly compressed and inferior appendage slightly concave at tip, thus appearing a bit bifid.

Distribution and dates.—CANADA: N. B., N. S., Ont., Que.; UNITED STATES: Ala., Conn., Ky., La., Maine, Mass., N. H., N. J., N. Y., N. C., Tenn.

April 4 (N. C.) to July 26 (Que.).

Genus SOMATOCHLORA Selys 1871

These are graceful clear-winged species. Their coloration is dark brown, generally with bronzy green reflections. There are no pale stripes on the front of the thorax, and the usual two on the sides vary greatly in size and spread from species to species. There is a narrow pale ring around the apex of abdominal segment 2, and in the males there are generally two pale areas beside the auricle on that segment.

The venation of the wings is open. Triangles are two-celled and supertriangles one-celled in both fore and hind wings, with two rows of cells in the trigonal area beyond. There is a single crossvein underneath the stigma, and a wide vacant space on each side of it. There is a single bridge crossvein. In the hind wing are two cubito-anal crossveins, generally five antenodals (rarely four, and sometimes six). There are two rows of cells in the second postanal interspace (*y*); the cells are very unequal, those of the first row being very much longer vertically than those of the marginal row.

There is a keel on the under side of the front and hind tibiae of the male, but none on the middle tibiae.

This is our largest genus of Cordulines. It is also one of the best known, having been monographed with great care and thoroughness and copiously illustrated by Dr. E. M. Walker in his volume, *The North American Species of the Genus Somatochlora* (Univ. Toronto Studies, Biol. Ser. I, pp. 1–202, 1925), a work indispensable to anyone who would make further studies of this genus. It is nearly as well represented in the northern parts of the Old World.

The nymph is a sprawler, somewhat elongate of body, and more or less hairy. The head is twice as wide as long. The eyes are rather small and but little elevated at the front. Behind them the head narrows to a nearly straight hind border. The mask-shaped labium is armed with from six to nine lateral setae and from eight to fifteen mental setae. The eight to ten teeth on the opposed edges of the lateral lobes are rather deeply crenate.

The abdomen is little wider than the head, more or less oval, without well-marked dorsal or lateral carinae. It is widest on segment 6, most narrowed thereafter on 9. Short and sharp lateral spines may be present

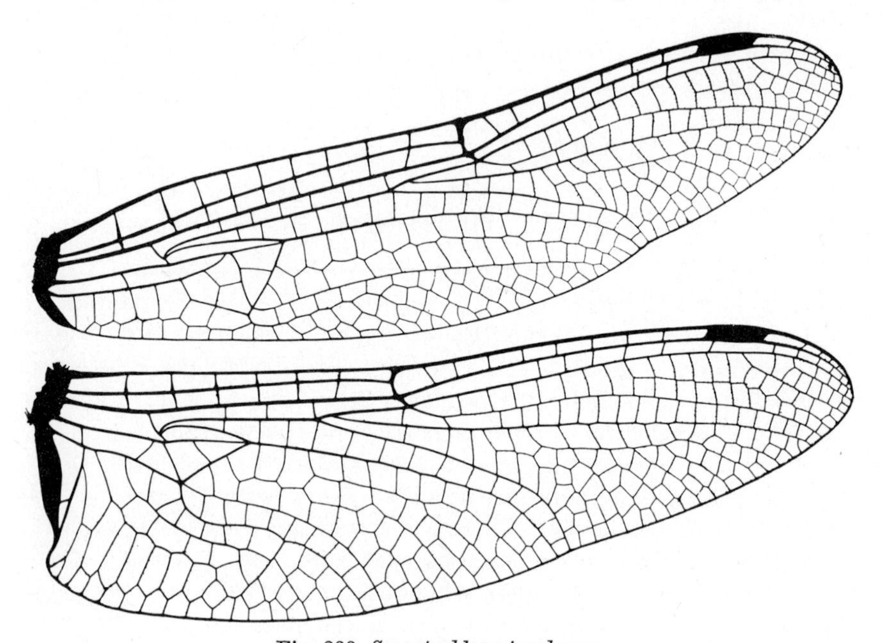

Fig. 233. *Somatochlora tenebrosa.*

on 8 and 9. Dorsal hooks may or may not be present on 3 to 8. The five caudal appendages are of nearly equal length.

The nymphs are found mostly in mucky edges of woodland streams and bogs.

KEY TO THE SPECIES*

Adults

1—Labrum yellowish, often with black band across front border; postclypeus mostly yellow 2
—Labrum black; if partly yellow, no black band across front border 10

2—Tibiae yellow externally; coloration hardly metallic; side stripes on thorax very long and wide; Southern .... **georgiana**
—Tibiae black externally; coloration strongly metallic 3

* Key to the genera of Cordulines is on pp. 347 and 349.

3—Lateral thoracic pale stripes entirely wanting .... **linearis**
—These stripes present 4

4—Abdominal segment 2 with single large lateral spot of yellow before auricle .. 5
—This segment with additional lateral spots of yellow 6

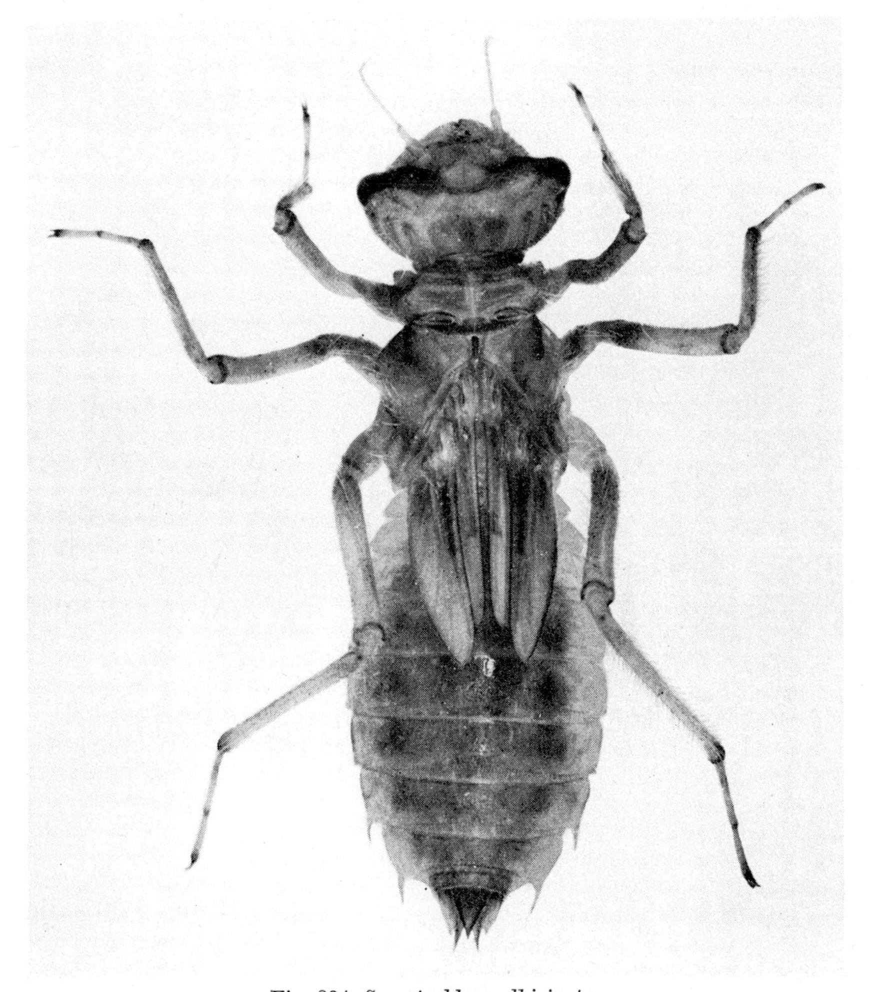

Fig. 234. *Somatochlora albicincta.*

5—Body with all pale markings dull in color .... **tenebrosa**
—Body with all markings bright in color .... **ensigera**

6—First lateral thoracic stripe narrower than second, and angulated or interrupted in middle .... **filosa**
—First stripe not so; more regular 7

7—Dorsum of 10 with large yellow spot in male (female unknown) .... **calverti**
—Dorsum of 10 with no large yellow spot 8

8—Second lateral stripe both longer and wider than first. . . . . **ozarkensis**
—Second stripe not both longer and wider than first 9

9—Second lateral stripe shorter than first. . . . . **hineana**
—Second stripe about equal to first. . . . . **provocans**

10—Middle segments of abdomen with narrow apical rings of white; postclypeus black; only one pale stripe on side of thorax (the first). 11
—No such rings of white present. 13

11—Inferior caudal appendage of male quadrangular, nearly parallel to its widely forking tip; caudal appendages of female, viewed from side, straightish below and decidedly convex on upper margin. . . . . **cingulata**
—Inferior caudal appendage of male elongate-triangular, strongly tapering to single upturned tip; caudal appendages of female about equally convex on upper and lower margins. 12

12—Superior caudal appendages of male, viewed from above, parrallel-sided externally for most of their length; subgenital plate of female extends to rearward less than half the length of sternum of segment 9. . . . . **albicincta**
—Superior caudal appendages of male, viewed from above, strongly divergent in basal fifth, then parallel; subgenital plate of female extends to rearward more than half the length of sternum of 9. . . . . **hudsonica**

13—Second lateral pale stripe of thorax wanting. 14
—Second stripe present. 17

14—Superior caudal appendage of male not recurved at tip; subgenital plate of female about as long as ninth sternite, entire at apex and scoop-shaped **franklini**
—Superior appendage of male strongly recurved at tip; subgenital plate of female much shorter than sternum of 9. 15

15—Third anal interspace (*z*) covered by dark brown spot. 16
—This interspace hyaline, with dark brown restricted to adjacent membranule . . . . **sahlbergi**

16—Rear of head with fringe of orange brown hair; subgenital plate of female entire or nearly so, a bit compressed and projecting downward. . . **whitehousei**
—Rear of head with fringe of whitish hair; subgenital plate of female bilobed and not projecting downward. . . . . **septentrionalis**

17—Each side stripe on thorax represented by round spot. . . . . **minor**
—Each side stripe represented by elongated spot. 18

18—First side stripe longer and narrower than second. **walshii**
—First side stripe not longer but yet narrower than second. 19

19—Both side stripes bright yellow and sharply defined. 20
—Both side stripes dull yellow and ill defined. 21

20—Middle abdominal segments with pale spots . . . . **forcipata**
—These segments with no pale spots. . . . . **elongata**

21—Abdominal segment 2 blackish, with single pale spot on each side. . . . **incurvata**
—Abdominal segment 2 with two larger pale areas on each side. 22

22—Tips of superior caudal appendages of male recurved; subgenital plate of female triangular in profile, taper-pointed and erect. . . . . **williamsoni**
—Tips of superior caudal appendages of male not recurved; subgenital plate of female broad, less pointed. 23

23—Facial lobes of postclypeus yellowish . . . . **kennedyi**
—Facial lobes of postclypeus black . . . . **semicircularis**

NYMPHS

1—Dorsal hooks present 2
—Dorsal hooks absent 8

2—Dorsal hooks falciform, acute, the last projecting beyond middle of segment 10 4
—Dorsal hooks low knobs, rather blunt, not falciform, the last not projecting to middle of 10 3

3—Dorsal hooks on 4 to 9 . . . . **ozarkensis**
—Dorsal hooks on 5 or 6 to 9 . . . . **williamsoni**

4—Superior caudal appendage of male flat above, anteapical tubercle (apex of developing inferior appendage of adult) not at all elevated 5
—Median appendage of male longitudinally concave, anteapical tubercle slightly elevated . . . . **tenebrosa**

5—Lateral spines at least half as wide as long; lateral appendages shorter than segments 9 and 10 inclusive; median appendage not acuminate . . **linearis**
—Lateral spines less than half as wide as long; lateral appendages as long as 9 and 10 inclusive; median appendage acuminate, with very slender tip 6

6—Hind femora 8 mm. long; hind tibiae 9 mm. long; lateral spines of 9 one-sixth as long as its margin . . . . **elongata**
—Hind femora 7 mm. long or less; hind tibiae 8 mm. long or less; lateral spines of 9 one-third as long as its margin 7

7—No trace of dorsal hook on 3; dorsal hook on 4 minute, less than one-third as long as hook on 5; lateral setae ten . . . . **walshii**
—Dorsal hook on 3 distinct; dorsal hook on 4 large, almost as long as hook on 5; lateral setae eleven to thirteen . . . . **minor**

8—Labium extending laterally over inner edge of eyes; lateral appendages of male with outer margin not regularly arcuate, more or less sinuate; lateral spines, when present, on both segments 8 and 9 9
—Labium extending laterally not quite to inner edge of eyes; lateral appendages of male with outer margin regularly arcuate; lateral spines, when present, confined to segment 9 13

9—Lateral spines, though often very minute, on 8 and 9 11
—No trace of lateral spines or denticles on 8 and 9 10

10—Lateral setae nine or ten; abdomen with distinct markings . . . . . **septentrionalis**
—Lateral setae six or seven (rarely eight); abdomen without distinct markings . . . . **whitehousei**

11—Length of body less than 25 mm.; hind femora less than 7 mm. long; abdomen without median dorsal prominences; median appendages of male without lateral knobs; coloration uniform 12
—Length of body more than 25 mm.; hind femora more than 7 mm. long; abdomen with median series of slightly elevated prominences; median appendage of male with distinct knob on each side (developing furcate apex of inferior appendage); coloration not uniform . . . . **cingulata**

TABLE OF SPECIES

ADULTS

Species	Hind wing	*Face*		*Thorax*		*Abdomen*[4]		Distr.
				Side stripes[3]				
		Color labrum[1]	*Post-clypeus*[2]	*Front*	*Rear*	*White rings*	*Yellow spots*	
albicincta	28-33	metallic	black	bright	0	+	0	N
calverti	36-37	brown	yellow	bright	bright	0	0	Fla.
cingulata	33-41	metallic	black	dull	0	+	0	N
elongata	34-38	metallic	black	bright	bright	0	0	N, E
ensigera	33-35	brown	yellow	bright	bright	0	0	C
filosa	36-45	brown	yellow	bright	bright	0	0	E
forcipata	29-33	metallic	black	bright	bright	0	+	N
franklini	25-30	metallic	black	dull	0	0	0	N
georgiana	32-34	brown	yellow	bright	bright	0	0	SE
hineana	40-42	brown	yellow	bright	bright	0	0	Ohio
hudsonica	30-34	metallic	black	bright	0	+	0	N
incurvata	31-37	metallic	black	dull	dull	0	+	NE
kennedyi	29-34	metallic	black	dull	dull	0	0	N

sahlbergi	32-33	metallic	black	bright	0	0	0	Alaska
semicircularis	27-32	metallic	black	dull	dull	0	±	W
septentrionalis	26-30	metallic	black	dull	0	0	0	N
tenebrosa	34-41	brown	yellow	dull	dull	0	0	E
walshii	25-34	metallic	black	bright	bright	0	+	N
whitehousei	26-30	metallic	black	dull	0	0	0	N
williamsoni	35-40	metallic	black	dull	dull	0	+	NE

[1]Color of labrum: metallic (black) or brown; latter generally with narrow black front border.

[2]Approximate color of postclypeus: yellow or black.

[3]Vividness of yellow side stripes of thorax: bright or dull; 0 means these stripes lacking.

[4]Abdomen sometimes has narrow white rings at joinings of segments, and sometimes yellow spots on dorsum or sides posterior to segment 3.

12—Lateral spines minute, those of segment 9 from one-twelfth to one-sixth as long as remainder of segment margin.......................... **hudsonica**

—Lateral spines longer, those of 9 from one-fifth to one-third as long as remainder of segment margin.................................. **albicincta**

13—Fringe of hair on hind margins of abdominal segments not forming dorso-lateral tufts; lateral spines normally present on 9; lateral setae seven or eight **14**

—Fringe of hair on hind margins of 6 to 9 forming conspicuous dorso-lateral tufts; lateral spines absent from 9 or represented by very minute denticles; lateral setae nine or ten................................ **forcipata**

TABLE OF SPECIES

NYMPHS

Species	*Total length*	*Setae*[1]		*Lat. spines*[2]		*Dorsal hooks*[3]
		Lat.	*Ment.*	*On 8*	*On 9*	
albicincta	20-24	5-6	11-12	2	3	0
cingulata	26-28	5-7	10-13	3	4	0
elongata	23-24	6-7	11-12	2	3	+
forcipata	19-20	9-10	12-15	0	0?	0
franklini	17	7-8	13	0	1	0
hudsonica	24-25	5-6	10-13	2	2	0
kennedyi	21	9	12-13	0	1	0
linearis	22	6-7	11-12	2	2	+
minor	21-23	6-8	11-13	2	4	+
ozarkensis	23	7-8	12-13	2	3	+
semicircularis	21-22		10-13	0	var.	0
septentrionalis	19-20	7-8	11-13	0	0	0
tenebrosa	20	8	11-12	2	2	+
whitehousei	20-22	6-7	9-10	0	0	0
williamsoni	23-25	8	11-12	2	3	+
walshii	21	7	9-10	2	3	+

[1]Number of lateral and mental setae on labium.

[2]Length of lateral spines on abdominal segments 8 and 9, expressed in tenths of total length of lateral margin of which they form a part.

[3]Middorsal hooks on abdomen: present (+) or absent (0).

14—Fringe of hair on middle section of hind margins of 7 and 8 forming a thick tuft dorsally .... **kennedyi**

—Fringe of hair on hind margins of 7 and 8 not forming median tuft......... **15**

15—Fringe of hair on hind margins of 8 and 9 equal, or almost equal, to median length of segment; body length 17 mm., hind femur 5.25 mm. ...... **franklini**

—Fringe of hair on hind margins of 8 and 9 much less than median length of segment; body length 21–22 mm., hind femur 5.8–6.0 mm. ..... **semicircularis**

Nymphs unknown: **calverti, ensigera, filosa, georgiana, hineana, incurvata, provocans,** and **sahlbergi.**

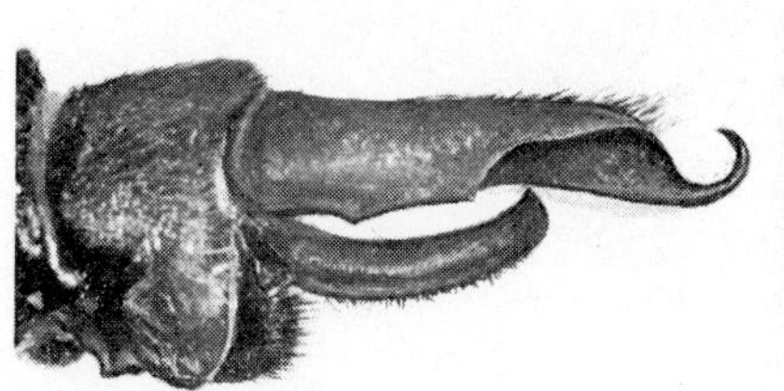
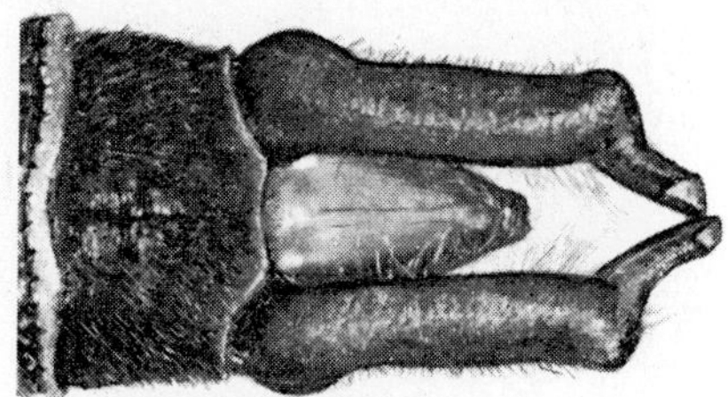

Fig. 235. *Somatochlora albicincta.*

Somatochlora albicincta Burmeister

1839. Burm., Handb., p. 847 (in *Epophthalmia*).
1906. Mrtn., Coll. Selys Cordulines, p. 28 (figs.).
1925. Walk., N. Amer. Somatochlora, p. 167 (figs.).
1927. Garm., Odon. of Conn., p. 231 (figs.).
1929. N. & H., Handb., p. 189 (figs.).

Length 45–52 mm.; abdomen 33–41; hind wing 28–33.

A widely distributed Northern species of rather stocky form, with a color pattern that is enlivened by a white-ringed abdomen. Sides of bronze green frons broadly yellow. Occiput brown, with a dense fringe on its rear margin of stiff tawny hairs. Thorax thinly hairy, with broad prominences of its side showing shining green through the hairs. Membrane of wings slightly tinged with brown.

Abdomen hairy at base and on swollen basal segments, and around edges of genital pocket in male. It becomes greatly contracted on segment 3 and bare; as it then widens to rearward it gradually becomes more hairy again. Caudal appendages of male hairy on inner side, but bare on their slender recurving tips. A narrow pale apical ring on abdominal segment 10, divided by black on middorsal line.

Inhabits slow streams and boggy places in Far North. Males patrol edges of water, flying usually about a foot above surface. Female releases her eggs at surface of water by tapping surface film with tip of her abdomen as she flies.

Distribution and dates.—Alaska; Labrador; CANADA: Alta., B. C., Man., Nfld., NW. Terr., Ont., Que., Yukon; UNITED STATES: N. H., N. Y., Oreg., Wash.

June 25 (Alaska) to September 1 (Ont.).

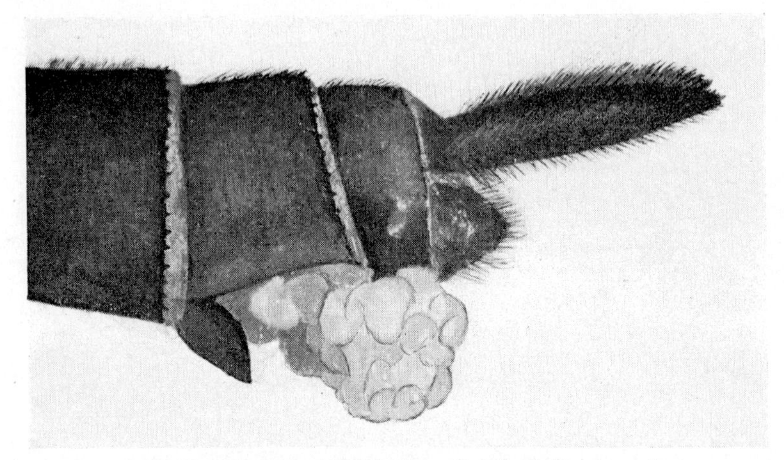

Fig. 236. *Somatochlora albicincta*, lateral view of end of female abdomen carrying a clutch of eggs, ready for release.

Somatochlora calverti Williamson and Gloyd

1933. Wmsn. & Gloyd, Occ. Pap. Mus. Zool. Univ. Mich., 262:1–7 (figs.).

Length 50–52 mm.; abdomen 36–40; hind wing 36–37.

A rare, light-colored Florida species; only male has as yet been described. Face largely yellow up to a brown bar on edge of frons. Above this bar, entire top of frons metallic bluish green, sides pale yellow. Vertex violet; occiput black. Rear of head black, with a thin bordering fringe of whitish hairs.

Thorax brown, with brilliant green and violet reflections. Top of crest and interalar area pale yellow. Two pale yellow side stripes unusually long and plain; in addition, smaller markings of same pale color both before and between them: a small one on front of thorax on each side, half length and pointed upward; several small streaks in vicinity of spiracle; and another far above, beside subalar carina. Legs blackish. Wings hyaline, with a light tinge of brown in membrane, paler toward base; costa and adjacent antenodal crossveins and stigma yellowish brown.

Abdomen dark brown on basal segments, marked with a broad bar of yellow on dorsum of segment 2. Three unequal pale streaks run separately low down on sides, last one almost covering broad genital lobe in male. It is confluent with lower end of narrow apical ring of yellow on 2. Basal triangles on 3 both long and very pale. Remainder of abdomen blackish, with little luster, except for a squarish middorsal apical pale spot on segment 10, and narrow apical rings of yellow in intersegmental membrane, especially bright on 8 and 9. Hairs on these segments light amber; hairs on appendages black.

Distribution and dates.—UNITED STATES: Fla.

August 25.

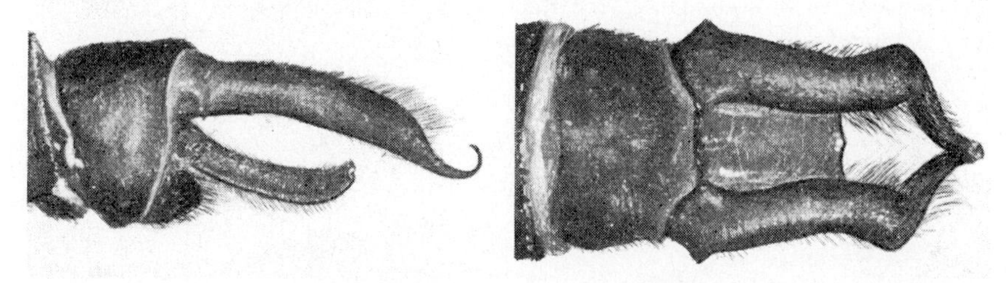

Fig. 237. *Somatochlora cingulata.*

Somatochlora cingulata Selys

1871. Selys, Bull. Acad. Belg., (2)31:302 (reprint, p. 68) (in *Epitheca*).
1906. Mrtn., Coll. Selys Cordulines, p. 23.
1925. Walk., N. Amer. Somatochlora, p. 182 (figs.).
1927. Garm., Odon. of Conn., p. 233 (figs.).
1929. N. & H., Handb., p. 190 (figs.).

Length 54–68 mm.; abdomen 40–50; hind wing 33–41.

Another Northern species, also with white-ringed abdomen. It appears to differ from others of genus in having more abundant crossveins, there being oftenest six antenodals in hind wings, and frequently a second bridge crossvein and likewise under stigma. Male unique in having double notch in apex of broad inferior caudal appendage. Triangle of yellow on each side of bronzy green-capped frons nearly meets its fellow of opposite side on middle line of postclypeus in front. Anteclypeus pale yellow; postclypeus brown. Rear of brown occiput nearly bare.

Tawny hair that thinly clothes thorax is rather long, and continues so both above and below on swollen basal abdominal segments. Wings hyaline, with tawny costa and stigma; membranule at their base pale, its basal half white.

Abdomen shaped as in *albicincta,* but less hairy to rearward, and with area of paler colors more extended.

Inhabits ponds and small boggy-bordered lakes; flies and oviposits much as does *albicincta*.

Distribution and dates.—Labrador; CANADA: Alta., B. C., Man., N. B., Nfld., N. S., Ont., Que.; UNITED STATES: N. H.

June 23 (Alta.) to September 1 (Ont.).

Somatochlora elongata Scudder

Syn.: saturata Hagen (no desc.)

1861. Hagen, Syn. Neur. N. Amer., p. 138 (no desc.) (as *Cordulia saturata*).
1866. Scudder, Proc. Boston Soc. Nat. Hist., 10:218 (in *Cordulia*).
1906. Mrtn., Coll. Selys Cordulines, p. 23 (figs.).
1925. Walk., N. Amer. Somatochlora, p. 70 (figs.).
1927. Garm., Odon. of Conn., p. 220 (figs.).
1929. N. & H., Handb., p. 193 (figs.).

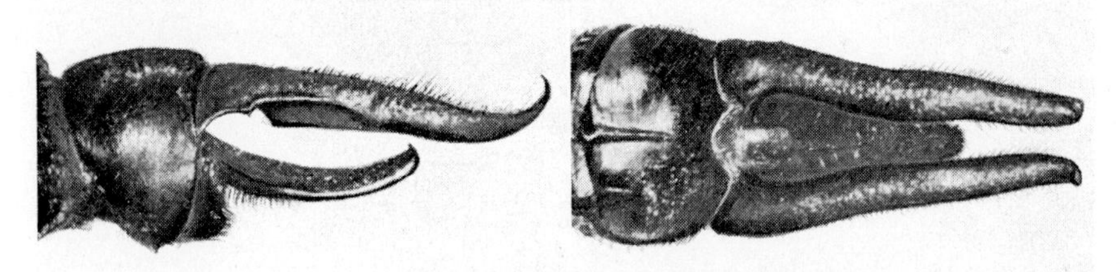

Fig. 238. *Somatochlora elongata.*

Length 52–62 mm.; abdomen 41–48; hind wing 34–38.

A large brown Northeastern species, with a very slender blackish abdomen. Frons bears a cap of shining bronzy green, surrounded except in rear by a broad belt of yellow. Labrum shining greenish black, with two dots or a bar of yellow at its base. Nearly bare occiput brown.

Thorax thinly clad with whitish hairs through which yellow stripes of sides show plainly. Prominences of thorax bronzy at front and pure green on sides. Wings slightly but uniformly tinged with brown; veins and stigma brown.

Abdomen very slender, broadly marked with yellow on lower half of each side of the three swollen basal segments. Portion of abdomen beyond contracted long segment 3 very gradually widened, its constituent segments shortened; all are black, with greenish reflections. Genital lobe of male black and densely fringed with stiff erect brown hairs. Top side of long and slender superior appendages of male hairy out to recurved tips. Scoop-shaped, downthrust subgenital plate of female bare; beyond its tip, end segments and their appendages clothed with brown hair.

Distribution and dates.—CANADA: N. B., N. S., Ont., P. E. I., Que.; UNITED STATES: Maine, Mich., Minn., N. H., N. Y., N. C., Pa., Wis.

June 17 (N. C.) to September 6 (N. S.).

Somatochlora ensigera Martin

Syn.: charadraea Williamson

1906. Mrtn., Coll. Selys Cordulines, p. 29 (fig.).
1925. Walk., N. Amer. Somatochlora, p. 86 (figs.).
1929. N. & H., Handb., p. 187 (figs.).

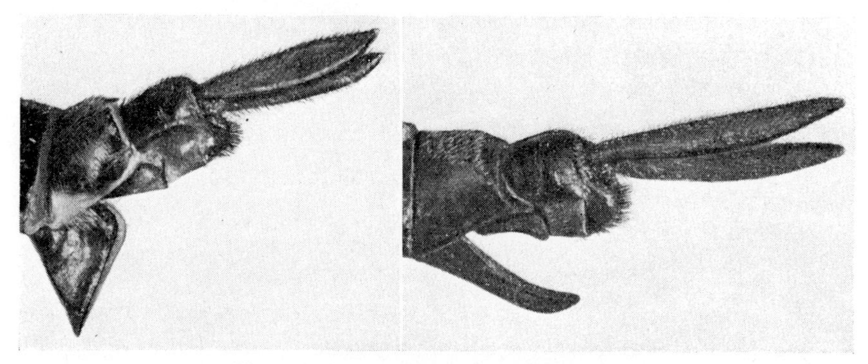

Fig. 239. *Somatochlora elongata; S. filosa.*

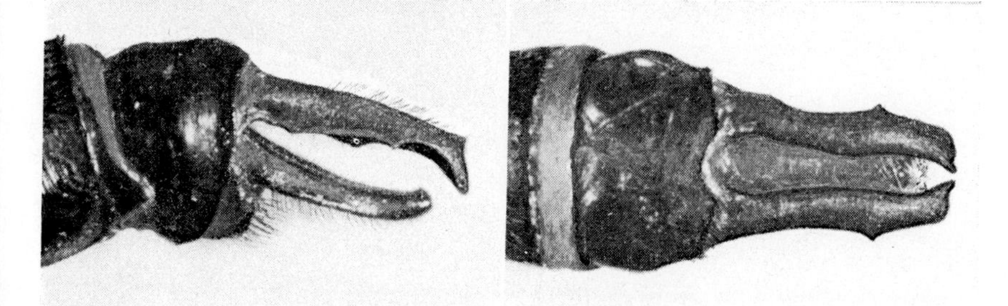

Fig. 240. *Somatochlora ensigera.*

Length 48–51 mm.; abdomen 35–38; hind wing 33–35.

An inland species of medium size and well-defined markings. Orange labrum with black front border and median basal spot. Frons has a broad cap of metallic blue green; its side spots of yellow narrow forward to meet or nearly meet on middle line in front. Occiput shining brown.

Thorax scantily hairy. Side stripes of thorax and of abdominal segment 2 bright yellow, bordered by black. Thorax dark reddish brown in front, with yellow carina; it becomes bluish toward crest and between yellow stripes on sides. Wings hyaline, touched with amber yellow at extreme base. Costa ochre yellow; stigma blackish; veins brown.

Abdomen brown, much swollen on segment 2; two large antero-lateral spots, in male a smaller postero-lateral spot runs down and half covers genital lobe. Of two usual elongate lateral triangles on segment 3, lower one is much longer. Remainder of abdomen and caudal appendages black.

This species nearly like *linearis* but smaller and more brightly marked with yellow. Caudal appendages of male more slender and more nearly parallel in their apical third. Ovipositor of female longer, straighter, more slender.

Frequents small woodland streams. Female deposits her eggs on damp clay at water's edge. Males fly at a low level over shoals.

Distribution and dates.—CANADA: Man., Sask.; UNITED STATES: Colo., Ind., Iowa, Mont., Okla.

June 14 (Sask.) to July 24 (Man.).

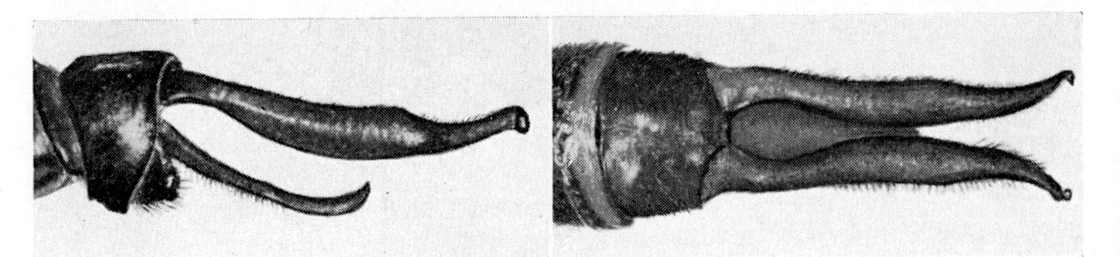

Fig. 241. *Somatochlora filosa.*

Somatochlora filosa Hagen

1861. Hagen, Syn. Neur. N. Amer., p. 136 (in *Cordulia*).
1906. Mrtn., Coll. Selys Cordulines, p. 22 (figs.).
1925. Walk., N. Amer. Somatochlora, p. 112 (figs.).
1929. N. & H., Handb., p. 188 (figs.).

Length 54–66 mm.; abdomen 42–52; hind wing 36–45.

One of largest species of genus, and probably the most southerly in its distribution. A very slender broad-winged species, with rather open-meshed venation. Face broadly yellow from blackish band that borders labrum in front up to metallic blue green cap that covers more than top of frons. Front of bilobed vertex also bluish green. Middle of brown occiput paler.

Two side stripes of thorax well developed: first stripe long and narrow, angulated or interrupted in middle; second shorter and broader. Wings hyaline, with an area near rear of stigma sometimes lightly tinged with brown, especially in female. Hind wings of unusual breadth.

Abdomen long and slender, especially segments 3 and 4. Sides of basal segments marked with three vertical stripes, last of which runs down in male to cover a large part of genital lobe on 2. Narrow intersegmental half rings of pale yellow on shortened end segments; elsewhere shining black, including long and slender caudal appendages of male.

Easily recognized by large size, peculiar first lateral thoracic stripe, form of tips of superior appendages of male, and long sled-runner-like ovipositor of female.

Distribution and dates.—UNITED STATES: Ala., D. C., Fla., Ga., Ill., Ky., Md., Miss., N. J., N. C., Pa., Tenn., Va.

June 2 (Tenn.) to December 21 (Fla.).

Fig. 242. *Somatochlora forcipata; S. franklini.*

Somatochlora forcipata Scudder

Syn.: chalybea Hagen (no desc.)

1861. Hagen, Syn. Neur. N. Amer., p. 138 (no desc.) (as *Cordulia chalybea*).
1866. Scudder, Proc. Boston Soc. Nat. Hist., 10:216 (in *Cordulia*).
1906. Mrtn., Coll. Selys Cordulines, p. 25 (figs.).
1925. Walk., N. Amer. Somatochlora, p. 134 (figs.).
1927. Garm., Odon. of Conn., p. 228 (figs.).
1929. N. & H., Handb., p. 192 (figs.).

Length 43–51 mm.; abdomen 34–39; hind wing 29–33.

Another rather small and darkly colored Northeastern species. Face dark, with black labrum, black and yellow postclypeus, and dark brown frons that has metallic green reflections above. Sides of frons broadly yellowish. Occiput blackish, covered with short erect brown hair. Black rear of head with dark brown hair fringe around rear margin.

Thorax clad in a rather thick growth of brownish hairs. In front and on sides, dark reddish brown, with blue and green reflections. Pale stripes of sides appear as two conspicuous, similar, roundish-oval spots of pale yellow. Legs black, with front femora paler beneath. Wings hyaline, with black veins and a dark yellow stigma.

Abdomen very hairy on swollen basal segments, blackish above and marked below middle of sides with two dimly defined pale spots, rear one of which runs down a little on prominent genital lobe in male. Apical ring on segment 2 conspicuous; upper and lower basal spots on 3 rather short. Remainder of abdomen greenish black, with a dull luster. A row of small roundish lateral spots on 4 to 8, these spots diminishing in size to rearward.

Inhabits small woodland streams and pools.

Distribution and dates.—Labrador; CANADA: Alta., Man., N. B., Nfld., NW. Terr., N. S., Ont., Que.; UNITED STATES: Maine, Mich., N. H., Vt.

May 27 (Que.) to August 27 (Que.).

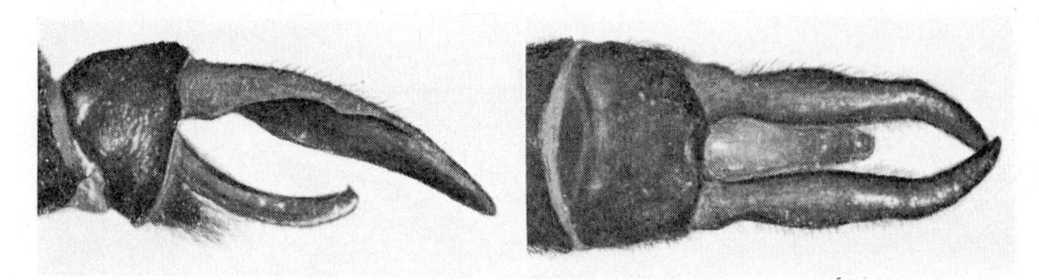

Fig. 243. *Somatochlora franklini.*

Somatochlora franklini Selys

Syn.: macrotona Williamson

1861. Hagen, Syn. Neur. N. Amer., p. 138 (no desc.) (in *Cordulia*).
1878. Selys, Bull. Acad. Belg., (2)45:195 (reprint, p. 17) (in part) (in *Epitheca*).
1906. Mrtn., Coll. Selys Cordulines, p. 25.
1925. Walk., N. Amer. Somatochlora, p. 117 (figs.).
1929. N. & H., Handb., p. 195 (figs.).

Length 44–54 mm.; abdomen 33–43; hind wing 25–30.

A small Northern transcontinental species. Face bronzy green, with three large yellow spots: one covers anteclypeus, other two cover sides of frons. Top of low vertex green. Occiput brown, its rear edge fringed with rusty hairs.

Thorax mostly pale, with unusually narrow strips of green between sutures showing through a rather heavy covering of rusty hairs. Only first stripe shows as an oval spot, but an extra side spot of yellow covers spiracle. Two cross stripes of yellow cover front and rear margins of dorsum of prothorax. Basal half of front femur yellowish. Wings rather short and narrow, with tawny costa and stigma; a touch of brown in anal angle covers triangle of male. Membranule pale.

Base of abdomen bulbous, with segment 3 abruptly constricted beyond its swelling. Segment 2 bears a narrow apical ring of yellow, more conspicuous than its other scanty pale markings. Genital lobe of male black and densely hairy. Abdomen in female with line of yellowish streaks low down on sides of middle segments; line wide on base of 3, narrow on 4 to 6, and wider again and yellower on genitalia at end of abdomen.

Distribution and dates.—Labrador; CANADA: Alta., B. C., Man., N. B., Nfld., NW. Terr., Ont., Que., Sask.; UNITED STATES: Maine, Mich., Minn.

May 23 (Ont.) to first week in September (Canada, Whts.).

Somatochlora georgiana Walker

1924. Root, Ent. News, 35:320 (*Somatochlora* sp. ?).
1925. Walk., N. Amer. Somatochlora, p. 98 (figs.).
1929. N. & H., Handb., p. 186.

Length 48–49 mm.; abdomen 34–37; hind wing 32–34.

A large Southeastern species, lacking usual metallic coloration of genus. Face dull yellow save for a very narrow black line on front margin of labrum.

Thorax dull brown in front and scantily hairy. Yellow stripes of side pale and very long, second one also unusually wide. Legs brown, yellowish on outer face of tibiae. Wings hyaline, with brown venation including costa and stigma. There may be only four antenodal crossveins in hind wing.

Abdomen dark brown, with scant metallic greenish reflections; basal segments dull yellowish; appendages brownish. Segment 2 bears usual pale apical ring, and another broader one that is duller yellow. Ovipositor of female broadly exposed, compressed, triangular as seen from sides, with equal sides and pointed obliquely downward.

Easily recognized by large extent of yellow on face, lack of metallic coloration, and yellow color of tibiae on their outer face.

Distribution and dates.—UNITED STATES: Ala., Ga., S. C.

July 6 (Ga.) to July 30 (S. C.).

Somatochlora hineana Williamson

1931. Wmsn., Occ. Pap. Mus. Zool. Univ. Mich., 225:1–8 (figs.).

Length 58–63 mm.; abdomen 42–49; hind wing 40–42.

A little-known midland species of large size. Face mostly yellow, including labrum and lower half of front of frons, above which frons bears a well-defined cap of dark metallic green. Occiput ochre yellow.

Thorax brown in front, with green and blue reflections. On the dark sides, yellow stripes well defined, especially at lower ends; second wider and shorter than first.

Abdomen dark brown, broadly marked with paler on its swollen basal segments. Pale apical ring on segment 2 interrupted on middorsal line. Long pale basal triangles on side of 3 large but ill defined. Beyond 3, remainder of abdomen and appendages dull brown.

Allied to *tenebrosa* but larger, and side stripes of thorax brighter yellow.

Distribution and dates.—UNITED STATES: Ohio.

June 7 to July 4.

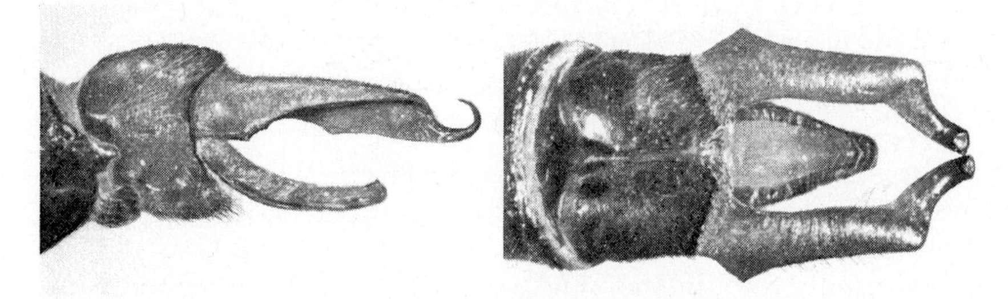

Fig. 244. *Somatochlora hudsonica.*

Somatochlora hudsonica Hagen

1871. Hagen, in Selys, Bull. Acad. Belg., (2)31:301 (reprint, p. 67) (in *Epitheca*).
1906. Mrtn., Coll. Selys Cordulines, p. 27 (figs.).
1925. Walk., N. Amer. Somatochlora, p. 176 (figs.).
1929. N. & H., Handb., p. 190 (figs.).

Length 50–54 mm.; abdomen 36–40; hind wing 30–34.

A wide-ranging Canadian species, allied to *albicincta* and *cingulata,* and having middle abdominal segments narrowly ringed with white. Frons and vertex blackish, with metallic green and blue reflections. Ochre yellow spots on sides of frons convergent or conjoined at middle of postclypeus. Occiput covered all over its upper surface with short, soft brownish hairs. Rear of head fringed heavily with whitish hairs.

Thorax in front bright green, with a distinct interrupted cross streak of yellow before crest from side to side. This is present but small and indistinct in *albicincta.* Also, a streak of yellow on under side of front femur. Wings hyaline, more or less tinged with yellow at base of anal area.

Abdomen much as in *albicincta,* but abdominal segment 10 wholly black above. Genital lobe of male metallic green, fringed with rusty hairs.

Distribution and dates.—Alaska; CANADA: Alta., B. C., Man., NW. Terr., Ont., Sask., Yukon; UNITED STATES: Colo.

June 12 (NW. Terr.) to September 28 (Ont.).

Somatochlora incurvata Walker

1918. Walk., Can. Ent., 50:365 (figs.).
1925. Walk., N. Amer. Somatochlora, p. 142 (figs.).
1929. N. & H., Handb., p. 194 (figs.).

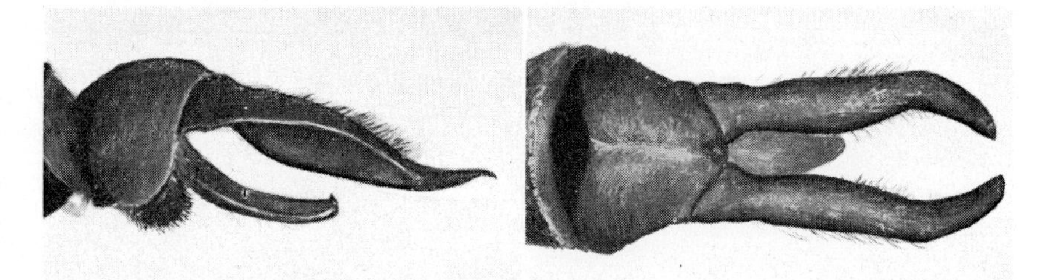

Fig. 245. *Somatochlora incurvata.*

Fig. 246. *Somatochlora incurvata; S. kennedyi.*

Length 49–59 mm.; abdomen 38–47; hind wing 31–37.

A little-known species from northern Michigan. Face very dark, with a shining black labrum, a very yellow anteclypeus, and a brown postclypeus with a yellow lower border. Frons tricolored, with a cap of shining metallic green, surrounded below by a band of cinnamon brown, and below that a yellow band formed by meeting of two large yellow spots of sides. Vertex shining and thinly hairy. Occiput and rear of head bare and black in male; in female, a pale area about tubercle on rear margin of eye, and a thin marginal fringe of whitish hair across from side to side.

Thorax thinly clad with long whitish hairs; brown in front, with strong green reflections; clouded on side, with pale stripes obscure. Wings hyaline or very lightly tinged with brown, more deeply tinged in female.

Abdomen brown, paler on basal segments. Pale apical ring on segment 2 rather wider than usual. Pale spots above and below auricle sometimes connected. Basal lateral triangles on 3 shorter than usual; beyond them, small basal tawny yellow spots on 5 to 8 (largest on 6, minute on 8) on middle segments only in male, and on all remaining segments in female.

Closely related to *forcipata* but larger and differing in form of genitalia.

Distribution and dates.—CANADA: N. S., Ont.; UNITED STATES: Mich. July 19 (N. S.) to October 15 (Canada, Whts.)

Somatochlora kennedyi Walker

1892. Harvey, Ent. News, 3:116 (in part) (as *Cordulia forcipata*).
1925. Walk., N. Amer. Somatochlora, p. 125 (figs.).
1929. N. & H., Handb., p. 196 (figs.).

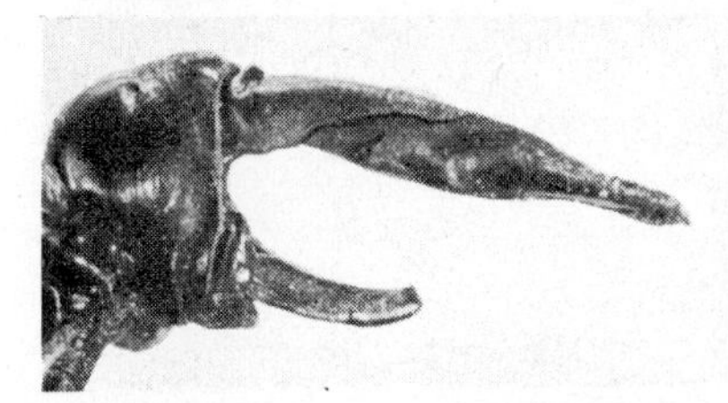

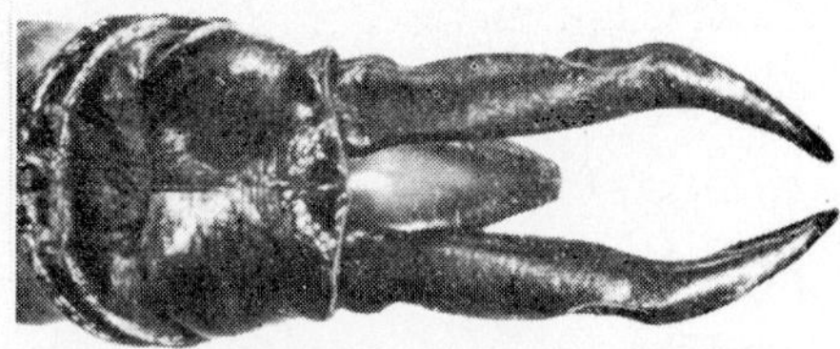

Fig. 247. *Somatochlora kennedyi.*

Length 47–55 mm.; abdomen 35–42; hind wing 29–34.

A hoary species of wide distribution. Face hairy. Frons wears a blackish cap having metallic green reflections, at either side of which are areas of dull yellow. Vertex black, with greenish overcast. Occiput dark brown.

Front of thorax metallic green, with bluish reflections, clouded above and below with reddish brown. Pale stripes on sides ill defined and wide; first one abbreviated below and followed by a suggestion of a long slender irregular middle stripe; usual second lateral expanded into a broad terminal band. Legs black, paler basally. Wings hyaline; costa yellow; stigma tawny.

Abdomen long and stout, with markings on swollen and hairy basal segments obscure, narrow apical ring on segment 2 being best defined. In male, middle band on 2 runs down on genital lobe. Remainder of abdomen blackish dorsally, with greenish reflections.

Related species are *franklini* and *forcipata,* from both of which it is distinguished by forcipate appendages of male and long scooplike ovipositor of female.

Frequents shallow boggy ponds.

Distribution and dates.—CANADA: Man., N. B., NW. Terr., Ont., Que.; UNITED STATES: Maine, Mass., Mich., N. H., N. Y.

May 28 (Que.) to August 5 (Ont.).

Somatochlora linearis Hagen

Syn.: procera Selys

1861. Hagen, Syn. Neur. N. Amer., p. 137 (in *Cordulia*).
1906. Mrtn., Coll. Selys Cordulines, p. 21 (figs.).
1925. Walk., N. Amer. Somatochlora, p. 91 (figs.).
1927. Garm., Odon. of Conn., p. 225 (figs.).
1929. N. & H., Handb., p. 186 (figs.).

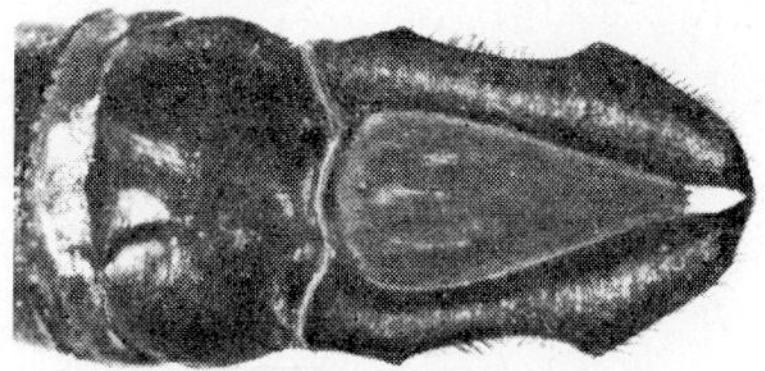

Fig. 248. *Somatochlora linearis.*

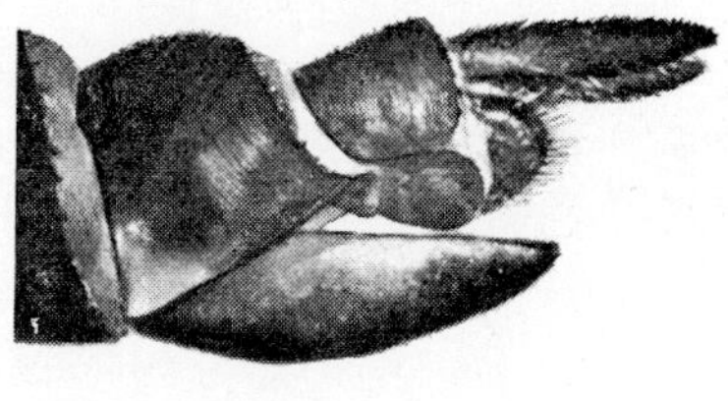

Fig. 249. *Somatochlora linearis; S. ozarkensis.*

Length 58–68 mm.; abdomen 44–55; hind wing 39–47.

One of handsomest Eastern species, long-legged, broad-winged, and unique in its entire lack of yellow stripes on thorax. Face mainly dull yellow, darker on labrum; top of frons and vertex metallic blue. Occiput shining brown.

Thorax shining brown, with greenish reflections, paler beneath. Legs black. Wings broad and shining, with iridescent reflections, black veins and costa, and a rich venational pattern.

Abdomen very long and slender, shining brown, with greenish sheen. Segment 2 has a conspicuous pale apical annulus and a large yellow spot below middle of blackish sides; beyond 3, segments black, shining, nearly bare. Genital lobe of the male black, broad, blunt, nearly bare.

This species near *ensigera,* but larger and with markedly less development of color pattern.

Distribution and dates.—UNITED STATES: Ala., Conn., Fla., Ga., Ill., Ind., Ky., La., Miss., Mo., N. J., N. Y., N. C., Ohio, Okla., Pa., S. C., Tenn., Va.

June 1 (Ind.) to October 3 (Ga.).

Fig. 250. *Somatochlora minor.*

Somatochlora minor Calvert

1898. Calv., Ent. News, 9:87 (fn.).
1898. Harvey, Ent. News, 9:86.
1925. Walk., N. Amer. Somatochlora, p. 62 (figs.).
1927. Garm., Odon. of Conn., p. 222 (figs.).
1929. N. & H., Handb., p. 191 (figs.).

Length 42–50 mm.; abdomen 29–38; hind wing 30–34.

A blackish Northern species, with notably short abdomen in male. Blackish, with brilliant metallic green reflections. Face black, with a large crescentic yellow spot covering anteclypeus, and another roundish one on each lateral face of frons. Top of frons, also vertex, brilliant dark metallic green. Top of occiput and rear of head shining black, with gray hairs all over former, and with a fringe of longer hair all around rear margin of head. On hind border of compound eye, an area around rear tubercle is yellowish.

Thorax in front mostly green and coppery. Thinly clad with soft hair that is long and white at front, shortens and darkens upward, and becomes black along carina and crest. Sides of thorax shining green, with usual stripes reduced to two well-defined subequal roundish spots. Hyaline wings have a touch of golden yellow in anal triangle of male at base of hind wings. Legs black, with under side of femora of front and middle ones yellowish.

Abdomen short, especially in male, and somewhat more depressed and widened in middle than in other species of genus. Shiny blackish basal segments broadly marked with yellow. On segment 2, one spot above, another below, level of auricle; spots join to form a long yellow crescent in female. Segment 3 with usual long basal triangles above and below on each side; beyond 3, remainder of abdomen and appendages black.

Distribution and dates.—CANADA: Alta., B. C., Man., N. B., N. S., Ont., P. E. I., Que., Sask.; UNITED STATES: Conn., Maine, Mich., N. H., N. Y., Wyo.

June 6 (Ont.) to third week in September (Canada).

Somatochlora ozarkensis Bird

1933. Bird, Occ. Pap. Mus. Zool. Univ. Mich., 261:1–7 (figs.).
1933. Wmsn. & Gloyd, Occ. Pap. Mus. Zool. Univ. Mich., 262:1, 5 (fig.).
1936. Pritchard, Ent. News, 47:99 (figs., nymph).

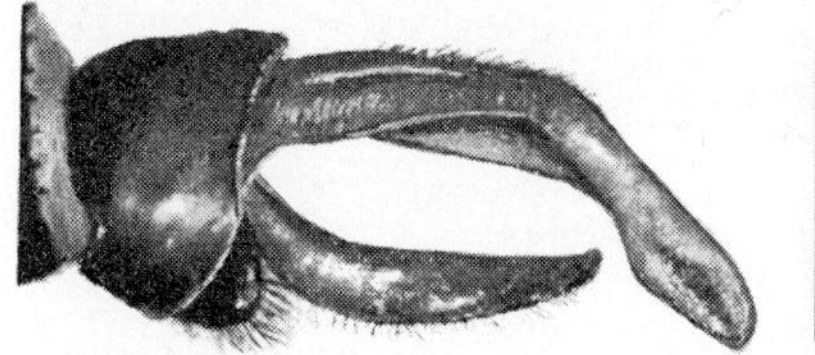
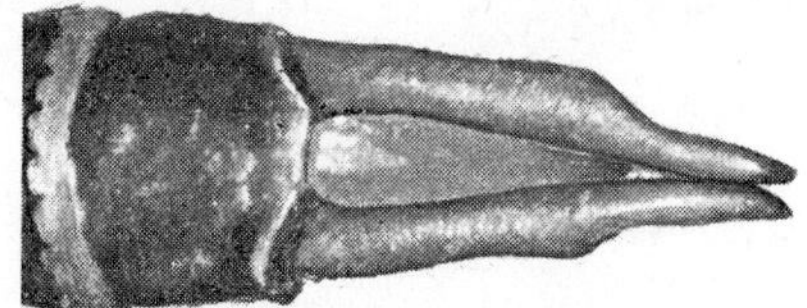

Fig. 251. *Somatochlora ozarkensis.*

Length 51–55 mm.; abdomen 37–42; hind wing 34–38.

A slender inland species, with metallic coloration marked by brilliant blue and coppery reflections. Face pale. Frons has a squarish cap of metallic blue green that fades to shining brown in depth of middle furrow. Vertex brown, paler on upper angles. Occiput blackish, shining; posterior fringe of head composed of long whitish hairs.

Front of thorax brown, lighted with metallic blue. Carina and edges of yellow crest black. Sides brilliant metallic blue, especially on wider upper surfaces. Pale yellow side stripes wide and variable, with added touches of yellow about spiracle below and next to wing roots above. Wings hyaline, touched with golden brown at base. Legs black, pale at base, extensively on first pair, less so successively on others.

Abdomen very slender and graceful in design. Usual markings on swollen basal segments whitish and prominent. Pale lower band on segment 2 a continuous crescent in female; in male it is interrupted; distal half runs down to cover about half of broad blunt genital lobe.

Inhabits woodland streams of Ozark Plateau.

Distribution and dates.—UNITED STATES: Mo., Okla.

June 10 (Okla.) to August 7 (Mo.).

Somatochlora provocans Calvert

1903. Calv., Ent. News, 14:39.
1906. Mrtn., Coll. Selys Cordulines, p. 29 (figs.).
1925. Walk., N. Amer. Somatochlora, p. 109 (figs.).
1929. N. & H., Handb., p. 188 (figs.).

Length 53–56 mm.; abdomen 37–43; hind wing 33–37.

An Eastern coastwise species of rather large size and brilliant coloration. Face mainly yellowish up to metallic green cap that covers top and

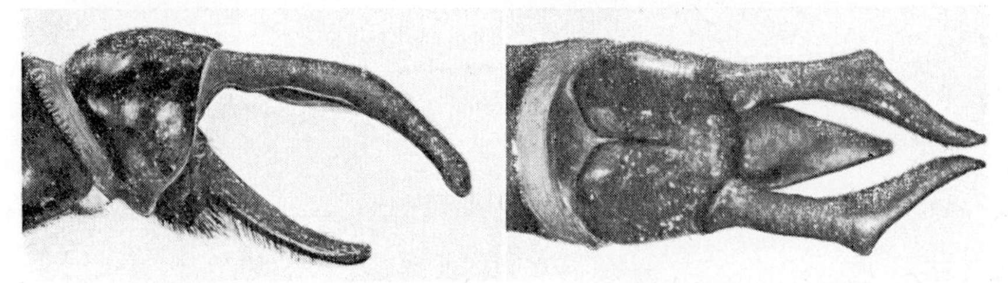

Fig. 252. *Somatochlora provocans.*

part of front of frons, except for narrow black lateral margins on labrum. Vertex and occiput brown, with greenish reflections. Around rear of head, a thin fringe of rather inconspicuous hair.

Thorax brown in front, with green reflections, with or without a pair of short frontal stripes of yellow that are divergent downward, pointed upward, and reach up only to half height. Green sides with usual two yellow stripes well developed, subequal, and more or less pointed at ends. An added sinuous yellow streak close behind spiracle. Wings hyaline, with tawny costa and brown stigma. Legs black beyond their pale bases; keel on hind femur of male pale whitish.

Abdomen long and slender, hairy at base both above and below. Sides of segment 2 bare and shining, with a conspicuous white apical ring, and with three yellow spots in male, spots which in female become conjoined to form an inverted Y. Lower angle of yellow in male almost covers broadly oval genital lobe. Beyond usual pale spots on base of 3 on each side, abdomen is shining bare and black save for narrow apical half rings of yellow across apices of 8 and 9.

Distribution and dates.—UNITED STATES: Ala., Fla., Ga., Ky., Miss., N. J., Pa., Tenn.

June 28 (Ky.) to August 30 (Fla.).

Somatochlora sahlbergi Trybom

Syn.: walkeri Kennedy

1889. Trybom, Bihang till K. Svenska Vet. Akad., Handlingar, 15:7, 16, 20 (figs.).
1925. Walk., N. Amer. Somatochlora, p. 163 (figs.).
1929. N. & H., Handb., p. 195 (figs.).

Length 48 mm.; abdomen 35; hind wing 32–33.

A rather small dark hairy species of Far North. Face blackish, with bronzy green luster on frons, with pale anteclypeus and small yellow spot on each side of frons next to eye. Occiput black.

Thorax bright metallic green, with carina and edges of crest black, and a long dense covering of hairs. Side stripes obsolete except for a minute spot toward front which remains as a vestige of first stripe. Legs black; wings hyaline, with black venation; membranule at their base conspicuously white.

Abdomen stout and rather hairy, with only apical ring on segment 2 very evident. Auricle of male very hairy.

Distribution and dates.—Alaska; CANADA: NW. Terr.

July 6–19 (NW. Terr.).

Flight period "probably the same as *S. septentrionalis*" (Whts., 1948).

Somatochlora semicircularis Selys

Syn.: nasalis Selys

1871. Selys, Bull. Acad. Belg., (2)31:295 (reprint, p. 61) (in *Epitheca*).
1906. Mrtn., Coll. Selys Cordulines, p. 26 (figs.)
1925. Walk., N. Amer. Somatochlora, p. 145 (figs.).
1929. N. & H., Handb., p. 193 (figs.).

Length 47–52 mm.; abdomen 34–40; hind wing 27–32.

A brilliantly metallic green species of Northwest. Face hairy and black except for pale anteclypeus and large yellow spots that cover sides of frons. Top of frons and vertex bright metallic green. Occiput black, covered with short black hair. Rear of head black, fringed with long white hair.

Thorax heavily clothed with tawny hair, through which comes a shine of brilliant green on all bulging surfaces; on sides, dim outline of usual two stripes and a smaller pale streak beside spiracle. Legs black; wings hyaline, lightly tinged with smoky brown.

Abdomen marked on segment 2 with dull yellow spots and an interrupted apical ring. Beyond usual pale basal triangles on 3, segments black and hairy above, tawny on flat venter, and marked with tawny basal lateral spots on 4 to 8 in male. In female, spots tend by coalescence to form a lateral stripe. These markings fairly well seen in younger specimens; tend to disappear with age.

Distribution and dates.—Alaska; CANADA: Alta., B. C.; UNITED STATES: Calif., Colo., Oreg., Utah, Wash., Wyo.

June 6 (B. C.) to September 6 (Colo.).

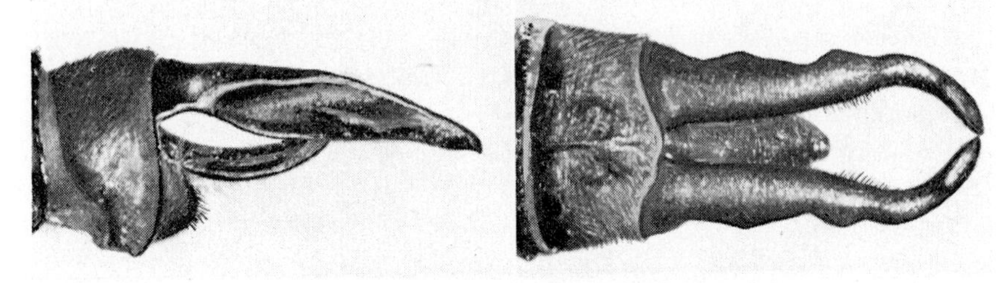

Fig. 253. *Somatochlora semicircularis.*

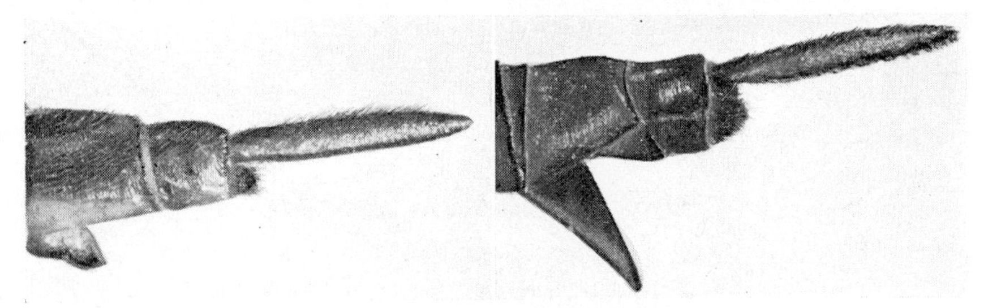

Fig. 254. *Somatochlora semicircularis; S. tenebrosa.*

Somatochlora septentrionalis Hagen

1861. Hagen, Syn. Neur. N. Amer., p. 139 (in *Cordulia*).
1925. Walk., N. Amer. Somatochlora, p. 160 (figs.).
1929. N. & H., Handb., p. 197 (figs. as *whitehousi*).
1941. Whts., Odon. of B. C., p. 534 (figs., nymph).

Length 39–48 mm.; abdomen 29–37; hind wing 26–30.

A small, narrow-winged Canadian species. Face dark metallic green except for pale anteclypeus and a pair of large yellow spots on sides of frons. Occiput black and hairy above. Rear of head black, with a white hair fringe.

Thorax clothed all over with a rather dense growth of long white hairs. It is green in front, with a yellow crossbar on crest that is narrowly interrupted in middle by black of carina. Sides green, with dull pale markings. Rather short first stripe prolonged to rearward below subalar carina; second stripe interrupted, obscure, variable. Wings with a yellow costa and a tawny stigma. Legs black, front femora yellowish beneath. Trigonal interspace notably narrow in this species, with no increase in number of included rows of cells out to margin of wing. A blackish spot covers anal triangle of male.

Abdomen shining brown, with encircling black carinae on swollen basal segments, dark metallic green beyond these segments, tawny on flat venter. Female has a short subgenital plate, its margins fringed with tawny hair.

This species nearest allied to *whitehousei*, with which (and with *franklini*) it shares black spot covering hind-wing triangle of male.

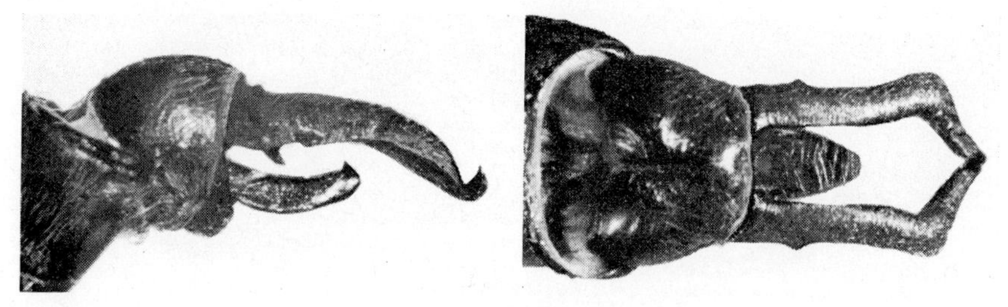

Fig. 255. *Somatochlora septentrionalis.*

Distribution and dates.—Labrador; CANADA: B. C., Man., Nfld., NW. Terr., Ont., Que.

June 29 (Ont.) to August 30 (Labrador).

Somatochlora tenebrosa Say

Syn.: tenebrica Hagen (no desc.)

1839. Say, J. Acad. Phila., 8:19 (in *Libellula*).
1906. Mrtn., Coll. Selys Cordulines, p. 24 (figs.).
1925. Walk., N. Amer. Somatochlora, p. 100 (figs.).
1927. Garm., Odon. of Conn., p. 227 (figs.).
1929. N. & H., Handb., p. 187 (figs.).

Length 48–64 mm.; abdomen 35–40; hind wing 34–41.

A twilight-flying species, slender and very handsome. Dark brown in general coloration, with green and coppery reflections on head and thorax. Face clothed with blackish hairs. Orange labrum, more or less heavily edged and crossed with black. Remainder of face up to middle of frons and around to eyes ochre yellow. High prominences and deep furrow on top of frons and whole of vertex black, with brilliant metallic reflections. Blackish occiput thinly clothed with erect hair. Fringe of hair around rear of head scanty and inconspicuous.

Thorax hairy in front and at rear, less so on sides; dark brown, with strong coppery reflections. Sides shining green on broader surfaces, with usual lateral stripes dull yellowish and rather inconspicuous; first stripe longer than second. Legs black, femora paler externally. Wings hyaline, sometimes with a faint basal tinge of amber; veins and stigma dark brown.

Abdomen long and slender, with sides of swollen basal segments shining brown. Segment 2 with large pale spots; hinder one in male runs down on rear side of genital lobe. Dorso-lateral spot on 3 short, ventral spot long and slowly tapering; remainder of abdomen black.

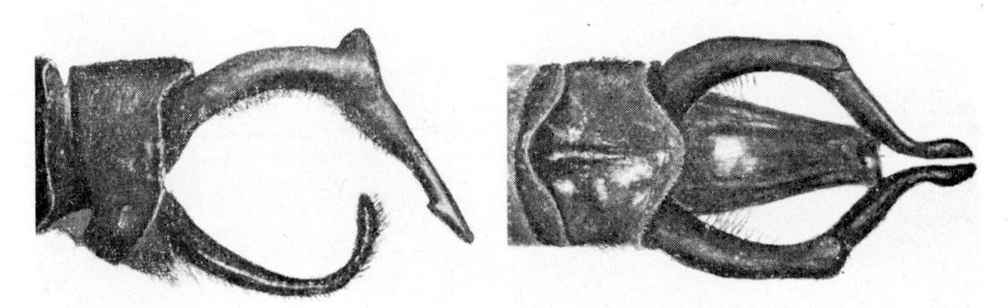

Fig. 256. *Somatochlora tenebrosa.*

This species has been known since the days of Thomas Say, "father of American entomology," who described it from Indiana. Easily recognized by elbowed and uniquely complicated form of caudal appendages of male; and by very long, strongly compressed, and downwardly directed ovipositor of female.

Distribution and dates.—CANADA: N. S., Ont., Que.; UNITED STATES: Ala., Conn., Ill., Ind., Iowa, Ky., Maine, Md., Mass., Miss., Mo., N. H., N. J., N. Y., N. C., Ohio, Pa., R. I., S. C., Tenn., Va.

June (N. C.) to October 6 (S. C.).

Somatochlora walshii Scudder

Broom-Tail

1866. Scudder, Proc. Boston Soc. Nat. Hist., 10:217 (in *Cordulia*).
1906. Mrtn., Coll. Selys Cordulines, p. 26 (figs.).
1925. Walk., N. Amer. Somatochlora, p. 55 (figs.).
1927. Garm., Odon. of Conn., p. 223 (fig.).
1929. N. & H., Handb., p. 192 (figs.).
1941. Walk., Can. Ent., 73:203 (figs., nymph).

Length 41–52 mm.; abdomen 28–40; hind wing 25–34.

A rather stout, bright-hued species; relatively short abdomen, terminated by abundantly bewhiskered caudal appendages. Face varicolored:

labrum black; anteclypeus yellow; postclypeus olivaceous, with blackish markings; frons with a squarish cap of brilliant green, surrounded in front and at side with a band of orange, and, outside that band, olivaceous next to eyes. Summit of vertex dull brown. Occiput brown, yellow behind; rear of eyes black, with a long fringe of tawny hairs.

Thorax hairy, shining green in front, with a coppery sheen that seems to emanate from underneath the green; sides similar but a shade lighter.

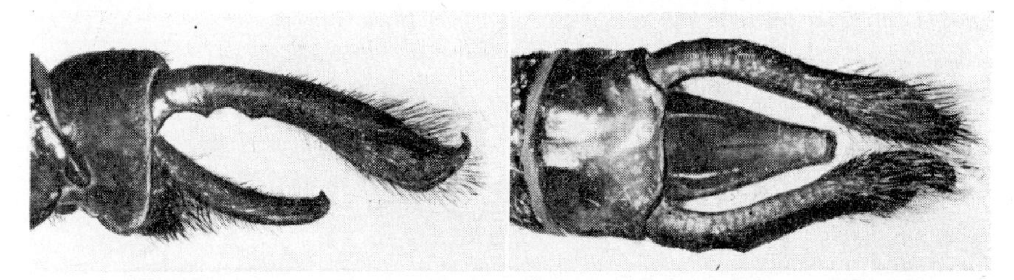

Fig. 257. *Somatochlora walshii.*

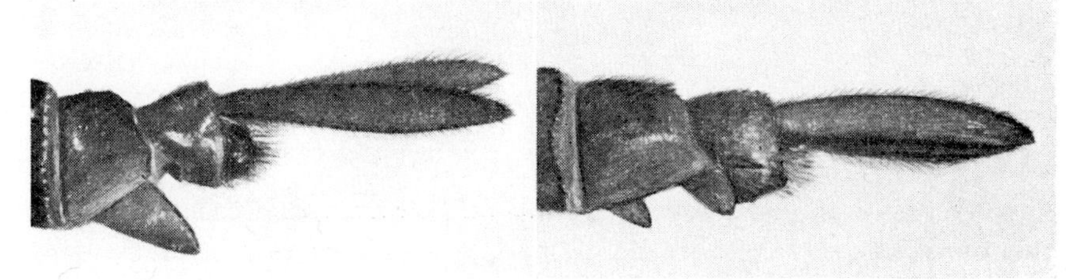

Fig. 258. *Somatochlora walshii; S. whitehousei.*

Two side stripes bright yellow, first longer and narrower than second; between the two are touches of yellow around spiracle. Wings hyaline, with a touch of yellow at their extreme bases. Legs black, femora of first pair pale beneath.

Much swollen segment 2 of abdomen brown at sides, with paired lateral yellow spots sometimes confluent with each other and with apical annulus. In male, yellow barely reaches genital lobe. Yellow triangles on 3 of moderate length; beyond them, abdomen black above, washed with clay yellow on ventral side of 4 to 9.

Distribution and dates.—CANADA: Alta., B. C., Man., N. B., Nfld., N. S., Ont., Que.; UNITED STATES: Conn., Maine, Mass., Mich., Minn., N. H., N. Y.

June 3 (Ont.) to September 3 (N. S.).

Somatochlora whitehousei Walker

1917. Kndy., Can. Ent., 49:234–235 (figs.) (as *septentrionalis*).
1925. Walk., N. Amer. Somatochlora, p. 154 (figs.).
1929. N. & H., Handb., p. 196 (figs. as *septentrionalis*).

Length 46–48 mm.; abdomen 33–38; hind wing 26–30.

Another hairy Northern species of rather stout form. Face mostly shining metallic green save for yellow anteclypeus. Rear of head black, with orange-brown bordering fringe of rather long hair.

Thorax dark olive, with brilliant metallic green reflections, or somewhat bluish on sides, where pale side stripes are short, wide, and ob-

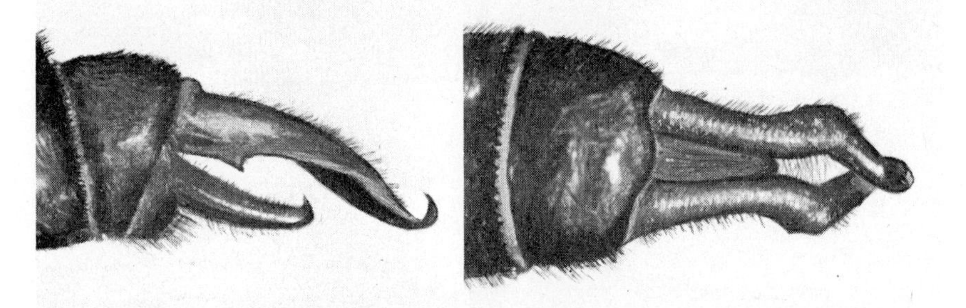

Fig. 259. *Somatochlora whitehousei.*

scure. Wings hyaline, with a tinge of yellow throughout membrane. A spot of dark brown covers anal triangle of hind wing in male. Legs black; front femora paler beneath.

Abdomen short, strongly constricted on segment 3, widened thereafter to 7, then narrowed again to end. Pale apical ring on 2 conspicuous. Genital lobe on 2 in male rather short and wide, very hairy.

Inhabits mossy bog pools.

Distribution and dates.—Labrador; CANADA: Alta., B. C., Man., Ont., Que., Sask.

June 12 (B. C.) to August 28 (Labrador).

Somatochlora williamsoni Walker

1901. Ndm., Bull. N. Y. State Mus., 47:499 (as *elongata*).
1907. Walk., Can. Ent., 39:70.
1925. Walk., N. Amer. Somatochlora, p. 78 (figs.).
1927. Garm., Odon. of Conn., p. 224 (figs.).
1929. N. & H., Handb., p. 194 (figs.).

Length 53–59 mm.; abdomen 40–48; hind wing 35–40.

A dark-colored Northeastern species. Face yellowish. Top of frons and vertex metallic blue. Labrum greenish black. Occiput brown, hair fringe at rear of head whitish.

Thorax thinly clothed with tawny hairs. Under the hair it is shining green, with strong reflections of blue and violet. Pale side stripes dull and often hardly discernible: first long and straight, second obscure.

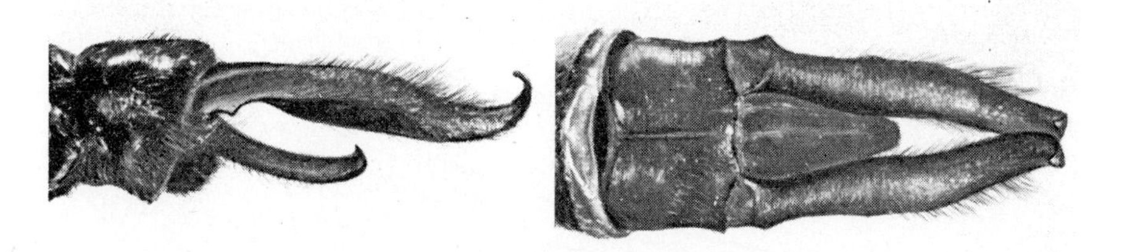

Fig. 260. *Somatochlora williamsoni.*

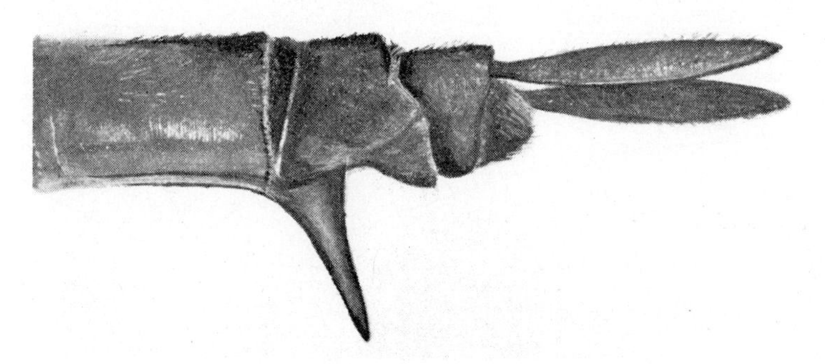

Fig. 261. *Somatochlora williamsoni.*

Legs black, with femora brownish basally. Wings hyaline, sometimes yellowish in membrane at base.

Abdomen hairy about base, becoming bare and shining brown on side of segment 2, where there are three lateral yellowish spots more or less confluent, or obsolete. Genital lobe of male broad and black, with only a scanty fringe of short black hairs. Beyond usual basal triangles on 3, abdomen black. Sparse dorsal hairiness of middle segments increases in length and density to rearward.

Distribution and dates.—CANADA: Man., Ont., Que.; UNITED STATES: Conn., Maine, Mich., Minn., N. H., N. Y., Pa., Tenn.

June 19 (Man.) to September 14 (Ont.).

Genus CORDULIA Leach 1815

These are clear-winged dragonflies, with bronzy green color showing through a rather heavy coat of pale tawny hair. The thorax is hairy all over; the abdomen is bare and shining above after segment 2, but hairy beneath for its full length, and with a tuft of golden hairs on the dorsum of segments 1 and 2 between the hind wings. The abdomen is slenderest

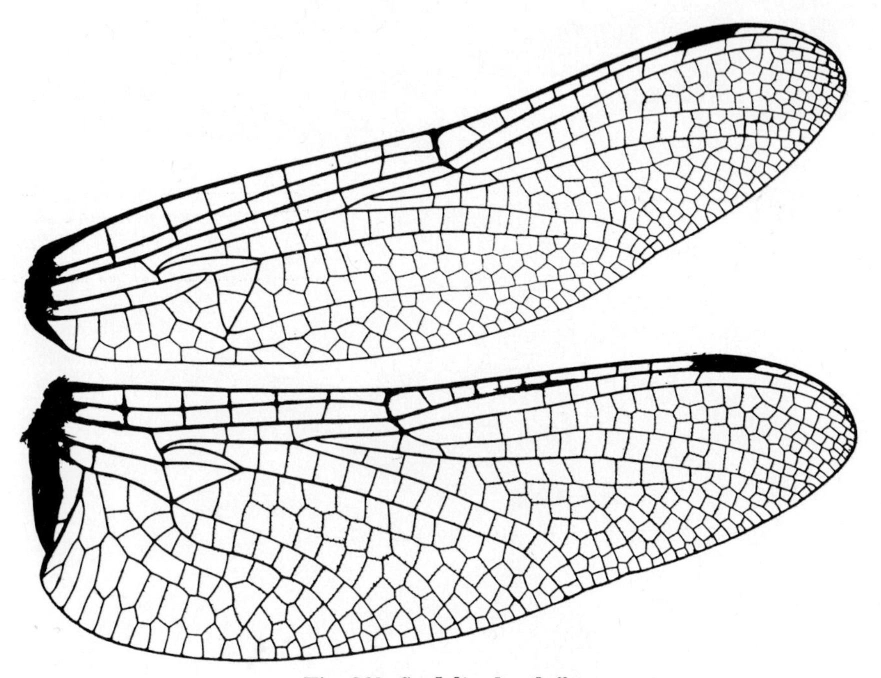

Fig. 262. *Cordulia shurtleffi.*

on segment 4, widening thereafter to 6, and narrowing a little thereafter to the end. The venation of the wings is as shown in the table of genera.

The nymph is stocky, somewhat cylindric, and fringed with rather coarse and sparse hairs, with much the appearance of *Libellula* nymphs. The distinctive characters of the nymphs are as shown in the table of genera. They lie sprawling amid the trash in shady nooks in the edges of bog ponds.

There is one Northern species in our fauna, and another at similar latitudes in the Old World.

Cordulia shurtleffi Scudder

1866. Scudder, Proc. Boston Soc. Nat. Hist., 10:217.
1901. Ndm., Bull. N. Y. State Mus., 47:502 (nymph).
1906. Mrtn., Coll. Selys Cordulines, p. 37 (figs.).
1927. Garm., Odon. of Conn., p. 214 (figs.).
1929. N. & H., Handb., p. 197 (fig.).

Fig. 263. *Cordulia shurtleffi.*

Length 43–50 mm.; abdomen 30–37; hind wing 29–32.

A fine bronzy green species. Top of frons and whole of vertex shining green, with violet reflections. Face variegated: labrum blackish, with a pair of enclosed tawny spots; anteclypeus whitish; postclypeus brown, sparsely covered with drooping white hairs; below green of frons, an encircling band of bright tawny or orange. Occiput hairy and blackish.

Thorax green, darkened only in sutures of sides and on carinae about wing roots. Legs black.

Abdomen black beyond swollen basal segments, where it is paler, with a tawny apical ring and an area of same color below auricle on segment 2 and a long basal triangle low down on sides of 3. Most easily recognized by broad inferior appendage of male, twice forked at tip.

Distribution and dates.—Alaska; Labrador; CANADA: Alta., B. C., Man., N. B., Nfld., NW. Terr., N. S., Ont., Que., Sask., Yukon; UNITED STATES: Calif., Colo., Conn., Maine, Mass., Mich., Nev., N. H., N. J., N. Y., Oreg., Pa., Utah, Wash., Wyo.

May 2 (B. C.) to August 26 (B. C.).

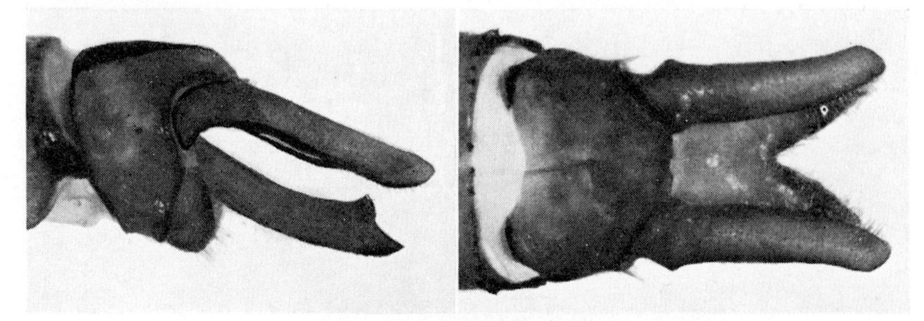

Fig. 264. *Cordulia shurtleffi.*

Genus DOROCORDULIA Needham 1901

These are elegant little dragonflies, with hairy thorax and smooth, shiny, bronzy green abdomen. The frons and vertex are shining metallic green. The eye-seam of the head is moderate, and the rather large bare occiput is hairy in the rear. The bulging frons is rounded to the sides and sunken at the base above in front of the middle ocellus.

The thorax is green, densely clothed in front with tawny hairs, darkened in the depths of the sutures and paler beneath. The legs are black; the wings, hyaline. The venation is open, with crossveins reduced to a minimum, there being no crossvein in either triangle or supertriangle (except occasionally in the female). The cells beyond the fore-wing triangle are reduced to a single row for a distance of three to five cells. There are two (often three in the female) rows of cells behind the anal loop in the second anal interspace (y). The marginal cells in the base of that interspace are very long.

The abdomen is narrowed before the middle segments and then widened to rearward, with the segments of this second widening somewhat depressed.

The nymph is as shown in the table of genera.

The genus is North American, and contains two known bog-inhabiting species.

KEY TO THE SPECIES

ADULTS

1—Rear of occiput yellowish; end segments of abdomen very moderately widened, lacking middorsal carina; tips of superior appendages of male blunt, parallel; subgenital plate of female divided by a shallow rounded notch .... **lepida**

—Rear of occiput blackish, hairy; segments of abdomen spatulately widened, with low sharp middorsal carina; tips of superior appendages of male divergent and somewhat sharply pointed; subgenital plate of female divided by a narrow cleft.. **libera**

NYMPHS

1—Low dorsal hook on abdominal segment 9.............................. **lepida**

—No dorsal hook or ridge on 9.. **libera**

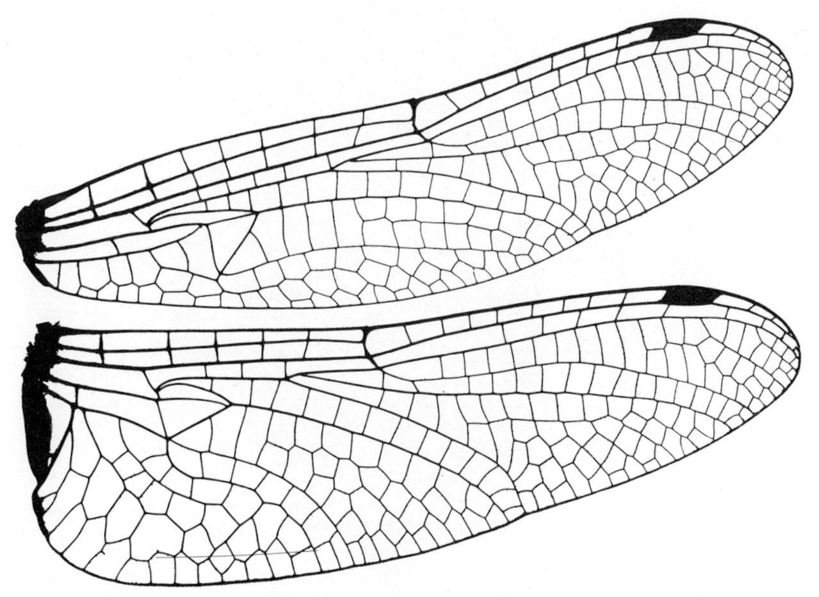

Fig. 265. *Dorocordulia libera.*

Dorocordulia lepida Hagen

1871. Hagen, in Selys, Bull. Acad. Belg., (2)31:264 (reprint, p. 30) (in *Cordulia*).
1901. Ndm., Bull. N. Y. State Mus., 47:506 (figs.)
1927. Garm., Odon. of Conn., p. 237.
1929. N. & H., Handb., p. 199 (fig.).

Length 37–43 mm.; abdomen 25–28; hind wing 26–29.

A dainty little Corduline of coastwise northeastern distribution. Face bronzy green above a whitish clypeus. Large yellow areas on sides of

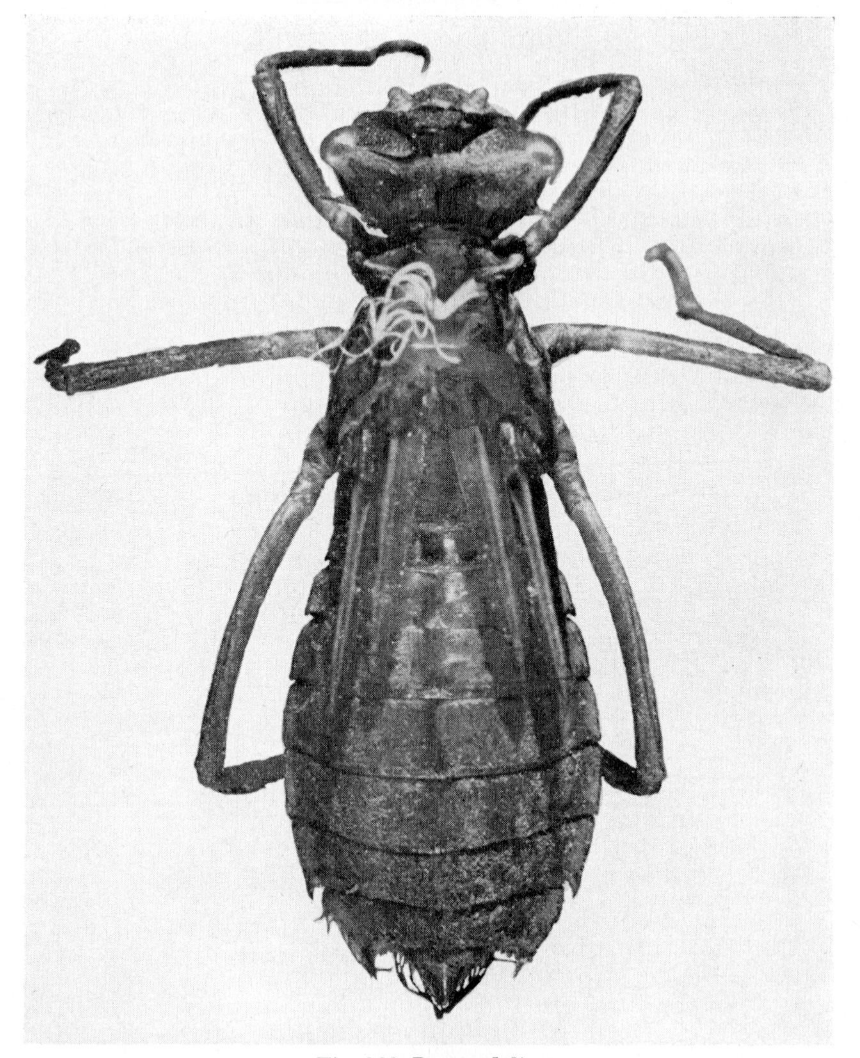

Fig. 266. *Dorocordulia.*

frons. Vertex more or less yellowish at summit, and occiput at its middle and in rear above.

On basal segments of abdomen, a large yellow area on side of segment 2 and another on base of 3, beyond which it is black when fully mature. Teneral female specimens have yellowish basal lateral areas on 4 to 7.

Distribution and dates.—CANADA: N. B., N. S.; UNITED STATES: Conn., Maine, Md., Mass., N. H., N. J., N. Y.

June 11 (Maine) to August 31 (New England).

Dorocordulia libera Selys

1871. Selys, Bull. Acad. Belg., (2)31:263 (reprint, p. 29) (in *Cordulia*).
1901. Ndm., Bull. N. Y. State Mus., 47:505 (figs.).
1927. Garm., Odon. of Conn., p. 237.
1929. N. & H., Handb., p. 198 (fig.).

Length 28–29 mm.; abdomen 28–31; hind wing 28–31.

A charming little bronzy green species, similar to preceding, but with lower part of face a little more yellowish, and yellow areas on sides of

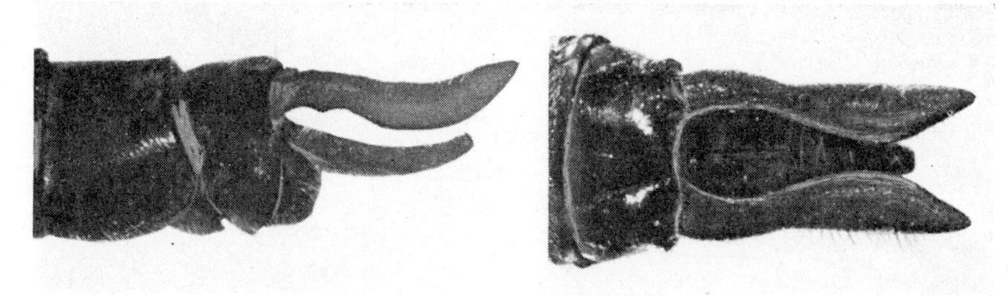

Fig. 267. *Dorocordulia lepida.*

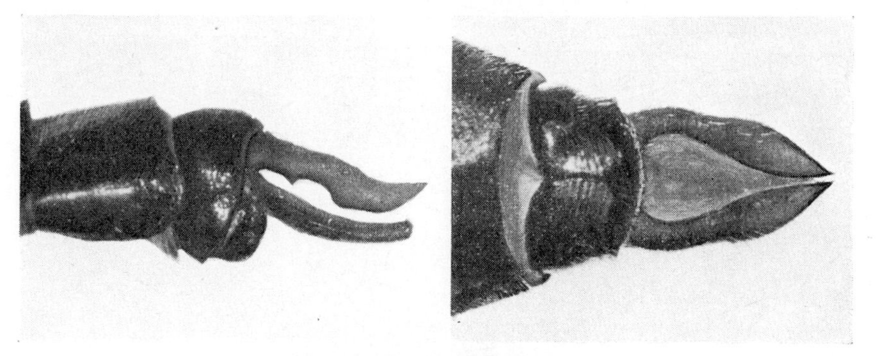

Fig. 268. *Dorocordulia libera.*

frons and on swollen basal abdominal segments less extensive. Occiput hardly yellow at all. These differences are so minor that determinations are best made by comparisons of specimens with figures of male appendages herewith presented. Spatulate dilated end segments of abdomen less marked in female than in male.

Common about borders of small upland lakes, where it flies in midsummer during sunshine, gaily in and out of little bays and over shallows. Nymphs inhabit edges of water, commonly under overhanging turf, and clamber up projecting roots and stumps to transform. Usually leave their cast-off skins less than a foot above water.

Distribution and dates.—CANADA: N. B., N. S., Ont., Que.; UNITED STATES: Conn., Ind., Maine, Mass., Mich., Minn., N. H., N. J., N. Y., Ohio, Vt., Wis.

May 4 (Mass.) to August 29 (New England).

Genus WILLIAMSONIA Davis 1913

This is a genus of two rare, small brownish Cordulines that are without metallic coloration except on top of the head in mature males. The

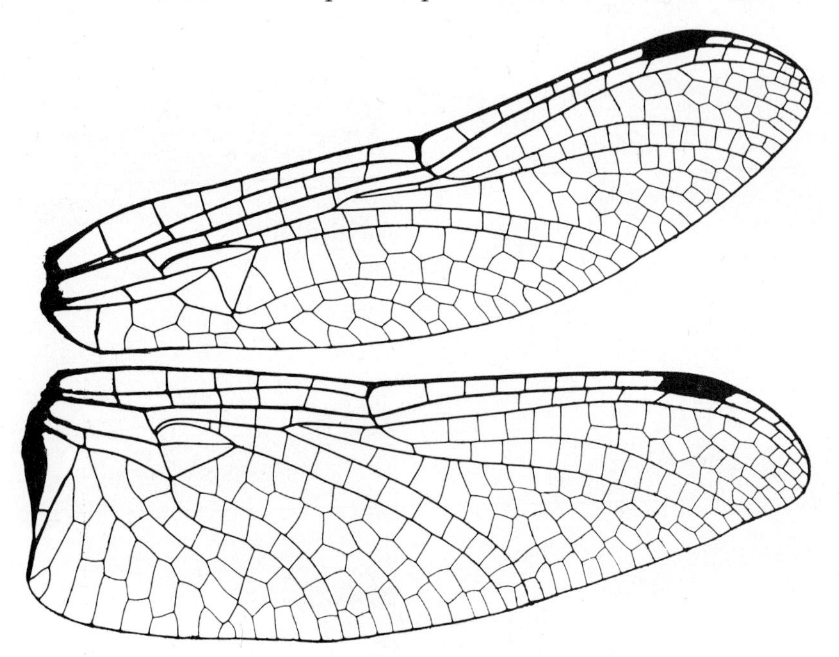

Fig. 269. *Williamsonia fletcheri.*

thorax is brown, with legs blackish beyond their pale bases. A keel on the under side of the tibiae is lacking in the female and on the middle legs of the male; it is present and about half length on the fore and hind tibiae of the male, half length on the fore and longer on the hind tibiae.

The wings are hyaline, with only slight flavescence at the extreme base. There is a single row of cells in the trigonal interspace for a distance in both fore and hind wings. The anal loop is small, almost without a "toe," but with three cells within the loop along its "sole," as in most Corduline genera. The cells of the loop on the side next to the gaff are much larger than those on the other side. Veins A2 and A3 are strongly convergent at their bases, narrowing the patella greatly at its upper end. The anal triangle is two-celled and well developed, though the wing border is almost rounded. Tornus and patella are in long contact.

The abdomen is brown, with more or less broken rings of yellow across the middle segments; the under parts and appendages are blackish. It is short and thinly hairy, cylindric, with the usual swollen basal segments, and with segment 3 considerably constricted in the male. There are large auricles on 2 in the male. The long subgenital plate of the female is divided for two-thirds of its depth into two parallel lobes with strongly convergent points.

The nymph is still unknown (1953).

KEY TO THE SPECIES

ADULTS

1—Face mostly yellow or light greenish.............................. **lintneri**
—Face blackish .... **fletcheri**

Williamsonia fletcheri Williamson

1923. Wmsn., Can. Ent., 55:96 (figs.).
1929. N. & H., Handb., p. 200 (figs.).
1941. Walk., Trans. Roy. Canad. Inst., 23:247.
1943. Mtgm., Ent. News, 54:1–4.
1944. Borror, Can. Ent., 76:145.

Length 35 mm.; abdomen 22–24; hind wing 22–23.

Very similar to *lintneri,* darker in coloration, with somewhat denser venation. Head much darker above; labium duller yellow. Synthorax black above.

Abdomen black dorsally, including caudal appendages. Latter in male more widely forcipate. Lobes of subgenital plate of female oblong in outline and almost as long as ninth sternite.

"A glacial relict inhabiting muskegs.... Its season very early and brief."—Walker (1941).

Distribution and dates.—CANADA: Man., N. B., Ont., Que.; UNITED STATES: Maine, Mich.

May 19 (Ont.) to July 1 (N. B.).

Williamsonia lintneri Hagen

1878. Hagen, in Selys, Bull. Acad. Belg., (2)45:187 (reprint, p. 9) (in *Cordulia*).
1890. Hagen, Psyche, 5:371 (figs.).
1912. Davis, Bull. Brooklyn Ent. Soc., 8:93 (fig.).
1923. Wmsn., Can. Ent., 55:98.
1929. N. & H., Handb., p. 200 (figs.).
1940. Davis, Ent. News, 51:61.
1943. Mtgm., Ent. News, 54:1–4.

Length 34 mm.; abdomen 21–23; hind wing 22–23.

Smallest of our Cordulines, a delicate little bog-inhabiting species of very local distribution. Head small; labium and labrum bright yellow; face greenish. Rounded frons above yellow, vertex and occiput brown, latter fringed with whitish hairs.

Thorax brown, thinly clad with soft hairs, its front paler toward crest. Sides darker in humeral and femoral pits and at leg bases.

Abdomen with pale orange markings on apex of segments 2 to 9. Lobes of subgenital plate of female lanceolate in outline.

"The orange ring on each abdominal segment makes the insect particularly easy of identification in the field."—Howe (1923).

Distribution and dates.—UNITED STATES: Mass., N. J., N. Y., R. I.

April 30 (Mass.) to June 1 (Mass.).

Subfamily LIBELLULINAE Selys

This dominant group of anisopterous dragonflies is of world-wide distribution. It includes nearly half of our North American genera. In it is found the utmost variety of form and coloration, but metallic coloration is less common than in the Cordulinae. The eyes are very large and always (in ours) in contact on the top of the head, often broadly so, by a long eye-seam.

The hind border of the prothorax is sometimes raised in a large, bilobed, hair-fringed plate. The synthorax is marked by a short sigmoid curvature of its humeral suture at the level of the humeral pit. The abdomen is sharply triquetral along its middle segments. It is in this group that wing venation attains the most marked differences between fore and hind wings. As told in our key to families and subfamilies (p. 63), the difference lies mainly in the relative position and shape of the triangles: the main triangle (T) far out beyond the arculus in the fore wing, retracted to the arculus in the hind. There is no basal triangle in the hind wing of the male.

In addition there are other venational subfamily features. There is a great development of planates in the wing, especially of the radial and median planates, with an apical planate added, but never a trigonal planate. The foot-shaped anal loop here reaches its highest development (fig. 272). It is open below in *Nannothemis*. It is complete with a toe in all our other genera. Both antenodal and postnodal crossveins of the first and second series tend to become matched in crosswise position. The trigonal interspace often contains more than two rows of cells; sometimes there is only one row. The Libellulinae have gone farthest in fusion of the sectors of the arculus at their departure from the arculus into a long common stalk.

The fore-wing subtriangle is greatly enlarged, and at its greatest development is so altered in appearance that it is hardly recognizable. The anal vein is broken, and turned sharply rearward, far out of its original

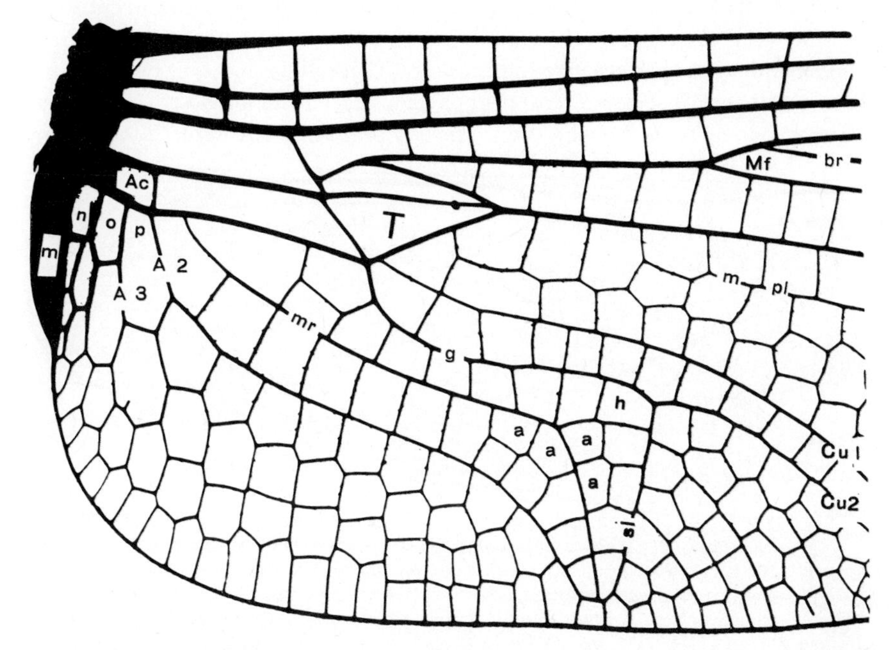

Fig. 270. Base of hind wing of *Lepthemis vesiculosa,* male. Note that rounded hind angle of wing has no basal male triangle and no tornus, but has a distinctly footshaped anal loop complete with knee, heel, and toe; an extra long gaff (*g*), a single heel cell (*h*), four interpolated ankle cells (*a, a, a, a*), and five cells bordering on sole (*sl*). Four cell rows between vein A2 and hind-wing margin. There arises between first and second paranals (*n* and *o*) what may be considered a fourth anal vein.

This portion of wing includes beginning of a median planate (*m pl,* better shown in whole wing of *Lepthemis,* p. 556). Middle fork (Mf), anal crossing (Ac), and principal veins as shown in introductory fig. 7 (p. 17).

Note that first anal interspace (*x*) lies wholly within anal loop. It is divided full length by a midrib (*mr*); second anal interspace (*y*) below cell *p* widens out unilaterally to rearward; third interspace (*z*) begins with three cells in vertical row below cell *o,* and continues in irregular cells that dwindle in size to inner wing border. Membranule (*m*) black in this species.

direct course. The cubito-anal crossvein, which always forms one side of the subtriangle, is so far aslant and so long that it does not look like a crossvein, but resembles a continuation of the anal vein, extended more or less in line with the front side of the triangle. A glance at the subtriangle of *Nannothemis* (fig. 274, p. 435) will help the reader to understand how these shifts have come to pass.

Figure 271 displays the distinctively Libelluline features in the outer part of the wing beyond the nodus. These are practically alike in fore and hind wings.

Figure 272 shows the high degree of differentiation in the greatly expanded anal area in *Libellulinae*.

These dragonflies are predominantly of perching habits. They do not hang by their feet except at night: they perch on long thin legs. Best known among them are the pond species that skim through the air along

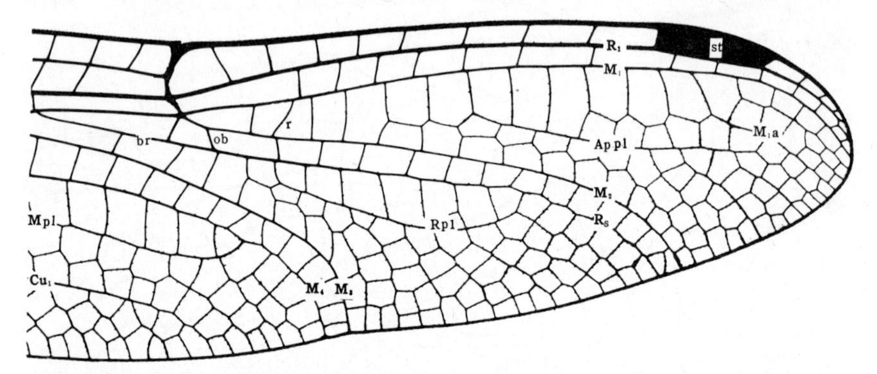

Fig. 271. Venation of *Tauriphila australis*, fore wing, showing planates in relation to principal veins: Ap pl, apical planate; R pl, radial planate; M pl, median planate. Note looped-up outer ends of two latter. Little reverse crossvein, *r* slanted as if to counterbalance upturned inner end of radial planate, is peculiar to subfamily Libellulinae.

the shores, or perch on reed tips and twigs, with evident preference for places that afford a wide outlook for discovery of food and mates and for avoidance of enemies. Their long, thin, neatly bunched legs enable them to rest easily on the side of vertical stems.

The outstanding work on adults of this subfamily is by F. Ris, *Catalogue systematique et descriptif* of the Libellulinae of the de Selys Collection.

The nymphs are for the most part stocky, and very similar to the nymphs of Cordulinae, so like them indeed that no single character has yet been found that will separate all of them. They are easily distinguished from nymphs of Macrominae by being less depressed of body, and by the absence of the pyramidal horn on the front of the head. For the most part, genera will be readily recognized by the photographs hereinafter presented.

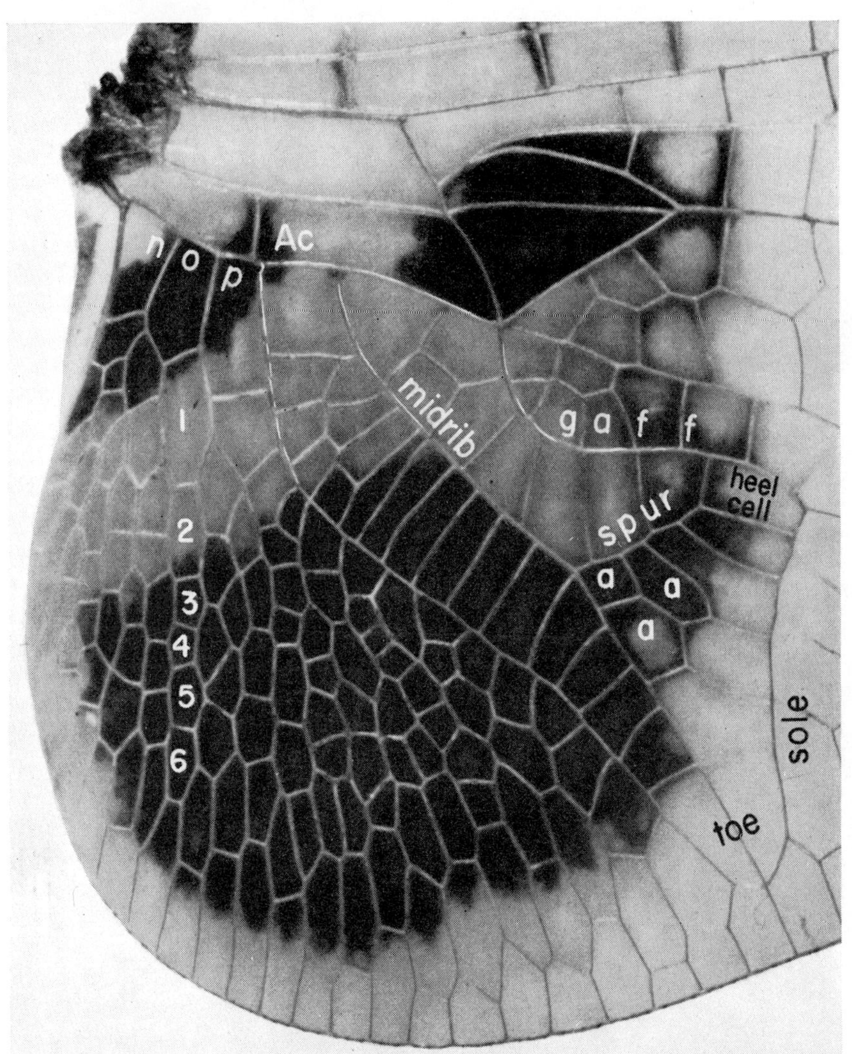

Fig. 272. Base of hind wing of *Celithemis amanda,* showing principal venational characters in expanded anal area in Libellulinae. Ac, anal crossing; *n, o, p,* first three paranal cells; three other paranals fall within anal loop. Loop is bounded by veins A1 and A2, as always; its parts are named on figure, except for *a, a, a,* the ankle cells, which in other species may lie on either side of midrib. A more or less constant column of cells standing between cell *o* and marginal row bears numbers 1 to 6.

TABLE OF GENERA

ADULTS

Genera	*Hind wing*	*Cells*					*Crossveins*		*Vn. Cul h.w. base*	*Distri-bution*
		Sub T	*T*	*Rows bynd.*	*Paranals f.w.*	*Paranals h.w.*	*At bridge*	*Under stigma*		
Brachymesia	32-36	3	2	3	5+3	2+3	1	2	down	Tex., W.I.
Brechmorhoga	34-43	3	2	2	6+2	3+3	1	2-3	down	SW
Cannacria	34-42	3	2	3	5+3	2+3	1-2	2	down	SE, W.I.
Cannaphila	30-31	3	2	2	6+2	3+2	1	3-4	up	*
Celithemis	23-34	1-3	2-4	2-4	4-6+1-3	3+2-4	1-2	1-3	down	E
Dythemis	29-38	3	2	3	5+2	3+3	1	2-3	down	S, W.I.
Erythemis	28-38	3	2-3	2-3	6+2	3+3	1	2	up	General
Erythrodiplax	17-34	1-4	1-3	2-4	5+1-3	3+2-3	1	2	var.	E,S, W.I.
Idiataphe	28-30	1	1	2	6+1	3+2	1-2	2	down	*
Ladona	26-34	2-3	2	2-3	5-7+2	3+2-3	2-4	2-4	down	N, E, S
Lepthemis	39-40	3	2	3	6+2	3+3	1	2	up	S, W.I.
Leucorrhinia	18-32	1-3	1-2	1-3	4-5+1-2	3+1-2	1	2	down	NE, NW
Libellula	31-51	3-14	2-5	3-5	5-7+2-3	3-4+3-5	3-4	2-5	down	General

Nannothemis	10-15	1	1	2	4-5+1	3+2	1	2	down	E
Orthemis	37-44	3-5	2	3	5+2-3	3+3-4	1	5-6	down	S, W.I.
Pachydiplax	30-42	3	2	3	5+2	3+3	1	0-1	up	General
Paltothemis	43-46	3	2	3	6+2	3+3	1-2	2-3	down	SW
Pantala	36-45	4-5	2	3	6-7+2-3	3+3	1	2	down	General
Perithemis	14-23	1-2	1-2	2	4+1	2+3	2-3	3	down	E,S, W.I.
Planiplax	28-34	3	2	3	5+2	2+3	0-1	2-3	down	*
Plathemis	30-38	5-8	3	4	5-6+2-4	3+3-4	2-4	3-5	down	E,S,W
Pseudoleon	30-35	3-4	2-3	4	6+2-3	3+2-3	1	2	down	W
Scapanea	35-38	3	2	3	6+2	3+3	1	3-4	down	*
Sympetrum	18-31	3	2	3	5+2	3+2-3	1	1	down	General
Tarnetrum	26-30	3	2	3	5+2	3+2-3	1	1	down	General
Tauriphila	36-38	3-4	2	3	6+2	3+3	1	2	down	*
Tholymis	36-39	3	2	3	6+2	3+3	1-2	2-3	var.	*
Tramea	38-47	4-6	2-3	3-4	6+2-3	3+3	1	2	down	E,S,W,W.I.

*Neotropical.

KEY TO THE GENERA

Adults

1—Fore-wing triangle with front side broken, making a quadrangle of it; anal loop of hind wing incomplete, open to rearward .... **Nannothemis**
—Fore-wing triangle usually normal, three-sided; anal loop usually complete, and more or less foot-shaped 2

2—Antenodal crossveins of both wings with row of roundish brown spots and pattern shown in fig. 284 .... **Pseudoleon**
—Crossveins with no such coloration 3

3—Vein M2 waved (undulation slight in **Brechmorhoga** and **Macrothemis**) 4
—Vein M2 smoothly curved 14

4—Hind wing narrow, with two cubito-anal crossveins, and with vein Cu1 arising from outer side of triangle; West Indian .... **Cannaphila**
—Hind wing wider at base, and with no such combination of characters 5

5—Vein Cu1 in hind wing arising from hind angle of triangle 6
—Vein Cu1 arising from outer side 19

6—Wings with several bridge crossveins (except in most **Libellula semifasciata**) 7
—Wings with a single bridge crossvein 9

7—Fore-wing triangle of three or more cells 8
—Fore-wing triangle two-celled .... **Ladona**

8—Arculus at or very close to second antenodal crossvein; fore-wing triangle straight or very slightly convex on outer side; abdomen of male bare on under side of segment 1; abdomen of female slowly tapered to rearward on side of middle segments .... **Libellula**
—Arculus nearer midway between first and second antenodal crossveins; fore-wing triangle distinctly convex on outer side; abdomen of male with pair of stout processes on under side of segment 1; abdomen of female with side margins of middle segments parallel .... **Plathemis**

9—Stigma very long, surmounting five or six crossveins .... **Orthemis**
—Stigma moderately long, surmounting two to four crossveins 10

10—Hind wing with two cubito-anal crossveins .... **Pantala**
—Hind wing with a single cubito-anal crossvein 11

11—Fore wing with two rows of cells beyond triangle 12
—Fore wing with three rows of cells beyond triangle 13

12—Fore-wing subtriangle two-celled .... **Macrothemis**
—Fore-wing subtriangle three-celled .... **Brechmorhoga**

13—Two to four parallel rows of cells between vein A2 and marginal row at hind angle of hind wing .... **Dythemis**
—Three parallel rows of cells in this space; West Indies .... **Scapanea**
—Four or five very irregular rows in this space .... **Paltothemis**

14—Midrib of anal loop nearly straight or very slightly bent at ankle; postnodal crossveins of second series, and first crossvein under stigma strongly aslant; reverse vein also strongly aslant (except in **Planiplax**) ... 15
—Midrib of anal loop more angulated; crossveins and reverse vein much less if at all aslant 19

15—Fore-wing triangle of two to four cells, and usually with three or four rows of cells beyond in trigonal interspace; radial planate often subtends two rows of cells . . . . **Celithemis**

—Fore-wing triangle of one cell (except in some **Perithemis**) and with two rows of cells beyond in trigonal interspace (except in **Planiplax**); radial planate subtends a single row of cells 16

16—Fore-wing triangle with inner side about as long as front side; more than a single bridge crossvein . . . . **Perithemis**

—Inner side much longer than front side; usually one bridge crossvein only . . . 17

17—Fore-wing triangle of two cells, with three rows of cells beyond; subtriangle of three cells; reverse vein not strongly aslant; two paranal cells before anal loop . . . . **Planiplax**

—Fore-wing triangle of one cell, with two rows beyond; reverse vein strongly aslant 18

18—Fore-wing subtriangle of one cell; two paranal cells before anal loop . . **Idiataphe**

—Fore-wing subtriangle of two cells; three paranal cells before anal loop **Macrodiplax**

19—Wings with more than a single bridge crossvein . . . . **Micrathyria**

—Wings with one bridge crossvein only 20

20—Wings with a single crossvein or none under stigma 21

—Wings with two or more crossveins under stigma (except in **Miathyria**) 23

21—Wings with triple-length vacant space before single remote crossvein under stigma . . . . **Pachydiplax**

—Wings with a single crossvein more proximal and adjacent spaces more normal 22

22—Radial planate subtends two rows of cells . . . . **Tarnetrum**

—Radial planate subtends a single row of cells (except in **S. madidum**) **Sympetrum**

23—Fore-wing triangle with inner side less than twice as long as front side; stigma short and thick, about twice as long as wide, at least in male; face white . . . . **Leucorrhinia**

—Fore-wing triangle with inner side more than twice as long as front side; stigma about three times as long as wide; face not white 24

24—Hind wing with two paranal cells before anal loop 25

—Hind wing with three paranal cells before anal loop 26

25—Hind wing with six antenodal crossveins; median planate subtends a single row of cells; black on abdomen covers segments 8 and 9 . . . . **Brachymesia**

—Hind wing with seven or eight antenodal crossveins; median planate subtends two rows of cells; black on abdomen covers segments 4 to 9 . . **Cannacria**

26—Toe of anal loop incomplete or resting on hind-wing margin . . . . **Tholymis**

—Toe complete, but not quite reaching hind margin, there being a row of small marginal cells between 27

27—Wings with stigma trapezoidal, front side distinctly longer than rear; some double-length cells above apical planate, reaching from planate to M1 28

—Wings with front and rear sides of stigma about equal in length; apical planate poorly developed and with no double-length cells above it 30

28—All cells above apical planate double length and in a single row; one crossvein under stigma . . . . **Miathyria**
—About half of these cells in a single row, then a double row; two crossveins under stigma **29**

29—Fore wing with three rows of cells in trigonal interspace . . . . **Tauriphila**
—Fore wing with four rows of cells in trigonal interspace . . . . **Tramea**

30—Fore wing with five paranal cells before subtriangle . . . . **Erythrodiplax**
—Fore wing with six (only rarely five) paranal cells before subtriangle **31**

31—Radial planate subtends two rows of cells; base of midrib of anal loop four times as far from triangle as from vein A2 . . . . **Lepthemis**
—Radial planate subtends a single row of cells; base of midrib three times as far from triangle as from vein A2 . . . . **Erythemis**

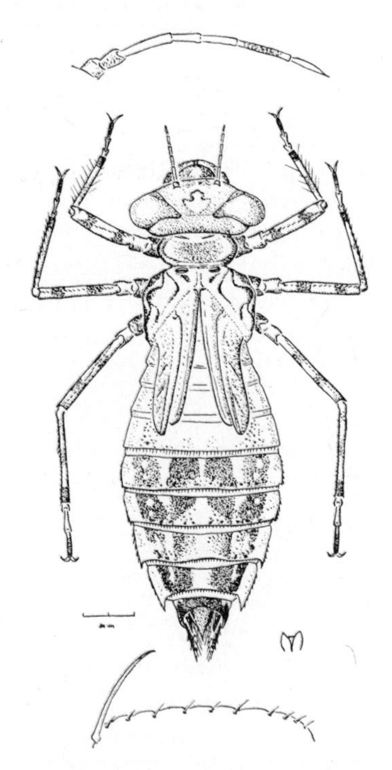

Fig. 273. Nymph of *Micrathyria hageni*. Above it, an antenna enlarged; below, end of lateral lobe of labium showing slender movable hook, and marginal row of obsolescent teeth with spinules set singly between teeth. (Drawings by Dr. May Gyger Eltringham.)

NYMPHS

1—Eyes capping fronto-lateral part of head (fig. 296); abdomen long and tapering **2**
—Eyes lower, more broadly rounded and more lateral in position; abdomen usually ending more bluntly **5**

2—Margin of median lobe of labium smooth........................... **Libellula**
—Margin of median lobe of labium crenulate................................ 3

3—No dorsal hooks on middle abdominal segments.................... **Orthemis**
—Dorsal hooks present on middle abdominal segments........................ 4

4—No dorsal hook on 8... **Plathemis**
—Dorsal hook present on 8.. **Ladona**

5—Inferior abdominal appendages strongly decurved at tip.................. 6
—These appendages straight or nearly so.................................. 7

6—No lateral spines on abdomen; lateral setae of labium seven to nine.. **Erythemis**
—Minute lateral spine on 9; lateral setae eleven or twelve............ **Lepthemis**

7—Dorsal hooks on some abdominal segments................................ 8
—No dorsal hook on any abdominal segment................................ 23

8—Dorsal hook on 9.. 9
—No dorsal hook on 9 17

9—Dorsal hooks cultriform, the series in lateral view like teeth of a circular saw .... **Perithemis**
—Dorsal hooks more spinelike or low and blunt........................... 10

10—Dorsal hooks long and laterally flattened............................. 11
—Dorsal hooks short and thick... 14

11—Abdomen broadly depressed, little longer than wide.............. **Tauriphila**
—Abdomen about twice as long as wide.................................... 12

12—Superior abdominal appendage, seen from above, slightly longer than its basal width; lateral abdominal appendage more than half as long as inferior appendages .... **Brachymesia**
—Superior abdominal appendage about twice as long as its basal width........ 13

13—Length when grown 21 mm. or more; tip of hind-wing case extends to rearward about halfway across abdominal segment 6................. **Cannacria**
—Length when grown 20 mm. or less; tip of hind-wing case extends about halfway over 7.. **Idiataphe**

14—Teeth on lateral lobe of labium large................................ 15
—Teeth on lateral lobe obsolete..................................... **Dythemis**

15—Lateral setae six... **Macrothemis**
—Lateral setae seven to ten... 16

16—Dorsal hooks high and conspicuous............................. **Brechmorhoga**
—Dorsal hooks low, ridgelike....................................... **Scapanea**

17—Dorsal hook on 8.. 18
—No dorsal hook on 8... 21

18—Lateral setae seven, mental setae nine to eleven.................. **Miathyria**
—Lateral setae nine to twelve, mental setae twelve to eighteen............. 19

19—Superior abdominal appendage as long as, or nearly as long as, inferiors.... 20
—Superior abdominal appendage much shorter than inferiors........... **Sympetrum**

20—Mental setae sixteen to seventeen; lateral spines of segment 8 as long as middorsal length of that segment........................... **Macrodiplax**
—Mental setae ten to fifteen; lateral spines of 8 much shorter than its middorsal length .... **Leucorrhinia**

21—Lateral spines of 9 much longer than its middorsal length.......... **Celithemis**
—Lateral spines of 9 not longer than its middorsal length................... 22

TABLE OF GENERA

NYMPHS

Genera	*Total length*	*Labium*			*Abdomen*					
		Setae[1]		*Teeth size*[2]	*Dorsal hooks*[3]	*Lat. spine*[4]		*Appendages*[5]		
		Lat.	*Ment.*			*On 8*	*On 9*	*Lat.*	*Sup.*	*Inf.*
Brachymesia	21-23	7-8	10-13	high	3-10	3	4	7	10	10
Brechmorhoga	22-24	7-9	14-15	high	2-9	5	5	6	10	10
Cannacria	21-27	6-10	8-11	high	3-10	4-7	5-9	4	9	10
Celithemis	13-21	7-10	10-14	obs.	4-7	6-12	18-20	4-5	7-8	10
Dythemis	17-18	6-7	9	obs.	3-9	5	7	8	10	10
Erythemis	15-17	7-9	11-12	0	0	0	0	6	10	10
Erythrodiplax	12-20	6-11	8-14	obs.	0-7	3-7	4-8	5-8	6-10	10
Idiataphe	17-20	6	9-11	high	3-10	5	12	2	10	10
Ladona	20-24	6	0-3	obs.	4-8	6	5	4	10	10
Lepthemis	16-18	11-12	16-17	0	0	0	2	6	10	10
Leucorrhinia	17-18	9-12	10-15	obs.	0-8	3-7	6-9	4-5	9	10
Libellula	20-28	5-11	12-13	low	0-8	6	6	3-7	10	10
Macrodiplax	16-20	10	16-17	low	3-8	10	14	4	10	10

Orthemis	20-23	9	3-4	high	0	5	6	4	10	10
Pachydiplax	20-21	10	10-13	low	0	6	12	4	6	10
Paltothemis	21-23	7-9	11-13	high	2-6	5	7	6	10	10
Pantala	26-28	14-15	16-18	high	0	11	14	8	12	10
Perithemis	15-18	5	9-10	high	3-9	6	6	7	10	10
Plathemis	22-25	8-10	8-11	high	3-6	4	5	6	10	10
Pseudoleon	20-21	8	11	0	0	4	5	9	10	10
Scapanea	20-22	7-9	9-10	high	2-9	4	5	5	10	10
Sympetrum	13-18	9-12	11-18	low	0-8	3-9	3-12	4-7	6-10	10
Tarnetrum	18-21	9-16	12-16	0	0	0-1	1-6	5-7	9	10
Tauriphila	16-19	8	9-10	med.	3-9	8	9	5	10	10
Tramea	24-27	10-11	13-15	low	0	12	12	8	8	10

[1]Raptorial setae on lateral lobe of labium (Lat.) and on each side of mentum (Ment.).

[2]Height of teeth on opposed edges of lateral lobes: high, medium (med.), obsolete (obs.), wanting (0).

[3]Middorsal hooks present on abdominal segments numbered.

[4]Relative length of spines terminating lateral margins of segments 8 and 9 expressed in tenths of middorsal length of their respective segments.

[5]Relative length of caudal appendages: lateral (Lat.), superior (Sup.), and inferior (Inf.), expressed in tenths of length of inferior pair.

22—Teeth on lateral lobe of labium obsolete........................**Erythrodiplax**
—Teeth on lateral lobe of labium conspicuous......................**Paltothemis**

23—Lateral spines of 9 not longer than its middorsal length..................24
—Lateral spines of 9 much longer than its middorsal length.................28

24—Small species: length 10 mm.; lateral setae six; end of abdomen truncated and very hairy...**Nannothemis**
—Larger species: length 12–20 mm.; lateral setae more than six..............25

25—Lateral spines of 8 very minute or wanting.......................**Tarnetrum**
—Lateral spines of 8 at least half as long as those of 9......................26

26—Abdomen abruptly rounded to tip; segment 10 annular, almost included in 9; caudal appendages projecting little beyond ventral margin of 9..**Pseudoleon**
—End of abdomen not so..27

27—Northern species ....**Leucorrhinia**
—Southern species ....**Erythrodiplax** and **Micrathyria**

28—Lateral spines on 8 about half as long as on 9...................**Pachydiplax**
—Lateral spines on 8 nearly as long as on 9...............................29

29—Superior abdominal appendage as long as, or longer than, inferiors....**Pantala**
—Superior abdominal appendage shorter than inferiors.................**Tramea**

Nymphs unknown: **Cannaphila, Planiplax,** and **Tholymis** (see p. 590).

Genus NANNOTHEMIS Brauer 1868

Syn.: Aino Kirby

These are the smallest of North American anisopterous dragonflies. Dumpy little fellows, they are less than an inch long, clear-winged, and with a shape of triangle and of anal loop that at once distinguish them from everything else in our fauna. The triangle of the fore wing appears to be four-sided, the front side being broken about midway of its length. The anal loop is incomplete at its end, with the midrib generally running out to the wing margin through the open distal end of the loop. Other venational characters are as stated in the table of Libelluline genera (p. 426).

The single known species is confined to the eastern states, and is very local in occurrence. Being a weak flyer it is easily taken with a net when found. It forages about little stagnant pools in marshy places, making short flights to capture the small dipterous insects that hover there, and perching much of the time on grass stems in the sunshine.

The nymph, when grown, is easily recognizable by its small size and long wing sheaths that reach abdominal segment 6, by the long apical fringes of hair on 9, and by the little thornlike incurving lateral spines on 8 and 9. Other characters will be found in the table for nymphs of Libelluline genera (p. 432).

The peculiar habitat of this nymph was described by its discoverer, the late Mr. R. J. Weith (1901), as follows:

Using a dragnet, I explored the shore and grass-fringed margin of the lake, near where the imago is found, but without success. These places yielded lots of other nymphs, but no *Nannothemis*. Then, collecting in those places in the marsh where the water is from one to three feet in depth among the rushes and sedges proved fruit-

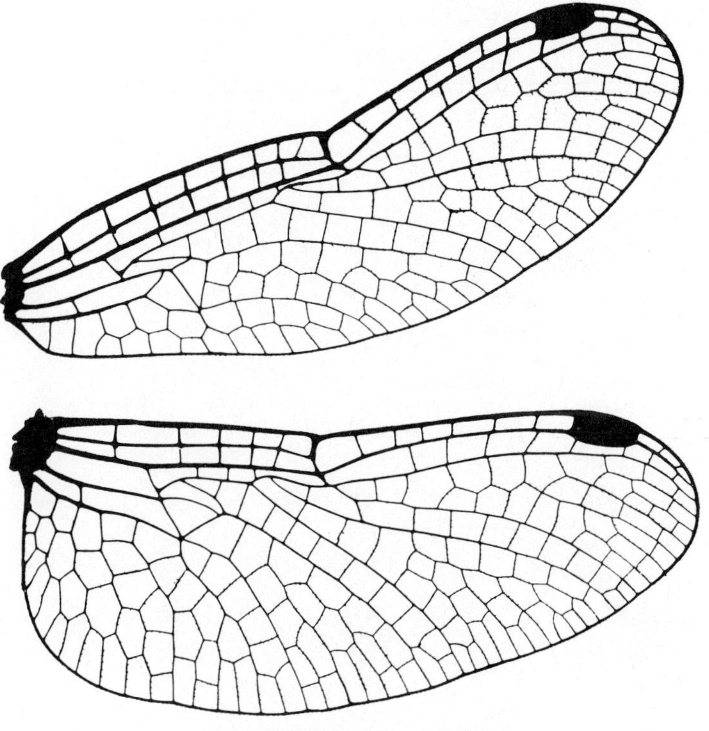

Fig. 274. *Nannothemis bella.*

less also. This convinced me that the home of the coveted nymph must be the almost dry marsh-land, with here and there a hole with a few inches of water in it. The holes were too small to allow the use of a net: I had to dip the water out with my hands. In them I was surprised to find a great number of Libellula nymphs, among which were two that proved to be the nymphs desired.

Not being able to find any more in these holes, I then searched thoroughly the debris which had been deposited on the marsh during high water, and which still lay in many places covered by a few inches of water. Here I found I could collect in an hour eighteen to twenty-five of them. But it was very trying; for, removed from the water, the nymph clings closely to grass or debris of exactly its own color, and does not stir even after letting this dry: so it is hard to see, and everything has to be picked over very carefully.

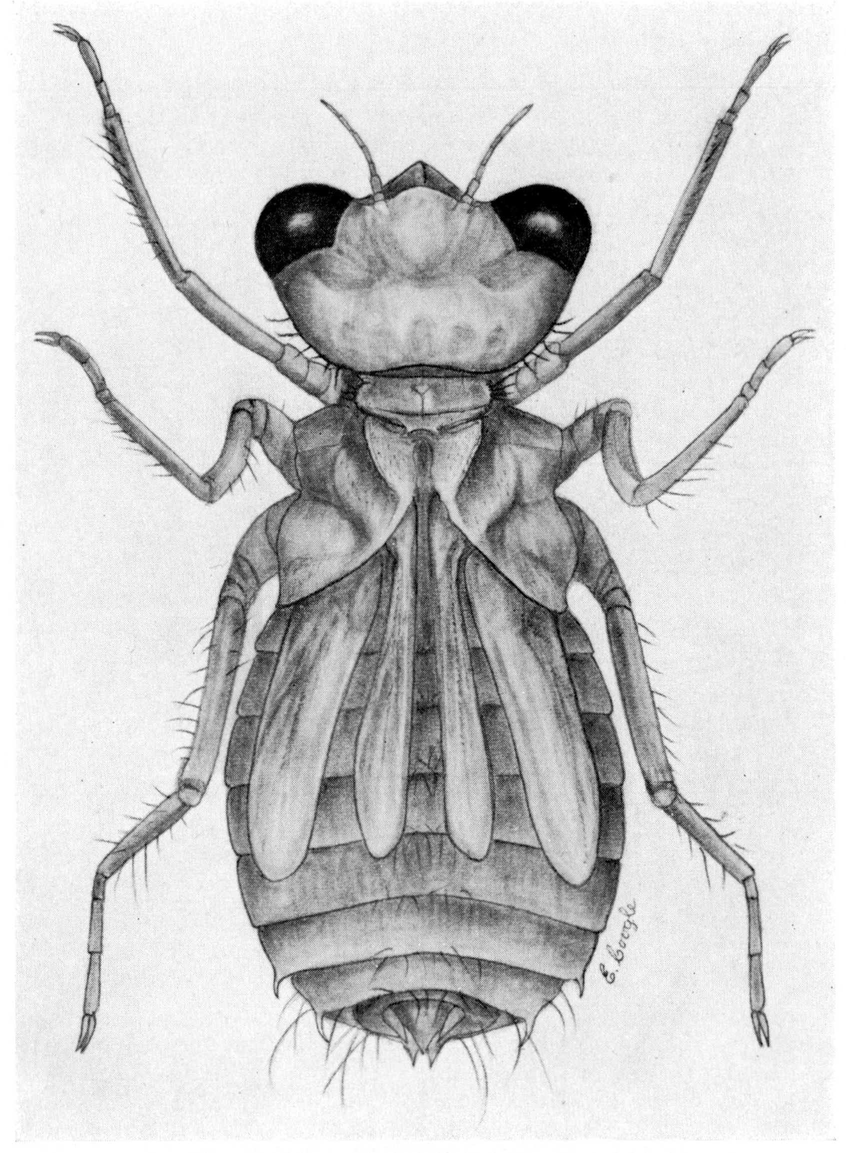

Fig. 275. *Nannothemis bella.* (Drawing by Esther Coogle.)

This little dragonfly has retained two very primitive characters: (1) the fore-wing triangle is small, its front-side crossvein little longer than neighboring crossveins; (2) its anal loop is incomplete. Yet it is highly specialized in two characters: (1) sectors of the arculus are low down on the arculus and long-stalked in both fore and hind wings; (2) the hind-wing triangle is retracted quite to the arculus.

Nannothemis bella Uhler

Syn.: puella Kirby

The Blue Bell

1857. Uhler, Proc. Acad. Phila., p. 87 (in *Nannophya*).
1867. Packard, Amer. Nat., 1:311 (fig.).
1893. Calv., Odon. of Phila., p. 260.
1901. Ndm. & Weith, Can. Ent., 33:252 (figs., nymph).
1911. Ris, Coll. Selys Libell., p. 388 (fig.).
1929. N. & H., Handb., p. 204.

Length 18–20 mm.; abdomen 12–13; hind wing 10–15.

A delicate little clear-winged dragonfly, subject to great changes in appearance of body with age. Marked with black and yellow in teneral specimens in a pattern that may become entirely obscured under a powdery coat of pruinose blue with age, especially in males. At first, ground color is yellow. Middorsal triangle of black on front of thorax. Three lateral sutures on its sides bear very irregular black stripes that diminish in width to rearward; ragged edges of these stripes often conjoined to cover sides in a marbled pattern, especially in males. Face white, with labrum and lower margin of postclypeus black, and all thinly covered with hairs. Vertex metallic black. Eyes barely meet on top of head.

Small thorax, thinly clad with rather long whitish hairs. Legs black. Wings clear in male, but often tinged in their membrane with amber brown as far out as triangles. Arculus long-stalked, equally in fore and hind wings. Stigma blackish touched with yellow, especially at proximal end. Single bridge crossvein placed unusually far to proximal side of subnodus. Only two crossveins under stigma.

Abdomen very slender, especially in middle segments, and sharply ridged on its three lateral carinae. Basal segments swollen slightly; genitalia on their under side very prominent. Erect inner branch of male hamule clawlike; outer one a low and somewhat triangular blade, partly hidden by long straplike genital lobe. Sides of this lobe as well as its end fringed with long hairs. Straight and slender caudal appendages of male yellow toward their ends on their upper sides.

Abdomen of female stouter and its color pattern less obscured by pruinosity. Predominant yellow at its base reduced to a triradiate band on segment 2, to smaller proportions on 4, 5, and 6, and to a single pair of minute yellow dots on 7; 8 to 10 black. Cerci pale. Scooplike subgenital plate extended to rearward to quite the length of segment 9; longer than caudal appendages.

Distribution and dates.—CANADA: Ont., Que.; UNITED STATES: Ala., Conn., Fla., Ga., Ind., Ky., La., Maine, Md., Mass., Mich., Miss., N. H., N. J., N. Y., N. C., Pa., S. C.

April 18 (Miss.) to September 5 (N. J.).

Genus PERITHEMIS Hagen 1861

Amber Wings

These attractive little dragonflies are easily recognized by the tints of amber and gold in their wings, which in the female are beautifully patterned in brown; by the very short and wide abdomen; and by some unique features of wing venation. The eyes are large and in contact for some distance on the top of the head. The hind lobe of the prothorax is broadly expanded and bilobed, and its margin is fringed with long thin hair. The wings are longer than the abdomen; their brown markings tend to run in two crossbands: one at the triangles and the other at the outer side of the nodus.

The wing venation is unique in the form of the triangle in the front wing and of the anal loop in the hind wing. The following characters appear to be common: the arculus situated before the middle of the space between the first and second antenodal crossveins; a terminal half-antenodal generally present; also two or three bridge crossveins, and three crossveins behind the stigma. Planates are all recognizable, though not very well defined at their ends. Triangles and subtriangles vary in the number of contained cells (one to three), but are constant in shape. The fore-wing triangle is about as long on the front as on the outer side. Its inner side is shorter than either its front or its outer sides, and is very strongly retracted, making with its base vein Cu an angle of less than 90°. The trigonal interspace (area between veins M4 and Cu1) narrows outward to the wing margin, though separated by three rows of cells nearly all the way out. The hind wing generally has five paranal cells: two before and three within the anal loop. The form of that loop suggests a toe-dancer's foot because of the long-pointed toe and the high-lifted heel. The midrib is nearly straight. The sole is longer than the gaff. The cells that fill the anal loop are rather large, and those of the area behind it are elongated toward the wing margin.

The abdomen is wider in the middle than at the ends, parallel-sided along the middle segments, and often somewhat depressed there. The caudal appendages of the male are of the usual simple Libelluline type. The genitalia of segment 2 are not prominent, but mostly included and not easily observed. The genital lobe is small, semioval, hardly longer

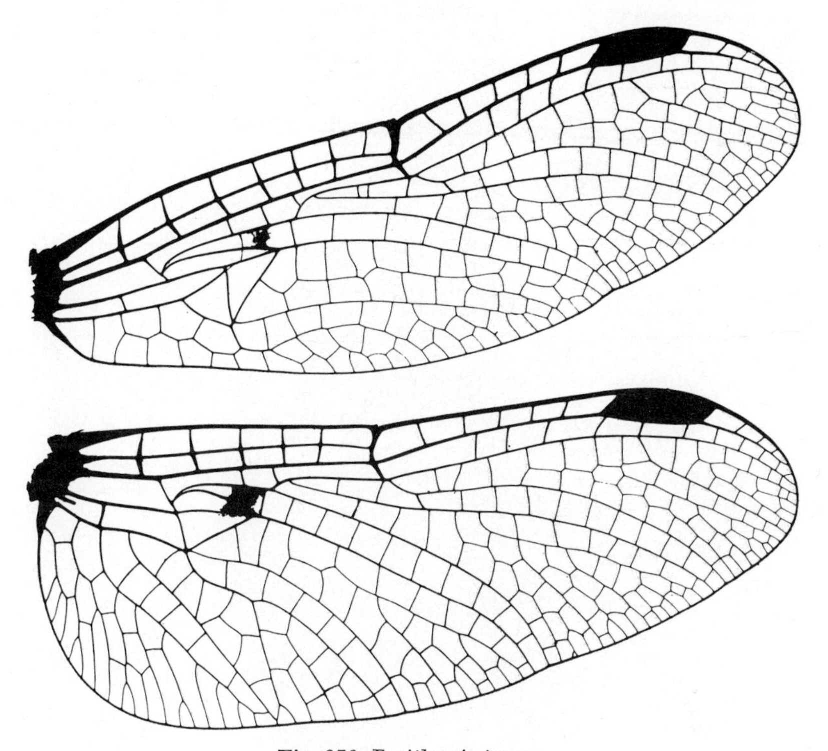

Fig. 276. *Perithemis tenera.*

than wide, and thinly clothed with pale hairs. The inner branch of the hamule is a rather long, slightly twisted, recurved hook that is hairy externally; the outer branch is practically wanting. The subgenital plate of the female is broad and short, not reaching halfway to the palps that stand near the middle of the ninth sternite. The plate is deeply divided, with outrolled blunt tips.

This is a Neotropical genus of about a dozen species, five of which occur within our range.

The outstanding work on the genus is the monograph by Dr. F. Ris, "A Revision of the Libelluline Genus Perithemis (Odonata)," *Misc. Pub. Mus. Zool. Univ. Mich.*, no. 21 (1930), pp. 1–60 (9 pls.).

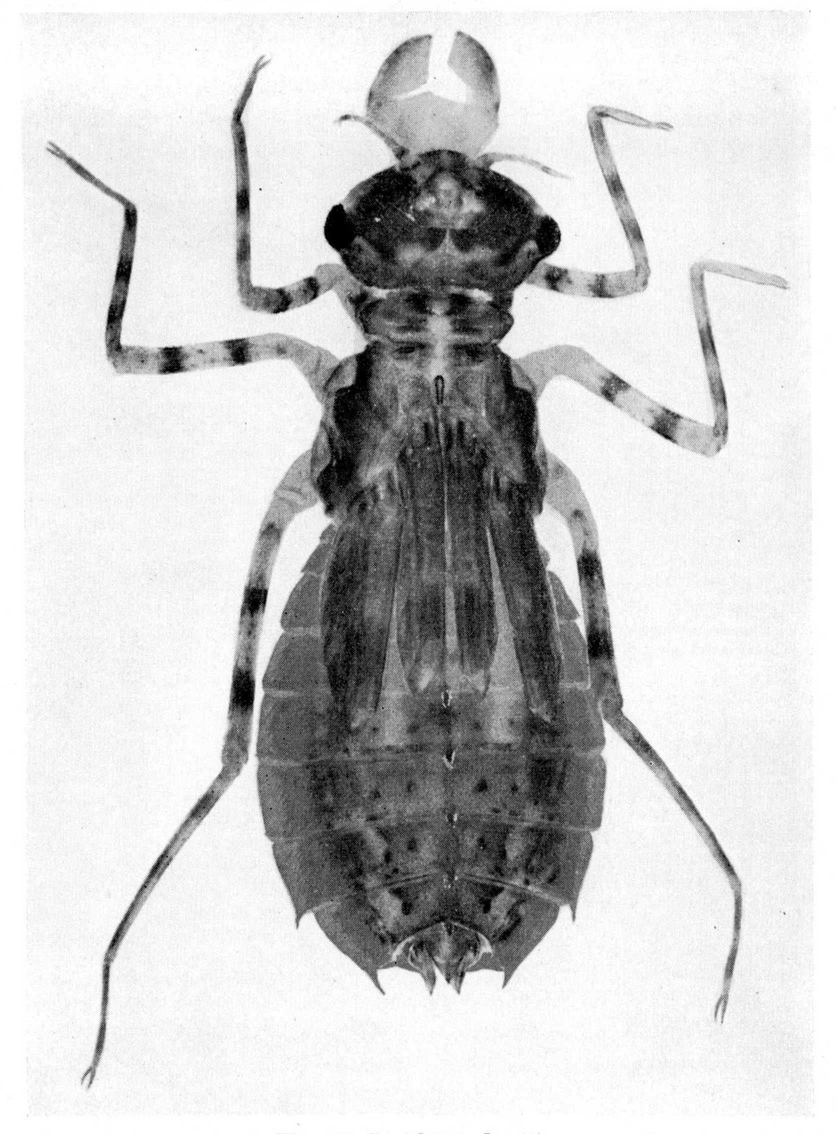

Fig. 277. *Perithemis domitia.*

The nymph is unique among our Libellulines in possessing a full series of cultriform high dorsal hooks on the abdominal segments, ending on segment 9. Its characters are as stated in the table for nymphs of the genera (p. 432).

Nymphs are known for only three species of this genus.

KEY TO THE SPECIES

ADULTS

1—Triangles and subtriangles of fore and hind wings without crossveins........ 2
—Some triangles or subtriangles divided by crossveins...................... 4

2—Sides of thorax light brown without markings, or only slightly clouded; wings of male golden yellow, venation and stigma red, usually without dark spots at triangles; wings of female with markings at triangles and also with a band at or before nodus, rarely slightly distal to it; Antillean **mooma**
—Sides of thorax, like dorsum, dark purplish brown with two lateral dull greenish bands, which may be interrupted; wings of male deep golden yellow, often with blackish markings at triangles; wings of female with markings not as in above; United States.............................. 3

3—Wings of male without brown spots, or with very small ones restricted to triangle and extreme base of hind wings; all four wings of female with two crossbands or clouds of brown; nodus in both sexes with brown not extending along costa all the way to stigma.........................**tenera**
—Wings of both male and female variable; both nodus and triangle with crossbands of brown; nodus with brown extending along costal interspace all the way to stigma in fore wings of female........................**seminole**

4—Hind-wing triangle with no crossvein...............................**domitia**
—Hind-wing triangle with crossvein....................................**intensa**

NYMPHS

1—Lateral setae of labium five; middorsal hook on abdominal segment 9 smaller than hook on 8*...**tenera**
—Lateral setae six; dorsal hook on 9 about same size as hook on 8........**domitia**

Nymphs unknown: **intensa, mooma.**

Perithemis domitia Drury

Syn.: iris Hagen, metella Selys, pocahontas Kirby

1773. Drury, Ill. Exot. Ins., 2:83 (pl. 45, fig. 4) (in *Libellula*).
1919. Calv., Trans. Amer. Ent. Soc., 45:372.
1929. N. & H., Handb., p. 206.
1930. Ris, Misc. Pub. Mus. Zool. Univ. Mich., 21:26 (figs.).

Length 22–25 mm.; abdomen 13–14; hind wing 16–19.

A reddish brown Antillean species; blackish legs, with contrasting stripe of light ochraceous yellow on outer face of tibiae. Venation of wings dark red. Top of vertex yellowish gray.

Adults fly low over water, never departing far from it. They dart about very swiftly and perch frequently on emergent twigs or grass stems. Males on meeting face to face in flight may dart upward to con-

* The Floridian **P. seminole** has been reared first by Dr. Byers (see Byers, 1930) and later by the senior author. Byers could not find that it differed in any way from the nymph of **tenera**, nor can we; which perhaps may indicate that it should be regarded as a variety of that species.

siderable heights, threatening each other, but return at once to low-level perches. Return again and again to same perch, even when driven from it by an unskillful stroke of net. Easily caught if collector waits close by a favorite perch with a net in readiness.

The senior author observed and reared this species at the pools in a mountain brook near San Juan de las Matas, Dominican Republic; also in roadside pools at lower elevations in Puerto Rico and in Cuba.

Distribution and dates.—ANTILLES: Cuba, Dom. Rep., Haiti, Jamaica, P. R.; also south to Colombia and Venezuela.

April 18 (Cuba) to September 24 (Cuba).

Perithemis intensa Kirby

Syn.: californica Martin

1889. Kirby, Trans. Zool. Soc. London, 12:326 (figs.).
1907. Calv., B. C. A., p. 311 (figs.).
1930. Ris, Misc. Pub. Mus. Zool. Univ. Mich., 21:19 (figs.).

Length 23–26 mm.; abdomen 15–16; hind wing 20–23.

A fine autumnal Southwestern species, more yellowish than preceding species. Face light yellow, top of vertex greenish gray. Front of thorax light golden brown, with no antehumeral pale stripe. Sides olivaceous or greenish, with short comma-like dark streaks in depths of lateral sutures. Wings brilliant orange yellow in male, without color bands and with only a dark spot at triangles of hind wings; stigma reddish. In females, usual crossbands vary from golden yellow to brown, not extending across to hind border of wing. Tibiae yellow ochraceous, with black spines; their outer sides more clear yellow.

Abdomen stout, compressed on basal segments 1 to 3, parallel-sided along middle segments, narrowed beyond 7 to end. Sides of 1 to 3 greenish; middle segments dull yellow, with fine dark markings ranged alongside middorsal carina; on 8 and 9 each side is marked by an oblique dark line.

Distribution and dates.—UNITED STATES: Ariz.; MEXICO: Baja Calif., Sonora; also southward in Mexico.

September 3 (Ariz.) to October 12 (Baja Calif.).

Perithemis mooma Kirby

Syn.: cloe Calvert, octoxantha Ris

1889. Kirby, Ann. Nat. Hist., (6)4:233.
1906. Calv., B. C. A., p. 314 (figs.).
1930. Ris, Misc. Pub. Mus. Zool. Univ. Mich., 21:21 (figs.).

Length 22–29 mm.; abdomen 12–16; hind wing 14–21.

A Neotropical species; wings of deep rich golden yellow, with red veins and stigma. In more highly colored female, crossbands of brown cover moderate areas at triangle and nodus and often entire area between, but brown does not reach rearward to wing margin. Outer third of wing generally clear except for stigma.

Face yellowish; vertex red-brown. Dorsum of thorax dull red-brown, with two abbreviated pale stripes. Sides of thorax olive green.

Abdomen laterally compressed on segments 1 and 2, becoming depressed and more or less fusiform on 5 to 7; dull yellowish brown, with wavy pale lines on sides of 4 to 9. Segment 10 blackish, marked at sides with yellow.

Distribution and dates.—ANTILLES: Jamaica; probably elsewhere in the Antilles; also south to Argentina.

December 5 to April 13.

Perithemis seminole Calvert

1907. Calv., B. C. A., p. 316.
1930. Byers, Odon. of Fla., p. 130 (fig., nymph).
1940. Ris, Misc. Pub. Mus. Zool. Univ. Mich., 21:17 (figs.).

Length 23–25 mm.; abdomen 13–15; hind wing 16–18.

Perhaps the handsomest member of the genus. Labrum and front of frons dull ochraceous to orange, with paler tints on clypeal area intervening. Top of vertex greenish gray. Front of thorax rusty brown; sides the same, with two oblique greenish yellow bands that may become clouded in their upper half in old specimens. Wings yellow, with venation somewhat lighter and stigma darker. Legs yellowish, with black spines.

Abdomen somewhat widened on middle segments, dull reddish brown, with sharply defined black carinae.

On Trout Creek, a sandy-bedded stream near Olga, Florida, the senior author found nymphs in transformation. They were clinging to the stems of spider lilies that grew at the brink of the slowly flowing water, only inches above its surface.

Distribution and dates.—UNITED STATES: Ala., Fla., Ga., N. C. (?).

March 4 (Fla.) to December 13 (Fla.).

Perithemis tenera Say

Syn.: chlora Rambur, tenuicincta Say

1839. Say, J. Acad. Phila., 8:31 (in *Libellula*).
1901. Ndm., Bull. N. Y. State Mus., 47:512 (figs., nymph).

1927. Garm., Odon. of Conn., p. 258.
1930. Ris, Misc. Pub. Mus. Zool. Univ. Mich., 21:14 (figs.).
1937. Mtgm., Ent. News, 48:61 (oviposition).

Length 20–24 mm.; abdomen 12–13; hind wing 17–19.

The one wide-ranging species of the eastern United States. A small brownish species, with amber-tinted and brown-spotted wings. Face yellowish, darker above, greenish across clypeus, brownish at sides and on top of frons. Labrum dull light yellow. Vertex and occiput brown. Thorax brown, densely clothed with soft brown hairs beneath which appear obscure stripes of yellow: pair on front parallel with carina, and wide apart; the two on the sides interrupted in middle portion, lower ends appearing as roundish spots of considerable size. Legs yellowish, with black spines. Wings tinted all over with amber yellow; stigma reddish. Wing pattern as shown in figure 276.

Abdomen short, stout, a little compressed at base, depressed toward middle and tapered toward end. Obscure brownish in color, paler laterally and beneath; appendages yellowish superiorly.

Flies low over surface of water, resting frequently on tops of low stems and twigs near shore. Perches horizontally with fore and hind wings often unequally lifted. Flight rather weak and vacillating. When over water, habitually avoids altitude of larger and stronger species, keeping down nearer surface. Very sensitive to cloudiness and moisture; seldom seen in flight except when sun is shining. Female sometimes held by male while ovipositing, but oftener is unattended, dropping her eggs on bits of floating dead pond scum by many successive dips made at nearly same spot. When a female, while ovipositing, is taken in hand and dipped to surface of water in a tumbler in imitation of her own act, she liberates ten to twenty eggs at each descent.

Nymph stands up well on its slender legs and crawls rather actively, and is clean and smooth of body surface. Seems to prefer stumps as a place of transformation. A lone stump in water's edge may be fairly sprinkled with cast skins, while there are none on emergent weeds round about.

Distribution and dates.—CANADA: Ont.; UNITED STATES: Ala., Conn., D. C., Ga., Ill., Ind., Iowa, Kans., Ky., La., Md., Mass., Mich., Miss., Mo., Nebr., N. J., N. Y., N. C., Ohio, Okla., Pa., R. I., S. C., Tenn., Tex., Va., Wis.; MEXICO: Coahuila.

April (Tex.) to October 7 (S. C.).

Genus PLANIPLAX Muttkowski 1910

Syn.: Platyplax Karsch

This is a small Neotropical genus of which one Mexican species falls within our limits. The frons and vertex are low and rounded. The eye-seam is short; the occiput is hairy. The prothorax is very broadly bilobed and bears a very long fringe of strongly radiating hairs.

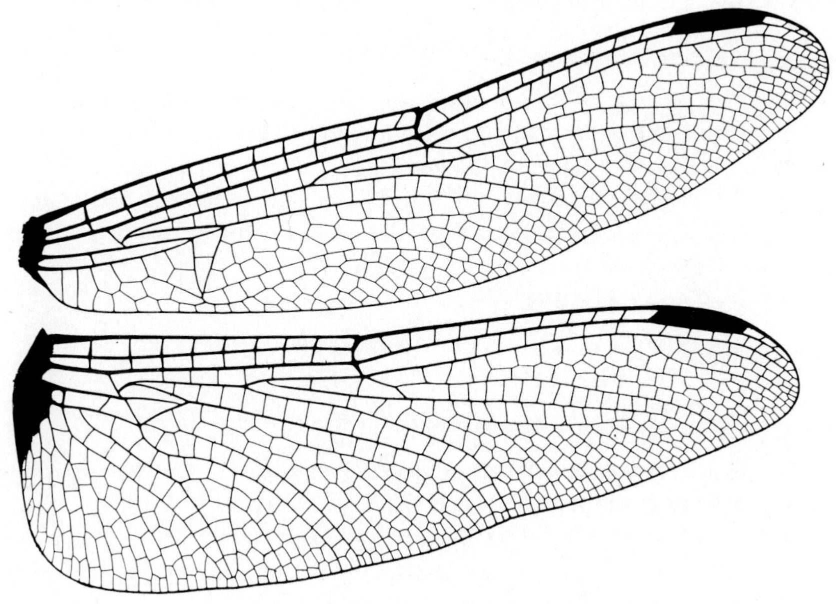

Fig. 278. *Planiplax erythropyga.*

The venation of the wings is as shown in our table of genera (p. 427), to which may be added that there are only two paranal cells in the hind wing outside the anal loop.

The regularly tapering abdomen is rather stout, somewhat depressed, and sharply triquetral, having sharp carinae. The caudal appendages are a little longer than segments 9 plus 10.

The nymph is unknown.

Planiplax sanguiniventris Calvert

1907. Calv., B. C. A., p. 327 (figs.) (in *Platyplax*).

Length 37 mm.; abdomen 21–26; hind wing 28–34.

A striking species by reason of strongly contrasting colors of abdomen: pruinose blue on basal segments, bright red beyond. Face blackish. Top of frons and vertex become metallic violet in old males. Occiput brown, low, and hairy.

Thorax rich brown under a heavy coat of blackish hair, becoming bluish pruinose in old males in front and on sides and back. Legs black, long, and spiny. Wings hyaline, with stigma yellowish or reddish. Membranule blackish, paler at extreme base. Touches of brown at extreme base of fore wings, in subcostal and cubital interspaces; in hind wings a large red-brown spot covers base outward halfway to first antenodal crossvein, to anal crossing, to triangle, and thence around beyond lower end of membranule.

Abdomen nearly uniform red-brown, becoming pruinose blue on back of segments 1 and 2; 3 to 10 and appendages bright red. In this last character it bears a superficial resemblance to *Erythemis peruviana,* which has pruinosity on basal segments but is in other characters very different.

Distribution and dates.—MEXICO: Baja Calif., Tamaulipas; also south to Guatemala.

June 23–27 (Tamaulipas).

Genus IDIATAPHE Cowley 1934

Syn.: Ephidatia Kirby

This is a Tropical American genus of two species, one Brazilian and the other West Indian; the latter is within our limits.

The nymph is recognized by the characters shown on page 431. It lives among the roots of emergent aquatic plants in the shoals of slightly brackish water.

Idiataphe cubensis Scudder

Syn.: amazonica Kirby, specularis Hagen

1866. Scudder, Proc. Boston Soc. Nat. Hist., 10:190 (in *Macromia*).
1936. Ndm. & Fisher, Trans. Amer. Ent. Soc., 62:108 (nymph).
1938. Garcia, J. Agr. Univ. Puerto Rico, 22:60 (fig.).

Length 36–41 mm.; abdomen 24–27; hind wing 28–30.

Frons low, with a wide furrow. Eye-seam long. Thorax rather frail and thinly hairy. Legs very slender, with long tibial spines. Network of wings rather wide-meshed, with third paranal cell especially wide on both fore and hind wings. No apical planate, but radial and median planates well developed, radial generally circumscribing and enclosing seven cells. Only two paranal cells within anal loop; midrib very slightly angulated.

Thorax blackish in front, clad with tawny hairs. Sides brownish olive, with wide blackish stripes on first and third lateral sutures.

Slender abdomen dark brown, becoming wholly blackish with age in male; female retains paler colors on enlarged basal segments and on basal half of segments 4 to 7, sometimes to 9.

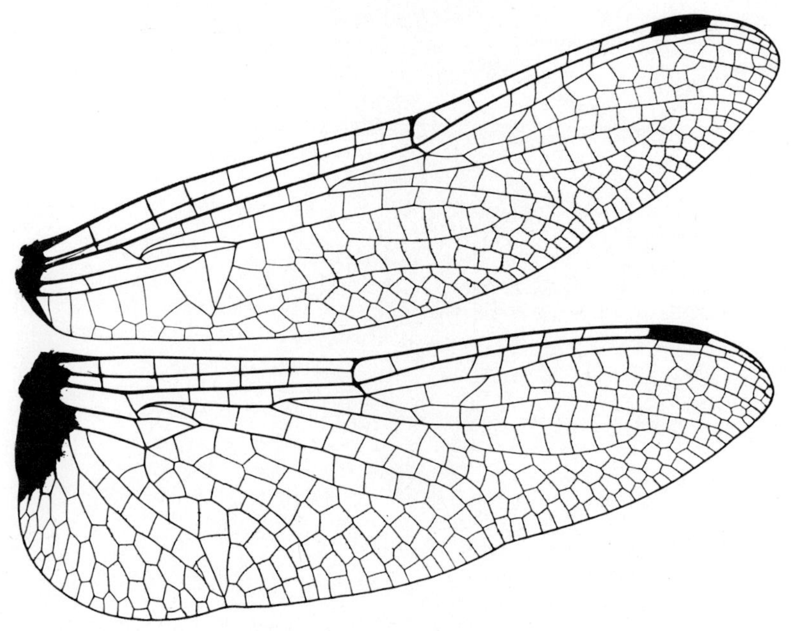

Fig. 279. *Idiataphe cubensis.*

Distribution and dates.—MEXICO: Tamaulipas; ANTILLES: Cuba, Jamaica, P. R.; also south to Colombia and Peru.

Apparently year-round; dates include December to March (Jamaica), and October (Cuba); Garcia reported exuviae in August in Puerto Rico.

Genus CELITHEMIS Hagen 1861

This genus of beautiful dragonflies is found only in eastern North America. The sexes are alike in the rich color pattern of their wings. The face is generally of some tint of yellow. The top of the head is crossed by a band of black that includes the middle ocellus. The eye-seam is rather long.

The thorax is not robust. The erect hind margin of the prothorax is broadly bilobed and fringed conspicuously with long pale hairs. The stripings of the thorax are of black on a pale ground, but by the darken-

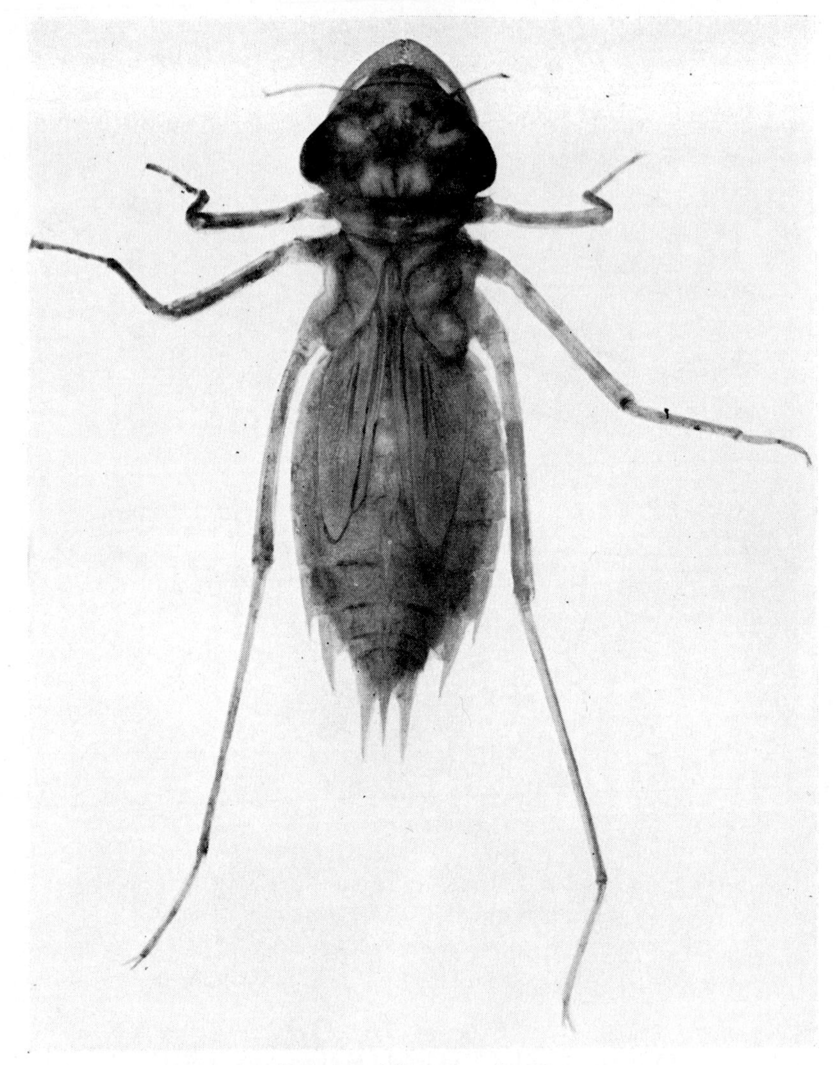

Fig. 280. *Idiataphe cubensis.*

ing of the ground color they become obscured with age. On the front of the synthorax a middorsal band of black is constantly present, variable in breadth and in connections at its ends. Antehumeral and humeral stripes are generally blended into one broad shoulder band. The lateral stripes generally roughly follow the sutures, but the midlateral one is usually abbreviated above. The slender legs are black, with a portion of the under side of the front femur pale.

The wings have a rather open venation. The base of the hind ones is broad, sustained by a rather large anal loop that has a nearly straight midrib and very prominent heel. The arculus is nearer to the first than to the second antenodal crossvein. There is generally but one bridge crossvein. The number of cells in the triangles is generally two in the fore wing and one in the hind, but the larger species often have more crossveins. There are five or six antenodal crossveins in the hind wing. There

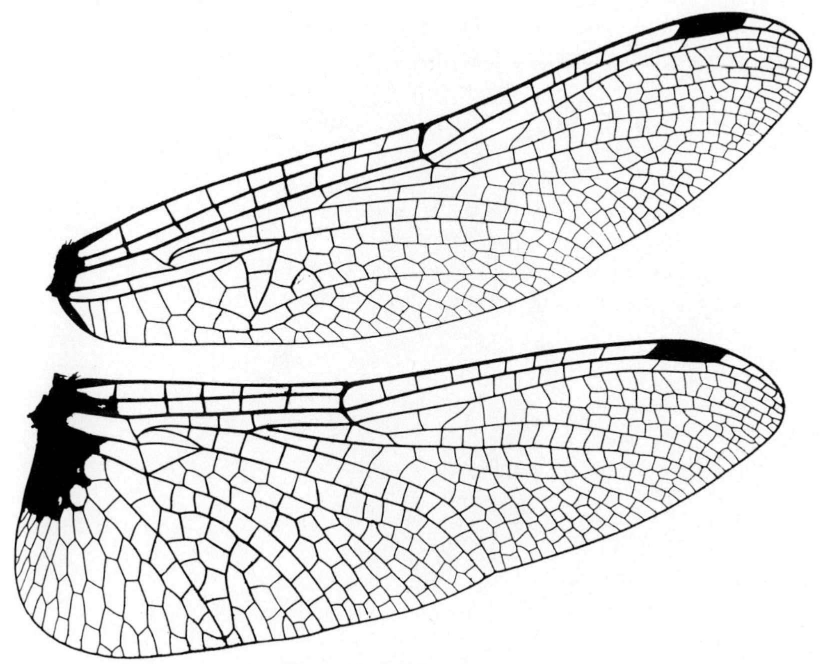

Fig. 281. *Celithemis verna.*

are three paranal cells before the anal loop in the hind wing. The radial planate is much better developed than the median in all wings.

The slender abdomen is a little compressed on the paler basal segments. The dorsal pale color is progressively reduced to a line of middorsal spots on segments 4 to 7; 8 to 10 are black.

The nymphs of this genus, so far as known, live in the weed beds of ponds and shoals. They clamber about over submerged vegetation or hide among the roots of creeping stems. They climb well but swim poorly. They are cleaner and less sprawling than most Libellulines. Like the adults, they are of a rather delicate build. At transformation they go no farther from the water than is necessary to find a suitable place—generally but a few inches above the surface of the water.

They are greenish, neatly patterned in pale brown and tan. They are distinguished by having rather short and spinelike dorsal hooks on segments 3 to 7, highest on 6. Lateral spines on 8 and 9 are slender, straight, and long, especially those on 9, the tips of which may attain the level of the tips of the long inferior caudal appendages.

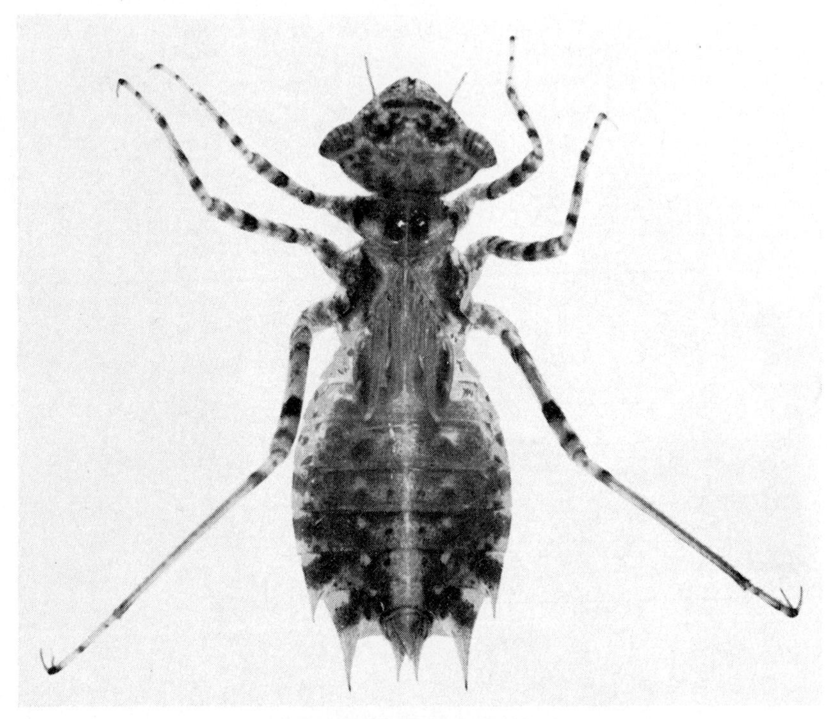

Fig. 282. *Celithemis fasciata.*

The number of raptorial setae on the labium is not constant for the species (seven to nine laterals, and ten to thirteen mentals); the dorsal hooks on the abdomen are much alike in all; therefore the species, so far as known, are best distinguished by the shape of the hind angles of the head.

The paper of first importance for the study of adults of this genus is the monograph by Williamson (1922), often cited below under species.

KEY TO THE SPECIES

ADULTS

1—Wings with large dark spots at or near nodus 2
—Wings with no large dark spots at or near nodus 5

2—Wing tips with no dark markings beyond stigma (sometimes smoky there); wing membrane yellow to orange.................................**eponina**
—Wing tips with dark edgings beyond stigma.................................. 3

3—Wings with roundish spot just beyond nodus, which generally does not touch costa ....**elisa**
—Wings with crossband just beyond nodus, which reaches costa and extends well to rearward 4

4—Fore wing with dark area before nodus extending to rearward across vein Cu1; anal loop generally with two or more ankle cells; radial planate generally subtends two rows of cells.................................**fasciata**
—Fore wing with dark area before nodus extending to rearward no farther than radial planate; anal loop generally with single ankle cell; radial planate generally subtends single row of cells.................**monomelaena**

5—Radial planate generally subtends two rows of cells.....................**verna**
—Radial planate subtends single row of cells.................................. 6

TABLE OF SPECIES

ADULTS

Species	Hind wing	Crossveins		Cells		Distr.
		Antenod.[1]	*Stigma*[2]	*Rad.pl.*[3]	*Ankle*[4]	
amanda	23-26	5	1	1	2-3	SE
bertha	24-28	5	2	1	1	Fla.
elisa	27-28	5-6	2	2-3	3-6	E, C
eponina	32-34	6	2-3	2-3	5-9	E, C
fasciata	29-32	6	2	2	1-3	SE
leonora	23-25	5	2	1	1	SE
martha	23-28	5	1-2	1	1	NE
monomelaena	25-28	5-6	2	1-2	1-3	NE, C
ornata	23-28	5	2	1	1	SE
verna	25-28	5	1-2	2	1	SE, C

[1]Number of antenodal crossveins in hind wing.

[2]Number of crossveins under stigma.

[3]Number of cell rows subtended by radial planate.

[4]Number of cells at ankle of foot-shaped anal loop that form no part of its two marginal rows.

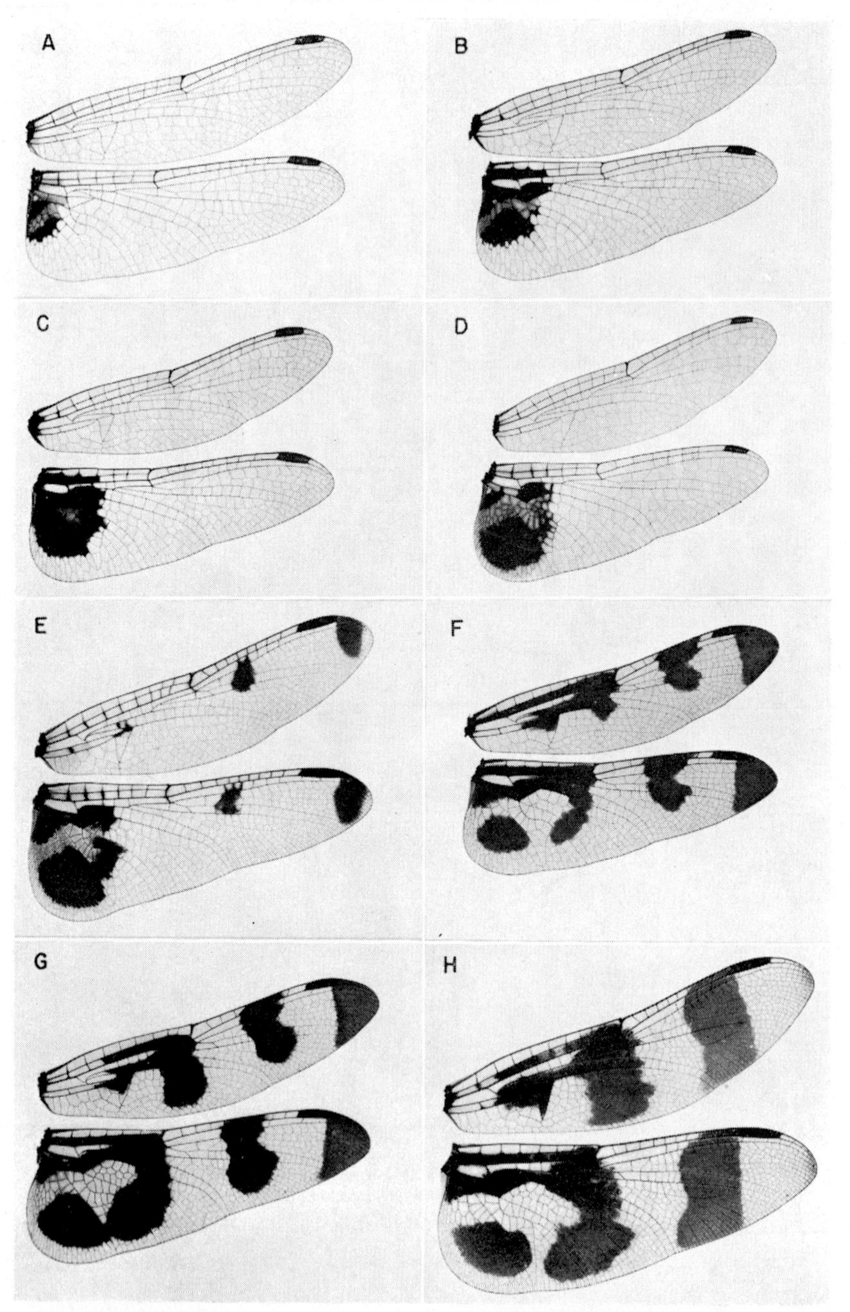

Fig. 283. A, *Celithemis bertha;* B, *C. ornata;* C, *C. martha;* D, *C. amanda;* E, *C. elisa;* F, *C. monomelaena;* G, *C. fasciata;* H, *C. eponina.*

6—Anal loop with two or three ankle cells; generally a single crossvein under stigma .... **amanda**
—Anal loop with a single ankle cell; generally two crossveins under stigma.... 7

7—Trigonal interspace of fore wings with two rows of cells at level of nodus...**martha**
—Trigonal interspace of fore wings with three rows of cells at level of nodus... 8

8—Cross-venation bright red in basal hyaline portion of wing of male; trigonal interspace of fore wings generally with one to three cells, reduced to two rows ....**bertha** and **leonora**
—Cross-venation yellowish, or brown to black, in basal hyaline portion of wing; trigonal interspace of fore wings generally with no rows of less than three cells...**ornata**

Nymphs

1—Eyes tapered postero-laterally to an acutely conic tip; lateral spines of segment 9 more than twice the middorsal length of 9, usually ending a little beyond level of tips of inferior caudal appendages (with normal degree of telescoping)... 2
—Eyes rounded laterally; lateral spines of 9 twice the middorsal length of 9, or less (except in *eponina*)... 3

2—Abdomen and legs with sharply contrasting color pattern; lateral setae usually nine, rarely ten; dorsal hook on segment 4 vestigial, on 5 much smaller than on 7..............................**fasciata** and **monomelaena**
—Abdomen light brown without definite color pattern, legs only indistinctly annulate with darker rings; lateral setae usually eight, rarely nine; dorsal hook on segment 4 high, on 5 about as large as on 7.........**ornata** and **verna**

TABLE OF SPECIES

NYMPHS

Species	*Total length*	*Eyes*[1]	*Setae of lab.*[2]		*Appendages*[3]		
			Lat.	*Ment.*	*Lat.*	*Sup.*	*Inf.*
amanda	13-15	round	7	*10-11*-12	5	7	10
bertha	15-17	round	7-*8*	11-14	4	8	10
elisa	15-17	round	7-*8*	11-13	4	7	10
eponina	19-21	round	7-9	10-13	5	8	10
fasciata	17-19	conic	9-10	12-14	5	8	10
martha?	17-18	round	7-9	12	4	8	10
monomelaena	15-17	conic	9	13	5	8	10
ornata	17-18	conic	*8*-9	12-13	4	7	10
verna	15-17	conic	8	11-12	4	7	10

[1]Contour of latero-posterior tip of eyes.

[2]Number of lateral and mental raptorial setae on labium.

[3]Relative length of lateral, superior, and inferior appendages, the last taken as 10.

3—Length of body 19 mm. or more; lateral spines of segment 9 more than twice the middorsal length of 9 . . . . **eponina**
—Length less than 19 mm.; lateral spines of 9 twice the middorsal length of 9, or less 4

4—Lateral setae seven; mental setae usually eleven; lateral appendages half as long as inferiors; dorsal hooks strongly developed, those on 6 and 7 subequal . . . . **amanda**
—Lateral setae usually eight, only rarely seven or nine; mental setae eleven to fourteen; lateral appendages about four-tenths as long as inferiors 5

5—Dorsal hooks almost obsolete on 4 and 7, subequal on 5 and 6 . . . . **?martha**
—Dorsal hooks all better developed, about as high on 4 as on 5 6

6—Superior abdominal appendage about eight-tenths as long as inferiors; inferiors shorter than lateral spines of segment 9; lateral spines of 9 noticeably divergent to their tips . . . . **bertha**
—Superior abdominal appendage about seven-tenths as long as inferiors; inferiors longer than lateral spines of 9; lateral spines of 9 project straight to rearward, or slightly incurved to tips . . . . **elisa**

Nymph unknown: **leonora.**

Celithemis amanda Hagen

1861. Hagen, Syn. Neur. N. Amer., p. 183 (in *Diplax*).
1912. Ris, Coll. Selys Libell., p. 728 (fig.).
1922. Wmsn., Occ. Pap. Mus. Zool. Univ. Mich., 108:3 (figs.).
1929. N. & H., Handb., p. 212 (fig.).

Length 27–31 mm.; abdomen 18–22; hind wing 23–26.

A trim little dragonfly, with spots of brown in a field of yellow at wing bases and with a yellow stigma. General pale color of head and thorax becomes brownish with maturity. Face yellow; labium also, but with a median black stripe that forms a narrow edging on convergent lateral lobes. Occiput brown.

Thorax brown in front, thinly clothed with pale hairs and overlaid with a broad urceolate middorsal stripe of black. Thorax becomes almost uniformly brown, its blackish side stripes evident only in deepest part of first and third lateral sutures. Wing pattern generally about as shown in figure 283.

Abdomen slender, regularly tapering rearward to end, mostly yellow on basal segments, the yellow quite overspreading segments 3 and 4, narrowing to rearward to form a middorsal line of yellow spots on 5, 6, and 7; 8 to 10 black; appendages, mainly black, become partly yellow in females or reddish in old males on upper side.

Distribution and dates.—UNITED STATES: Ala., Fla., Ga., Miss., N. C., S. C.

April 2 (Miss.) to November (Fla.).

Celithemis bertha Williamson

1922. Wmsn., Occ. Pap. Mus. Zool. Univ. Mich., 108:8 (figs.).
1929. N. & H., Handb., p. 213 (fig.).
1930. Byers, Odon. of Fla., p. 124.

Length 28–36 mm.; abdomen 18–24; hind wing 24–28.

A pretty little reddish brown species, with red-veined wings. Face yellowish, becoming rufous with age, including top of vertex. Thorax yellowish, becoming brownish with age, heavily striped with black. Wide middorsal black stripe on front of thorax spreads out laterally along collar. Unequal black stripes on the three lateral sutures, middle one incomplete above and narrowed below level of spiracle, but widened in its middle portion. Wings clear beyond brown basal markings; veins reddish, especially before nodus, and in the little yellow that lies in midst of largest basal spot of hind wing.

Abdomen blackish beyond a pallid base that becomes reddish with age. Segments 3 and 4 mainly yellowish. These colors restricted to middorsal spots on 5 to 7. Caudal appendages blackish but paler on upper side.

In shallow sand-bottomed lakes of central ridge of southern Florida this species haunts zone of thin maiden-cane a little way out from shoreline. Flies rather constantly in outermost fringe of that zone, clinging lightly to tallest leafy tips as they sway in wind, practically outside *eponina* territory and in deeper water.

Distribution and dates.—UNITED STATES: Fla., Ga.

March 31 to December 3.

Celithemis elisa Hagen

1861. Hagen, Syn. Neur. N. Amer., p. 182 (in *Diplax*).
1901. Ndm., Bull. N. Y. State Mus., 47:515 (nymph).
1912. Ris, Coll. Selys Libell., p. 725 (fig.).
1922. Wmsn., Occ. Pap. Mus. Zool. Univ. Mich., 108:3.
1927. Garm., Odon. of Conn., p. 287 (fig.).
1929. N. & H., Handb., p. 210 (fig.).

Length 29–34 mm.; abdomen 18–22; hind wing 27–28.

A slender yellowish or reddish brown species. Face yellow, clear yellow at first, becoming olive yellow with age; thinly clad with short blackish hair; vertex and occiput yellow also.

Thorax dull yellow, becoming brownish, with a rather small middorsal blackish triangle on its front. Sides more yellow than brown, with brown disposed in narrow and diffuse stripes on three lateral sutures.

Wings have a diffuse roundish spot just beyond nodus that generally does not extend forward to touch costa. Wing membrane exposed between bands is hyaline; that exposed between basal spots is tinged with yellow. Large basal spot on hind wings becomes black with age; veins that traverse, red. Radial planate subtends two rows of cells. Crossveins of wing somewhat less numerous than in *eponina,* which this species most closely resembles.

Slender abdomen mostly yellow on basal segments 1 to 4, and mostly black beyond them, with all carinae shining black. Usual interrupted line of pale saddle marks on 5 to 7, with line on 7 generally well defined; 8 to 10 and caudal appendages black, becoming reddish on dorsal side in old males.

Female similar in coloration but with more yellow, especially along lateral margins of abdomen in its basal half.

Haunts bulrush beds in standing fresh shallow water.

Distribution and dates.—CANADA: N. S., Ont., Que.; UNITED STATES: Ala., Conn., Fla., Ga., Ill., Ind., Kans., Ky., Maine, Md., Mass., Mich., Minn., Miss., Mo., N. H., N. J., N. Y., N. C., Ohio, Okla., Pa., R. I., S. C., Tenn., Tex., Va., Wis.

April 4 (Miss.) to October 7 (S. C.).

Celithemis eponina Drury

Syn.: camilla Rambur, lucilla Rambur

1773. Drury, Ill. Exot. Ins., 2:86 (pl. 47, fig. 2) (in *Libellula*).
1901. Ndm., Bull. N. Y. State Mus., 47:514 (nymph).
1912. Ris, Coll. Selys Libell., p. 724 (fig.).
1914. Whedon, Minn. State Ent. Rep., p. 100.
1920. Howe, Odon. of N. Eng., p. 149 (fig.).
1922. Wmsn., Occ. Pap. Mus. Zool. Univ. Mich., 108:3.
1929. N. & H., Handb., p. 209 (fig.).

Length 36–42 mm.; abdomen 23–28; hind wing 32–34.

A very familiar pond species, largest of genus, easily recognized by its size and by having wing membrane entirely suffused with yellow and heavily marked with brown in both sexes. Labium pale yellow; face darker, and with age may become blackish or brownish, with coppery reflections. Top of vertex yellowish; top of occiput brown.

Synthorax yellowish, striped with black: a broad middorsal triangle on front, and stripes on three lateral sutures, midlateral stripe incomplete above, other two conjoined at both ends. Sometimes side stripes are conjoined by an additional irregular black band across spiracle. Wings banded after general pattern shown in figure 283. Middle band often connected with more basal spots. Veins yellow, becoming red at full maturity. Crossveins of wings more numerous than in other species, there being often more than a single bridge crossvein and more than one cubito-anal, and often from seven to nine cells interpolated at ankle between two regular rows of cells in anal loop. True course of cubital vein around rear of subtriangle in fore wing often obscured by irregularities of cell arrangement, and triangle of that wing may be composed of three, four, or five cells. Three to five rows of cells occupy trigonal interspace. Radial planate subtends two or sometimes three rows of cells. Slender legs, black beyond their bases, except for half of under side of front femora.

Abdomen long and slender, somewhat compressed on basal segments; on each side, a broad lateral band of yellow. An interrupted middorsal line of yellow streaks on segments 3 to 7. Dark caudal appendages mostly yellow on upper side. Female colored much as in male, with more breadth of yellow and with an extra edging of it on lateral carinae of abdomen as far to rearward as 8.

Frequents borders of ponds and neighboring grassy slopes; sometimes, when foraging, is carried far from water by wind. In its flight is a flutter, suggestive of flight of a butterfly. Female in ovipositing is held by male; both are apt to be seen on windy days when other species are in shelter, dipping to crests of foaming waves, far out from shore. Transformation occurs in early morning, often on stumps about a foot above surface of water.

Whedon (1914) records these observations on adults in Minnesota:

A few were taken . . . along a bay filled with cat tails, bulrushes and sedges . . . they were present in great numbers on the gravel flats, notwithstanding that the day was very dull and a steady drizzle of cold rain falling. They were covered with glistening rain drops which were shaken from their wings as they fluttered from perch to perch.

In bright weather they were much more agile and quite difficult to capture. When in copulation they would ascend fifty or sixty feet and dart off over the lake for a time. During windy days, and it was very windy when it was bright, they seemed to delight in battling with the gale and in clinging like weathervanes to tallest weed stalks, their wings half set.

Distribution and dates.—CANADA: Ont.; UNITED STATES: Ala., Conn., Fla., Ga., Ill., Ind., Iowa, Kans., Ky., La., Md., Mass., Mich., Minn., Miss., Mo., Nebr., N. J., N. Y., N. C., Ohio, Okla., Pa., R. I., S. C., Tenn., Tex., Va., Wis.; ANTILLES: Cuba.

Year-round in Fla.

Celithemis fasciata Kirby

1889. Kirby, Trans. Zool. Soc. London, 12:326 (fig.).
1912. Ris, Coll. Selys Libell., p. 726.

1922. Wmsn., Occ. Pap. Mus. Zool. Univ. Mich., 108:3.
1929. N. & H., Handb., p. 211 (fig.).
1942. Wstf., Ent. News, 53:127.

Length 36–38 mm.; abdomen 24–26; hind wing 29–32.

A very dark and handsome Southeastern species, very clear-winged between well-defined blackish crossbands. Labrum black; opposed inner edges of lateral lobes of yellow labium margined with dark brown. Olive yellow face above labrum early becomes smudged with black; top of frons and vertex become shining black, with metallic violet reflections. A little streak of fawn color on outer posterior margin of eye.

Thorax, though brightly striped with brown on a yellow ground in teneral specimens, becomes wholly black in maturity, or dimly shows a narrow antehumeral T-shaped stripe of yellow before broad humeral band of black. Behind humeral black band is more yellow with two well-marked stripes: midlateral one tapering upward, third lateral downward; from upper end of latter a branch runs to rearward to base of abdomen.

Wings of well-defined pattern. A tinge of yellow in membrane only within clearings in midst of basal brown spots. Veins, at first yellow, become blackish with age; stigma brown.

Abdomen mostly shining black; even basal segments, at first greenish yellow, at full maturity become more black than yellow. There remain, however, middorsal saddle marks on segments 5 to 7; 8 to 10 and caudal appendages black. An interrupted midlateral band of pale spots on 2 to 5, rapidly dwindling in size of spots to rearward; below them along lateral carina is usual edging of paler color.

Distribution and dates.—UNITED STATES: Ala., Fla., Ga., Ind., Ky., La., Miss., N. C., Okla., S. C., Tex.

May 5 (Fla.) to October 1 (Ga.).

Celithemis leonora Westfall

1952. Wstf., Fla. Ent., 35:109 (figs.).

Length 27–33 mm.; abdomen 17–21; hind wing 23–25.

A dainty little species, very similar structurally to *C. bertha*, of which it is here considered a variety, but in typical specimens is distinguished at once by presence of a large roundish brown spot near apex of all four wings. This spot, however, while generally large, is variable in size and may be small in one or both pairs of wings. It lies close behind outer corner of stigma and may often extend almost to wing margin at that level.

Distribution and dates.—UNITED STATES: North Fla., Ga., N. C.

June 7 (Fla.) to October 2 (Fla.).

Celithemis martha Williamson

1922. Wmsn., Occ. Pap. Mus. Zool. Univ. Mich., 108:5 (figs.).
1927. Garm., Odon. of Conn., p. 288.
1929. N. & H., Handb., p. 212 (fig.).

Length 25–33 mm.; abdomen 16–20; hind wing 23–28.

A blackish Northeastern species, closely allied to *ornata*. Face dull yellow, early becoming blackish. Top of frons and vertex become shining black, with metallic reflections. Thorax densely clothed with tawny hairs. Front almost covered by a dark middorsal band, so broad and so conjoined at its ends with black humeral area that only a dim roundish pale spot remains on each side of front. Side stripes very obscure, and are early lost, the whole then becoming dull black. Wings hyaline beyond basal spots, with blackish veins and tawny stigma. Pattern of brown basal markings generally somewhat as shown in figure 283.

Abdomen mainly blackish, with considerable pale areas on basal segments and usual middorsal line of spots on middle segments, similar to those of *amanda* but smaller. All beyond segment 7 black.

Species at once distinguishable from *ornata* by its darker color and by having but two rows of cells in trigonal interspace where that species has three rows.

Distribution and dates.—CANADA: N. S.; UNITED STATES: Maine, Md., Mass., N. H., N. J., N. Y., Pa.

June 26 (Maine) to September 7 (N. J.).

Celithemis monomelaena Williamson

1910. Wmsn., Ohio Natural., 10:155 (figs.).
1912. Ris, Coll. Selys Libell., p. 726 (fig.).
1920. Howe, Odon. of N. Eng., p. 87 (fig.).
1922. Wmsn., Occ. Pap. Mus. Zool. Univ. Mich., 108:3.
1929. N. & H., Handb., p. 211 (fig.).
1934. Leonard, Occ. Pap. Mus. Zool. Univ. Mich., 297:1 (fig., nymph).

Length 30–36 mm.; abdomen 19–26; hind wing 25–28.

Another slender Northeastern species, similar to *fasciata,* having a black labrum and above it an olive brown face. Front of thorax varies from fawn color to black. Sides brown and yellow, about half and half; crossed obliquely behind broad humeral band by narrower second and third lateral stripes; these two become confluent upward, but are very variable in extent and connections. Wings patterned as in *fasciata,* with bands of brown narrower, with intervening membrane clear. A clear area within largest basal spot lightly tinged with yellow.

Abdomen similar to that of *fasciata,* with middorsal reddish streaks of segments 4 to 7 diminishing in size to rearward and somewhat better defined than in that species.

Distribution and dates.—CANADA: Ont.; UNITED STATES: Conn., Ind., Mass., Mich., Mo., N. J., N. Y., Ohio, Okla., Wis.

June 18 (Ohio) to September 7 (N. J.).

Celithemis ornata Rambur

Syn.: pulchella Burmeister

1842. Rbr., Ins. Neur., p. 96 (in *Libellula*).
1912. Ris, Coll. Selys Libell., p. 727 (fig.).
1922. Wmsn., Occ. Pap. Mus. Zool. Univ. Mich., 108:7 (figs.).
1924. Root, Ent. News, 35:321.
1929. N. & H., Handb., p. 212 (fig.).
1930. Byers, Odon. of Fla., p. 127.

Length 33–35 mm.; abdomen 23–25; hind wing 23–28.

A dainty little yellowish dragonfly, clear-winged except for some brown streaks at wing base. Face clear yellow on frons and labrum, with intervening clypeal area darker and tinged with olive. Labium yellow; broad longitudinal median band of black covers inner edges of both lateral lobes. Occiput yellow.

Thorax yellow, striped with brown. A very broad middorsal stripe covers most of front; it is almost interrupted by sharp lateral notches just beneath crest, and is confluent below at collar with black stripes of sides. Usual three side stripes well developed; humeral very broad, midlateral expanded to form a quadrangle of black on middle of side, narrowed again below spiracle. It is thus of unusual breadth in its middle portion. Often conjoined with complete third lateral at its upper end. Wings hyaline, with more or less of a basal tinge of yellow and with brown markings often as shown in figure 283.

Abdomen blackish, with much yellow overspreading basal segments and running out to rearward in a line of middorsal spots on segments 3 to 7; 8 to 10 and appendages black.

This frail little species flits gaily about edges of ponds from one leaf tip of arrowhead or bulrush to another, or forages widely over adjacent fields, preferably over open fields with sparse low vegetation.

Distribution and dates.—UNITED STATES: Ala., Fla., Ga., La., Miss., N. J., N. C., S. C., Va.

January 24 (Fla.) to September 27 (Fla.).

Celithemis verna Pritchard

1935. Pritchard, Occ. Pap. Mus. Zool. Univ. Mich., 319:6 (adult and nymph).
1942. Wstf., Ent. News, 53:127.

Length 32–35 mm.; abdomen 21–23; hind wing 25–28.

Another blackish species, with only basal wing markings. Face yellowish olive, with black labrum. Labium mostly yellow, its middle stripe running only part way out along opposed edges of lateral lobes. Top of frons and vertex dull brown, becoming shining black with age. Occiput brown.

Front of thorax blackish, usual middorsal stripe being expanded at both ends and conjoined there to black of humeral stripe, with only a dim circular spot left between on each side. Behind broad humeral, second and third lateral stripes are dim and irregular, more like dark clouds, and do not conform closely to sutures.

Membrane of wings hyaline or tinged with yellow only at base; no brown areas on front wings or, at most, little brown spots covering first antenodal and anal crossing. Pattern of brown in hind wing of male generally about as shown in figure 281. Radial planate subtends two rows of cells; median planate, a single less perfectly developed row.

The junior author found this species locally common near Hendersonville, N. C., but difficult to capture, for it kept well out from land near the outer fringe of shore vegetation and was very alert and active. Only mating pairs could be taken with a net. A slingshot was used to get single males.

Distribution and dates.—UNITED STATES: Ala., Fla., Ga., Ky., N. C., Okla., S. C.

April 28 (Okla.) to July 26 (Ky.).

Genus PSEUDOLEON Kirby 1889

This genus contains a single ornate Southwestern species. The head is of moderate size. The frons is low, without cross carina, but with well-developed furrow. The eye-seam is short, and vertex and occiput are correspondingly large. The eyes in life are streaked up and down with alternate stripes of brown and tan. The vertex is high and double-tipped. The occiput is low and bare.

The prothorax is bare, with no marked development of its free hind border. The synthorax is stout, roughened on the surface, and sparsely beset with short stiff hairs. The wings are beautifully patterned. Their venation is as shown in the table of genera (p. 427). They are notable for the very long stigma, and for the slant of the few crossveins in the second postnodal interspace before it that simulate brace veins. The anal loop is very broad, with a spur vein from the ankle supporting a very large heel and with several interpolated ankle cells on both sides of the midrib. The cells in and about the triangles are inconstant in number and often unsymmetrical in right and left wings. The triangle of the hind wing is notably concave on the outer side. Vein M2 is strongly

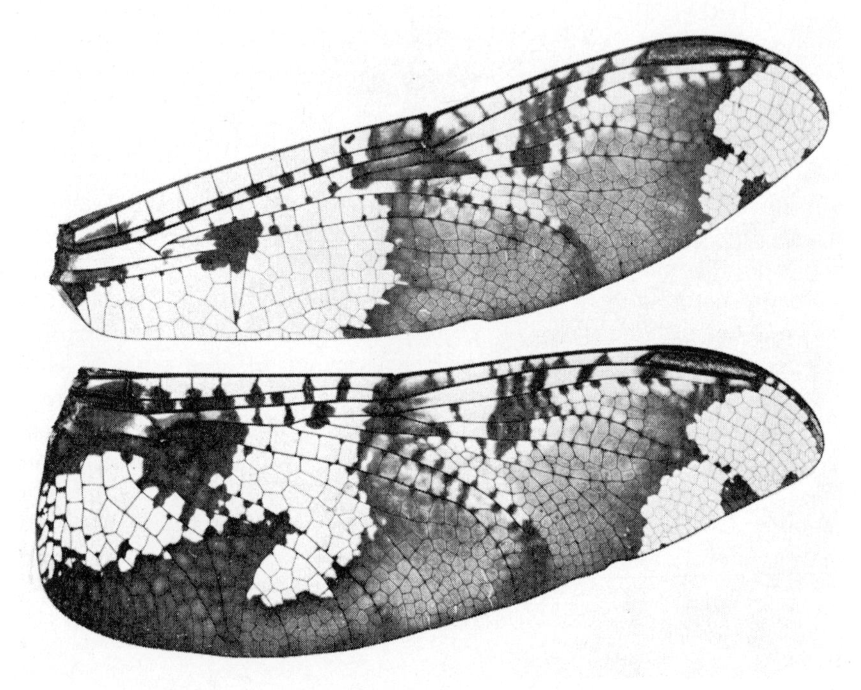

Fig. 284. *Pseudoleon superbus.*

undulate. The cells of the trigonal interspace are disposed in four more or less irregular rows.

The abdomen is moderately stout, tapered regularly from front to rear; triquetral, with dorsal and lateral carinae strongly developed and the sides between them flattened. The closely parallel ventral carinae also are strongly developed, and on each segment are produced to rearward in a spinulose triangular point. Segments 2 and 3 have an extra transverse submedian carina. The caudal appendages of the male are stout, the inferior one being very broad and cordate in outline as viewed from beneath. The genitalia of segment 2 are moderately prominent, the inner branch of the hamule being lifted high, the low-lying outer branch about four times as broad as the inner, and lying close against the back-

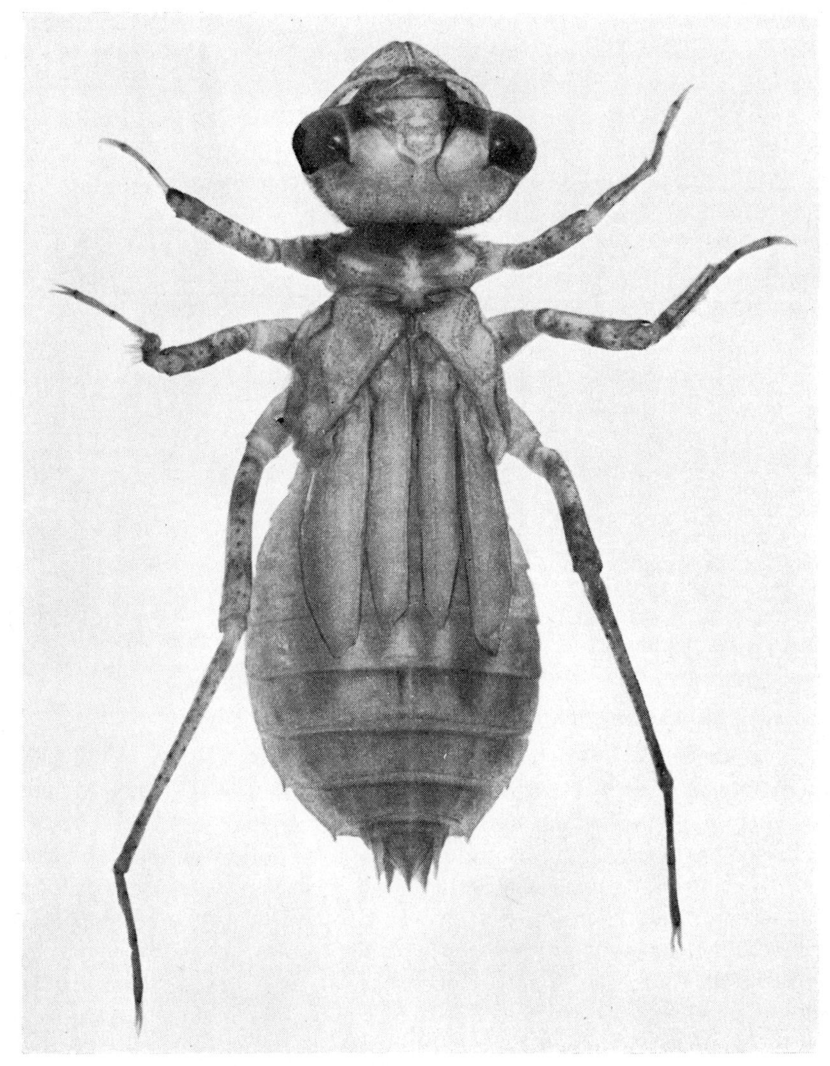

Fig. 285. *Pseudoleon superbus.*

wardly directed and hairy genital lobe. The third joint of the penis is very long and club-shaped; the fourth joint, small and almost included in the obliquely truncated tip of the third. The peduncle of the penis is somewhat hoodlike. It has a hollow front into which the tip of the penis is folded between elevated and hairy lateral lobes on each side, recalling conditions in *Gomphus*.

The nymph is easily recognized by the labium, which has its thin lateral lobes besprinkled with brown dots externally. Lateral setae are eight; mental setae, about eleven. Opposite the base of the mentals on the outer edge of the mentum is a line of smaller, less regular setae, and beyond these at the outermost angle is a crossrow of three or four additional; the edge of the median lobe is smooth and bare. The teeth of the lateral lobe are obsolete, merely indicated by about fifteen single spinules.

The abdomen is widest in the middle, slowly narrowed on segments 8 and 9, abruptly on 10, with no dorsal hooks at all. Lateral spines on 8 and 9 are similar, sharp, straight, on 9 half as long as the middorsal length of 9, on 8 a little shorter. Segment 10 is very short, almost included in the apex of 9. Appendages are sharply pointed, superior and inferiors about equal in length, equilateral triangular; laterals are more slender and a trifle shorter.

Pseudoleon superbus Hagen

1861. Hagen, Syn. Neur. N. Amer., p. 148 (in *Celithemis*).
1889. Kirby, Proc. Zool. Soc. London, 12:274 (fig.).
1895. Calv., Proc. Calif. Acad. Sci., (2)4:518 (figs.).
1906. Calv., B. C. A., p. 214.
1911. Ris, Coll. Selys Libell., p. 528.
1929. N. & H., Handb., p. 213.
1937. Ndm., J. Ent. Zool. Claremont, Calif., 29:107 (figs., nymph).

Length 38–45 mm.; abdomen 24–28; hind wing 30–35.

Adult elegantly decorated on both body and wings. Face yellowish in teneral specimens, but becomes heavily smudged with black in more mature specimens. Occiput yellow.

Thorax has a thin line of long erect hairs on collar, behind which, on front, matt surface is thinly besprinkled with short stiff brown hairs. Ground color yellow, marbled at first in a somewhat vermiculate pattern of fine brown lines; later the brown predominates; at last, in old specimens, whole thorax becomes black. Legs, at first light brown, also become blackish, except for outer face of tibiae, which continue pale.

Wings as shown in our figure, but very variable in extent of their dark coloration. A wide blackish band across both wings between nodus and stigma, which may or may not be connected with another large spot of black nearer base of hind wing; smaller spots on triangles that may be connected with latter large spot; also, little flecks of brown on antenodal crossveins of second series, and others scattered along whole margin of hind wing from tip to base. Spread of these markings varies with individual, and blackness deepens with age. Often take on beautiful purplish brown tints. Membrane between colored areas very clear and brilliantly iridescent.

Abdomen unique in pattern: a series of yellow V marks with wavy margins on dorsum of segments 3 to 7. V's open to rearward and spread from margin to margin on each segment. Twin streaks of yellow border the black middorsal carina in midst of V's, and continue farther back on 4 to 9. At other end of abdomen a wash of yellow overspreads sides of 3, and yellow is continued in small streaks on 4 and 5. In old males nearly entire abdomen becomes shining black.

The following observations on habits of this species were made by Mr. Robert Flock in Sabina Canyon, Arizona:

> A female was observed fluttering about some grass and willow roots that were floating in the water. She would flutter along a few inches above the roots and then would drop down and thrust the end of the abdomen into the water. The refuse and algae which had collected at the exit of the pond seemed to be a favorite place for egg-laying. Sometimes eggs were laid in the shallow parts of the pond where there was no debris.
>
> While laying eggs the female was often pursued by the male. Then she would suddenly cease her fluttering flight and dart swiftly away. Occasionally the male would catch her and they would fly attached for a short time. Much time was spent in wild pursuit flights: flights of male and female; of male and another male; or of a female followed by two or more males: long flights with swooping, wheeling and all kinds of maneuvers over the pond or more often over the adjacent countryside.
>
> These dragonflies are rather hard to catch, as they do not follow a regular beat. They usually sit on the rocks with the abdomen up and the wings down all ready for flight.

Mr. Flock found adults of this species "in fair abundance" in the lower part of Sabina Canyon (alt. about 3,000 ft.) on July 11, but no nymphs. No more adults were seen there after August 8. Farther up the canyon (alt. about 4,000 ft.) he found both grown nymphs and adults on September 25 in a cooler part of the canyon. Adults were still flying there on October 3.

Distribution and dates.—UNITED STATES: Ariz., Calif., Tex.; MEXICO: Baja Calif., Nuevo León, Tamaulipas; also south to Guatemala (Mtk.).

January (Tamaulipas) to October 3 (Ariz.).

Genus MACRODIPLAX Brauer 1868

This is a Tropical genus of two species, one Old World, the other American; the latter within our southern limits. The furrow on the frons is deep; the eye-seam is long; the vertex is broad. The thorax is robust, though small. The hind wings are broad.

The genus is readily recognizable by this combination of venational characters: in the fore wing the paranal cells are very large, with the marginals alongside them very small; the wide triangle points to rearward, and is followed by two rows of cells. There is no apical planate; the radial planate circumscribes and encloses five cells in both fore and hind wing.

The abdomen is short. It regularly diminishes in girth beyond the middle of segment 3, being tapered to the end.

The nymphal characters are as shown in our table of Libelluline nymphs (p. 432).

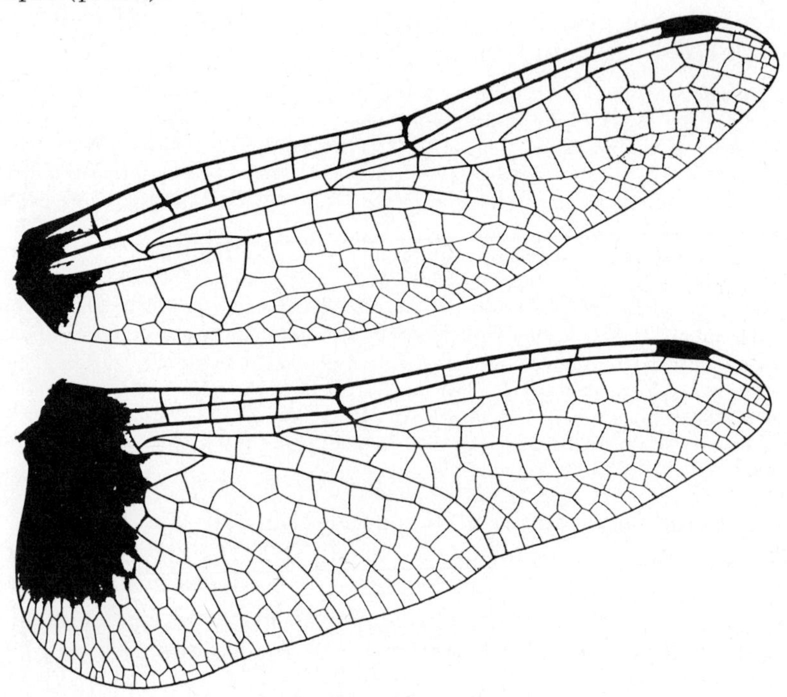

Fig. 286. *Macrodiplax balteata.*

Macrodiplax balteata Hagen

1861. Hagen, Syn. Neur. N. Amer., p. 140 (in *Tetragoneuria*).
1913. Ris, Coll. Selys Libell., p. 1038 (fig.).
1929. N. & H., Handb., p. 252.
1936. Ndm. & Fisher, Trans. Amer. Ent. Soc., 62:109 (fig., nymph).
1945. Ndm., Bull. Brooklyn Ent. Soc., 40:109.

Length 37–42 mm.; abdomen 27–29; hind wing 32–35.

A fine little Southern species, with a band of brown across bases of both fore and hind wings. Face olive brown, with labrum darker. Black crossband between yellow-topped frons and vertex quite wide. Occiput olivaceous, clothed behind with tawny hairs.

Front of thorax reddish golden brown, clothed with long gray-brown hairs. Sides more olivaceous, with irregular blackish stripes on three lateral sutures, all broadly connected at their lower ends above leg bases.

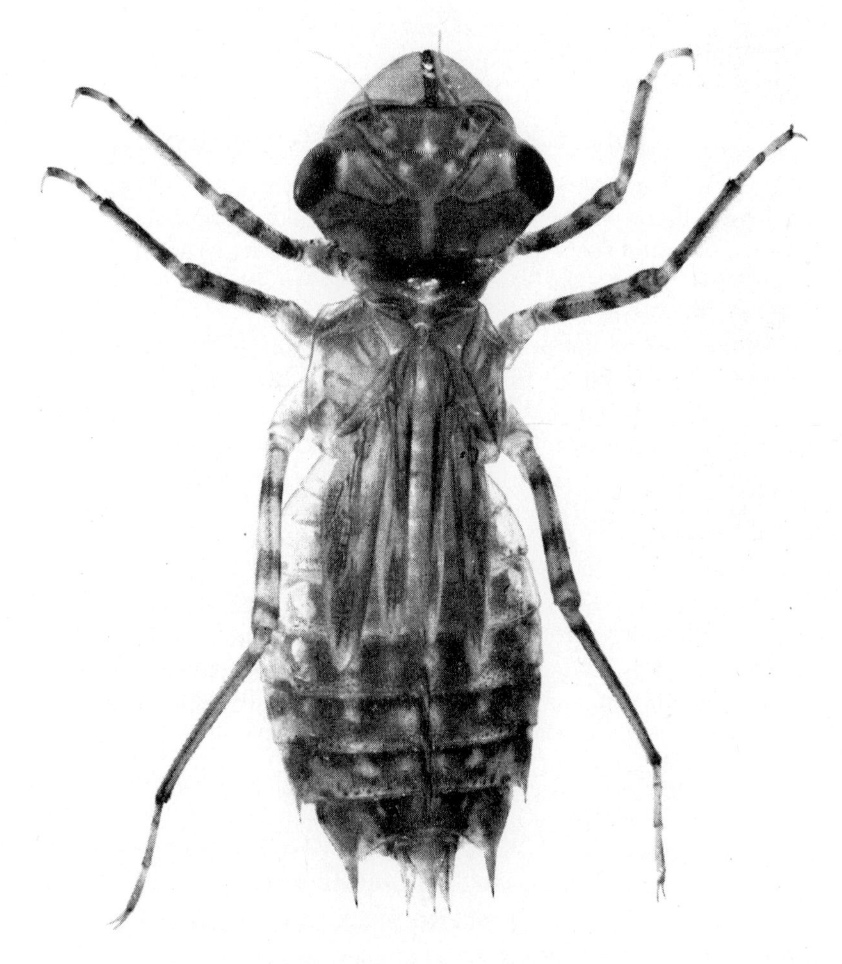

Fig. 287. *Macrodiplax balteata.*

Legs yellowish at base, black beyond. Wings pointed at tip and very broad basally. Basal brown band extends outward about as far as anal crossing in fore wing, out to triangle in hind wing, where it is then rounded to rearward. Stigma trapezoidal in shape, with its front side longer. No extra ankle cells in anal loop; its midrib is nearly straight.

Abdomen regularly tapers from apex of segment 2 to end. Becomes almost wholly blackish beyond paler basal segments.

Nymphs of this species to be found in slightly brackish permanent pools along southern coast of Florida.

Distribution and dates.—UNITED STATES: Fla., La., Miss., N. C., Tex.; ANTILLES: Cuba, Haiti, Jamaica.

Year-round.

Genus CANNAPHILA Kirby 1889

This is a genus of several Neotropical species, two of which occur within our limits. They are middle-of-the-road dragonflies, relatively unspecialized, and seemingly lacking in any single distinctive character. The head is broad. The vertex in front is divided by a wide median notch. The eye-seam is short.

The hind lobe of the prothorax is low and simple. The legs are short and spiny. The tooth on the under side of the tarsal claws is very remote from the tip. The hind wings have little widening at the base. The stigma is large. The arculus is at or beyond the second antenodal crossvein, and there is hardly any clearance of crossveins in the usual vacant spaces. Vein M2 is very slightly undulate. Veins M3 and M4 are smoothly curved and nearly parallel toward their tips. The anal loop is simple, with small development of the toe, and in the space between loop and hind-wing border is a single row of about four large hexagonal patellar cells. The hind wings are but little broader than the fore wings.

The abdomen is moderately swollen at base and parallel-sided thereafter. The caudal appendages of the male are hardly as long as segments 9 and 10. Segment 8 of the female has roundly expanded lateral margins.

The nymph is unknown.

Cannaphila funerea Carpenter

Syn.: angustipennis Rambur

1897. Carpenter, Sci. Proc. Roy. Dublin Soc., 8:434 (figs.) (in *Misagria*).
1910. Ris, Coll. Selys Libell., p. 295.

Length 37 mm.; abdomen 25–26; hind wing 30–31.

A narrow-winged species of moderate size, generally brown. Face pallid, bright yellow about mouth; tops of frons and vertex become metallic green in old males.

Thorax chocolate brown in front, with a yellow carina and edgings of black on crest. On sides are some narrow streaks of yellow in teneral specimens, generally darkening at maturity. Legs brown, becoming blackish. Wings hyaline, with brown veins and stigma. Venational characters as shown in our table (p. 426), to which may be added: intermedian crossveins 5/4; no half-antenodal before nodus; patella at inner base of anal loop very large, sometimes consisting of two large cells.

Abdomen cinnamon brown, with all carinae black; some streaks of yellow dorsally on basal segments in male; wider dull yellow areas on

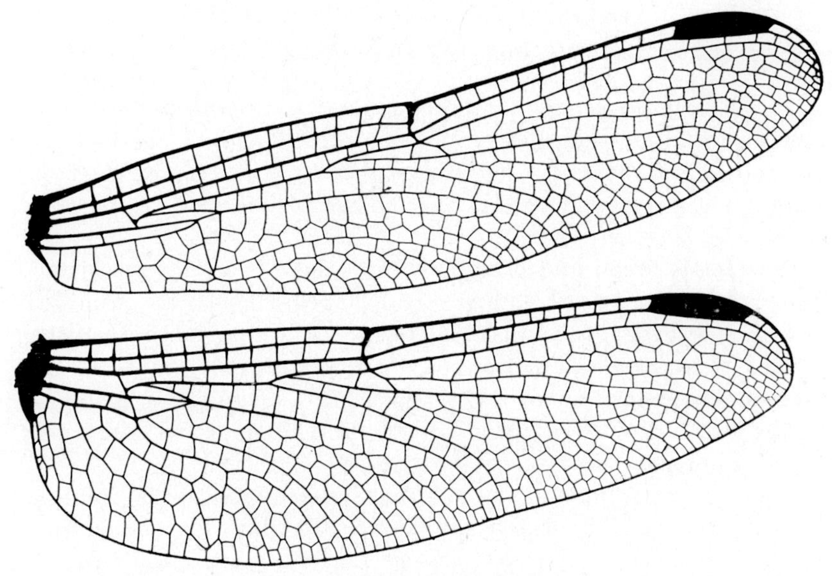

Fig. 288. *Cannaphila insularis.*

3 to 10 in female. Leaflike expansions at sides of 8 brown concolorous with sides.

Distribution and dates.—UNITED STATES: Tex.; ANTILLES: Cuba; also from Honduras to Brazil.

May 15 (Cuba) to June 28 (Isle of Pines).

Cannaphila insularis Kirby

1889. Kirby, Trans. Zool. Soc. London, 12:306.
1906. Calv., B. C. A., p. 239.
1910. Ris, Coll. Selys Libell., p. 295.

Length 42–44 mm.; abdomen 29–30; hind wing 30–31.

This and *funerea* are probably varieties of one species, and are so treated by Ris in his monograph (p. 294). He distinguishes the two.

1—Labium wholly yellow (exceptionally with a small spot of black) .... *funerea*
—Labium with a broad quadrangular spot of black covering its middle lobe and adjacent edges of the lateral lobes; hind wing a little broader, as is also the end of its anal loop .... *insularis*

Distribution and dates.—ANTILLES: Dom. Rep., Haiti, Jamaica. March and April (Haiti).

Genus ORTHEMIS Hagen 1861

Syn.: Neocysta Kirby

This is a Neotropical genus of about a dozen species, only one of which comes within our range. The species are rather large, the body stout, and the coloration prevailingly brownish. Only the top of the frons in mature males becomes metallic. The abdomen of the male (in our species) is red.

The face is broad and prominent, and there is a long eye-seam. The top of the frons is more or less shield-shaped, rimmed all around front and sides with a low carina, and the sides of the rim are convergent toward the base. Within the rim the top surface is bare and slopes toward the center, where there is a narrow longitudinal groove. The vertex also is divided longitudinally by a forking groove into a pair of rounded humps.

The legs are strong and spiny. The wings are hyaline. The venation is as stated in the table of genera (p. 426), to which may be added: the apical sector (M1a) is short, arising far out toward the end of the stigma; just before its base in the radial interspace is a close-set series of double-length crossveins, elongated crosswise to the axis of the wing; the median planate is weakly developed, discernible but zigzagged; the gaff of the anal loop is longer than the sole.

The abdomen is long and tapering (in our species) to the end, triquetral in cross section from the middle onward, and with flat sides. The genitalia of abdominal segment 2 are moderately prominent. In the female the sides of 8 are dilated at the ends into large and conspicuous semicircular flaps.

The nymph is rather elongate, with subcylindric and slowly tapered abdomen. Coarse scurfy hairs surround the vestigial dorsal hooks on the middle abdominal segments. The eyes broadly overspread the sides of the face, and rise in an elevated rounded prominence above the general base of the head.

Other characters are as shown in our table (p. 433).

Orthemis ferruginea Fabricius

Syn.: discolor Burmeister, macrostigma Rambur

1775. Fabr., Syst. Ent., p. 423 (in *Libellula*).
1906. Calv., B. C. A., p. 234 (fig.).
1910. Ris, Coll. Selys Libell., p. 282.
1928. Calv., Proc. Calif. Acad. Sci., (2)4:520 (figs.).
1929. N. & H., Handb., p. 216.

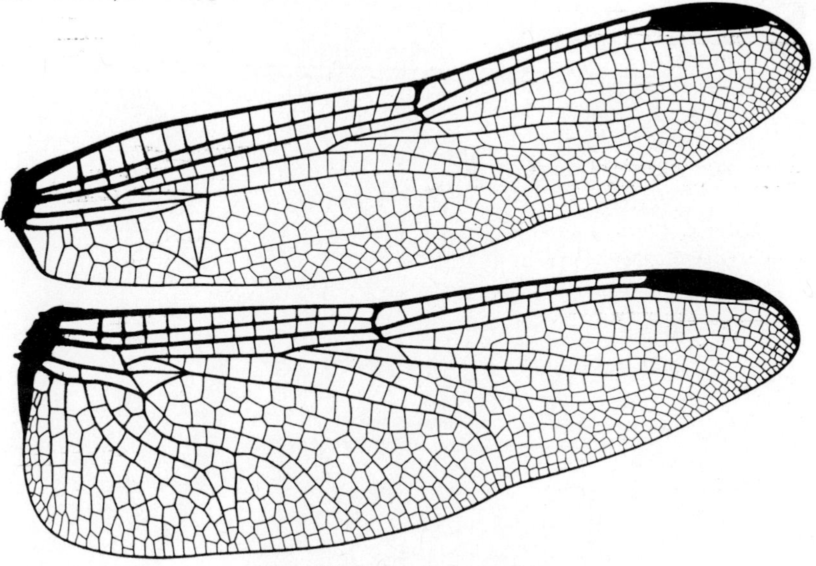

Fig. 289. *Orthemis ferruginea.*

Length 52–55 mm.; abdomen 36–39; hind wing 37–44.

A dashing red species, high-perching, strong in action, and swift at dodging. General color at first brown, with whitish streakings and edgings, and black carinae. Top of head in mature male becomes wholly metallic black, with violet reflections. Thorax and abdomen become red, with a purplish overcast. Dim stripings on thorax, but they are only irregular pale streaks left over between darker shadings that spread from depths of sutures. On collar is a long erect fringe of pallid hairs.

Abdomen nearly uniform in color, being darkened along all carinae. Female much lighter on slopes of sides of middle segments. Appendages pale.

Distribution and dates.—UNITED STATES: Ala., Ariz., Fla., Miss., N. Mex., Okla., Tex., Utah; MEXICO: Baja Calif., Chihuahua, Coahuila, Nuevo León, Sonora, Tamaulipas; ANTILLES: Cuba, Dom. Rep., Jamaica, P. R.: also south to Uruguay and Chile.

Year-round southward.

Fig. 290. *Orthemis ferruginea.*

Genus LADONA Needham 1897

These dragonflies are of moderate size, brownish of body, and clear-winged except for basal brown spots. The face is pale, becoming much darkened with age. The frontal prominence of the face is low, rounded, and without top marginal carina or longitudinal furrow. The vertex is shallowly emarginate. The occiput is small, brown, and bare.

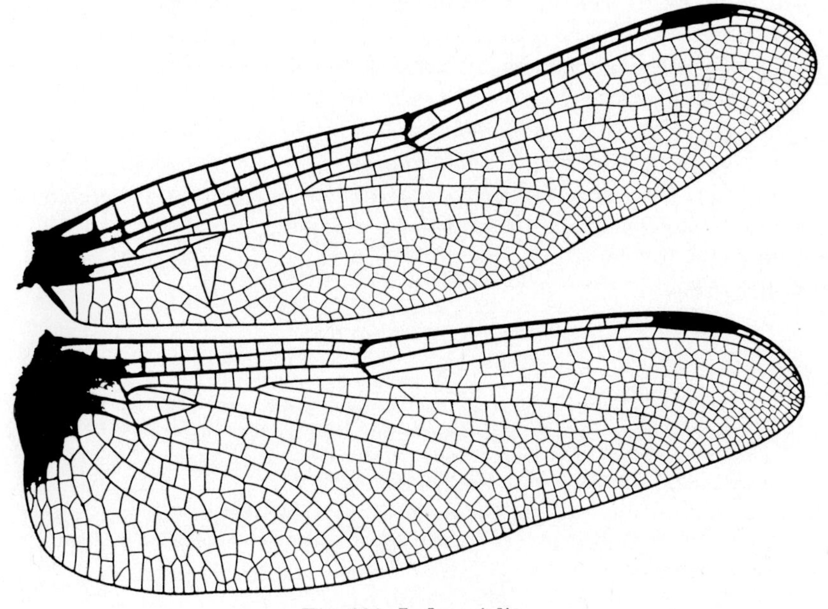

Fig. 291. *Ladona julia.*

The disc of the prothorax is narrowed rearward to an erect flat marginal lobe. The synthorax is rather thinly hairy. It lacks definite side stripes, but is variously clouded in shades of brown. The wings are sometimes lightly tinged with yellow in their membrane. The membranule is white.

To the chief characters of wing venation shown in our table of genera (p. 426), these may be added: vein M1a arises before the middle of the stigma; there is a crossvein in the supertriangle; three rows of cells are subtended by the radial planate and one by the median in both fore and hind wings. Behind vein A2 in the anal area are three well-defined rows of cells.

The abdomen regularly tapers to rearward all the way, and is very moderately enlarged on the basal segments. All carinae are black. A broad middorsal band of brown on segments 3 to 7 widens and darkens in color to rearward, with paler areas on the sides and brown again along the side margins. Segments 8 and 9 are black; 10 and the appendages, paler. The subgenital plate of the female is merely the roundly notched margin of the eighth sternite, little or not at all produced to rearward.

This is a small genus of about five species and subspecies: three of them occur in North America, the other two in Europe.

The nymphs are slender and thin-legged, with a long taper to the end of the abdomen. The body is neatly patterned in green and brown. On the lateral lobe of the labium are six lateral setae. The caudal appendages are very long, fully as long as segments 9 and 10 together, and very sharp. The end segments of the abdomen tend to a triquetral form, especially in *deplanata*, being more ridged than rounded on the dorsal side.

Two of our three species we have reared: *deplanata* and *julia; deplanata* is smaller, more sharply triquetral on segments 9 and 10, and has no mental setae (or sometimes a single very thin and inconspicuous one); *julia* has three.

KEY TO THE SPECIES

Adults

1—Wing base with brownish area divided lengthwise by clear strip that occupies midbasal space; foot-shaped anal loop with no interpolated ankle cells between its two marginal rows; sometimes only a single paranal cell in two-celled subtriangle of fore wing.......................... **deplanata**

—Wing base with brown continuous across midbasal space; anal loop with extra ankle cells; two paranal cells in three-celled subtriangle of fore wing.. 2

2—Front of synthorax entirely pale save for narrow black middorsal carina, becoming wholly pruinose white at maturity.......................... **julia**

—Front of synthorax with wide stripe of pale brown bordering black middorsal carina on each side, reducing pruinose white area by half....... **exusta**

Ladona deplanata Rambur

The Little Corporal

1842. Rbr., Ins. Neur., p. 75 (in *Libellula*).
1897. Ndm., Can. Ent., 29:144 (nymph).
1910. Ris, Coll. Selys Libell., p. 259 (in *Libellula*).
1929. N. & H., Handb., p. 217.

Length 34–38 mm.; abdomen 21–24; hind wing 26–29.

A handsome little species, generally recognizable by clear streak through brown of basal spots in wings, and by its smaller size. Fully mature males have a light dusting of pruinosity rather uniformly distributed over front of thorax and whole length of black abdomen. Two blackish dots on dorsum of middle abdominal segments show through

Fig. 292. *Ladona julia.*

pruinose coat at full maturity. Female retains a paler color; large yellowish spots along sides of abdomen remain a continuous series all the way back to segment 9, becoming brighter on 7 and 8. Caudal appendages of female may become rusty red.

Adults frequent emergent vegetation in shoal water and perch on low stems and on adjacent gently sloping banks. They fly rather haltingly, mainly from one low perch to another, and are quite easy to capture with a net.

Nymph of species distinguished by lack of (or weak development in) mental setae on labium.

Fig. 293. *Ladona exusta.*

Distribution and dates.—UNITED STATES: Ala., Fla., Ga., Ky., La., Miss., N. J.(?), N. Y.(?), N. C., Okla., S. C., Tex.

January 4 (Fla.) to June 29 (N. J.).

Ladona exusta Say

1839. Say, J. Acad. Phila., 8:29 (in *Libellula*).
1909. Wlsn., Proc. U. S. Nat. Mus., 36:655, 657.
1910. Ris, Coll. Selys Libell., p. 257 (in *Libellula*).
1920. Howe, Odon. of N. Eng., p. 71 (in *Libellula*).
1927. Garm., Odon. of Conn., p. 246 (in *Libellula*).
1929. N. & H., Handb., p. 218.

Length 37–46 mm.; abdomen 22–26; hind wing 31–33.

A slightly larger species than *deplanata,* with scanty pruinosity on thorax and abdomen even in fully mature males. Occasional males have number of rows of cells in trigonal interspace reduced to two for a short distance, as in that species, and lack ankle cells interpolated in anal loop; but generally, full complement of crossveins for genus is present.

Subgenital plate of female has a deeper median notch, widely V-shaped. Tops of its two divisions reach rearward only to level of end of segment 8. This plate in *deplanata* narrower, very shallowly notched, and projects to rearward beyond general level of end of segment.

Distribution and dates.—CANADA: N. S.; UNITED STATES: Conn., Ga., La., Maine, Mass., N. H., N. J., N. Y., N. C., Vt.

May 19 (New England) to July 28 (New England).

Ladona julia Uhler

1857. Uhler, Proc. Acad. Phila., p. 88 (in *Libellula*).
1905. Osburn, Ent. News, 16:195.
1910. Ris, Coll. Selys Libell., p. 258 (in *Libellula*).
1920. Howe, Odon. of N. Eng., p. 71 (in *Libellula*).
1927. Garm., Odon. of Conn., p. 246 (in *Libellula*).
1929. N. & H., Handb., p. 218.

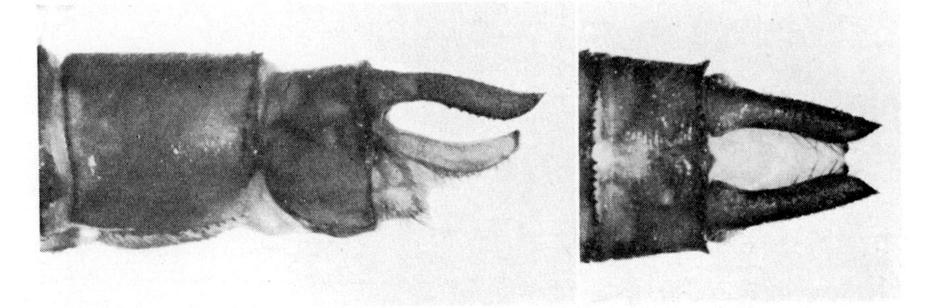

Fig. 294. *Ladona julia.*

Length 41–45 mm.; abdomen 29–31; hind wing 28–34.

This species a little larger than *exusta,* and decidedly more robust; also, darker in general coloration, and much whiter in pruinosity of top of thorax and of abdominal segments 2 to 4; beyond 4 there is generally little whiteness. Contrasts of these colors, therefore, make this species more conspicuous than the others. It is also more widely ranging, having greater strength of wing. Generally has more crossveins in wing, with those at ankle adding ankle cells on both sides of midrib of anal loop.

Female with broad front of thorax olivaceous, no pruinosity, and only carinae black. Sides dull and patternless. Hamules of male stouter than in *deplanata* in both inner and outer branches, with inner branch a little less elevated, its tip less turned to rearward.

Distribution and dates.—CANADA: B. C., Man., N. S., Ont., Que., Sask.; UNITED STATES: Conn., Ind., Maine, Mass., Mich., Minn., N. H , N. Y., N. Dak., Ohio, Pa., Wash., Wis.

May 15 (Ind.) to September 10 (Ind.).

Genus LIBELLULA Linnaeus 1758

Syn.: Belonia Kirby, Eolibellula Kennedy, Eurothemis Kennedy, Holotania Kirby, Leptetrum Newman, Neotetrum Kennedy, Syntetrum Kennedy

These are the best known of our larger dragonflies. They are to be seen coursing through the air over fresh-water ponds in summertime everywhere. They are showy insects. The head is rather large, the eye-

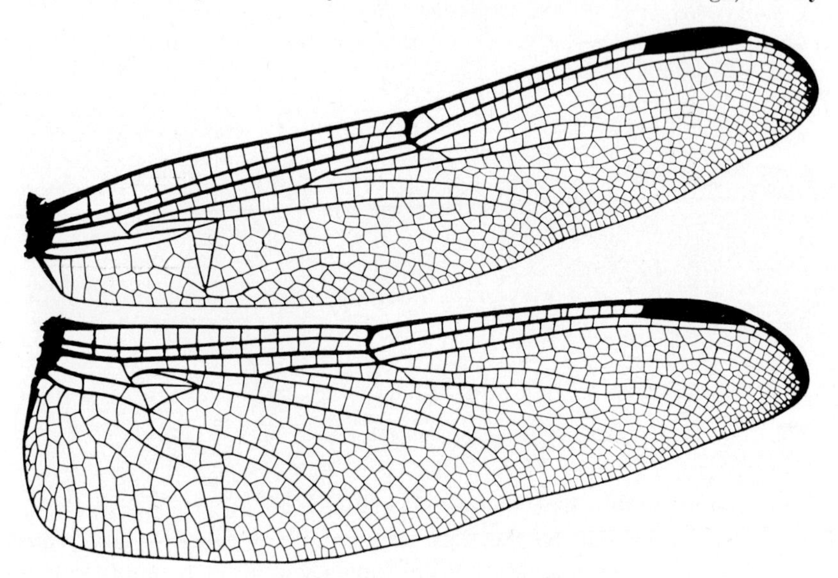

Fig. 295. *Libellula incesta.*

seam on top short. The frons is not very prominent, but has a rather deep longitudinal furrow.

The thorax is very robust, and is more or less densely clothed with hair. The stripings of the thorax are indistinct even in teneral specimens, and always become darkened and obscured with age, and in the males often overcast with a dense pruinosity. The hind lobe of the prothorax is rather narrow and low-lying, and its margin is entire and bare.

The wings are long and in general rather densely veined. In most species they are beautifully patterned in brown, with added tints of red, gold, and yellow in the membrane. The venation varies, but the following characters seem to be common. The arculus is much nearer to the second antenodal crossvein than to the first. The sectors of the arculus (veins M1–M3 and M4) are in contact, but not solidly fused at the base in the fore wing, slightly consolidated in the hind. The crosswise-elongated triangle of the fore wing is narrow, its front side being generally less than half as long as the inner side. The triangle of the hind

Fig. 296. *Libellula vibrans.*

wing is smaller, more nearly equilateral, and two-celled. Vein Cu1 springs from its hind angle, not from its outer side. There are two to four bridge crossveins. The three planates—apical, radial, and median—are constantly present and well developed, and they all generally subtend two rows of cells (the apical, sometimes three in the fore wing; the median, sometimes only one in the hind wing). The anal loop is long and rather narrow and sinuous, often with but two included rows of cells; its gaff is equal to or longer than its sole in all our species except *luctuosa*. The venation of the large fan-shaped second anal interspace (*y*) takes on chiefly two patterns as it spreads to the hind-wing margin: either it is filled with forking branchlets of the second anal vein, or its cells are lined up in more or less regular parallel columns, mainly without forks in their course to the hind margin. The columns are double in *luctuosa, odiosa,* and *saturata*. Variations in other characters are shown in the table.

The abdomen is stout, more or less depressed, its sides regularly tapering from front to rear in both sexes. Paler colors overspread the swollen basal segments in most species. In the midst of the black on each side there is generally a line of large pale spots on 1 to 7; a similar line of spots is often present on the under side of the abdomen, on each side of the mid-ventral groove. Thereafter the color deepens to rearward. Segments 8 to 10 are generally wholly blackish. The paler colors are more extensive in the female. They may be wholly obscured with age in the male, often with pruinosity overspreading and changing the black to white.

The nymphs of this genus are hairy creatures that sprawl amid the bottom silt in ponds and streamside pools. The form is rather elongated, tapering slowly to the rear. The eyes are very prominent, capping the anterior angles of the head. The front border of the middle lobe of the labium is smooth and not distinctly crenulate. Dorsal hooks and lateral spines of the abdomen are variously developed, as indicated in the accompanying table.

The genus is mainly Holarctic in distribution, being found throughout the Temperate Zone of the Northern Hemisphere. Most of its species are found in the United States, where they form a significant element of the aquatic carnivorous population.

The principal authoritative work on the adults of this genus is that of Ris (1901 to 1908), to which reference is made under every species heading that follows. Ris's work was followed by Kennedy's study of the penis (*Ent. News*, 33:33–40, 1922). There is as yet no comprehensive treatment of the nymphs of this genus, though nearly all the species have been reared. We have ventured to present their characters, so far as yet known, in a nymph table which follows.

KEY TO THE SPECIES*

Adults

1—Basal third of both fore and hind wings covered full width by blackish band; wing tips clear .... **luctuosa**

—Wings without blackish markings, or when present not completely covering so great an area **2**

* Key to the genera of Libellulines is on p. 428.

2—Front fourth (costo-radial area) of both wings covered full length by diffuse brownish band strongly tinged with amber yellow; stigma whitish, heavily margined with black; wing tips of female generally blackish; sides of thorax creamy white crossed by narrow midlateral brown stripe...... **flavida**

—Not so 3

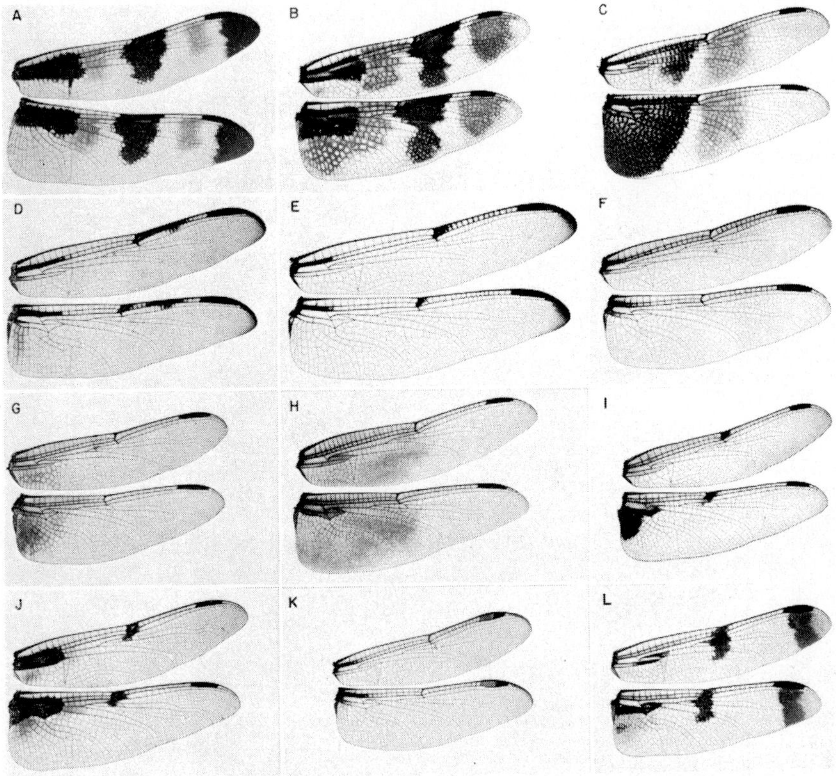

A B C D E F G H I J K L

Fig. 297. A, *Libellula pulchella;* B, *L. forensis;* C, *L. luctuosa;* D, *L. axilena;* E, *L. vibrans;* F, *L. flavida;* G, *L. croceipennis;* H, *L. saturata;* I, *L. quadrimaculata;* J, *L. nodisticta;* K, *L. cyanea;* L, *L. semifasciata.*

3—Wings throughout their membrane yellow, orange, or red, with no dark marking in definite color pattern.. 4

—Wings clear, or with more or less definite pattern in brown.............. 7

4—Wings dark red or reddish; Western.. 5

—Wings yellow (becoming bright red in old males of **L. auripennis**); Eastern.... 6

5—Wings with reddish color extending outward to stigma; basal cubital space, triangle, and supertriangle in hind wing dark reddish brown; generally one cubito-anal crossvein; stigma 4.5–5.5 mm. long.................. **saturata**

—Wings with reddish color extending only to nodus; basal cubital space with no brown tint; generally two cubito-anal crossveins; stigma 5.5–6.5 mm. long .... **croceipennis**

TABLE OF SPECIES—ADULTS

Species	Hind wing	Cells in triangles[1]					Prnls. h.w.[2]		Cvs. under stig.[3]	Distr.
		Fore wing			Hind wing		Anal loop			
		T	Sub T	Rows	T	Rows	Before	In		
auripennis	39-43	3-4	6-8	3-4	2	2-3	3-4	3-4	4-5	E, S
axilena	43-49	3-5	5	3-4	2	2	3-4	3-4	4-5	S
comanche	35-44	3-4	3-6	3-4	2	2-3	3-4	2-3	3-4	SW
composita	35-37	2-3	3-4	3	2	2	3	3	2-3	W
croceipennis	37-45	4-5	8-10	3-4	2	3	3-4	3-4	3-5	SW
cyanea	33-35	2-3	3-6	3-4	2	2-3	3	3-4	4-5	E, C, S
flavida	38-40	3-4	3-7	3	2	2	3	3-4	4-5	E, S
forensis	38-40	5	6-8	3	2	2-4	3-4	3-4	3-5	W
incesta	36-42	2-3	3-4	3	2	2	3-4	3	4-5	NE, C, S
luctuosa	38-40	4-5	6-9	4	2	2-3	3-4	4-5	3-4	NE, C, S
needhami	39-44	3-4	5-6	3-4	2	3	3-4	3-4	4-5	E, S
nodisticta	38-41	3-4	4-5	3-4	2	3-4	3-4	3	2-3	W
pulchella	42-46	3-4	4-6	3-4	2	3	3-4	3-4	2-4	G
quadrimaculata	33-36	2-3	3-6	3-4	2	2-3	3-4	3-4	2-3	W, C, E
saturata	41-45	3-5	9-14	4-5	2-4	3-4	4	3-4	3-4	W, C
semifasciata	31-37	3-4	4-7	3-5	2	2-4	3-4	3-4	3-5	NE, S
vibrans	48-51	3-4	3-7	3-4	2	2-3	3-4	3-4	3-4	E, S

[1]Number of cells composing triangle and subtriangle of front and hind wing, and number of cell rows beyond triangle (in trigonal interspace). Prevailing numbers italicized.

[2]Number of paranal cells before and in anal loop in hind wing.

[3]Number of crossveins under stigma.

6—Costa somewhat bicolored, dark before nodus, yellow beyond; hind tibiae light brown, with black spines; light-colored band on sides of thorax overlaps humeral suture; wing veins pale red or yellow; tip of red vesicle of vertex pale; hind wing generally with four paranal cells before anal loop .... **needhami**

—Costa all one color; hind tibia black, its spines all black; light-colored band on sides of thorax stops at humeral suture; wing veins all red; vesicle of vertex all red; hind wing generally with three paranal cells before anal loop .... **auripennis**

7—Costa white out to stigma; short band of saffron yellow across wing bases; no crossvein in either fore-wing supertriangle or hind-wing triangle; Western desert .... **composita**

—Not so in color 8

8—Stigma bicolored, brown and yellow, half and half.......... 9

—Stigma not bicolored.......... 10

9—Hind wing about 35–44 mm.; Western.......... **comanche**

—Hind wing about 33–35 mm.; Eastern.......... **cyanea**

10—Wings with wide crossband of brown at nodus.......... 11

—Wings with no crossband of brown at nodus.......... 13

11—Wings with both basal and nodal light brown spots double, bordered with golden yellow .... **semifasciata**

—Wings with spots dark brown or black, and not gold-bordered.......... 12

12—Wing tips brown beyond stigma.......... **pulchella**

—Wing tips hyaline.......... **forensis**

13—Hind wing with large brown spot at base, traversed by white crossveins and connected by streak of yellow in membrane with small brown spot at nodus .... **quadrimaculata**

—Not so marked.......... 14

14—Trigonal interspace of fore wings generally with four rows of cells; both wings with conspicuous basal black spots reaching out to triangles... **nodisticta**

—Generally three rows of cells; smaller markings of brown.......... 15

15—Wings black-tipped; face white.......... **vibrans**

—Wings hyaline; face not white.......... 16

16—Wings with basal brown streaks reaching out well toward triangles; generally an isolated black streak between nodus and stigma.......... **axilena**

—Basal streaks much smaller or wanting; rarely any color between nodus and stigma .... **incesta**

Nymphs

1—Head squarish seen from above, not narrowed behind eyes; six deeply cut, sharply serrate teeth on lateral lobes of labium; single raptorial seta on each side of mental lobe.......... **composita**

—Head narrowed behind eyes; ten or eleven shallow, broadly truncated teeth on lateral lobes; mental setae nine or more.......... 2

2—No middorsal hooks on abdomen.......... 3

—Dorsal hooks present, at least on some middle segments.......... 4

TABLE OF SPECIES—NYMPHS

Species	Length	Setae[1]		Dorsal hooks[2]		Lat. spines[3]		Caudal append.[4]		
		Lat.	Ment.	On segs.	Max.	On 8	On 9	Lat.	Sup.	Inf.
auripennis	27	5	5+5	3-8	6,7,8	4	4	4	10	10
axilena	27	5	6,7+4	3-8	6,7,8	4	3	5+	10	10
comanche	27	6	5+4	0	0	2	1	6	10	10
composita ?	20-21	7-8	1+0	3-5	4	0±	0±	3	10	10
croceipennis	26	7-9	6+8	3,4-8	7	0±	0±	7	10	10
cyanea	22	5-6	6+3	4-8	6,7,8	4	4	5	10	10
forensis	27	5-7	5+4-6	4-6	5	2	1	7	10	10
incesta	26	5	5,6+4	4-8	6,7,8	4	4	4+	10	10
luctuosa	25	7	5+5	4-8	6	4	4–	5–	10	10
needhami	28	5	5+5	4-8	6,7,8	4–	4–	5	10	10
pulchella	26	9	7+6	3-7	5	2–	2–	5	10	10
quadrimaculata	26	7-8	5+5	3-8	5,6,7	1	1	7	10	10
saturata	28	8-10	5,7+11	0	0	0±	0+	7	10	10
semifasciata	24	7	8+8	3-8	8	3	2	5	10	10
vibrans	27	5	6+5	4-8	8	4	4	5–	10	10

[1]Number of raptorial setae of labium: on lateral lobe and on each side of mentum: an outer group of longer setae in close alignment plus a continuing line of lesser setae, dwindling in size, paler, and more difficult to see, running down into hollow at center of mentum; lesser setae less constant.

[2]Middorsal line of prominences on abdominal segments 3 to 8, whatever their shape (max.: dorsal hook of maximum size on segment indicated).

[3]Length of lateral spine on abdominal segments 8 and 9 expressed in tenths of total length of lateral margin of segment, of which it is a part.

[4]Relative length of caudal appendages, laterals, superior, and inferiors, the last taken as 10.

3—Lateral setae of labium six; short lateral spines on abdominal segments 8 and 9 .... **comanche**
—Lateral setae of labium eight to ten; no lateral spines .... **saturata**

4—Dorsal hooks lacking on one or two antepenultimate abdominal segments 5
—Dorsal hooks regularly present on segments 4 to 8 6

5—Lateral setae of labium five, *six*, or seven .... **forensis**
—Lateral setae nine .... **pulchella**

6—Lateral setae seven or more 7
—Lateral setae five (sometimes six in **cyanea**) 10

7—Lateral spine of segment 8 three-tenths or four-tenths as long as lateral margin of which it is a part; lateral abdominal appendages about half as long as inferiors 8
—Lateral spine of 8 not more than one-tenth as long as lateral margin of which it is a part; lateral abdominal appendages about seven-tenths as long as inferiors 9

8—Largest dorsal hook on 8; a dorsal hook on 3; mental setae of labium about sixteen .... **semifasciata**
—Largest dorsal hook on 6; no dorsal hook on 3; mental setae about ten **luctuosa**

9—Lateral spines vestigial on 8 and 9; mental setae about fourteen ... **croceipennis**
—Lateral spines of 8 and 9 form at least a tenth of lateral margin of which they are a part; mental setae ten .... **quadrimaculata**

10—Superior abdominal appendage notably decurved to tip 11
—This appendage straight or very slightly decurved at extreme tip 12

11—Length of body 22 mm.; lateral abdominal appendages half as long as inferiors .... **cyanea**
—Length of body 26 mm.; lateral appendages less than half as long as inferiors .... **incesta**

12—Superior caudal appendage, viewed directly from above, little longer than width of extreme base 13
—Superior appendage longer, more slender, subacuminate, at least one and a half times as long as width of base 14

13—Lateral caudal appendages a little less than half as long as inferiors .... **vibrans**
—Lateral appendages a little more than half as long as inferiors .... **axilena**

14—Lateral appendages about half as long as inferiors .... **needhami**
—Lateral appendages less than half as long as inferiors; all appendages a little more slender, and dorsal hooks a little less hairy toward their bases **auripennis**

Nymphs unknown: **flavida, nodisticta.**

Libellula auripennis Burmeister

Syn.: costalis Rambur, jesseana Williamson

1839. Burm., Handb., p. 861.
1910. Ris, Coll. Selys Libell., p. 273.
1929. N. & H., Handb., p. 222.
1943. Wstf., Trans. Amer. Ent. Soc., 69:17 (fig.).

Length 51–58 mm.; abdomen 34–40; hind wing 39–43.

A beautiful species, with golden wings and yellow or reddish stigma. Face yellow, becoming reddish with age; likewise, prominent biconic vertex. Narrow black preocellar cross stripe of head widened before each lateral ocellus. Occiput brown above. A thin fringe of white hairs across rear of head.

Thorax dark brown, becoming rusty in front and paler on sides and beneath, with two obscure lateral yellowish bands: one just posterior to humeral suture, other band midlateral. Some fully mature males acquire a distinctly purplish tint on top of head and body. Wings tinted throughout their membrane. Stigma may become red. Hind tibiae blackish, including spines. No black stripes on sides, only little narrow black pits in depths of lateral sutures and a narrow black rim around spiracle.

Long abdomen tapers regularly to rearward, mainly yellowish at first, with a middorsal black band that widens and darkens in color to rearward as far as segment 8, then narrows and pales a little on 9, and much more on 10; appendages pale, becoming rufous in male. Segment 8 of female with no leaflike lateral expansions, only a slight rearward extension of outer posterior angle.

Distribution and dates.—UNITED STATES: Ala., Conn., Fla., Ga., La., Md., Mass., Miss., Mo., N. J., N. Y., N. C., Ohio, Pa., R. I., S. C., Tenn., Tex.; MEXICO: Tamaulipas.

February 23 (Fla.) to October 8 (S. C.).

Libellula axilena Westwood

Syn.: leda Say

1837. Westwood, edit. Drury, Ill. Exot. Ins., 2:96 (pl. 47, fig. 1).
1861. Hagen, Syn. Neur. N. Amer., p. 156.
1903. Ndm., Bull. N. Y. State Mus., 68:273 (nymph).
1910. Ris, Coll. Selys Libell., p. 269.
1929. N. & H., Handb., p. 228.

Length 60–62 mm.; abdomen 39–41; hind wing 43–49.

A very fine big species, patterned in black and white. Face obscure yellowish, brighter at sides, but all darkening with age to black, and tinged with rich metallic purple on labrum and top of frons.

Thorax at first brown and yellow, chocolate brown in front, with usual middorsal yellow line becoming black and then pruinose blue. Sides of thorax brown along wing roots and just above legs, with a wide intervening band of yellow or olive that extends entire length of side from collar in front to basal segments of abdomen. Half-length stripe of brown on hindmost lateral suture extending down from subalar carina.

Legs black, with yellowish bases. Wings marked with black about as shown in our figure, with membrane hyaline. Venation as shown in our table (p. 482).

Abdomen very long and slender, especially in male. Enlarged basal segments pale, with middorsal and lateral stripes of brown. Intervening yellow stripe between two brown ones extends to rearward full length of abdomen, narrowly interrupted on each segment by transverse carina of segment. Yellow becomes duller on middle segments and brighter again on 8. In female, a narrowly semilunar expansion of side margin extends to rearward a little beyond tip of segment in a rounded lobe at its posterior end. Segment 10 more or less yellowish; caudal appendages black.

Distribution and dates.—UNITED STATES: Ala., Fla., Ga., Ky., La., Miss., N. Y., N. C., Okla. (?), S. C.

March 26 (Fla.) to September 26 (N. Y.).

Libellula comanche Calvert

Syn.: flavida Hagen

1907. Calv., Ent. News, 18:201.
1910. Ris, Coll. Selys Libell., p. 272.
1929. N. & H., Handb., p. 223.

Length 47–55 mm.; abdomen 31–34; hind wing 35–44.

A grayish Southwestern species, becoming densely pruinose at full maturity. Face cream color except for a black stripe on front border of labrum. Top of vertex and front half of occiput also creamy white. Antennae black.

Thorax brown in front, surface almost devoid of hair behind fringed collar, and besprinkled all over with black prickles. Brown divided by a middorsal whitish stripe, and at its upper margin spreads laterally a little way across humeral suture. Sides of thorax broadly creamy white, with an upper border of brown along subalar carina, a lower one above leg bases, and a diffuse brown stripe covering third lateral suture. On middle suture, a short stripe from below extends upward to cover spiracle. These stripes so conjoined below as to surround a pale spot behind spiracle.

Legs black beyond middle of femora. Wings subhyaline, or more often deeply tinged with brown, especially in female, more or less brown at tips. Costa black. Before base of hind wings, subalar crest, sharply triangular, is unusually prominent. In costal strip of wings, crossveins yellowish; stigma bicolored, white with a brown outer end. Deeper tinge of yellow in costal stripe sometimes extends over whole wing.

Abdomen rather stout, depressed to triquetral form, all its carinae black. Basal segments olivaceous, and on them begins the darkening in color that continues to end of abdomen, with only segment 10 paler in color on dorsal side. Usual lateral pale band wide and runs whole length of abdomen. In old males, entire color pattern of body may disappear under a dense coat of pruinosity, but inferior caudal appendage remains yellow.

"Dragonflies of this species sit mostly on green bushes at the banks and fly, mostly over the surface of the water, hunting and chasing each other, and (to the disappointment of the collector) disturbing other species that might easily be taken but for their continual interference."—F. G. Schaupp.

Distribution and dates.—UNITED STATES: Ariz., Calif., Mont., Okla., Tex., Utah; MEXICO: Chihuahua, Sonora.

June 10 (Utah) to September 11 (Calif.).

Libellula composita Hagen

1873. Hagen, Rep. U. S. Geol. Surv. Terr., 6(3):728 (in *Mesothemis*).
1910. Ris, Coll. Selys Libell., p. 267.
1917. Kndy., Proc. U. S. Nat. Mus., 52:625.
1929. N. & H., Handb., p. 227.

Length 42–49 mm.; abdomen 30–32; hind wing 35–37.

Another white-faced species, with a saffron yellow stain across base of both fore and hind wings. Whitish ridge behind eyes conspicuously crossed with three bars of black. Top of vertex and occiput white. Eyes creamy white in life, with a pearly luster.

Middorsal white band of thorax very wide, extending up and over and completely covering crest. Even ridge of crest white, though with a serrulate edging of black prickles. A pair of wide brown stripes cover about half of front of thorax. Sides white, with conspicuous black stripes on first and third lateral sutures, and with a half stripe between that reaches up to cover spiracle. The three stripes are conjoined above leg bases, enclosing a crescent-shaped white spot behind spiracle. Legs pale at base and halfway out on outer side of femora and black beyond; elsewhere black. Wings hyaline, with a bright yellow costa, brown stigma, and black veins. Saffron stain of wing bases extends outward to about arculus and backward beyond tip of white membranule.

Abdomen black, with three interrupted rows of large quadrate pale spots on each side: one midlateral on segments 1 to 3, one inferior on 1 to 4 or 5, and one superior on 3 to 8; spots of all rows narrowing to rearward. Distal half of abdomen in male all black, including inferior caudal appendage. Sometimes a trace of orange on dorsum of 10. Lateral margin of 8 in female scarcely expanded, only slightly convex and a little thickened.

Distribution and dates.—UNITED STATES: Ariz., Nev., Utah, Wyo.

June 16 (Utah) to July 21 (Utah).

Libellula croceipennis Selys

Syn.: uniformis Kirby

1868. Selys, C. R. Soc. Ent. Belg., 11:67.
1906. Calv., B. C. A., p. 212.
1910. Ris, Coll. Selys Libell., p. 276.
1929. N. & H., Handb., p. 222.

Length 55–57 mm.; abdomen 32–37; hind wing 37–45.

A large pale red species, closely resembling *L. saturata* in stature and color. Body all red save black spines on legs and midventral black stripe on abdomen. Yellow color of wing membrane overspreads wing only a little way outward beyond triangle, then narrows and is continued outward along costal strip only, fading insensibly beyond and behind stigma. No streaks of brown alongside midbasal space or elsewhere on wings.

Venation of wings as shown in table of species. Often two cubito-anal crossveins in hind wing and a single row of cells subtended by median planate. Stigma generally about 6 mm. long; in *saturata* about 5 mm. Fewer ankle cells in anal loop than in *saturata,* often a single one on proximal side of midrib.

Nymph distinguished from that of *saturata* by presence of more lateral setae arming lateral lobe of labium, and vestigial lateral spines on abdominal segments 8 and 9.

Distribution and dates.—UNITED STATES: Okla., Tex.; MEXICO: Baja Calif.; also south to Costa Rica.

June 21 (Tex.) to October 4 (Tex.).

Libellula cyanea Fabricius

Syn.: bistigma Uhler, quadrupla Say

1775. Fabr., Syst. Ent., p. 424.
1910. Ris, Coll. Selys Libell., p. 272.
1920. Howe, Odon. of N. Eng., p. 70.
1927. Garm., Odon. of Conn., p. 244.
1929. N. & H., Handb., p. 223.

Length 41–46 mm.; abdomen 29–32; hind wing 33–35.

A relatively small Eastern species, with bicolored stigma and some basal streaks of brown in wings. Much like Southwestern *comanche,* but with a blackish face that becomes very black and shiny on frons in old males.

Front of thorax brown, with usual middorsal pale stripe. Top of crest also whitish, within black carinae. Sides of thorax white, traversed by a blackish stripe on third lateral suture. This stripe widens below to leg bases. Clear wings have a well-developed but short blackish streak in base of subcostal interspace, and yellow tinge in membrane along whole nodal area; stigma yellow and brown, half and half.

Abdomen similar in color pattern to *comanche,* but early develops a dense pruinosity on surface that hides pattern completely. Old males black, including caudal appendages. In female, lateral margin of segment 8 expanded a little to a somewhat crescentic form.

Distribution and dates.—UNITED STATES: Ala., Conn., Del., Ga., Ind., Kans., Ky., Maine, Md., Mass., Mich., Miss., Mo., N. H., N. J., N. Y., N. C., Ohio, Okla., Pa., S. C., Tenn., Tex., Va.

April (N. C.) to September 10 (Pa.).

Libellula flavida Rambur

Syn.: plumbea Uhler

1842. Rbr., Ins. Neur., p. 58.
1903. Ndm., Bull. N. Y. State Mus., 86:274 (very immature nymph).
1910. Ris, Coll. Selys Libell., p. 271.
1920. Howe, Odon. of N. Eng., p. 70.
1929. N. & H., Handb., p. 224.

Length 48–51 mm.; abdomen 32–34; hind wing 38–40.

An East Coast species, recognizable by two large pale areas on each side of thorax. Face yellowish in female, blackish in male, and shining, with yellow anteclypeus and yellow line in fronto-clypeal suture. Vertex yellowish, with black in front at base. Occiput brown.

Thorax brown in front, with a middorsal stripe of yellow that is soon darkened with brown; at full maturity entire front becomes wholly pruinose blue in male. Sides of thorax largely yellow, with a narrow stripe on third lateral suture dividing yellow into two large spots; foremost spot trapezoidal in shape, other spot triangular. Legs blackish, paler to knees. Wings deeply tinged with yellow along front, with two streaks of brown in yellow area that borders midbasal space. Long stigma, sometimes distinctly bicolored with yellow or orange and brown. Wings of female tipped with brown from level of stigma outward.

Abdomen at first extensively pale on enlarged basal segments and along side, but becomes black with age. Middorsal band of black covers middle segments, ending on 9; 10 and base of caudal appendages yellow. Thorax of male heavily pruinose in front, lightly on pale sides, and to varying degrees on 1 to 4, and again on 8 and 9 of abdomen.

Distribution and dates.—UNITED STATES: Ala., Fla., Ga., Md., Mass., Miss., Mo., N. J., N. Y., N. C., Pa., S. C., Tex.

March 16 (Miss.) to September 28 (Ga.).

Libellula forensis Hagen

1861. Hagen, Syn. Neur. N. Amer., p. 154.
1910. Ris, Coll. Selys Libell., p. 265.
1927. Seemann, J. Ent. Zool. Claremont, Calif., 19:33.
1929. N. & H., Handb., p. 226.

Length 49–51 mm.; abdomen 31–33; hind wing 38–40.

A Far Western species, with a full-width nodal crossband of brown on all wings. Resembles Eastern *L. pulchella,* from which it differs most markedly in lacking brown color of wing tips. Face dull yellow, densely clothed with black hair. Two yellow spots at its sides and two more on top of frons. Black of preocellar band prolonged forward in bottom of frontal furrow. Rear of head black, with two yellow spots at hind margin of eye.

Thorax dull brown, clothed with tawny hair. All carinae black. On each side of thorax, two short longitudinal yellow or yellowish streaks, one in front of and one at rear of spiracle. Legs black. Wings patterned as shown in figure 297. Their venation as stated in table.

Rather stout abdomen, brown or black, with a wide longitudinal yellow stripe on each side, interrupted only by black lines of transverse carinae of segments. Appendages black. Segment 8 of female has a very slight expansion of its side margins, a mere low convexity, which does not extend to rearward beyond level of apex of segment.

Distribution and dates.—CANADA: B. C.; UNITED STATES: Ariz., Calif., Colo., Mont., Nebr., Nev., Oreg., Utah, Wash., Wis. (?).

June 3 (B. C.) to September 16 (B. C.).

Libellula incesta Hagen

1861. Hagen, Syn. Neur. N. Amer., p. 155.
1910. Ris, Coll. Selys Libell., p. 270.
1920. Brimley, Ent. News, 31:138.
1927. Byers, Odon. of Fla., p. 113.
1929. N. & H., Handb., p. 227.

Length 50–52 mm.; abdomen 32–34; hind wing 36–42.

This species very black in both male and female when fully mature, generally without wing markings. In teneral specimens, face yellowish, but in older ones darker, and top of head becomes shining metallic black.

Thorax brown in front, clothed with short dark pubescence, and with only vestiges of usual middorsal pale line. Soon becomes blackish and later in old males acquires a beautiful purplish pruinosity. Sides of thorax lack definite stripes, and early become blackish.

Legs black. Wings clear, with black veins and stigma, often with a touch of brown on wing tip beyond stigma. More rarely a little brown streak will appear in basal subcostal space or between nodus and stigma.

Abdomen in teneral specimens has usual pale stripe along each side, but it early becomes almost wholly black. Segment 8 in female has a low crescentic widening of lateral margin, very black, with a vestige of yellow of lateral stripe bordering its inner side.

This trim "tailor-made" species may be distinguished from *axilena* by its smaller size. In *incesta* there are generally three paranal cells before anal loop; in *vibrans*, generally four.

Distribution and dates.—CANADA: Ont.; UNITED STATES: Ala., Ark., Conn., Fla., Ga., Ill., Ind., Ky., La., Maine, Mass., Mich., Miss., Mo., N. H., N. J., N. Y., N. C., Ohio, Okla., Pa., R. I., S. C., Tenn., Tex., Va., Wis.

February 20 (Miss.) to October 7 (S. C.).

Libellula luctuosa Burmeister

Syn.: basalis Say

The Widow

1839. Burm., Handb., p. 861.
1861. Hagen, Syn. Neur. N. Amer., p. 152.
1910. Ris, Coll. Selys Libell., p. 263.
1920. Howe, Odon. of N. Eng., p. 71.
1927. Garm., Odon. of Conn., p. 248.
1929. N. & H., Handb., p. 221 (fig.).
1940. Ferguson, Field and Lab., 8:6 (oviposition).

Length 42–50 mm.; abdomen 28–31; hind wing 38–40.

A dark brown species, with a broad black band of crepe across basal third of both fore and hind wings. Face and top of head pale at first, darkening with age and becoming black on all prominent surfaces.

Thorax brown and somewhat hairy, with hairs thin and pale. At first a broad middorsal stripe of yellow that later becomes obscured, whole front of thorax darkening with age to a deep brassy brown in female, to jet black in male; at full maturity becomes overspread with pruinose white, save only bare black middorsal carina. No stripes on sides of thorax.

Legs black. Crepe bands of wings widest at front, where they extend well toward or to nodus, and are narrowed and rounded to rear. Included paler fenestrate areas along costa and before triangles. Wing tips often touched with smoky brown, especially in female. Venation as shown in table of species. This species unique in having sole of anal loop distinctly longer than gaff.

The abdomen slowly tapers in both sexes from segment 2 to end. Five longitudinal stripes cover dorsum: three black ones on dorsal and lateral carinae, and two along lateral carinae, dorsal one widening and deepening in color to rearward. Intervening yellow stripes on sides broken only by transverse black carinae. Whole dorsum becomes black with age, and more or less pruinose in male. Subgenital plate of female a short scoop, with a very wide U-shaped median emargination, directed obliquely downward.

Very common pond species. Flies steadily over water, resting occasionally on reed tips. Not very hard to capture with a net. In cool of evening, adults may be found hanging by their feet to drooping twigs of shrubbery near by. Female usually oviposits unattended by male. At transformation, cast-off skins of nymphs commonly left sticking to clustered grass stems a few yards from ponds.

Distribution and dates.—CANADA: N. S., Ont., Que.; UNITFD STATES: Ala., Conn., D. C., Ga., Ill., Ind., Iowa, Kans., Ky., Maine, Md., Mass., Mich., Minn., Miss., Mo., Nebr., N. J., N. Mex., N. Y., N. C., Ohio, Okla., Pa., R. I., S. C., S. Dak., Tenn., Tex., Vt., Va., Wis.; MEXICO: Chihuahua.

April 24 (Miss.) to September 16 (Ind.).

Libellula needhami Westfall

1920. Howe, Odon. of N. Eng., p. 69 (in part as *L. auripennis*).
1922. Wmsn., Ent. News, 33:13 (in part as *L. auripennis*).
1927. Garm., Odon. of Conn., p. 243 (in part as *L. auripennis*).
1943. Wstf., Trans. Amer. Ent. Soc., 69:22 (figs.).

Length 53–56 mm.; abdomen 35–38; hind wing 39–44.

This species very similar to *auripennis* in stature and general coloration, but paler, more yellowish, and with a more deeply tinted strip of color along rear of vein R1. Distinguished by having costa faintly bicolored but distinctly darker on antenodal than on postnodal portion. Paranal cells proximal to anal loop four in *needhami*, sometimes three; more often three in *auripennis*. Legs also rather uniformly light brown, contrasting with blackish color of tibial spines. Lateral pale band on thorax extends forward across humeral suture, and does not stop at it, as in *auripennis;* stigma may be yellow or orange, but not red.

Distribution and dates.—UNITED STATES: Ala., Conn., Del., Fla., Ga., La., Md., Mass., Miss., N. J., N. C., R. I., Tex., Va.; MEXICO: Tamaulipas; ANTILLES: Cuba.

February 20 (Fla.) to September 22 (Fla.).

Libellula nodisticta Hagen

1861. Hagen, Syn. Neur. N. Amer., p. 151.
1910. Ris, Coll. Selys Libell., p. 264.
1917. Kndy., Proc. U. S. Nat. Mus., 52:608.
1929. N. & H., Handb., p. 226.

Length 46–52 mm.; abdomen 32–34; hind wing 38–41.

A stout-bodied Western species, brown spotted with yellow, thorax becoming wholly pruinose with age. Face yellowish, with clay yellow labrum bordered in front with black. Deeply furrowed frons shining black above, but remains yellow at sides. Double-peaked vertex and occiput shining black. Black rear of head crossbarred with yellow, its long hair fringe white.

Thorax brown at first, with obscure middorsal pale band and black-edged carina and crest; well clothed with pallid hairs. Sides brown, black in depths of lateral sutures, washed with yellow on lower bulges between—all becoming obscured by pruinosity with age. Legs black. Wings spotted with brown, as shown in figure 297. Veins and stigma black, very clear wing membrane becoming pruinose behind basal spots of brown.

Abdomen blackish, increasingly so to rearward. Usual diffuse bands of yellowish spots both on sides and underneath end on segment 8; 9, 10, and caudal appendages black.

In the morning the individuals of this species were easily captured while seated on brush and weeds in the sunny openings along the stream. A female observed ovipositing flew about 2 feet above the water and made several quick swings downward, tapping the water with her abdomen just once for each swing.—Kennedy (1917).

Distribution and dates.—UNITED STATES: Calif., Colo. (Hagen, 1874), Mont., Nev., Utah, Yellowstone; also from Mexico to Venezuela.

June 24 (Utah) to September 17 (Calif.).

Libellula odiosa Hagen

1861. Hagen, Syn. Neur. N. Amer., p. 152.

Length 50–52 mm.; abdomen 31–33; hind wing 41–43.

Probably to be regarded as a variety of *luctuosa*, similar in coloring and in color changes with age, but a little larger and more robust, and with some small differences in wing venation. In hind-wing triangle a crossvein generally present. Aside from this, larger and stronger wings appear to have fewer cells and crossveins, especially in anal loop, where interpolated "ankle cells" run from three to six in *odiosa,* from nine to fourteen in *luctuosa.*

Distribution and dates.—UNITED STATES: Tex.

July 5 to August 13.

Libellula pulchella Drury

Syn.: bifasciata Fabricius, confusa Uhler, versicolor Fabricius

Ten Spot

1770. Drury, Ill. Exot. Ins., 1:115 (pl. 48, fig. 5).
1910. Ris, Coll. Selys Libell., p. 265.
1914. Whedon, Minn. State Ent. Rep. for 1913, p. 99.
1920. Howe, Odon. of N. Eng., p. 73.
1927. Garm., Odon. of Conn., p. 249.
1927. Seemann, J. Ent. Zool. Claremont, Calif., 19:33.
1929. N. & H., Handb., p. 224 (fig.).

Length 52–57 mm.; abdomen 33–36; hind wing 42–46.

A fine familiar species, large and strong-flying, with wings conspicuously spotted. Face obscure brownish, paler in middle and yellowish across top of frons. Vertex brown, with jet black covering its front side. Occiput yellow, with brown outer angles.

Thorax brown, well clothed with a thin white pubescence. On sides, two yellowish stripes: first lies across middle suture above spiracle; second extends to rearward, followed by yellow spots in line under swollen basal segments of abdomen.

Legs blackish, paler at base. Wings spotted with brown, as shown in figure 297. In male at maturity, ten chalky white spots develop on wings between spots of brown: two on each fore wing and three on hind, one of latter behind triangle. Membranule white. Scanty pruinosity of body in male developed on basal third of abdomen and on pale areas of sides of thorax.

Abdomen brown, with two interrupted stripes running entire length and a narrow pale streak bordering middorsal carina. Caudal appendages brownish, becoming black with maturity. Abdomen of female tapers all the way back to tip (a character that will distinguish it from similarly spotted female of *Plathemis*). Segment 8 has no leaflike lateral expansion, but only a short, rounded, lobelike projection to rearward of lateral margin.

A wide-ranging species. Wherever it occurs it is sure to be in evidence, flying much and resting little during hours of sunshine. It sometimes darts threateningly near, but is too wary to be caught easily. It flies horizontally several feet above the water. Female oviposits unattended, usually over submerged green vegetation in bays and shoals, striking the water, at points wide apart, with a vigor which often sweeps up a drop that falls back to surface with a splash a foot farther ahead. Cast skins are left at transformation several feet from shore, usually on grass or weeds.

Distribution and dates.—CANADA: B. C., Man., N. B., N. S., Ont., Que., Sask.; UNITED STATES: Ala., Calif., Colo., Conn., Fla., Ga., Ill., Ind., Iowa, Kans., Ky., La., Maine, Md., Mass., Mich., Minn., Miss., Mo., Nebr., Nev., N. H., N. J., N. Y, N. C., Ohio, Okla., Oreg., Pa., R. I., S. C., Tenn., Tex., Utah, Vt., Va., Wash., W. Va., Wis.

May 9 (Ohio) to October 14 (La.).

Libellula quadrimaculata Linnaeus

Syn.: maculata Harris, praenubila Newman, quadripunctata Fabricius, ternaria Say

1758. Linn., Syst. Nat., p. 543.
1910. Ris, Coll. Selys Libell., p. 251.
1920. Howe, Odon. of N. Eng., p. 70.
1927. Garm., Odon. of Conn., p. 251.
1929. N. & H., Handb., p. 226.

Length 42–46 mm.; abdomen 27–29; hind wing 33–36.

A Holarctic species of wide distribution. Face yellow; labrum bordered and traversed by black. Top of vertex yellow; occiput yellow, with brown outer angles.

Thorax tawny, with a dense covering of hair of same color. On lateral sutures, narrow black lines are conjoined at their lower ends, full length on first and third sutures, abbreviated above on middle one, and surrounded at spiracle by variable yellow spots. Wings patterned as shown in figure 297; rather conspicuously spotted in both fore and hind wings at base and nodus. The two spots connected by a yellow streak that fills subcostal space and then runs on along costa to wing tip. Large triangular spot on hind wings heavily overspread with brown; brown area traversed by whitish veins. Stigma brown.

Abdomen hairy at base only; olive brown on enlarged basal segments, mostly blackish on tapering apical segments, with a narrow streak of yellow running lengthwise just above lateral margin. Broad middorsal pale band of base dwindles in breadth to rearward and ends on segment 6. The little pruinosity developed by fully mature males in this species is mainly on back of 6 and 7 of abdomen. Segment 8 of female without lateral expansions.

Distribution and dates.—Alaska; Labrador; CANADA: Alta., B. C., Man.,

N. B., Nfld., NW. Terr., N. S., Ont., P. E. I., Que., Sask.; UNITED STATES: Ariz., Colo., Conn., Idaho, Ill., Ind., Iowa, Maine, Mass., Mich., Minn., Nebr., Nev., N. H., N. J., N. Mex., N. Y., N. C. (?) (Ris), Ohio, Oreg., Pa., S. Dak., Utah, Vt., Wash., Wis., Wyo.; also from northern Europe and Asia.

May 7 (B. C.) to September 29 (B. C.).

Libellula saturata Uhler

1857. Uhler, Proc. Acad. Phila., p. 88.
1904. Ndm., Proc. U. S. Nat. Mus., 27:705 (fig., nymph).
1910. Ris, Coll. Selys Libell., p. 274.
1923. Ndm., J. Ent. Zool. Claremont, Calif., 16:129.
1927. Seemann, J. Ent. Zool. Claremont, Calif., 19:34.
1929. N. & H., Handb., p. 222.

Length 52–61 mm.; abdomen 34–40; hind wing 41–45.

A big red Southwestern species; occurs also in warm springs farther north. Red all over, including legs and wing veins. Only eyes and occiput brown.

Thorax brownish red, clothed with short pubescence of same color, and without stripes. Wings broadly flavescent as far out from base as stigma, that color fading farther out and to rearward, with color deepening to two brown streaks that lie at sides of midbasal space in hind wings, and generally are connected at end of that space to cover triangles. Red on back of abdomen becomes brighter to rearward and on appendages, especially in mature males.

Considerable difference between sexes in number of crossveins in wing, especially in number of interpolated cells in anal loop. In males these cells may be so numerous as to form two added rows for a distance along midrib; female with but a few cells at ankle.

Flight rather heavy and lumbering, but well sustained.

Hairy nymph squats amid black ooze on bottom of stagnant pools; when ready to transform, crawls only a little way out of water.

Distribution and dates.—UNITED STATES: Ariz., Calif., Colo., Idaho, Kans., Mo., Mont., Nev., Okla., Tex., Utah, Wyo.; MEXICO: Baja Calif., Chihuahua, Coahuila, Nuevo León, Sonora; also southward in Mexico.

May 18 (Calif.) to October 2 (Ariz.).

Libellula semifasciata Burmeister

Syn.: hersilia Blanchard, maculata Rambur, ternaria Say (female)

1839. Burm., Handb., p. 862.
1910. Ris, Coll. Selys Libell., p. 266.

1920. Howe, Odon. of N. Eng., p. 73.
1927. Garm., Odon. of Conn., p. 252.
1929. N. & H., Handb., p. 224.

Length 45–48 mm.; abdomen 28–30; hind wing 31–37.

A rather slender species, with prettily spotted amber-tinted wings. Shining face waxy yellow, paler across middle or reddish in old males. Vertex yellow, becoming olivaceous. Antennae black. Occiput yellow and very hairy on rounded rear surface.

Thorax brown in front and yellowish on sides, with hairline black stripes in first and third lateral sutures, and an oblique pale band behind each suture. Legs yellowish at base and brown beyond, with black tarsi. Wings as shown in figure 297. Long stigma yellow, becoming rufous at maturity, bordered by black. Wing membrane richly tinged with yellow, broadly as far out as triangle, and narrowly along costa thereafter out to wing tip.

Abdomen regularly tapering, with a wide middorsal pale stripe as far out as segment 6, and a narrower lateral one low down on each side. Segments 7 to 9 and basal half of 10 black. Caudal appendages black-tipped. Segment 8 of female has a very slight widening of lateral margins.

Distribution and dates.—CANADA: Ont.; UNITED STATES: Ala., Conn., D. C., Fla., Ga., Ill., Ind., Ky., La., Maine, Md., Mass., Mich., Miss., N. H., N. J., N. Y., N. C., Ohio, Pa., R. I., S. C., Tenn., Tex., Va., Wis.

April 1 (S. C.) to August 16 (New England).

Libellula vibrans Fabricius

1793. Fabr., Syst. Ent., 2:380.
1910. Ris, Coll. Selys Libell., p. 268.
1920. Howe, Odon. of N. Eng., p. 70.
1927. Garm., Odon. of Conn., p. 253.
1929. N. & H., Handb., p. 228 (fig.).
1938. Camp, New Orleans Acad. Sci. (nymph abstract).

Length 56–63 mm.; abdomen 39–43; hind wing 48–51.

Our largest *Libellula,* an elegant insect with long gauzy black and white wings. Face white, with tawny labium and labrum, both narrowly edged with black. Frons white above, with its basal third and all vertex black. Occiput brown, with a thin fringe of pale hairs at its rear.

Thorax at first pale brown in front, with usual middorsal yellow stripe, but in old males becomes densely pruinose from black carina out laterally to and over humeral suture. Sides olive, with black subalar carinae, from which descend two little streaks of black in depths of first and third lateral sutures. Low down on sides, another longitudinal line of irregular black markings lies just above bases of legs. Legs black beyond their pale bases. Wings clear, with beautiful black markings, shown in figure 297. Brown on wing tips very variable in extent.

Abdomen gradually tapers, long and slender, especially in male. Usual middorsal black stripe begins on segment 3, widens progressively, ends on 9 or 10. Another brown stripe extends along each side margin from base to 8 or 9. Caudal appendages black. Segment 8 of female has lateral margin extended below level of yellow stripe in a thin plate of low crescentic form, jet black in color. This plate projects to rearward in a rounded lobe beyond level of apex of segment.

Distribution and dates.—CANADA: Ont.; UNITED STATES: Ala., Conn., Fla., Ga., Ill., Ind., Ky., La., Md., Mass., Miss., N. J., N. Y., N. C., Ohio, Okla., Pa., S. C., Tenn., Tex., Va., Wis.

April 1 (Fla.) to December 8 (Miss.).

Genus PLATHEMIS Hagen 1861

White Tails

This genus is remarkable for the striking difference in appearance between male and female, and for the still more remarkable resemblance of the female of *P. lydia* to the female of *Libellula pulchella.* The genus is very similar to *Libellula* in most characters, as will be seen by consulting the table of genera (p. 426). It has the maximum of undulation of vein M2 in all wings, and of curvature of veins A3 and A4 at their base. It differs strikingly in having on the under side of abdominal segment 1 of the male a large forked midventral process; and in the female, a short, wide, parallel-sided abdomen.

The nymph also is like that of *Libellula,* except that the head is widest *behind* the eyes, and the front border of the median lobe of the labium is crenulate.

KEY TO THE SPECIES

Adults

1—Median crossband of wings of male uniform dark brown; wing tips of female brown .... **lydia**
—Median crossband of wings of male divided by strip of paler color; wing tips of female hyaline .... **subornata**

Nymphs

1—Dorsal hooks on segments 2 to 6, highest on 5; all sharp, thornlike .... **lydia**
—Dorsal hooks on 2 to 6, highest on 4; all blunt and hairy .... **subornata**

Plathemis lydia Drury

Syn.: trimaculata DeGeer

1770. Drury, Ill. Exot. Ins., 1:112 (pl. 47, fig. 4) (in *Libellula*).
1867. Packard, Amer. Nat., 1:310 (fig.) (in *Libellula*).
1901. Ndm., Bull. N. Y. State Mus., 47:537 (fig., nymph).
1910. Ris, Coll. Selys Libell., p. 261 (in *Libellula*).
1917. Kndy., Bull. Univ. Kansas, 18:141 (pl. VII).
1920. Howe, Odon. of N. Eng., p. 74.
1927. Garm., Odon. of Conn., p. 255.
1929. N. & H., Handb., p. 229 (figs.).

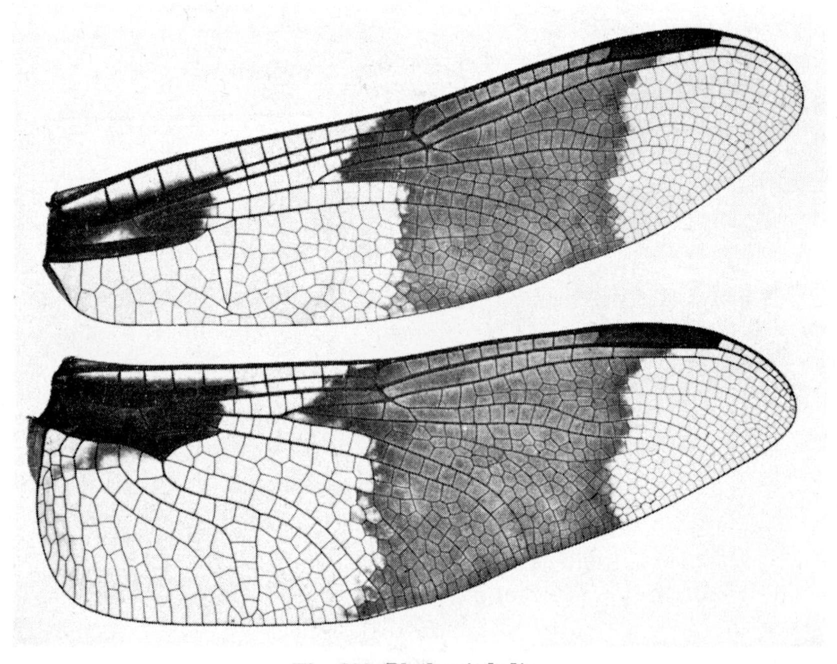

Fig. 298. *Plathemis lydia.*

Length 42–48 mm.; abdomen 25–27; hind wing 30–35.

A very common and widely distributed species. Face yellowish to olivaceous, thickly beset with short stiff hairs, and becoming darker with age. Labium pale and concolorous. Top of frons, vertex, and occiput shining brown.

Thorax brown, beset with microscopic prickles intermixed with short scattered hairs. Sides of thorax brown, blackish below, and marked with two longitudinal whitish or yellowish stripes: first stripe extends

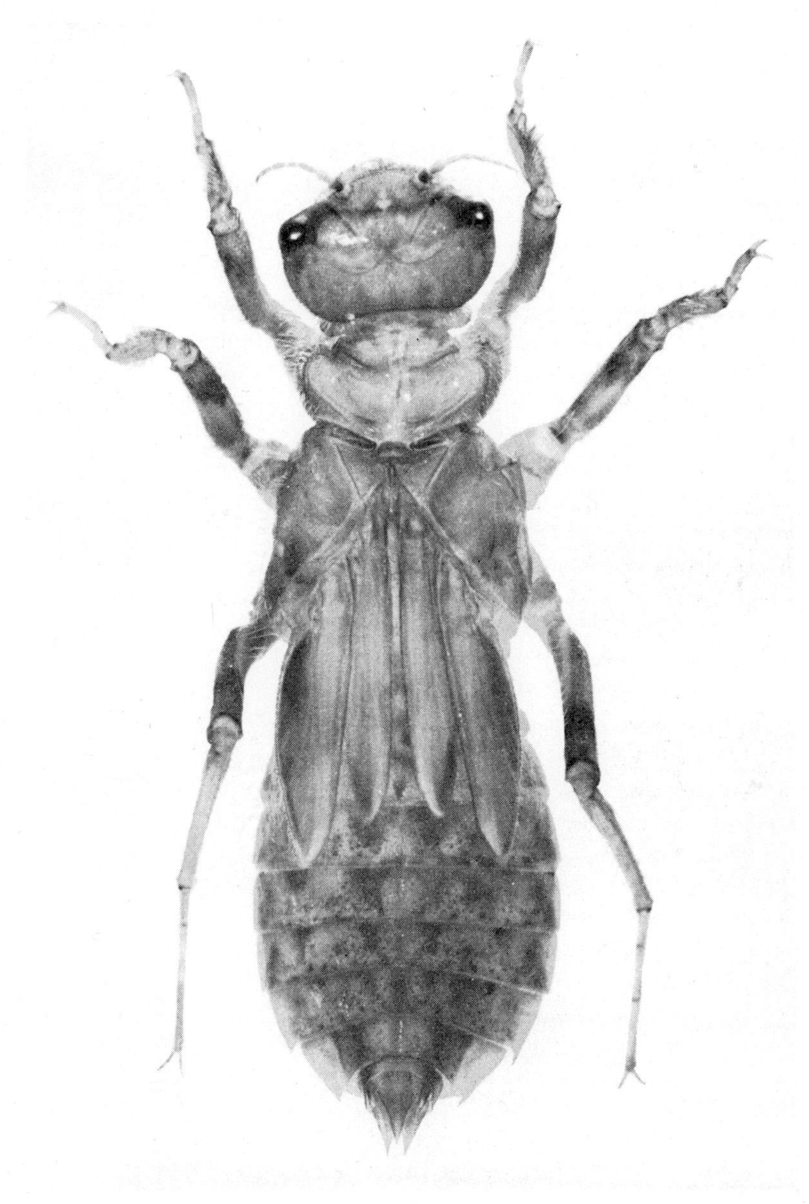

Fig. 299. *Plathemis lydia*.

from rear of prothorax backward and upward to end at top of mid-lateral suture, and is often broken into a line of two or three spots; second stripe extends entirely across metepimeron; both yellowish at front ends, where they are surrounded by black.

Abdomen of male triquetral, rather short and thick, with short mid-dorsal and lateral carinae; tapers regularly to end, becomes densely pruinose white at maturity (whence the common name). Hind wings acquire a patch of pruinose white beyond large basal brown spot. In female, abdomen strongly depressed, parallel-sided, with a conspicuous row of oblique yellow spots on each side of segments 3 to 8. These two characters, as well as smaller size, serve to distinguish it from female of *L. pulchella.* Prominent midventral inferior process on segment 2 in male is divided into two blunt-tipped arms by a full-depth U-shaped median notch.

This species often found associated with *L. pulchella.* The two often fly the same beat about the borders of ponds, making sudden dashes hither and yon, not infrequently meeting face to face in mid-air and stopping to hover threateningly with a great rustle of wings, then dashing away again. During hours of sunshine their time is divided between hovering over the surface of the water and perching on shore. The males come to rest with head low and tail elevated, moving the wings forward and downward into a drooping position by a succession of jerks. When old and pruinose they seem to prefer whitish perches. They will sometimes settle on the collector's white net or on his white shirt or straw hat.

The females slip in and out of sheltered places along the shore to lay their eggs. Hovering low over the water, marking time, they drop down to touch the surface many times in nearly the same place, liberating twenty-five to fifty eggs at each descent. They go deeper into the nooks and reedy shoals than does *L. pulchella.*

Distribution and dates.—CANADA: B. C., N. B., N. S., Ont., Que.; UNITED STATES: Ala., Ark., Calif., Colo., Conn., Fla., Ga., Idaho, Ill., Ind., Iowa, Kans., Ky., La., Maine, Mass., Mich., Minn., Miss., Mo., Nebr., N. H., N. J., N. Mex., N. Y., N. C., Ohio, Okla., Pa., R. I., S. C., Tenn., Tex., Utah, Vt., Va., Wash., W. Va., Wis.

April 18 (Miss.) to October 16 (Tenn.).

Plathemis subornata Hagen

1861. Hagen, Syn. Neur. N. Amer., p. 149.
1905. Calv., B. C. A., p. 205.
1906. Wmsn., Ent. News, 17:351 (figs.).
1910. Ris, Coll. Selys Libell., p. 263 (in *Libellula*).
1929. N. & H., Handb., p. 230.

Length 40–51 mm.; abdomen of male 25–31; abdomen of female 22–25; hind wing 32–38.

A Western species, similar to preceding in most respects, but differs in having a median blackish stripe on labium, which leaves only outer half of each lateral lobe yellow; in lacking brown wing tips in female; and in having nearly entire area between wing spots pruinose in old males.

Prominence on ventral side of abdominal segment 2 in male is divided only halfway to base by a wide V-shaped notch. Spots on 3 to 8 brighter yellow, matched by an almost equally conspicuous row on under side of same segments.

Appears to frequent mainly swales and seepage pools in desert and semidesert areas. Nymphs clamber a few inches up from water in thick beds of aquatic sedges and grasses to transform.

Distribution and dates.—CANADA: B. C.; UNITED STATES: Ariz., Calif., Colo., Kans., Nebr., Nev., N. Mex., Tex., Utah; MEXICO: Chihuahua, Sonora.

April (Calif.) to October 16 (Tex.).

Genus MICRATHYRIA Kirby 1889

This is a genus of mainly Neotropical dragonflies, more than twenty-five of them, five of which are found within our limits. They are mostly small species of blackish coloration and hyaline unspotted wings. The face is pale; the eye-seam is short.

The hind lobe of the prothorax is large and erect, more or less bilobed, and fringed with long hairs. The venation of the wings, while fairly constant in each species, covers an unusually wide variability in characters that elsewhere have been most used for distinguishing genera: a half-length antenodal crossvein may be present or absent; also, a second cubito-anal in the hind wing; and the fore-wing triangle and subtriangle may have one, two, or three included cells, according to species. A character that will distinguish this genus from all others in our fauna of similar size and aspect is the presence of two or three bridge crossveins, all its nearly allied genera having only one. Another significant character is that vein Cu1 springs from the outer side of the triangle in the hind wing.

The abdomen is long and slender, moderately widened at one or both ends.

The best account of adults of the genus as a whole is found in Ris (*Coll. Selys Libell.*).

The nymphs have been described for few species. For those few, see Needham's *Micrathyria* paper of 1943.

KEY TO THE SPECIES

Adults

1—Fore-wing subtriangle one-celled; five paranal cells; hind wing 22 mm. or less . . . . **debilis**
—Fore-wing subtriangle two- or three-celled; six or seven paranal cells; hind wing 24 mm. or more 2

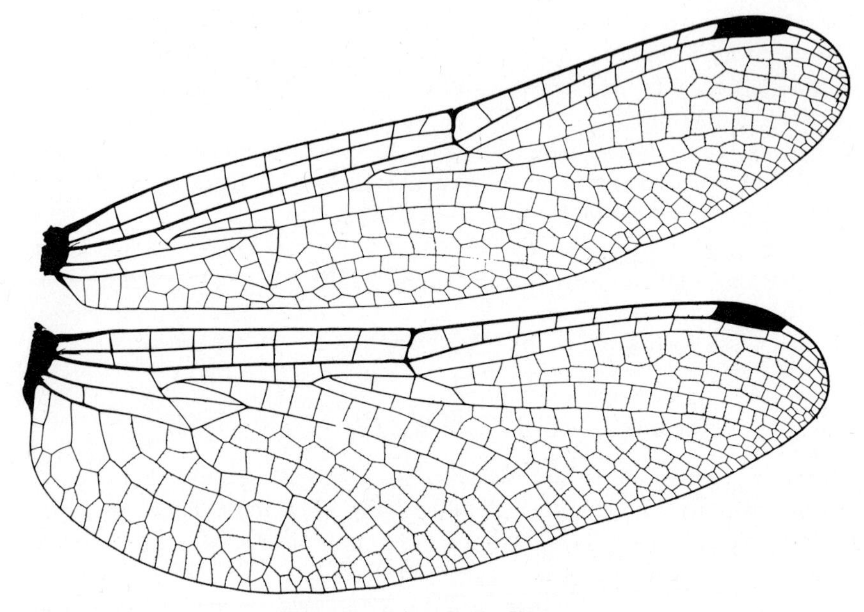

Fig. 300. *Micrathyria didyma.*

2—Trigonal interspace of hind wing with no full-width cells, series starting with two rows; sides of thorax with three unbranched stripes . . . . . . **didyma**
—That interspace with one or two full-width cells, followed by two rows; at least one side stripe of thorax forked at upper end 3

3—Fore-wing subtriangle two-celled; six paranal cells . . . . **aequalis**
—Fore-wing subtriangle three-celled; seven paranal cells 4

4—Antenodal crossveins seven . . . . **hageni**
—Antenodal crossveins six . . . . **dissocians**

Nymphs

1—Venter of abdomen pale, with double row of brown dots or dashes 2
—Venter obscure, without obvious pattern 3

2—Venter with minute round dots; lateral setae of labium nine or ten . . . . **aequalis**
—Venter with large transverse dashes; lateral setae eleven or twelve . . . . . . **didyma**

3—Dorsum of abdomen pale basally, blackish beyond middle . . . . **hageni**
—Abdomen clouded, with a pair of submedian bands of brown . . . . **dissocians**

Nymph unknown: **debilis.**

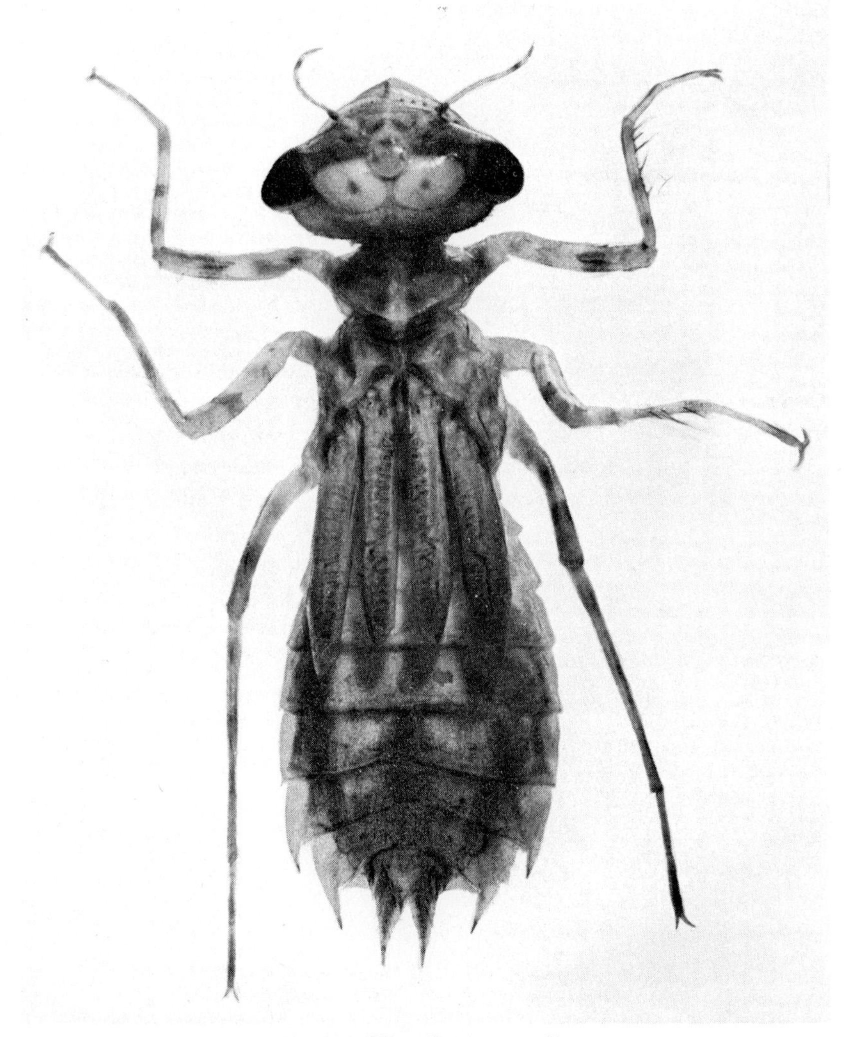

Fig. 301. *Micrathyria aequalis.*

Micrathyria aequalis Hagen

Syn.: septima Selys

1861. Hagen, Syn. Neur. N. Amer., p. 167 (in *Dythemis*).
1895. Calv., Proc. Calif. Acad. Sci., (2)4:543 (figs.).
1906. Calv., B. C. A., p. 229.
1943. Ndm., Ann. Ent. Soc. Amer., 36:187 (nymph).

Length 28–33 mm.; abdomen 16–22; hind wing 21–25.

A small blackish species, with top of head shining metallic green. Face pale, besprinkled with short black hairs. Occiput black, with white hair on hind margin.

Front of thorax black, with yellow stripes, becoming wholly black with age in old males. While teneral there is a pair of incomplete yellowish stripes that extend downward and broaden to cover middle coxae, and a pale cross streak of same color along lower margin of crest. Sides of thorax paler, with three diffuse brown stripes on the three lateral sutures, extending downward in a narrow antehumeral stripe. At their confluent lower ends are some yellow spots about leg bases. Wings hyaline, their membrane shining.

Abdomen slender, a little widened at both ends in both sexes. A broad pale band on sides of segments 1 to 3, crossed by narrow lines of black on encircling carinae. A large and conspicuous yellow spot on each side of dorsum of 7, and a much smaller one on 6. In female, these are the two end spots of a continuous lateral band of pale spots. Caudal appendages black.

Female does not drop her eggs into water during flight, but settles upon a floating leaf, bends abdomen downward in a U shape, and with its tip well under leaf fastens eggs in a single layer to under side of leaf.

Distribution and dates.—MEXICO: Baja Calif.; ANTILLES: Cuba, Dom. Rep., Haiti, Jamaica, P. R.; also from Martinique, and from Mexico south to Ecuador.

Apparently year-round.

Micrathyria debilis Hagen

1861. Hagen, Syn. Neur. N. Amer., p. 168 (in *Dythemis*).
1906. Calv., B. C. A., pp. 223, 229 (figs.).
1911. Ris, Coll. Selys Libell., p. 447.

Length 24–26 mm.; abdomen 15–19; hind wing 18–22.

A little black species, at once distinguished by smaller size and reduced wing venation. Face yellowish white; vertex of male steel blue.

Front of thorax blackish, clothed with ashen hair and pruinose in male; sides brassy black. Obsolescent dull yellowish streaks extend divergently upward from base of middle and hind legs. Legs black; wings hyaline. Venation notably scanty, with only one cell in fore-wing triangle; subtriangle with only two cells flanking inner margin; five paranal cells.

Abdomen cylindric on middle segments, widened again on 7 and 8; black, with a pair of large half-length semioval yellow spots on dorsum of 7; 9, 10, and caudal appendages black.

Distribution and dates.—MEXICO: Tamaulipas; ANTILLES: Cuba; also south to Guatemala.

June (Tamaulipas) to September 12–23 (Cuba).

Micrathyria didyma Selys

Syn.: dicrota Hagen, phryne Rambur, poeyi Scudder

1857. Selys, in Sagra, Hist. Cuba, Ins., 7:453 (in *Libellula*).
1861. Hagen, Syn. Neur. N. Amer., p. 166 (as *Dythemis dicrota*).
1906. Calv., B. C. A., pp. 221, 223.
1911. Ris, Coll. Selys Libell., p. 430.
1932. Klots, Odon. of P. R., p. 38 (figs.).
1943. Ndm., Ann. Ent. Soc. Amer., 36:189 (nymph).

Length 35–41 mm.; abdomen 22–29; hind wing 25–33.

Largest of our species, with densest wing venation. Differs from our other species in having generally two cells in triangle of fore wing and three in subtriangle. Face whitish. Top of frons covered with a metallic spot of shining green. Vertex also green; occiput has touches of yellow.

Front of synthorax brown, with a yellow middorsal carina, becoming metallic green in old males. While teneral it shows a pair of abbreviated dorsal stripes of yellowish green, a very narrow antehumeral stripe, with upper end turning inward beneath crest. Sides greenish, with three oblique brown stripes on three lateral sutures, diffuse and conjoined at upper and lower ends. A dark area below encloses one or more yellow spots above leg bases. Legs black; wings hyaline.

Abdomen slender and parallel-sided. Usual lateral pale spots on broad sides of basal segments, greenish in color, and a large and conspicuous pale spot on 7. Female has a continuous band of lateral spots on 3 to 7. Another very distinctive character is the very large genital hamule, its exposed portion long, bent far forward, truncated and serrated on end.

Distribution and dates.—MEXICO: Baja Calif., Tamaulipas; ANTILLES: Cuba, Dom. Rep., Haiti, Jamaica, P. R.; also south to Ecuador.

January 11 (Jamaica) to October (Baja Calif.). Probably year-round.

Micrathyria dissocians Calvert

1906. Calv., B. C. A., pp. 222, 226 (figs.).
1911. Ris, Coll. Selys Libell., p. 540.
1932. Klots, Odon. of P. R., p. 39 (nymph).

Length 37–40 mm.; abdomen 21–28; hind wing 24–29.

Adult males become heavily pruinose blue on bases of wings, on back between wings, and on abdominal segments 1 to 3. Face whitish; frons brown, becoming metallic bluish green in old males. Occiput black.

Thorax brown in front, with yellowish green carina, narrow and divergent middorsal antehumeral stripes, with their upper ends prolonged beneath crest. Sides of thorax yellowish, with forking and variously conjoined bands of brown on three lateral sutures. From third stripe a branch runs to rearward. Legs black, with a streak of green on under side of front femur. Wings tinged with yellowish in their membrane toward costal border, with a small flavescent basal spot in hind wings. Following combination of venation characters seems distinctive of species: fore-wing triangle one-celled, subtriangle three-celled, and two or three rows of cells in trigonal interspace.

Abdomen blackish, pale sides of basal segments with greenish markings, brown sides generally with pale spots on 4 and 5, and again a large one on 7. Segments 8 to 10 and caudal appendages black.

Distribution and dates.—ANTILLES: Cuba, Dom. Rep., Jamaica, P. R.; also from Mexico.

January (P. R.) to July 28 (P. R.).

Micrathyria hageni Kirby

1889. Kirby, Trans. Zool. Soc. London, 12:368.
1911. Ris, Coll. Selys Libell., p. 438.
1932. Klots, Odon. of P. R., p. 41.
1943. Ndm., Ann. Ent. Soc. Amer., 36:187 (nymph).

Length 33–35 mm.; abdomen 19–24; hind wing 25–29.

A slender blackish pond species, with abdomen moderately widened at both front and rear ends. Face yellowish. Top of frons metallic blue in male, tawny in female.

Front of thorax brown, with a yellow-edged carina and a rather wide pale stripe on each side that does not reach up to crest. A crooked pale stripe at humeral suture, and two cross streaks on each side, one within crest and other close under it. Sides of thorax greenish, with three oblique stripes of brown, middle one generally forked above and conjoined with its neighbors at top. Legs blackish beyond their short basal segments, except for pale under side of front femur. Wings have a flavescent tinge that fades toward end; in base of hind wings, a diffuse brownish spot reaches hardly halfway to anal crossing.

Abdomen slender and cylindric on middle segments and moderately widened on segments 7 to 9; blackish, with a line of pale spots on each side: greenish spots on basal segments, progressively diminishing in size; very large bright yellow on 7, covering most of dorsum. In female, spots larger, continuing subequal all the way back, with a smaller spot on 8. Segments 9, 10, and caudal appendages black in both sexes. In old males, color pattern much obscured by pruinosity.

Distribution and dates.—UNITED STATES: Tex.; MEXICO: Baja Calif., Tamaulipas; ANTILLES: Cuba, Dom. Rep., Jamaica, P. R.; also south to Costa Rica.

Apparently year-round; dates include February (Jamaica), July (Tex.), and October (Baja Calif.).

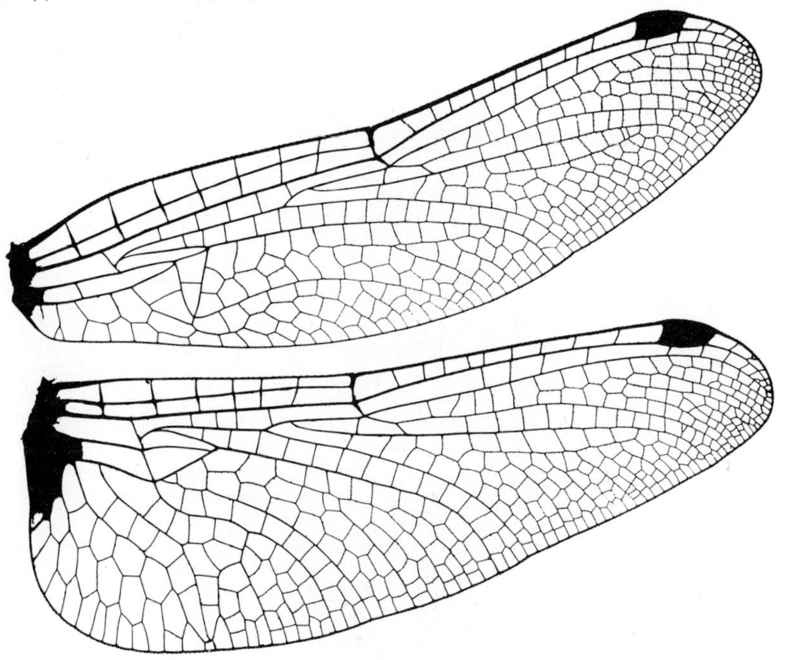

Fig. 302. *Leucorrhinia intacta.*

Genus LEUCORRHINIA Brittinger 1850

White-faces

These small, low-flying, white-faced dragonflies are of predominantly northward distribution. They are clear-winged save for some small brown streaks at the wing roots. The face is white, thinly beset with short black hairs; the labrum is yellowish, and the top of the head mainly deep black. The males are larger than the females.

The ground color of the thorax is red or dull yellow or tawny, heavily but diffusely marked with black; these markings are obscured by a dense growth of long silky hair. On the front the middorsal black stripe is bordered by two broad pale parallel bands that converge upon it at their upper ends. The sides of the thorax are clouded heavily with black along the full length of the first and third lateral sutures, and parts of the midlateral one with varying cross connections. The legs are black. The wings are hyaline, for the most part clear and shining, sometimes tinged with smoky brown outward for a distance from the base, especially in the females.

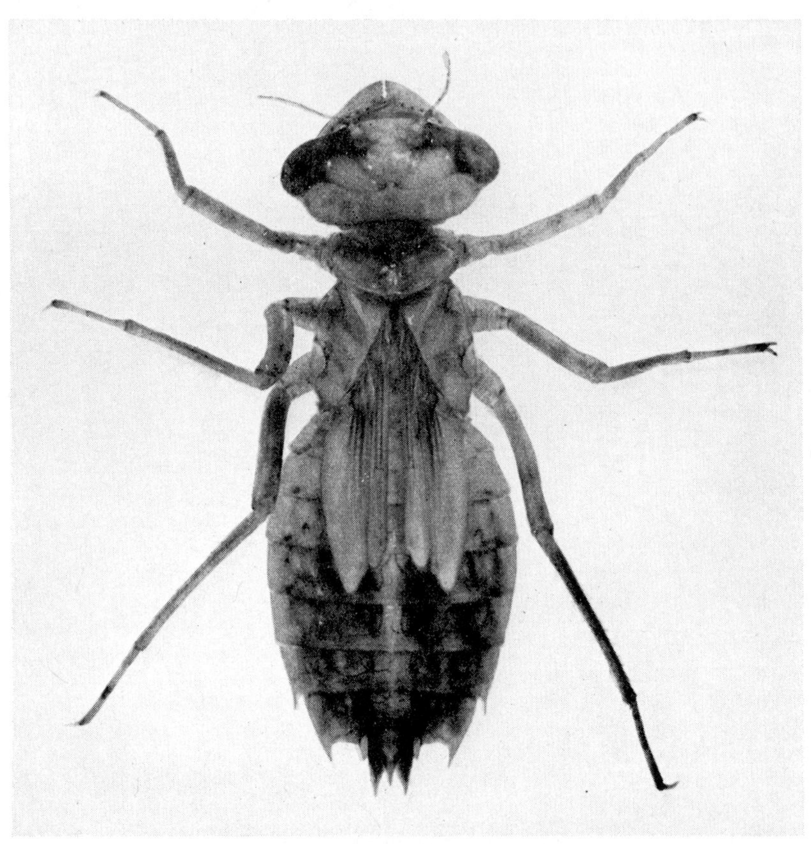

Fig. 303. *Leucorrhinia glacialis.*

The most distinctive venational characteristics of the genus are the enlarged heel and the short toe of the anal loop, and the short thick stigma, which in the males is only about twice as long as wide; in the females, longer. The midrib of the anal loop often appears forked at the ankle, where one or more ankle cells are interpolated between the two rows. Other characters are two crossveins under the stigma in all wings, the first one often aslant like a brace vein; radial planate well developed, median planate weakly developed; oftenest there is a transverse terminal row of four cells within the clubfoot-shaped anal loop along its sole.

The abdomen is mostly pale on the moderately swollen basal segments and blackish beyond them. It is brightly marked with red in the males; more extensively with yellow in the females. On the more slender middle segments is an interrupted middorsal row of pale spots, which may dwindle, segment by segment, until they disappear to rearward; 9 and 10 and the caudal appendages are black.

The nymphs are climbers in the midst of green waterweeds. They are clean, pale nymphs, green in life, daintily patterned in brown. On the ventral side of the abdomen occur these color characters: two rows of paired brown dots or else three longitudinal brown bands. The genus is often recognizable by these marks alone. The slender legs are faintly ringed with brown. In the armature of the labium are nine to twelve lateral setae and ten to fifteen mental setae. The lateral spines of the abdomen are well developed on segments 8 and 9, much longer on 9 than on 8. The superior caudal appendage is nearly as long as the inferiors.

Both nymphs and adults often abound in sphagnum pools and other boggy places. The adults fly low and keep near shore or near floating vegetation. They are rather easy to approach when found perching, but they can make a quick getaway and are artful at dodging a net.

The fading of specimens soon after they are taken is very disappointing to the collector. During the first week or two of adult life, before age and pruinosity have dimmed their fine colors, these are very beautiful insects. One who sees only preserved specimens would hardly expect this, for faded browns have then replaced the ruby red color of the males and the bright yellow of the females.*

The genus is Holarctic, represented in North America principally in the Canadian life zone. The most important papers dealing with the genus as a whole are those of Ris (*Coll. Selys Libell.*, 14:701–721, 1912) and Hagen (1890) for adults, and the several papers of Walker (1913, 1914, 1916) for American nymphs. (Hagen and Walker are cited under the different species.)

KEY TO THE SPECIES

Adults

1—Radial planate subtends two rows of cells (in at least some wings for a distance of two or more cells) . . . . **glacialis**
—Radial planate subtends single row of cells **2**

* "I well remember with what delighted surprise I greeted my first living specimen. It was a male with a brilliant red body phalerate with jet black, a flavescent tinge beyond the basal markings of the wings, a rich red-brown stigma with a touch of yellow on the costa either side of it, and a face with the whiteness and subopaqueness of fine china."—From the field notebook of the senior author.

2—Hind wing 29 mm. or more; ankle cells four to eight.................. **borealis**
—Hind wing less than 29 mm.; ankle cells one to four....................... 3

3—Middle abdominal segments black, at least in male....................... 5
—These segments marked with yellow in both sexes......................... 4

4—Middle abdominal segments bear middorsal line of narrowly linear yellow streaks .... **patricia**
—These segments bear line of broadly oval yellow spots............ **hudsonica**

5—Vein Cu1 in hind wing springs from hind angle of triangle or from very close to that angle.. **intacta**
—Vein Cu1 springs from outer side of triangle............................. 6

6—Trigonal interspace of fore wing with two rows of cells for a distance... **frigida**
—Trigonal interspace of fore wing with three rows of cells.............. **proxima**

TABLE OF SPECIES

ADULTS

Species	*Hind wing*		*Fore wing*		
	Length	*Antenodals*[1]	*Cu1*[2]	*Sub T*[3]	*Cell rows*[4]
borealis	29-32	6-7	opp.	3	3
frigida	21-24	5-6	up	1-2-3	2
glacialis	26-29	6-7	up	3	3
hudsonica	21-27	5-6-7	var.	1-2-3	2-3
intacta	23-25	5-6	var.	1-2-3	2-3
patricia	18-25	5-6	up	2-3	2
proxima	24-27	5-6	up	3	3

[1]Number of antenodal crossveins in hind wing.

[2]Point of origin of vein Cu1 in fore wing; from hind angle of triangle (opp.), or from outer side of triangle (up), or variable (var.).

[3]Number of cells in subtriangle of fore wing.

[4]Number of cell rows in trigonal interspace of fore wing.

NYMPHS

1—No dorsal hooks on abdominal segments 7 and 8; three wide longitudinal dark bands on under side of abdomen.................................. 2
—Dorsal hooks present on 7 and 8; no distinct band on under side of abdomen.. 4

2—Lateral spines of 9 pointing straight to rearward, with axes parallel..**hudsonica**
—Lateral spines of 9 distinctly convergent.................................. 3

3—Lateral spines of 8 small and close-laid to side margins of 9.......... **borealis**
—Lateral spines of 8 larger and distinctly outstanding................. **glacialis**

4—Tips of lateral spines of 9 about on a level with tips of inferior caudal appendages .... **frigida**
—Tips of spines of 9 distinctly shorter.. 5

5—Dorsal hooks on 7 and 8 slender; spines of 8 and 9 incurving........... **proxima**
—Dorsal hooks on 7 and 8 stouter; spines of 8 and 9 straight.............. **intacta**

Nymph unknown: **patricia.**

TABLE OF SPECIES

NYMPHS

Species	*Total length*	*Setae of labium*		*Lateral spines*[1]		*Caudal appendages*[2]		
		Lat.	*Ment.*	*On 8*	*On 9*	*Lat.*	*Sup.*	*Inf.*
borealis	18-21	10-11	13-14	3	6	4	9	10
frigida	15-16	9-10	10-13	7	9	4	9	10
glacialis	16-18	9-11	13-14	4	7	5	9	10
hudsonica	17-18	9-11	12-14	5	7	4	9	10
intacta	17-18	10-12	13-14	7	9	4	9	10
proxima	19-20	10-11	11-15	4	7	4	9	10

[1]Relative length of lateral spines expressed in tenths of length of their respective segments.

[2]Relative length of caudal appendages, lateral, superior, and inferior, expressed in tenths of length of inferiors taken as 10.

Leucorrhinia borealis Hagen

1890. Hagen, Trans. Amer. Ent. Soc., 17:231 (figs.).
1908. Mtk., Odon. of Wis., p. 113.
1916. Wălk., Can. Ent., 48:416 (figs., nymph).
1941. Whts., Odon. of B. C., p. 553.

Length 44–46 mm.; abdomen 18–20; hind wing 29–32.

Largest species of genus, with longest, broadest, and most continuous middorsal stripe of red on abdomen. Top of vertex whitish, middle of occiput pale. Thorax very hairy; more red than black in male; in female, more yellow. Costa yellow its full length; so also some adjacent cross-veins. Planates rather more equally developed than in other species; median less zigzagged. From swollen basal segments of abdomen a broad, nearly continuous band of red extends rearward so far that it covers segment 8 in male; 7 in female.

Distribution and dates.—Alaska; CANADA: Alta., B. C., Man., NW. Terr., Ont., Sask., Yukon; UNITED STATES: Wyo.

June 3 (Sask.) to July 15 (Man.).

Leucorrhinia frigida Hagen

1890. Hagen, Trans. Amer. Ent. Soc., 17:231 (fig.).
1908. Mtk., Odon. of Wis., p. 112.
1913. Walk., Can. Ent., 45:161–170 (figs., nymph).
1927. Garm., Odon. of Conn., p. 280.
1929. N. & H., Handb., p. 243.

Length 28–32 mm.; abdomen 18–21; hind wing 21–24.

A delicate little blackish species, with very little discoverable color pattern, even in teneral specimens. Vertex and occiput black. Thorax brown under a thin covering of whitish hair, with stripings evident only at front and in young specimens; later, all black. Basal markings of hind wings black, extending out far enough to cover first strong crossveins. In female, wing membrane tinged with brown out to triangles.

This species has scantiest venation, with all triangles and subtriangles generally of one or two cells, latter rarely of three cells. Bordering outer side of subtriangle in fore wing are only two cells, one in paranal row and one below it; in other species, two below it. Yet a half-antenodal crossvein is often present, sometimes also an extra bridge crossvein.

Adult male easily recognized by grayish pruinosity that covers swollen basal segments of abdomen. Beyond, middorsal line of pale spots almost lacking, reduced to short narrow lines on segments 5 or 6 and 7. Appendages black.

Distribution and dates.—CANADA: Man., N. S., Ont., Que.; UNITED STATES: Conn., Ind., Maine, Mass., Mich., N. H., N. J., N. Y., N. D., Pa., Vt., Va.

May (N. Y.) to August 19 (Ont.).

Leucorrhinia glacialis Hagen

1890. Hagen, Trans. Amer. Ent. Soc., 17:234 (figs.).
1927. Garm., Odon. of Conn., p. 281.
1927. Walk., Odon. of Can. Cordillera, Prov. Mus. Nat. Hist., Victoria, B.C., p. 15.
1929. N. & H., Handb., p. 243.
1941. Whts., Odon. of B. C., p. 555.

Length 34–35 mm.; abdomen 21–24; hind wing 26–29.

A ruby red and jet black species—in life and at maturity one of most beautiful of Odonata, but with most fugitive coloration. Reds fade to dull brown. Vertex black; occiput brown. Black stripe on front of thorax narrowed above to broadly triangular form. Cloud bands of first and third lateral sutures confluent above spiracle, and a long red stroke separates antehumeral stripe from humeral in their lower half. Female more broadly yellow where male is red. Wings hyaline, with rather abundant venation.

Abdomen predominantly shining black, but is more red than black on swollen basal segments. Beyond these, middorsal row of narrow spear-heads is reduced progressively to end on about middle of segment 7.

Distribution and dates.—CANADA: Alta., B. C., N. B., N. S., Ont., Que., Sask.; UNITED STATES: Calif., Maine, Mass., Mich., Nev., N. H., N. Y., Wis., Wyo.

May 19 (New England) to August 22 (B. C.).

Leucorrhinia hudsonica Selys

Syn.: hageni Calvert

1850. Selys, Revue Odon., p. 53 (in *Libellula*).
1861. Hagen, Syn. Neur. N. Amer., p. 180.
1914. Walk., Can. Ent., 46:375 (figs., nymph).
1890. Hagen, Trans. Amer. Ent. Soc., 17:233 (figs.).
1927. Garm., Odon. of Conn., p. 282.
1929. N. & H., Handb., p. 241.
1941. Whts., Odon. of B. C., p. 554.

Length 27–32 mm.; abdomen 18–20; hind wing 21–27.

A dainty little species, rather elegantly marked with red in male and with yellow in female. Top of head black, with touches of paler color on middle of vertex and occiput. A rather dense coat of blackish hairs covers front of thorax above and base of abdomen beneath. Middorsal thoracic black stripe contracted at upper end to a rectangular form resting on collar. Antehumeral and humeral stripes conjoined at both ends, but separated between by a pale streak and followed by three larger irregular pale areas farther to rearward. Wings hyaline, with veins brown; costa yellowish from nodus, becoming bright yellow about stigma in male.

Abdomen black, with usual predominantly red or yellow swollen basal segments. Low down on sides of 3 are patches of long whitish hair. Middorsal line of spots on 4 to 7 narrow and abbreviated. Radial and median planates almost equally well developed. Venation somewhat unstable, sometimes with an extra cubito-anal crossvein; or it may be a bridge crossvein or a half-antenodal.

Inferior caudal appendage of male notched at tip in a wide V, with upturned tooth-tipped outer corners. Subgenital plate of female extends to rearward a little more than halfway to palps of ninth sternum; cleft to base into two elongated, somewhat triangular plates, straight-edged and in contact along inner sides.

Distribution and dates.—Alaska; Labrador; CANADA: Alta., B. C., Man., N. B., Nfld., NW. Terr., N. S., Ont., P. E. I., Que., Sask., Yukon; UNITED STATES: Calif., Maine, Mass., Mich., Minn., Nebr., N. H., N. Y., Oreg., Pa., Utah, Wis.

May 7 (B. C.) to August 26 (B. C.).

Leucorrhinia intacta Hagen

1861. Hagen, Syn. Neur. N. Amer., p. 179 (in *Diplax*).
1890. Calv., Trans. Amer. Ent. Soc., 17:38 (figs.).
1901. Ndm., Bull. N. Y. State Mus., 47:517 (nymph).
1908. Walk., Ottawa Natural., 22:59 (figs.).
1916. Walk., Can. Ent., 48:417 (figs., nymph).
1927. Garm., Odon. of Conn., p. 282.
1929. N. & H., Handb., p. 242.
1941. Whts., Odon. of B. C., p. 555.

Length 29–33 mm.; abdomen 19–22; hind wing 23–25.

A common farm-pond species. Face all white down to yellow labrum. Labium black, with a large squarish pale spot on each lateral lobe. Top of black vertex and occiput partly pale. Hairs on face short and very sparse. Thorax densely clothed with tawny hairs in front and with pale-tipped hairs on sides and next to abdomen beneath. Pattern of black stripes of thorax very obscure, bands being wide, irregular, and broadly confluent.

Abdomen black, with usual pale areas on swollen basal segments. A conspicuous dorsal twin-spot on 7, isolated in male; but in female preceded by smaller spots on 4, 5, and 6. Segments 8, 9, 10, and appendages black.

Pale spot on 7 will generally distinguish this species from others of genus; further set off by extremely wide fork in which inferior appendage of male terminates—wider than spread of superiors; also by a pair of minute rounded spinulose lobes on rim of anterior lamina of genital pocket; also, in female by subgenital plate, which is not platelike at all, but is developed as two separate linear flaps that stand separate from each other by a distance almost equal to their own length. They do not reach backward to level of palps on segment 9.

Distribution and dates.—CANADA: Alta., B. C., Man., N. B., N. S., Ont., P. E. I., Que., Sask.; UNITED STATES: Calif., Colo., Conn., Idaho, Ill., Ind., Iowa, Ky., Maine, Mass., Mich., Minn., Nev., N. H., N. J., N. Y., N. Dak., Ohio, Oreg., Pa., R. I., S. Dak., Tenn., Utah, Vt., Wash., Wis.

April 18 (N. J.) to August 18 (Tenn.).

Leucorrhinia patricia Walker

1940. Walk., Can. Ent., 72:12 (figs.).
1942. Walk., Can. Ent., 74:74 (figs.).

Length 24–29 mm.; abdomen 17–21; hind wing 18–25.

Very similar to *hudsonica,* but with dorsal spots of middle abdominal segments reduced to short linear dashes or in part wanting.

Superior caudal appendages have their tips tilted up strongly, and the four or five little teeth on their under sides a little longer than in *hudsonica.*

Subgenital plate of female extends a little less than halfway out to genital palps on segment 9; its two lobes, roughly triangular and equilateral, arise well apart at their base.

Distribution and dates.—CANADA: Man., NW. Terr., Ont.

June 16 (NW. Terr.) to July 17 (NW. Terr.).

Leucorrhinia proxima Calvert

1890. Calv., Trans. Amer. Ent. Soc., 17:38 (fig.).
1890. Hagen, Trans. Amer. Ent. Soc., 17:236 (figs.).
1898. Harvey, Ent. News, 9:87 (variability).
1916. Walk., Can. Ent., 48:420 (figs., nymph).
1927. Garm., Odon. of Conn., p. 284.
1929. N. & H., Handb., p. 244.
1941. Whts., Odon of B. C., p. 555.

Length 33–36 mm.; abdomen 22–25; hind wing 24–27.

Lighter coloration somewhat more extended and hair on thorax somewhat less dense than in other species. Vertex and occiput black; anteclypeus sometimes dull brown.

Hind lobe of prothorax conspicuously fringed with long white hair. Pale stripes on front of thorax converge at upper ends to truncate squarish middorsal black stripe; at lower ends they are confluent to rearward with pale stripe between antehumeral and humeral black stripes. First and third lateral stripes complete; irregular oblique black band connecting them makes with them a big letter N on side.

Abdomen more red on back of swollen basal segments than in preceding species; in old males these segments become thinly pruinose on darker under side. Narrow middorsal spots of yellow or red reduced to cover little more than a part of carina on segments 4 to 7. Superior caudal appendages of male blunt and a little divergent at tips; straight, somewhat club-shaped; scarcely arched at all in their basal half; on the under surface is a line of from nine to eleven minute denticles. Stout bristles that taper to their slender tips beset segment 1 beneath; they also fringe erect edges of bilobed anterior lamina and end of genital lobe of 2. Inferior appendage parallel-sided and ends in a wide U-shaped emargination of tip. Subgenital plate of female so short as to be easily overlooked: a pair of rounded lobes, each more than twice as wide as high.

Walker (*Odonata of Can. Cordillera,* 1927, p. 15) states, "Generally distributed through the Canadian Zone across the continent and usually the commonest *Leucorrhinia*."

Distribution and dates.—Alaska; CANADA: Alta., B. C., Man., N. B., NW. Terr., N. S., Ont., Que., Sask., Yukon; UNITED STATES: Maine, Mass., Mich., Minn., N. H., N. Y., Wash., Wyo.

May 15 (B. C.) to August 22 (B. C.).

Genus ERYTHRODIPLAX Brauer 1868

This is a large genus of Neotropical dragonflies, a few of which enter our southern limits. They are mostly rather small, with varicolored wings that often differ markedly in extent of coloration in the two sexes. The changes in superficial appearance with age are great; in consequence, determination of species calls for careful discrimination of structural characters.

The face is pale in teneral specimens, but may become metallic black, or shining brown or red, in old ones. The frons is low and rounded, without transverse carina, and always beset with short blackish hairs. The hind margin of the prothorax rises in an erect lobe that is rounded on the margin, and but rarely slightly notched in the middle. It is fringed with long erect hairs. The synthorax, however much it may be striped or tinted in tenerals, becomes of one dark color in most mature specimens.

The wings are rather broad. Their venation has the following characters: arculus a little nearer to the second antenodal crossvein than to the first; triangle of the fore wing narrow, generally of two cells and followed by two or three rows of cells in the trigonal interspace; triangle of the hind wing generally of a single cell; stigma long, surmounting two crossveins; paranal cells of the fore wing seven, five before and two within the subtriangle (one within in *minuscula*); paranals in the hind wing five or six, three before the anal loop and two or three within it. There are one or more extra cells (ankle cells) within the anal loop in addition to its two regular rows, all on the outer side of its midrib.

The abdomen is rather stocky, slightly compressed on the basal segments, and variously tapered beyond. Segments 3 to 7 become concolorous in old males, but females retain some paler colors; generally there are paired dorsal saddle marks of red or yellow. The terminal segments 8 to 10 and all the longitudinal carinae are blackish.

The ultimate criteria for distinguishing species, here as elsewhere, are found in the structural differences in the genitalia of abdominal segments 2 and 3 of the male and in the subgenital plate of the female. These are minute differences, difficult to observe, and hardly need to be taken into account in naming our few species, for these differ in color and size, and are often well apart in range.

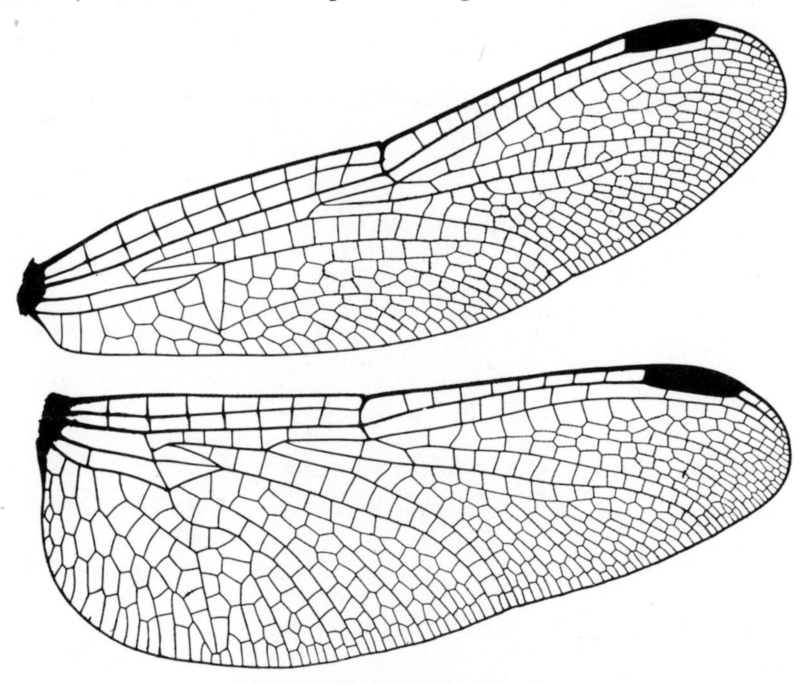

Fig. 304. *Erythrodiplax berenice.*

The nymphs are dwellers in tangles of submerged vegetation. They are of the most common Libelluline form type, differing from nymphs of *Celithemis* mainly in having more broadly rounded eyes, and shorter lateral spines on abdominal segment 9. Dorsal abdominal hooks may be present or lacking; when present they are little developed and occur on fewer segments. Best criteria for determining species are found in the armature of the labium.

The species show considerable variety. The big species (*funerea* and *umbrata*) stand apart from the others because of their more copious venation (see table of species); also because of the wide black bands across both fore and hind wings in mature males.

The latest and best work on the adults of this genus is by Donald J. Borror, *A Revision of the Libelluline Genus Erythrodiplax* (*Odonata*).

Borror lists fifty-one species and subspecies, of which but twelve are known to fall within our range limits. The nymphs of this genus are known for only six of our twelve species. For nymphs the best work is by Elsie Broughton Klots, *Insects of Porto Rico and the Virgin Islands,*

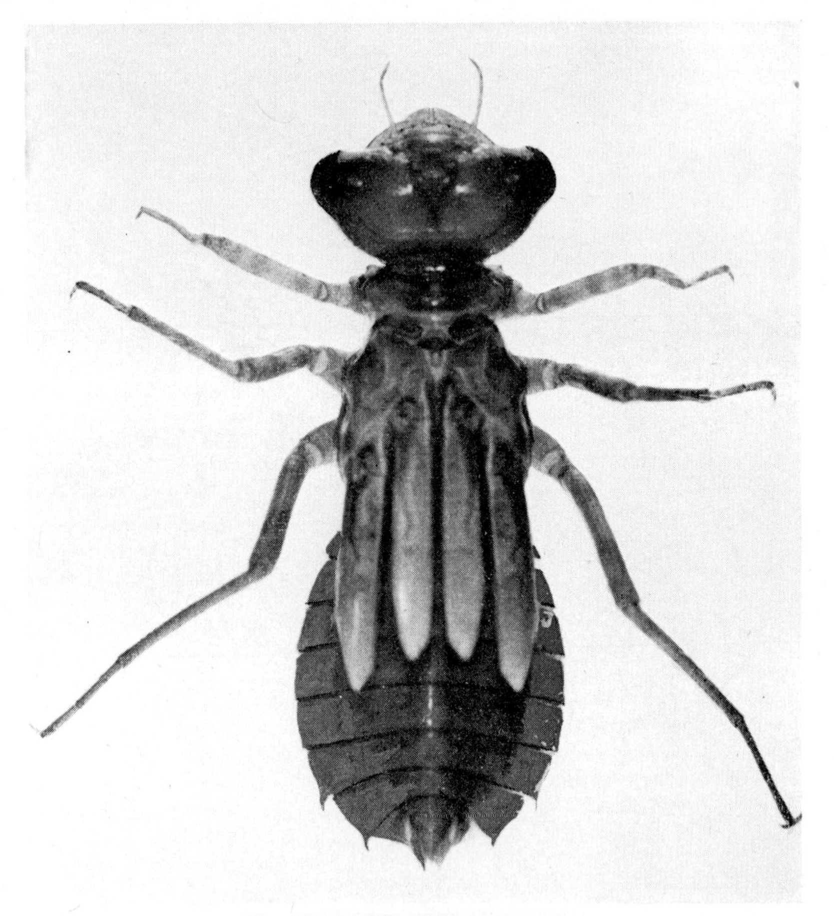

Fig. 305. *Erythrodiplax minuscula.*

Volume XIV, Part 1, "Odonata or Dragonflies," issued by the New York Academy of Sciences in 1932.

The accompanying key does not include the four Neotropical species that have been reported within our limits: *abjecta,* from Baja California, Mexico; *basalis* and *unimaculata,* from Jamaica; and *fusca,* from Tamaulipas, Mexico. We do not know these species. The following points of difference, gleaned from Borror's monograph of the adults of the genus, would seem to separate them.

The larger Jamaican *unimaculata* (hind wing 23–29 mm.) have large conspicuous brown spots at the base of both fore and hind wings. The other three have brown markings at base of hind wings only, with the brown not extending outward as far as the triangle.

Size will generally separate *abjecta* (hind wing 25–39 mm.) from the two following species. In *basalis* (hind wing 16–25 mm.) the adult male, viewed from the side, has yellow spots on the frons and behind the compound eye; *fusca* (hind wing 19–26 mm.) lacks these spots.

KEY TO THE SPECIES

Adults

1—Radial planate subtends two rows of cells.................................. 2
—Radial planate subtends single row of cells (occasionally two rows in **berenice**) 3

2—Median planate generally subtends two rows of cells; Southwestern....**funerea**
—Median planate subtends but a single row of cells....................**umbrata**

3—Vein Cu1 in fore wing arises from hind angle of triangle................... 4
—Vein Cu1 arises from outer side of triangle, well above hind angle.......... 7

4—Blackish species at maturity; trigonal interspace of fore wing with two rows of cells for a little way.. 5
—Brownish species at maturity; trigonal interspace of fore wing with three rows of cells 6

5—Fore-wing subtriangle generally three-celled; base of hind wing with large brown spot ....**justiniana**
—Subtriangle generally a single cell; hind-wing color restricted to extreme base ....**minuscula**

6—Southeastern in range; hind wing with basal brown spot generally extending outward to cover triangle...**fervida**
—Southwestern in range; hind wing with basal brown markings not extending outward to triangle..**connata**

7—Outer branch of posterior hamule of male extends straight to rearward alongside genital lobe..**berenice**
—Outer branch of hamule bent distinctly toward margin of genital lobe...**naeva**

Nymphs

1—Vestigial dorsal hooks on abdominal segments pointed upward............. 2
—Only a short ridgelike middorsal elevation on any segment, or no traces of dorsal hooks.. 3

2—Lateral setae of labium ten or eleven................................**berenice**
—Lateral setae eight or nine..**naeva**

3—A low vestige of middorsal hook on segment 5.........................**minuscula**
—No vestiges of dorsal hooks.. 4

4—Caudal appendages with lateral about half as long as inferiors.......**justiniana**
—Caudal appendages with lateral at least three-fourths as long as inferiors..... 5

5—Superior caudal appendage as long as inferiors.......................**funerea**
—Superior caudal appendage distinctly shorter than inferiors............**umbrata**

Nymphs unknown: **abjecta, basalis, connata, fervida, fusca,** and **unimaculata.**

TABLE OF SPECIES

ADULTS

Species	Hind wing	H. wing an. cvs.[1]	Paranal cells[2] f.w. / h.w.	Rows beyond T[3]	Hind wing		Distribution
					Cul[4]	Ankle cells[5]	
berenice	21-26	*6*, *7*, 8	5+2 / 3+2-3	2-3	up	1-2	E, S
connata	21-26	7, *8*, 9	5+2 / 3+3	3	down	1	SW, W.Ind.
fervida	22-28	7, *8*, 9	5+2 / 3+3	2-3	down	1	W.Ind.
funerea	25-34	8, 9, *10*	5+2 / 3+3	3	down	3	SW
justiniana	18-22	6, *7*	5+2 / 3+3	2	down	1	W.Ind.
minuscula	17-21	5, *6*	5+1 / 3+2-3	2	down	1	SE, C
naeva	23-27	6, *7*, 8	5+2 / 3+2-3	2-3	up	1	SE, W.Ind.
umbrata	25-33	8, *9*, 10	5+2 / 3+3	3	down	1-3	SE, C

[1]Number of antenodal crossveins in hind wing. Prevailing numbers italicized.

[2]Number of paranal cells of: fore wing and / hind wing, each counted in two groups as indicated by + sign: before + in subtriangle.

[3]Number of cell rows in trigonal interspace (discal rows) counted beyond triangle where complete rows are fewest.

[4]Point at which vein Cul in hind wing leaves triangle: at its hind angle (down) or part way up its outer side (up).

[5]Number of extra cells in anal loop included between its two marginal rows.

TABLE OF SPECIES

NYMPHS

Species	*Total length*	*Labium*				*Dorsal hooks*[5]	*Lat. spines*[6]		*Caudal appendages*[7]		
		Setae		*Teeth*[3]	*Spinls.*[4]		*On 8*	*On 9*	*Lat.*	*Sup.*	*Inf.*
		Lat.[1]	*Ment.*[2]								
berenice	14	9-10	10-11	11-12	3-4	6	3	4	7	8	10
funerea	20	9	11	13±	1	0	3	4	8	10	10
justiniana	15	10	12-13	7-8	1-2	0	5	6	5	6	10
minuscula	14	7	11	10	1	5	7	8	8	9	10
naeva	15-17	6	8-9	7	3	6	3	4	7	9	10
umbrata	16-17	10	13	11	2	0	3	4	6	9	10

[1]Number of raptorial setae on lateral lobe of labium.

[2]Number of raptorial setae on each side of median lobe of labium.

[3]Number of teeth on opposed edges of each lateral lobe.

[4]Number of spinules on single teeth near middle of row.

[5]Number of abdominal segment on which highest vestigial middorsal hooks occur.

[6]Relative length of lateral spines on segments 8 and 9, expressed in tenths of middorsal length of respective segments.

[7]Length of caudal appendages, lateral, superior, and inferior, expressed in tenths of length of inferior taken as 10.

Erythrodiplax berenice Drury

Syn.: histrio Burmeister

1770. Drury, Ill. Exot. Ins., 1:114 (pl. 48, fig. 3) (in *Libellula*).
1904. Calv., Ent. News, 15:174 (nymph).
1929. N. & H., Handb., p. 215.
1929. Shortess, Trans. Amer. Ent. Soc., 55:415–423.
1930. Byers, Odon. of Fla., p. 134.
1942. Borror, Revis. Erythrodiplax, p. 90 (figs.).
1943. Wright, Carnegie Inst. Misc. Pub., 435:125–142.

Length 32–35 mm.; abdomen 21–23; hind wing 21–26.

A brackish-water species. When teneral, may be zebra-striped in bright black and yellow, but it darkens progressively with age until it may become entirely black in male and nearly so in female. Face at first about equally black and yellow, labrum yellow with a black front border; sides of postclypeus and frons yellow, with front blackish. Top of frons and vertex brown and become shining black, with violet reflections.

On thorax a middorsal black stripe is divided by yellow of carina. Another black stripe precedes humeral, with yellow intervening. On sides, black stripes begin in sutures and spread and coalesce at their ends to form a big N mark over humeral area, and a coarse meshwork of black farther back; there is left of the pale color only a row of slender streaks along level of spiracle. Short thin white hair on under side of thorax. Wings hyaline, quite clear in male; often with a thin cloud of brown just beyond nodus, or across basal area, or both, in female.

Abdomen early becomes wholly black in male, but female retains something of large middorsal spots on segments 3 to 7, which may remain yellow or may become red, even as red as blood.

Distribution and dates.—CANADA: Que.; UNITED STATES: Ala., Conn., D. C., Fla., Ga., La., Maine, Md., Mass., Miss., N. J., N. Y., N. C., Pa., R. I., S. C., Tex., Va.; MEXICO: Baja Calif; also from the Bahamas and south to Salina Cruz, Mexico.

Year-round (Fla.).

Erythrodiplax connata Burmeister

Syn.: chloropleura Brauer, communis Rambur, fraterna Hagen. leotina Brauer, portoricana Kolbe

1839. Burm., Handb., p. 855 (in *Libellula*).
1906. Calv., B. C. A., p. 264.
1942. Borror, Revis. Erythrodiplax, p. 173 (figs.).

Length 28–30 mm.; abdomen 19–23; hind wing 21–26.

A brown Southwestern and Antillean species, almost clear-winged, with only a trace of amber in extreme base of fore wing and a yellowish brown spot in base of hind wing that fades out halfway to triangle in male; even less extended in female. Labrum black, with a narrow tawny front border. Face pale brown; top of head darker and, in male, with age the top becomes shining metallic black, with violet reflections.

Thorax pale brown, more or less clouded with darker brown, and paler on under side. Hairs on thorax pale brown.

Abdomen more or less tawny, with all carinae deep black in male, becoming wholly black and thinly covered with powdery blue, except for segment 10, which remains yellowish in both sexes. Caudal appendages black.

Distribution and dates.—UNITED STATES: Ariz., Tex.; ANTILLES: Cuba, Jamaica, P. R.; also south to Argentina.

Apparently year-round; dates include January 29 and March 4 (Jamaica), July 13 and August 23 (Tex.), October 31 (P. R.).

Erythrodiplax fervida Erichson

Syn.: distinguenda Rambur, erichsoni Kirby, fraterna Hagen (in part), justina Selys, pulla Burmeister

1848. Erichson, Insekten, in "Reisen in British Guiana, in den Jahren 1840–1844," by Richard Schomburgk, p. 584 (in *Libellula*).
1942. Borror, Revis. Erythrodiplax, p. 75 (figs.).

Length 32–36 mm.; abdomen 19–24; hind wing 22–28.

A Neotropical species, yellowish brown, becoming dark brown to black on top of frons when fully mature. Pale stigma becomes dark brown. Frons nonmetallic. Hyaline wings marked with a roundish basal brown spot. Spot faint in fore wing and extends outward only as far as second or third antenodal crossvein, but is larger in hind wing, extending out to cover triangle. Spot is yellow before it is brown, but never becomes black.

Large yellow side spots on segments 2 to 7 of abdomen become red, but are less well defined in male. Caudal appendages remain pale. Genital lobe of male squarely truncate on end.

Distribution and dates.—MEXICO: Tamaulipas; ANTILLES: Cuba, Dom. Rep., Haiti, Jamaica, P. R.; also south to Colombia.

Apparently year-round; dates include October 26 (Cuba), December 31, January 30, and April 14 (Jamaica); Wilson collected it in summer in Jamaica.

Erythrodiplax funerea Hagen

Syn.: affinis Kirby, tyleri Kirby

1861. Hagen, Syn. Neur. N. Amer., p. 158 (in *Libellula*).
1899. Calv., Proc. Calif. Acad. Sci., (3)1:398 (in *Trithemis*).
1911. Ris, Coll. Selys Libell., p. 483.
1929. N. & H., Handb., p. 215.
1942. Borror, Revis. Erythrodiplax, p. 114 (figs.).

Length 38–42 mm.; abdomen 20–33; hind wing 25–34.

Largest of our species; when mature, has broadest band of black on wings. Labrum black, with short fringe of tawny hair. Face blackish; top of head very black, but bulging rear of head behind eyes pale, coarsely marked with black blotches.

Teneral specimens have much olive color between thoracic stripings of black, but in old ones all becomes black. Hair on thorax thin and pale. Legs dark beyond their bases except for a single line of pale yellow on outer face of tibiae, also on sides of femora in female.

Nymph has superior caudal appendage as long as inferiors. Teeth on opposed edges of lateral lobes of labium so low that there is scarcely a notch between them.

Distribution and dates.—UNITED STATES: Ariz., Calif., Tex.; MEXICO: Sonora; also south to Ecuador.

July 18 (Ariz.) to August 27 (Sonora).

Erythrodiplax justiniana Selys

Syn.: ambusta Hagen

1857. Selys, in Sagra, Hist. Cuba, Ins., 8:451 (in *Libellula*).
1932. Klots, Odon. of P. R., p. 48.
1938. Garcia, J. Agr. Univ. Puerto Rico, 22:50, 59, 83, 90.
1942. Borror, Revis. Erythrodiplax, p. 155 (figs.).

Length 29–32 mm.; abdomen 15–20; hind wing 18–22.

Small Antillean species, with a big roundish spot on base of hind wing and a little touch of brown in extreme base of fore wing. Both wings otherwise hyaline. Spot in hind wing extends outward to cover triangle and backward to hind angle of wing.

Face brown (black in fully mature specimens), with edgings of paler along lower margin of labrum, postclypeus, and frons. All hair on thorax brown.

Abdomen becomes pruinose in mature males; in females, entirely tawny. Both sexes retain row of pale saddle marks on dorsum of segments 3 to 7; much more completely in female. In both, caudal appendages pale.

Distribution and dates.—ANTILLES: Cuba, Dom. Rep., Haiti, Jamaica. P. R.

Year-round.

Erythrodiplax minuscula Rambur

Topsy

1842. Rbr., Ins. Neur., p. 115 (in *Libellula*).
1904. Ndm., Proc. U. S. Nat. Mus., 27:709 (fig., nymph) (in *Trithemis*).
1908. Brimley, Ent. News, 19:135 (in *Diplacodes*).
1911. Ris, Coll. Selys Libell., p. 524.
1929. N. & H., Handb., p. 215.
1930. Byers, Odon. of Fla., p. 136.
1942. Borror, Revis. Erythrodiplax, p. 169 (figs.).

Length 25–27 mm.; abdomen 14–17; hind wing 17–21.

Our smallest species, as its specific name appropriately indicates. Face at first pale, with a touch of olive on top of frons. Hair of face yellowish. A broader area of olive apparent on front and on lateral prominences of thorax. No black stripes.

Abdomen becomes wholly black in male; both thorax and abdomen in old specimens become pruinose. In female, pale brown of abdomen gets black only in lines on carinae of segments 2 to 7, with a wash of black over 8 and 9; 10 and caudal appendages yellowish.

Distribution and dates.—UNITED STATES: Ala., Fla., Ga., Ind., Ky., La., Md., Miss., N. C., Ohio, Okla., Pa., S. C., Tex., Va., W. Va.

Year-round (Fla.).

Erythrodiplax naeva Hagen

1861. Hagen, Syn. Neur. N. Amer., p. 167 (in *Dythemis*).
1942. Borror, Revis. Erythrodiplax, p. 95 (figs.).

Length 31–34 mm.; abdomen 21–25; hind wing 23–27.

Blackish Antillean species, very similar in appearance to *berenice,* and of about same size. More of a touch of yellow in extreme base of wings in both sexes. Face and thorax reddish brown before they become black. Hair underneath thorax reddish brown. Legs reddish brown, with claws blackish.

Stigma a little more than 3 mm. long in *berenice;* a little less than 3 mm. in *naeva.* Often two rows of cells between veins M2 and Rs in *berenice;* appears to be always a single row in *naeva.* Caudal appendages of male whitish on upper side in *berenice;* black in *naeva.* Six antenodal crossveins in hind wing seems to be prevalent number in *berenice;* seven in *naeva.* But in this they overlap. The two seem not to overlap in range.

Nymph gives evidence of kinship with *berenice* in having teeth on lateral lobes of labium deeper cut and armed with more numerous spinules than in any other species.

Distribution and dates.—UNITED STATES: Fla. (Keys and Dry Tortugas); ANTILLES: Cuba, Jamaica, P. R.; also from Bahamas and south to Colombia and British Honduras.

Year-round; dates include December 22 (Jamaica), January 3 (Fla. Keys), February 24, April 13 (Jamaica), July and August (P. R.), August 25 (Dry Tortugas).

Erythrodiplax umbrata Linnaeus

Syn.: fallax Burmeister, flavicans Rambur, fuscofasciata Blanchard, montezuma Calvert, ruralis Burmeister, subfasciata Burmeister, tripartita Burmeister, unifasciata DeGeer

1758. Linn., Syst. Nat., p. 545 (in *Libellula*).
1911. Ris, Coll. Selys Libell., p. 484.
1928. Calv., Stud. Nat. Hist. Iowa Univ., 12(2):25 (nymph).
1929. N. & H., Handb., p. 215.
1942. Borror, Revis. Erythrodiplax, p. 105 (figs.).

Length 38–45 mm.; abdomen 23–33; hind wing 25–33.

Foregoing list of synonyms indicates extraordinary variableness of appearance in this species according to age and sex.

Species mainly olivaceous, and female remains so. Face largely olivaceous, but becomes shining metallic black in male, with a reddish spot on rear of vertex and another on middle of occiput.

Thorax pale brown, with a tinge of olivaceous, without black stripes except for edgings along carinae; does not become deep black even in male. Hair on its front stiff and brownish; underneath, pale and downy. Broad band upon wings of male slow in developing and, for a longer period than in other species, young teneral males look like females. Wings of females develop a thin wash of yellow along costo-radial strip and beyond stigma to wing tip that deepens on wing tip and fades out to rearward. Caudal appendages pale. Pruinosity develops on sides of thorax in mature males and on under side of rear half of abdomen.

Distribution and dates.—UNITED STATES: Ala., Fla., Ga., La., Ind., Ohio, Okla., Tex.; MEXICO: Nuevo León, Tamaulipas; ANTILLES: Cuba, Dom. Rep., Haiti, Jamaica, P. R.; also south to Argentina.

Year-round (Antilles).

Genus SYMPETRUM Newman 1833

Syn.: Diplax Charpentier

The Topers

This is a large genus of more than fifty species, about a dozen of which are North American. The genus belongs predominantly to the North Temperate Zone. Our species are mostly reddish in color (except

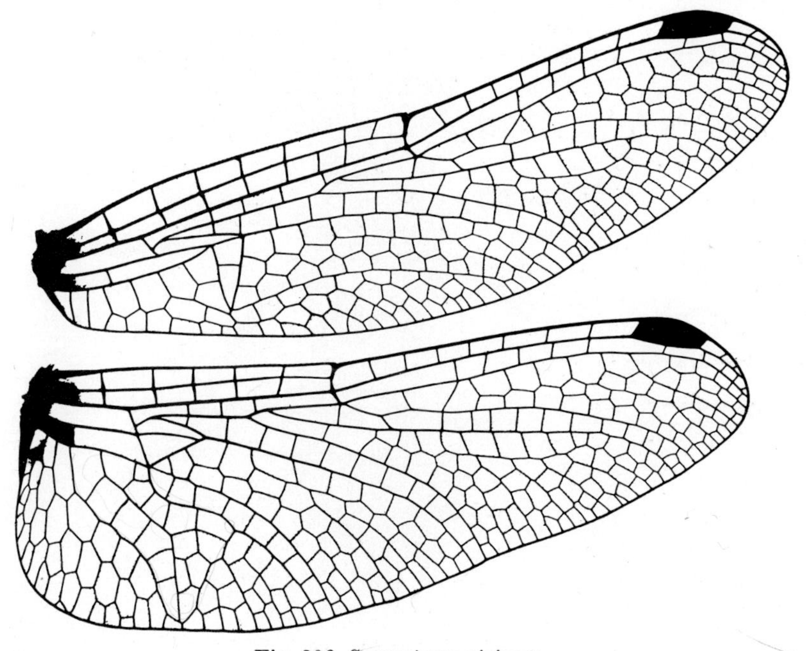

Fig. 306. *Sympetrum vicinum.*

danae), and autumn is their season of flight. The head is of moderate size; the frons low, with well-developed furrow; the vertex high and scarcely bilobed.

The prothorax has an elevated and broadly bilobed hind margin, fringed with long hair. The synthorax is thinly hairy. The legs are slender and the wings often strongly flavescent. The venation is as shown in our table for genera, except for the variations to be found in *S. madidum*.

The abdomen is slender and bare, especially in the male, compressed on the vertically expanded basal segments and parallel-sided beyond. Segment 4 has no extra cross carina.

These are mainly pond species that haunt the reedy shoals and forage over wet meadows and other low vegetation. They fly haltingly and rest frequently, and are among the easiest of dragonflies to capture with a net. But alas! The brilliant reds fade to dull browns in preserved specimens.

The nymphs are slender sprawlers that lie amid the bottom trash or

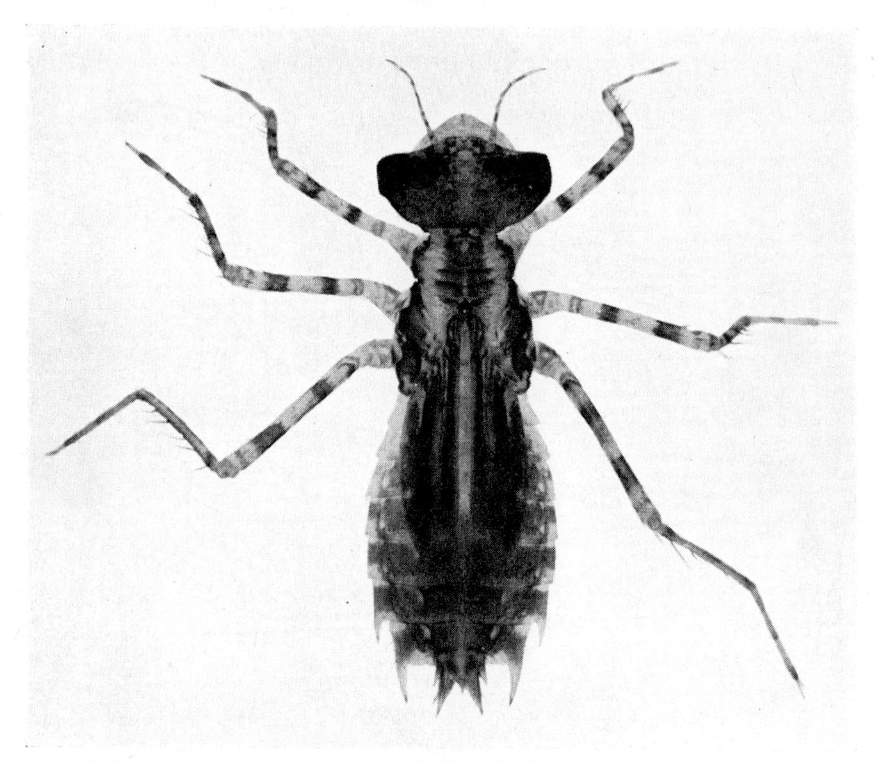

Fig. 307. *Sympetrum vicinum* (eyes accidentally compressed laterally).

clamber over it. They are protectively colored in patterns of dappled greens and browns. They vary greatly in the development of dorsal hooks and lateral spines (which are always present in our species) and in the numbers of raptorial setae on the thin labium. It is somewhat difficult to determine the species.

The most useful reference works for further study of this genus are those by Ris (*Coll. Selys Libell.*) for adults of the world; Walker (*Can. Ent.*, 1917) for North American nymphs; and more recent, lesser papers hereinafter cited under the several species headings.

KEY TO THE SPECIES

Adults

1—Radial planate in fore wing subtends two rows of cells.............. **madidum**
—Radial planate subtends a single row of cells.............................. 2

2—Superior appendages of male with a prominent inferior tooth; subgenital plate of female deeply bifid.. 3
—Superior appendages of male with no prominent inferior tooth (with denticles only); subgenital plate of female entire or a little emarginate...... 7

3—Tibiae yellow externally 4
—Tibiae black externally 5

4—Abdomen with black markings.................................. **ambiguum**
—Abdomen with yellowish brown markings......................... **pallipes**

5—Hamule of male bifid for a third of its visible length, outer (posterior) branch twice as stout as inner; tips of subgenital plate of female upturned against sternum of segment 9..................... **rubicundulum**
—Hamule of male bifid for not more than a fourth of its visible length, outer branch four times as stout as inner; tips of subgenital plate of female flat to rearward, diverging widely.................................... 6

6—Face of mature male cherry red; wing veins reddish................. **internum**
—Face of mature male white; wing veins not reddish; tips of subgenital plate of female extended flat to rearward, divergent at ends...... **obtrusum**

7—Wings tinged with yellow over basal half.............................. 8
—Wings clear, or yellow on costal strip, or only at extreme base..............11

8—Thorax with a streak of black before spiracle, and narrow lines of black in lateral sutures 9
—Thorax without these lateral markings...................... **semicinctum**

9—Wing base with yellow extending a little beyond nodus in both wings and deepening in color outward in both sexes, becoming brown or brownish in a crossband at its outer margin.............................. **fasciatum**
—No such crossband of brown present...................................10

10—Wings with yellow area very faint, almost lacking in female, and in fore wings of male; our palest species .... **californicum**
—Wings with yellow about equally developed in male and female; general color of body dark, becoming blackish in old males.............. **occidentale**

11—Legs more or less yellow...12
—Legs black 13

12—Tibiae entirely yellow .... **vicinum**
—Tibiae striped with black on sides.............................. **costiferum**

13—Superior abdominal appendages of male black; mature body black and yellow .... **danae**
—These appendages yellow or red; mature body black and brown or red.. **atripes**

Nymphs

1—Lateral and mental setae of labium generally equal in number...... **madidum**
—Lateral setae distinctly fewer than mental setae........................ 2

TABLE OF SPECIES

ADULTS

Species	*Hind wing*	*Prevailing color*					*Wings*[5]	*Distribution*
		Face[1]	*Tibiae*[2]	*Caudal app.*[3]		*Genital lobe*[4]		
				Sup.	*Inf.*			
ambiguum	26-28	white	yellow	yellow	yellow	pale	basal	E,S,C
atripes	23-30	olive	black	red	black	yellow	basal	W
costiferum	25-28	pale	pale	red	red	red	costal	NE,W
danae	20-27	black	black	black	black	yellow	basal	NE,W
internum	23-27	red	black	yellow	yellow	yellow	basal	NE,W
madidum	28-31	red	black	red	red	yellow	costal	W
obtrusum	20-29	white	black	red	red	red	basal	NE,W
occidentale	22-28	yellow	black	yellow	black	black	half	NW
pallipes	25-28	yellow	yellow	yellow	yellow	yellow	basal	W
rubicundulum	24-30	yellow	black	red	red	red	basal	E,W
semicinctum	18-23	yellow	black	yellow	black	red	half	E,C.
vicinum	21-23	red	yellow	red	red	red	basal	E,S,NW

[1]Prevailing color of face (mature males).
[2]Color of outer face of tibiae.
[3]Prevailing color of outer side of caudal appendages.
[4]Prevailing color of outer side of genital lobe.
[5]Location of colored area of wings.

2—Abdomen ends rather bluntly; lateral spines of segments 8 and 9 outstanding at sides, these spines fringed with stout spinules, tips of 9 attaining almost or quite the level of tips of inferior appendages.................. 3

—Abdomen more narrowed on 8 and 9; lateral spines of 8 and 9 shorter, inturned, following curvature of segments, and fringed with shorter spinules 4

3—Inferior caudal appendages stout, with blunt tips................**semicinctum**

—These appendages with slender acuminate tips........................**vicinum**

4—Lateral spines of 9 not less than two-fifths as long as whole lateral margin of segment; dorsal hooks well developed, slender, very acute, that of 7 nearly as long as middorsal line of its segment, that of 8 usually more than half as long as its segment.. 5

—Lateral spines of 9 not more than one-third as long as lateral margin of segment; dorsal hooks smaller and when well developed somewhat stouter and distinctly curved, that of 9 distinctly shorter than segment, that of 8 rarely half as long as segment.. 6

5—Base of mentum of labium distinctly wider than middle coxae and somewhat more than one-fourth of greatest width; lateral spines of 9 usually at least half as long as lateral margins of segment; lateral setae typically eleven ....**costiferum**

—Base of mentum of labium scarcely, if at all, wider than middle coxae and about one-fourth of greatest width; lateral spines of 9 two-fifths to one-half as long as lateral margins of segment; lateral setae typically ten..**pallipes**

6—Dorsal hooks present on 4 to 7 or 8, sometimes absent from 4 or 8; lateral spines of 9 about one-third as long as lateral margins of segment; lateral setae nine to eleven.. 7

—Dorsal hooks present on 5 to 7 only, vestigial, or absent altogether; lateral spines of 9 about one-fifth as long as lateral margins of segment; lateral setae usually eleven..**internum**

7—Lateral setae eleven; dorsal hooks vestigial or absent from segments 4 and 8 ....**danae**

—Lateral setae nine or ten; dorsal hooks somewhat larger, generally present, though small, on segments 4 and 8..............**obtrusum** and **rubicundulum**

Nymphs unknown: **atripes** and **occidentale.** Omitted: **ambiguum** (this key prepared before its nymph was known, but see table).

Sympetrum ambiguum Rambur

Syn.: albifrons Charpentier

1842. Rbr., Ins. Neur., p. 106 (in *Libellula*).
1900. Wmsn., Odon. of Ind., p. 177 (as *albifrons*).
1911. Ris, Coll. Selys Libell., p. 689.
1929. N. & H., Handb., p. 235 (figs.).
1943. Wright, Ecol. Monog., 13:493.
1946. Wright, J. Tenn. Acad. Sci., 21(1):135 (figs., nymph).

Length 36–38 mm.; abdomen 23–25; hind wing 26–28.

TABLE OF SPECIES

NYMPHS

Species	*Total length*	*Setae*[1]		*Dorsal hooks*[2]	*Caudal append.*[3]		
		Lat.	*Ment.*		*Lat.*	*Sup.*	*Inf.*
ambiguum	15-17	9	11-12	4-8 large	5	8	10
costiferum	14-15	*11*-12	13-*14*	4-8 spine-like	5	10	10
danae	14	*11*-12	13-*14*	5-7 or 0	5	8	10
internum		10-*11*		5-7 or 0	5	7	10
madidum	14-15	12	12	5-8 low	5	8	10
obtrusum	16	9-*10*	12-13	4-8 small	5+	7	10
pallipes	16-18	*10*-11	*13*-14	4-8 medium	6	8	10
rubicundulum	16-17	9-11	*12*-15	5-8 small; on 5 rud.	7	8	10
semicinctum	14-15	9	12	4-8	5	6	10
vicinum	13-15	*9*-10	12-13	4-8 medium	4	6	10

[1]Number of raptorial setae on one side, right or left, of labium. Prevailing numbers italicized.
[2]Middorsal hooks present on segments indicated, with size.
[3]Relative length of lateral, superior, and inferior caudal appendages.

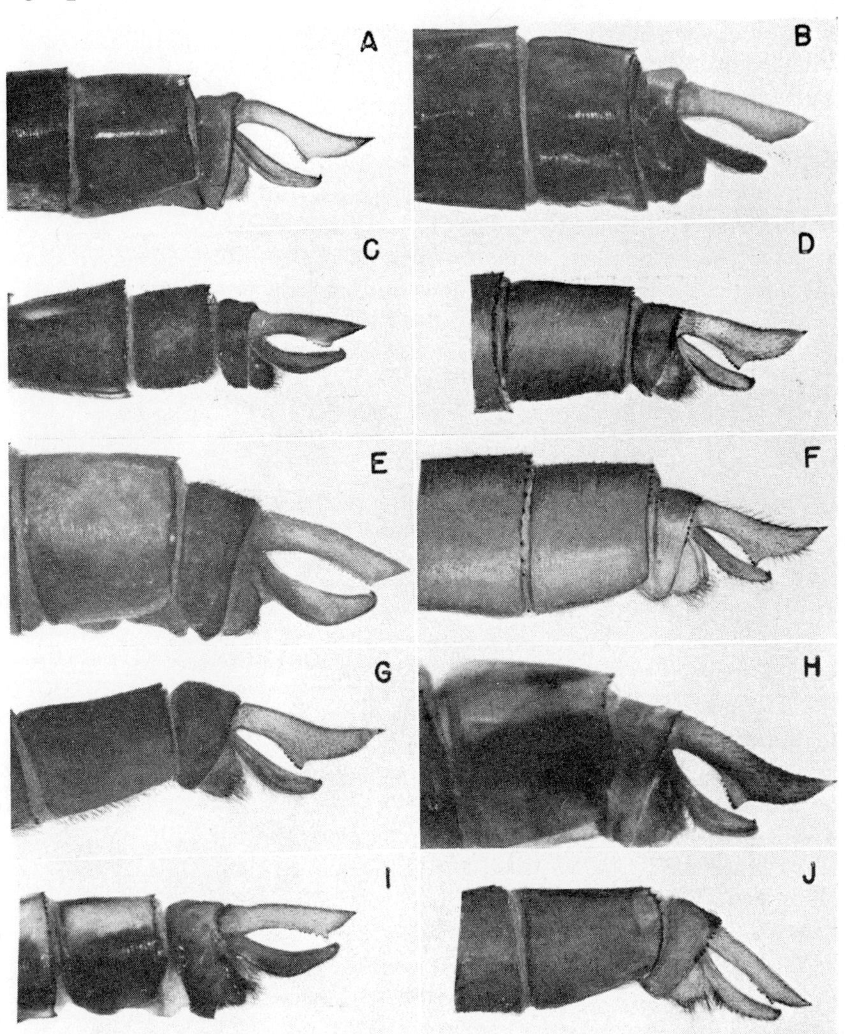

Fig. 308. A, *Sympetrum ambiguum;* B, *S. costiferum;* C, *S. danae;* D, *S. internum;* E, *S. madidum;* F, *S. obtrusum;* G, *S. pallipes;* H, *S. rubicundulum;* I, *S. semicinctum;* J, *S. vicinum.*

A white-faced, clear-winged species, with black-ringed abdomen. Top of frons and vertex yellow to China blue, with a rather wide black cross stripe intervening.

Broad front of thorax pale red brown, with only narrow black edgings to crest. Sides olivaceous, obscurely marked with irregular brown stripes on three lateral sutures. These brown stripes conjoined at their ends above legs and below wings.

Legs pale, with only inner face of tibiae, leg spines, and claws black. Wings hyaline, with faint flavescence in extreme base. Costa yellow; stigma also, but with thickened bordering veins jet black. There appear to be constantly six antenodal crossveins in hind wing. Apical planate undeveloped; radial planate circumscribes and encloses definitely a single row of five cells. Only two rows of cells in anal loop; no ankle cells, and, behind loop, only two rows of cells to marginal row.

Abdomen red in male, with broad and rather diffuse apical rings of black on segments 4 to 9; genital lobe and hamule pale on 2.

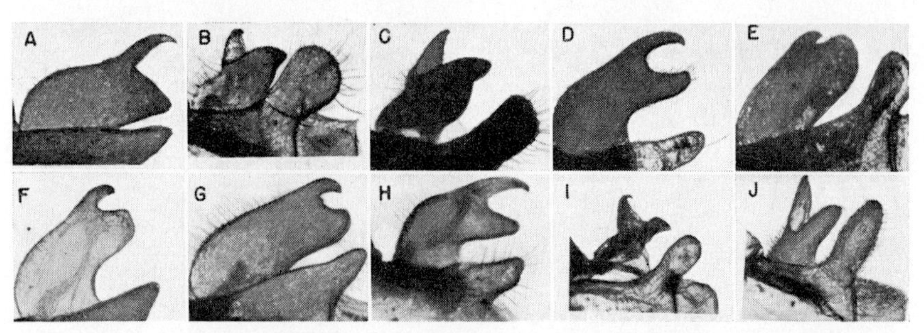

Fig. 309. A, *Sympetrum ambiguum;* B, *S. costiferum;* C, *S. danae;* D, *S. internum;* E, *S. madidum;* F, *S. obtrusum;* G, *S. pallipes;* H, *S. rubicundulum;* I, *S. semicinctum;* J, *S. vicinum.*

"At a glance both sexes of *albifrons* (*ambiguum*) may be recognized by the face: white below, shading into a clear China blue above."—Williamson (1900).

Distribution and dates.—UNITED STATES: Ala., Fla., Ga., Ill., Ind., Kans., Ky., Maine, Mass., Minn., Miss., Mo., N. J., N. C., Ohio, Okla., Pa., S. C., Tenn., Tex.

May 3 (Ala.) to November 10 (Fla.).

Sympetrum atripes Hagen

1873. Hagen, Rep. U. S. Geol. Surv. Terr., 7(3):588 (in *Diplax*).
1929. N. & H., Handb., p. 240 (figs.).

Length 32–40 mm.; abdomen 22–28; hind wing 23–30.

A small red Western species, with black legs. Face yellow and olive, thickly beset with short black hairs. Black band on ridge behind eyes crossed by bars of yellow.

Prothoracic dorsum marked with five purple spots; its rear lobe large and erect, bearing a dense fringe of long tawny hairs. Front of synthorax red or tawny; its sides olivaceous, with sinuous and interrupted streaks of black in depths of three lateral sutures. Three streaks conjoined above leg bases; second makes contact with third at points farther up toward its abbreviated upper end. Wings hyaline, with red veins and stigma, and with a touch of yellow in membrane at extreme base. In this species, as in *pallipes* and *costiferum,* triangle of hind wing appears to be stalked on vein M4, front and inner sides generally meeting before they reach that vein.

Abdomen pale on swollen basal segments, red beyond, and washed with black that deepens to rearward on sides of segments 3 to 10, black spreading upward to middorsal line on 8 and 9. On 2, hamules and genital lobe black. Superior caudal appendages of male red; inferior yellow. Female paler, with broader flavescent patches at wing bases, and with lower sides of 4 to 9 occupied by a continuous black band that covers their lower half.

Distribution and dates.—UNITED STATES: Calf., Colo., Nev., Oreg.

July 8 (Oreg.) to September 29 (Calif.).

Sympetrum californicum Walker

1951. Walk., Ent. News, 62:161 (figs.).

Length 28–37 mm.; abdomen 20–25; hind wing 22–28.

Described as a variety of *occidentale;* similar structurally but paler in color, especially in female, which has wings only faintly yellowish.

Distribution and dates.—UNITED STATES: Calif., Nev.

July 6 (Nev.) to August 28 (Nev.).

Sympetrum costiferum Hagen

1861. Hagen, Syn. Neur. N. Amer., p. 174 (in *Diplax*).
1901. Ndm., Bull. N. Y. State Mus., 47: 527 (fig., nymph).
1911. Ris, Coll. Selys Libell., p. 692.
1914. Wmsn., Ent. News, 25:456.
1920. Howe, Odon. of N. Eng., p. 81 (figs.).
1927. Garm., Odon. of Conn., p. 271 (figs.).
1929. N. & H., Handb., p. 239 (figs.).
1941. Whts., Odon. of B. C., p. 547.

Length 31–37 mm.; abdomen 21–26; hind wing 25–28.

A red species, with a costal strip of yellow in membrane of wings that doubtless suggested specific Latin name. Thickly hairy face crossed with alternating bands of yellow and olive. Black stripe on frons broad. Antennae black. Vertex yellow. Occiput brown. In female, face wholly whitish.

Thorax olive to red in front, with only edgings of carina and crest black. Dusky olivaceous sides with streaks of black in lateral sutures, irregular, often interrupted, quite inconstant in form and cross connections. All conjoined by black markings above leg bases.

Legs bicolored; outer side of tibiae and femora yellow; all else black. Wings shining hyaline, with veins red in costal strip. Golden or reddish stain in that strip tends to disappear in old males; and to broaden in some females until it overspreads greater part of hind wing. Stigma reddish, its bordering veins thickened and very black.

Abdomen quite hairy at base and moderately slender, with sides conspicuously black along whole length, color deepening to rearward and spreading upward to middorsal line on segments 8 and 9. Caudal appendages of male red, as are hamules and genital lobe of 2; latter widens distally from its base to a rounded and hairy tip.

Distribution and dates.—CANADA: Alta., B. C., Man., N. B., Nfld., NW. Terr., N. S., Ont., Que., Sask.; UNITED STATES: Calif., Conn., Maine, Mass., Mich., Mo., Nebr., Nev., N. H., N. Y., Oreg., R. I., Utah, Wash., Wis., Yellowstone.

July 7 (Nfld.) to November 1 (B. C.).

Sympetrum danae Sulzer

Syn.: scoticum Donovan

1776. Sulzer, Abgekürtze Geschichte, p. 169 (fig.) (in *Libellula*).
1911. Ris, Coll. Selys Libell., p. 646.
1917. Walk., Can. Ent., 49:217 (figs.).
1927. Garm., Odon. of Conn., p. 271.
1929. N. & H., Handb., p. 239 (figs.).

Length 21–23 mm.; abdomen 18–24; hind wing 20–27.

A little black and yellow species, ranging east and west over Northern Hemisphere. Common in northern Europe and Asia. First species of genus to have been described and named. Teneral specimens pale greenish olive marked with brown, but brown prevails with aging, and adult males become mostly black. At maturity, face greenish yellow, with entire labrum, a band across postclypeus, and all prominence of frons shining black. In female, face all greenish yellow save very black marginal line on front of labrum; at rear of frons on top of head a broad black cross stripe. In female, top of vertex remains pale.

Thorax dark brown or black in front, clad with golden hair. Sides of thorax fenestrate with black and yellow, yellow in three oblique rows of spots: foremost row of one to three large spots; middle row of three small ones; third row of three large ones, with hindmost next to abdomen. Legs very black beyond their pale bases. In female, yellow spots on sides larger and more confluent.

Wings hyaline, with a wash of gold on extreme base. Costa, some antenodal crossveins, nodus, subnodus, and stigma yellowish; other venation black. No apical planate and little development of a median one; single row of cells on radial planate not definitely delineated at outer end.

Abdomen black in mature male, with side spots of dull yellow showing through black on segments 2 to 4 and on 8 and 9. In female a yellow band develops by confluence and extension of these spots. Segment 10 and caudal appendages black. Often a thin golden cloud develops around nodus.

No trace of red on *danae.*

Distribution and dates.—Alaska; CANADA: Alta., B. C., Man., N. B., Nfld., NW. Terr., Ont., Que., Sask.; UNITED STATES: Calif., Colo., Maine, Mich., Nev., N. H., N. Y., Oreg., Utah, Wyo.

June 20 (B. C.) to October 10 (B. C.).

Sympetrum fasciatum Walker

1951. Walk., Ent. News, 62:162 (figs.).

Length 30–37 mm.; abdomen 19–24; hind wing 23–27.

A variety of *occidentale,* very similar in all structural characters but easily distinguishable by brown or brownish band on outer margin of yellow in both fore and hind wings. Toward base, wings paler.

Inhabits marshy places along small spring-fed streams.

Distribution and dates.—CANADA: Alta.; UNITED STATES: Ariz., Colo., Kans., Nebr., N. Mex., Okla. (?), S. D., Tex. (?), Utah, Wyo.

June 16 (Utah) to October 21 (N. Mex.).

Sympetrum internum Montgomery

Syn.: decisum Hagen

1911. Ris, Coll. Selys Libell., p. 688 (as *decisum*).
1933. Wmsn., Occ. Pap. Mus. Zool. Univ. Mich., 264:1–5 (figs.) (as *decisum*).
1941. Whts., Odon. of B. C., p. 550 (as *decisum*).
1943. Mtgm., Can. Ent., 75:57 (nomenclatural history).

Length 21–36 mm.; abdomen 20–23; hind wing 23–27.

A bright reddish species, with a yellowish red face and a marginal black line on front border of labrum. Vertex and occiput shining brown.

Thorax reddish brown, clad in hair of same color. No definite stripes, patternless in front and on sides. Legs black beyond their bases except for pale under side of front femur. Wings hyaline, with a wash of gold of variable extent in extreme base, generally not reaching first cross-veins. Two rows of cells above apical planate, one row each for radial and median planates. Generally six antenodal crossveins in hind wing, sometimes five. Anal loop with large heel, generally with one or two ankle cells on either side of midrib.

Abdomen cherry red, with a black band across dorsum of segment 1, an oblique wash of black across sides of 2, and large black subequal lateral triangles on sides of 4 to 8 or 9; 10 and appendages yellow. Caudal appendages of male yellow, superiors with black tips.

Williamson (1933, p. 5) says of this species that in flight it "moves in a golden reddish bit of haze due to the color of its venation. At rest its face at the distance of a dozen feet or more is almost cherry red, as contrasted with the white face of *obtrusum* and the yellowish brown face of *rubicundulum*."

Distribution and dates.—Alaska; CANADA: Alta., B. C., Man., N. B., Nfld., NW. Terr., N. S., Ont., P. E. I., Que., Sask.; UNITED STATES: Calif., Colo., Ind., Maine, Mich., Minn., Mo., Nev., N. H., N. Y., N. D., Oreg., Pa., Utah, Wash., Yellowstone.

June 15 (Ont.) to October 8 (B. C.).

Sympetrum madidum Hagen

Syn.: chrysoptera Selys, flavicostum Hagen

1861. Hagen, Syn. Neur. N. Amer., p. 174 (in *Libellula*).
1905. Osburn, Can. Ent., 16:194 (figs.).
1911. Ris, Coll. Selys Libell., p. 679 (figs.).
1929. N. & H., Handb., p. 234 (figs.).
1941. Whts., Odon. of B. C., p. 547.

Length 42–45 mm.; abdomen 25–30; hind wing 28–31.

A yellow-winged Northwestern species. Body red from head to tail; yellowish in female. Top of head all red save for narrow frontal basal stripe and black antennae.

Thorax dark red in front, pale red on sides, without black stripes, and with a very thin covering of reddish hair. Faint suggestion of a pair of oblique white stripes on each side of thorax.

Legs black. Wings subhyaline, their veins and stigma red. Wing membrane strongly tinged with yellow, but in varying degrees in different localities. Always a yellowish strip over nodal crossveins out to stigma. Often deeper yellow in anal area, sometimes overspreading inner half of both wings. Network of venation decidedly dense for a *Sympetrum*. Besides double row of cells on radial planate, sometimes two cells in hindwing triangle, two cubito-anal crossveins, and always a number of ankle cells within anal loop. First paranal cell very wide, its base on vein A being about as long as second and third together. Sometimes extra bridge crossveins (more than one regularly present), extra (and often unmatched) nodal crossveins, and more or less irregularity of cell arrangement in some other interspaces of wing. These mark species as somewhat more primitive than those which follow. However, its most constant venational characters are those given for *Sympetrum* in our table.

Abdomen red, paler and phalerate with black carinae on basal segments, brighter red thereafter to and including caudal appendages; yellowish on these parts in female, with some black markings on sides of segments 8 and 9.

"The even-textured red wings make the insect conspicuous to the collector . . ."—Whitehouse (1941).

Distribution and dates.—CANADA: Alta., B. C., Man., NW. Terr., Sask.; UNITED STATES: Calif., Colo., Mo., Mont., Nev., Oreg., Wash., Wyo.

June 7 (B. C.) to September 7 (Sask.).

Sympetrum obtrusum Hagen

1867. Hagen, Stettin. Ent. Ztg., 28:95 (in *Diplax*).
1911. Ris, Coll. Selys Libell., p. 685.
1914. Wmsn., Ent. News, 25:456.
1920. Howe, Odon. of N. Eng., p. 80 (figs.).
1929. N. & H., Handb., p. 236 (figs.).
1943. Mtgm., Can. Ent., 75:57.

Length 31–39 mm.; abdomen 22–26; hind wing 20–29.

Very similar to *internum;* to be distinguished from that species only by critical examination of male genitalia of abdominal segment 2. (See figs. 308 and 309.) Face white, with only free edge of labrum black. Vertex and occiput olivaceous.

Thorax reddish brown, with black only on edges of crest. No color pattern. Legs and wings about as in *internum,* with apparently a minor difference in number of antenodal crossveins in hind wing: generally six in *internum;* generally five in *obtrusum;* apical planate undeveloped.

Abdomen shining red beyond paler basal segments, with band of black triangles along side much as in *internum.*

Distribution and dates.—CANADA: Alta., B. C., Man., N. B., N. S., NW. Terr., Ont., P. E. I., Que., Sask.; UNITED STATES: Colo., Idaho, Ill., Ind., Iowa, Kans., Maine, Md., Mass., Mich., Minn., Nebr., N. H., N. J., N. Y., N. C., Ohio, Oreg., Pa., R. I., Utah, Wash., Wis., Wyo.

April 14 (Ind.) to October 19 (B. C.).

Sympetrum occidentale Bartenev

1911. Ris, Coll. Selys Libell., p. 692 (no name).
1915. Bartenev, Univ. Izviestija, 46:1.
1951. Walk., Ent. News, 62:156 (figs.).

Length 31–40 mm.; abdomen 21–25; hind wing 24–28.

A slender pale brownish species. Face brownish, with front half or more of labrum and top of vertex yellow.

Thorax pale brown, with narrow blackish stripes in its lateral sutures, stripes widening downward to a confluence at their lower ends. Yellow of basal half of wings deepens from a nearly colorless basal area to about the level of the nodus, then fades rather abruptly without forming any dark submarginal crossband.

Abdomen reddish above toward base, with a black side stripe that widens to rearward, segment by segment, until on 9 that black almost covers abdomen. Caudal appendages of abdomen of male almost as in *semicinctum* (fig. 308).

Distribution and dates.—CANADA: B. C.; UNITED STATES: Idaho, Oreg., Utah, Wash.

June 28 (Idaho) to October 6 (Idaho).

Sympetrum pallipes Hagen

1874. Hagen, Rep. U. S. Geol. Surv. Terr., 7(3):589 (in *Diplax*).
1905. Osburn, Can. Ent., 16:192 (figs.).
1911. Ris, Coll. Selys Libell., p. 688.
1914. Walk., Can. Ent., 46:373 (figs., nymph).
1929. N. & H., Handb., p. 235 (figs.).
1941. Whts., Odon. of B. C., p. 551.

Length 34–38 mm.; abdomen 22–26; hind wing 25–28.

A pale clear-winged Western species. Labrum, frons, and vertex clear yellow. Black on ridge behind eyes envelops three spots of yellow.

Front of thorax a light golden brown under a pale pubescence, through which are dimly seen three whitish stripes: a faint antehumeral; and two broader, oblique lateral stripes, behind first and third lateral sutures, respectively. Ground color of sides olivaceous.

Legs yellow, with black spines; tarsi and sides of femora more or less smudged with black. Wings hyaline, with only a touch of yellow stain in extreme bases. Venation sparse; veins pale; costa and stigma yellow, latter bordered by thick black veins.

Abdomen olivaceous to rufous, with apical carinae of middle segments blackish. Touches of jet black on hinge articulations of end segments. Caudal appendages of male yellow; also narrowly triangular genital lobe on segment 2. Subgenital plate of female swollen toward base and thereafter cleft into two acute triangular recurving lobes.

Distribution and dates.—CANADA: Alta., B. C.; UNITED STATES: Calif., Colo., Mont., Nebr., Nev., Oreg., Tex., Utah, Wash.

June 6 (B. C.) to September 30 (Oreg.).

Sympetrum rubicundulum Say

Syn.: assimilatum Uhler

1839. Say, J. Acad. Phila., 8:26 (in *Libellula*).
1901. Ndm., Bull. N. Y. State Mus., 47:524 (figs.).
1911. Ris, Coll. Selys Libell., p. 682.
1920. Wmsn., Proc. Ind. Acad. Sci., p. 103.
1927. Garm., Odon. of Conn., p. 272.
1927. Ndm., Utah Agr. Exp. Sta. Bull., 201:20.
1929. N. & H., Handb., p. 236 (figs.).
1933. Wmsn., Occ. Pap. Mus. Zool. Univ. Mich., 264:1 (figs.).

Length 33–34 mm.; abdomen 21–23; hind wing 24–30.

Another dainty little red species of wide distribution and often abundant occurrence. Face yellow; top of head more or less brownish.

Thorax reddish brown in front, pale olivaceous in female, olivaceous on sides without dark stripes. Legs black beyond basal segments. Wings hyaline, with generally a trace of flavescence in their bases.

However, in some individuals (formerly ranked as a separate species under name *assimilatum*) flavescence of wing bases extends almost as far out as in *semicinctum*, most often in females.

Abdomen slender in male, smooth and shining, pale red on enlarged basal segments, cherry red beyond, with a band of black triangles resting on lateral margin of segments, and reaching up on rear of each segment to middorsal carina. Caudal appendages red. Female yellowish where male is red.

Pale teneral yellowish specimens of this species begin fluttering up out of the grasses of shallows in the upper reaches of most fresh water ponds about the latter part of June. A month later, when they have attained their brilliant red and black coloration and have become more numerous, we find them scattered everywhere.

The female in ovipositing is accompanied by the male. He seems to direct the course and to assist in the flight. Together the two descend to touch the water many times in rapid succession in nearly the same place; then a short flight, and many more descents in another place.—From the field notebook of the senior author.

Distribution and dates.—CANADA: Ont., Que.; UNITED STATES: Colo., Conn., D. C., Ill., Ind., Iowa, Kans., Ky., Md., Mass., Mich., Minn., Mo., Nebr., Nev., N. H., N. J., N. Y., N. C., Ohio, Pa., R. I., S. Dak., Utah, Vt., Va., Wis., Wyo.

May 5 (Ohio) to October 29 (Ky.).

Sympetrum semicinctum Say

1839. Say, J. Acad. Phila., 8:27 (in *Libellula*).
1911. Ris, Coll. Selys Libell., p. 690.
1917. Kndy., Proc. U. S. Nat. Mus., 52:623.

1920. Howe, Odon. of N. Eng., p. 82.
1927. Garm., Odon. of Conn., p. 274.
1929. N. & H., Handb., p. 237 (figs.).
1933. Wmsn., Occ. Pap. Mus. Zool. Univ. Mich., 264:1 (figs.).
1951. Walk., Ent. News, 62:153 (figs.).

Length 24–31 mm.; abdomen 16–20; hind wing 18–23.

A species with half-banded wings, as its specific name indicates, a flavescent cloud covering them out to nodus. Colored area rounded externally, and deepest color lies toward outer margin. On fore wings, flavescence often greatly reduced or sometimes even wanting. Face yellow, becoming reddish above on frons.

Thorax reddish brown, paler and olivaceous on sides, and thinly clothed with brown pubescence. Narrow black lines on edges of crest and subalar carinae; also, black streaks in depths of lateral sutures and broader connecting ones above leg bases. Legs black beyond short basal segments.

Abdomen red its full length above, superior caudal appendages red, inferior black. In female a wide black band covers lower half of each side of abdomen, double on segments 1 to 3, with added small middorsal triangles of black on 7, 8, and 9. Segment 1 also black across dorsum.

Distribution and dates.—CANADA: N. B., N. S., Ont., Que.; UNITED STATES: Conn., Ill., Ind., Iowa(?), Maine, Md., Mass., Mich., Minn.(?), N. H., N. J., N. Y., N. C., Ohio, Pa., R. I., Tenn., Vt., Va., W. Va., Wis.

June 19 (Ont.) to October 23 (Mich.).

Sympetrum vicinum Hagen

1861. Hagen, Syn. Neur. N. Amer., p. 175 (in *Diplax*).
1911. Ris, Coll. Selys Libell., p. 693.
1913. Davis, J. N. Y. Ent. Soc., 21:27.
1920. Howe, Odon. of N. Eng., p. 82.
1926. Calv., Ecology, 8:185.
1927. Garm., Odon. of Conn., p. 275.
1929. N. & H., Handb., p. 238 (figs.).

Length 31–35 mm.; abdomen 21–22; hind wing 21–23.

A delicate little thin-legged autumnal species of wide distribution and great abundance in many places. Face clay yellow, later becoming red in male.

Thorax reddish on front and olivaceous on sides. Wings hyaline, with more or less yellow at base. Venation somewhat reduced and scanty. Radial planate subtends and encloses generally five cells only, sometimes four or six. Little or no development of a median planate.

Abdomen slender and bright red of a lighter hue than in associated species. Hamules of male as shown in our figure. Scooplike subgenital plate of female very wide after oviposition, and its lateral margins are then outspread to meet and join flaring sides of segment 9, forming a trumpet-shaped end to abdomen.

Distribution and dates.—CANADA: B. C., N. B., N. S., Ont., Que.; UNITED STATES: Ala., Colo., Conn., D. C., Ga., Ill., Ind., Iowa, Kans., Ky., Maine, Md., Mass., Mich., Minn., Miss., Mo., Nebr., N. H., N. J., N. Y., N. C., Ohio, Okla., Pa., R. I., S. C., Tenn., Tex., Va., Wash., Wis.

June 2 (Ky.) to December 13 (S. C.).

Genus TARNETRUM Needham and Fisher 1936

This small group of American species is closely allied to *Sympetrum*. Two species occur within our limits; two or three more are Neotropical. The head is rather broad, with well-developed frontal furrow but no frontal carina. The trigonal interspace is narrowed to the wing margin in the fore wing. In the hind wing the anal area is broad, with a wide anal loop, a broad angular heel, and a very moderately angulated midrib. The network of venation behind vein A4 is more copious than in *Sympetrum*. Other venational characters are as shown in our table of genera (p. 427).

The abdomen is rather stout, and segment 4 has an extra encircling carina on the dorsal side in addition to the regular apical carina. The caudal appendages of the male are short and nearly straight, and abruptly and obliquely truncate from below upward, giving but a short taper to the end. They are armed beneath with a long line of subequal denticles. The subgenital plate of the female is scoop-shaped, inclined downward, roundly excavate across its free border, the edge of which is outturned like the lip of a water pitcher.

The nymph is stocky and somewhat hairy, with laterally rounded eyes; the teeth are vestigial, on the lateral lobes of the labium. There are no dorsal hooks on the abdomen, and lateral spines are also wanting, except for a small vestige of one sometimes present on segment 9.

KEY TO THE SPECIES

ADULTS

1—Wings hyaline, or with only a faint wash of yellow at base; legs more black than yellow; hind wing with five antenodal crossveins............**corruptum**

—Wings with brown streaks in basal subcostal and cubital spaces, surrounded by a yellow area that covers most of basal half of wing; legs more yellow than black; hind wing with six antenodal crossveins................**illotum**

NYMPHS

1—Length of body about 21 mm.; lateral caudal appendages half as long as inferiors ....**corruptum**

—Length of body about 18 mm.; lateral caudal appendages two-thirds as long as inferiors ....**illotum**

Tarnetrum corruptum Hagen

1861. Hagen, Syn. Neur. N. Amer., p. 171 (in *Mesothemis*).
1901. Ndm., Bull. N. Y. State Mus., 68:271 (fig., nymph) (in *Sympetrum*).
1911. Ris, Coll. Selys Libell., p. 678 (in *Sympetrum*).
1915. Bethel, Ent. News, 26:119 (in *Sympetrum*).
1929. N. & H., Handb., p. 233 (figs.) (in *Sympetrum*).

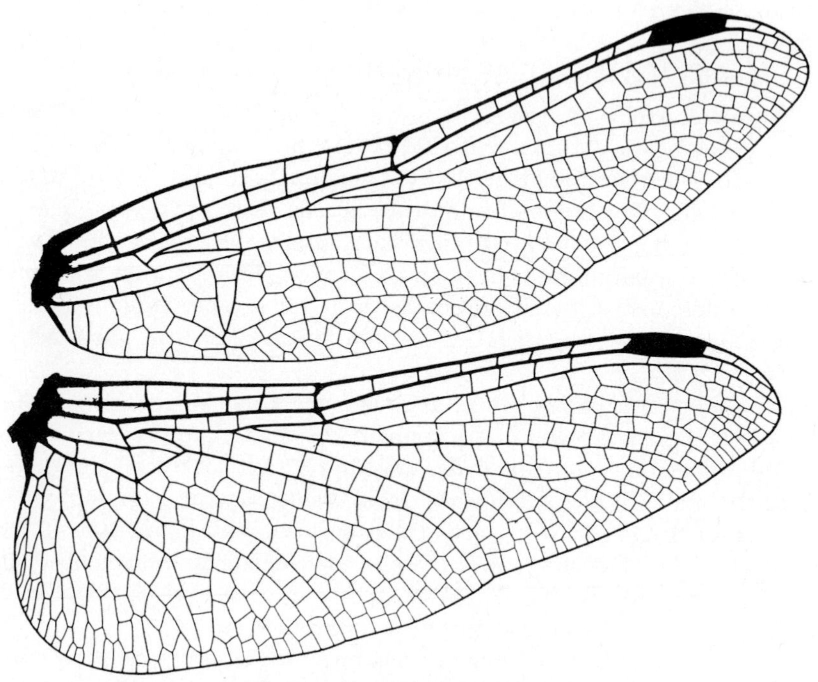

Fig. 310. *Tarnetrum corruptum.*

Length 39–42 mm.; abdomen 26–27; hind wing 29–30.

A wide-ranging species. In life it is patterned in soft pastel colors that fade sadly in dried specimens. Face yellowish, becoming bright red above in males; base line of frons rather narrow; top of vertex yellow or red.

Thorax stout, clothed with white hairs. Carina and crest very narrowly edged with black. Front of thorax grayish brown, sometimes with a trace of a pale stripe showing low down near collar on each side. Sides dark grayish olive, with two very oblique pale stripes that are conspicuous only at their lower ends. First oblique stripe lies midway of sides, second at rear. Both whitish in middle, overlaid at lower end by a conspicuous yellow spot, half encircled by black on lower side.

Abdomen olivaceous, with dark spots along sides of segments 3 to 9, forming a diffuse blackish band, and with middorsal black spots on 8 and 9. Caudal appendages black.

This long and late season species sometimes overwinters as an adult, and in North is taken on wing in early spring.

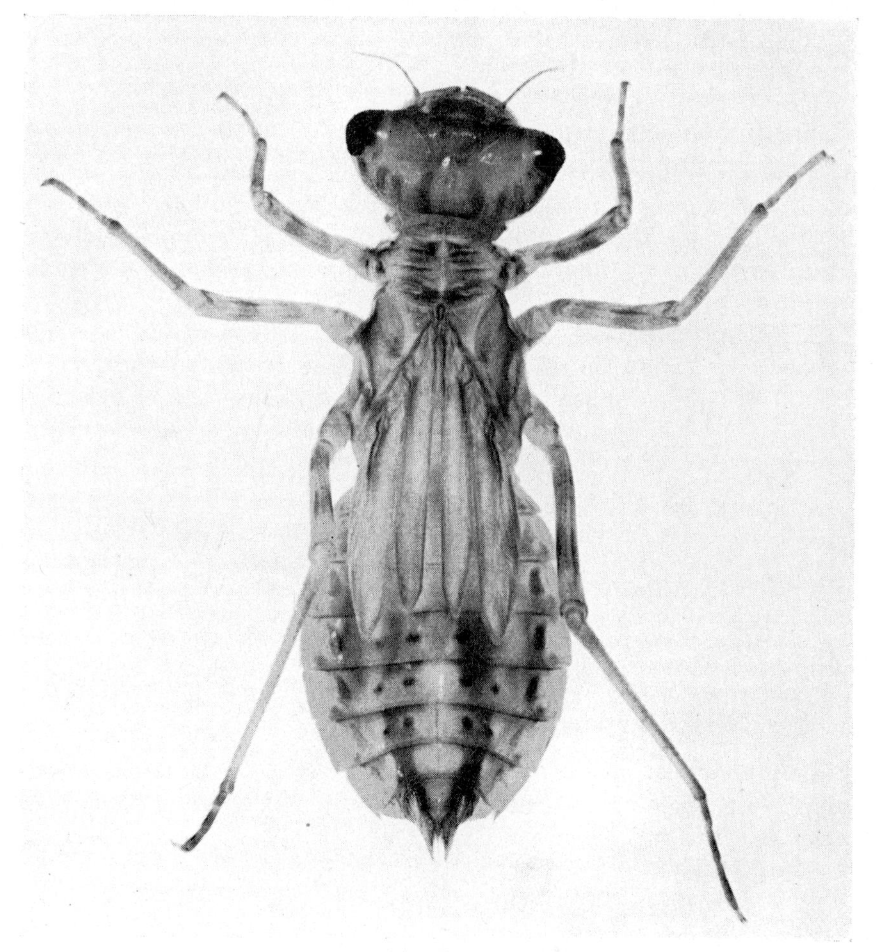

Fig. 311. *Tarnetrum corruptum.*

Distribution and dates.—CANADA: Alta., B. C., Man., Ont., Sask.; UNITED STATES: Ala., Ariz., Ark., Calif., Colo., Fla., Ga., Idaho, Ill., Ind., Iowa, Kans., La., Mich., Minn., Miss., Mo., Mont., Nebr., Nev., N. J., N. Mex., N. Y., Ohio, Okla., Oreg., Pa., S. C., S. Dak., Tenn., Tex., Utah, Wash., Wis., Wyo.; MEXICO: Baja Calif., Coahuila, Sonora, Tamaulipas; also south to Honduras, and from Asia and Sea of Okhotsk.

February (Calif.) to December 31 (Fla.).

Tarnetrum illotum Hagen

Syn.: gilvum Selys, virgulum Selys

1861. Hagen, Syn. Neur. N. Amer., p. 172 (in *Mesothemis*).
1901. Ndm., Bull. N. Y. State Mus., 47:521 (pl. 25, fig. 1, nymph) (in *Sympetrum*).
1911. Ris, Coll. Selys Libell., p. 676 (in *Sympetrum*).
1917. Kndy., Proc. U. S. Nat. Mus., 52:609 (in *Sympetrum*).
1929. N. & H., Handb., p. 233 (figs.) (in *Sympetrum*).

Length 38–40 mm.; abdomen 23–24; hind wing 26–28.

A reddish species, with two faint oblique and often interrupted stripes on sides of thorax, red hair thinly covering front and sides. Face and top of vertex and occiput all red. Red legs have black spines and claws. Wings conspicuously marked with basal brown streaks, as stated in preceding key.

Abdomen red, its lateral carinae narrowly hairlined with black, and a thin wash of black on sides of segments 7 to 10. Caudal appendages also red. Scooplike subgenital plate of female surpasses tip of 8 by about half the length of segment; roundly emarginate, with edge outrolled like lip of a water pitcher. It extends to, but does not cover, palps of elongated ninth sternite.

Usually the female of this species oviposits unaccompanied by the male but here I observed a pair working together. These copulated on the wing, then rested half a minute in copulation on a branch, when they flew about over the water, the male holding the female by the prothorax (head). The pair made tentative dives from an elevation of about two feet. After half a minute they dropped two inches above the water when with a swinging motion the female dipped her abdomen in the water about thirty times, after which they made a sudden upward flight and separated, each to seat itself on a twig.—Kennedy (1917).

Distribution and dates.—CANADA: B. C.; UNITED STATES: Calif., Nev., Oreg., Wash., Yellowstone; MEXICO: Baja Calif.; also south to Argentina.
March (Calif.) to November (Calif.).

Genus ERYTHEMIS Hagen 1861

Syn.: Mesothemis Hagen

These dragonflies are of moderate size and great diversity of appearance in different species and at different ages in the same species.

The legs are strong and spiny, with very long black spines on the tibiae, short and numerous small ones on the proximal part of the hind and middle femora, and three or four very large and stiff ones near the end of it.

The wings are hyaline; in some species, brown or flavescent at the base. The color patches, when present, are larger in the male than in the female. The venation is as shown in our table of Libelluline genera (p. 426). The antenodal crossveins in the hind wing vary in number from 7 to 11; the intermedian crossveins are 6–10/3–4. The crossveins under the stigma have 7, 8, or 9 crossveins before them in the second postnodal interspace. The apical, radial, and median planates subtend one to two, one, and one row, respectively, in both fore and hind wings. The rather

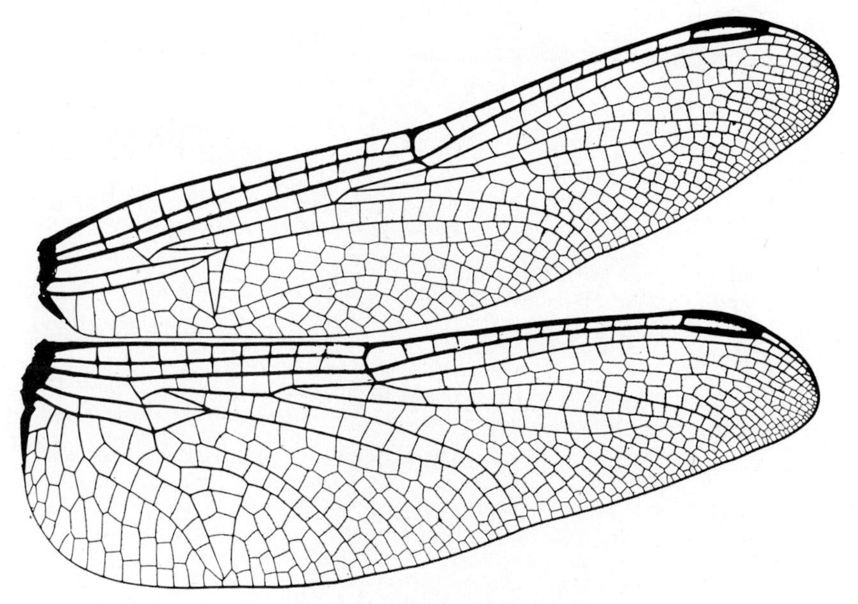

Fig. 312. *Erythemis simplicicollis.*

compact anal loop generally has a single ankle cell next to the prominent heel on the outer side of the midrib.

The abdomen is slender along its middle portion and sharply triangular in cross section.

The nymph of this genus is as shown in our table (p. 432). It is easily recognizable among allied forms (except *Lepthemis*) by the strongly decurved inferior caudal appendages, the very thick, stocky body, and the head with bulging green eyes. The eyes in life are striped with green and brown.

The most important single paper on the adults of this genus is that of Williamson (1923), to which reference is hereinafter made under each species heading. There is as yet no comparable paper on nymphs, and only two species, *plebeja* and *simplicicollis,* appear to have been reared to date.

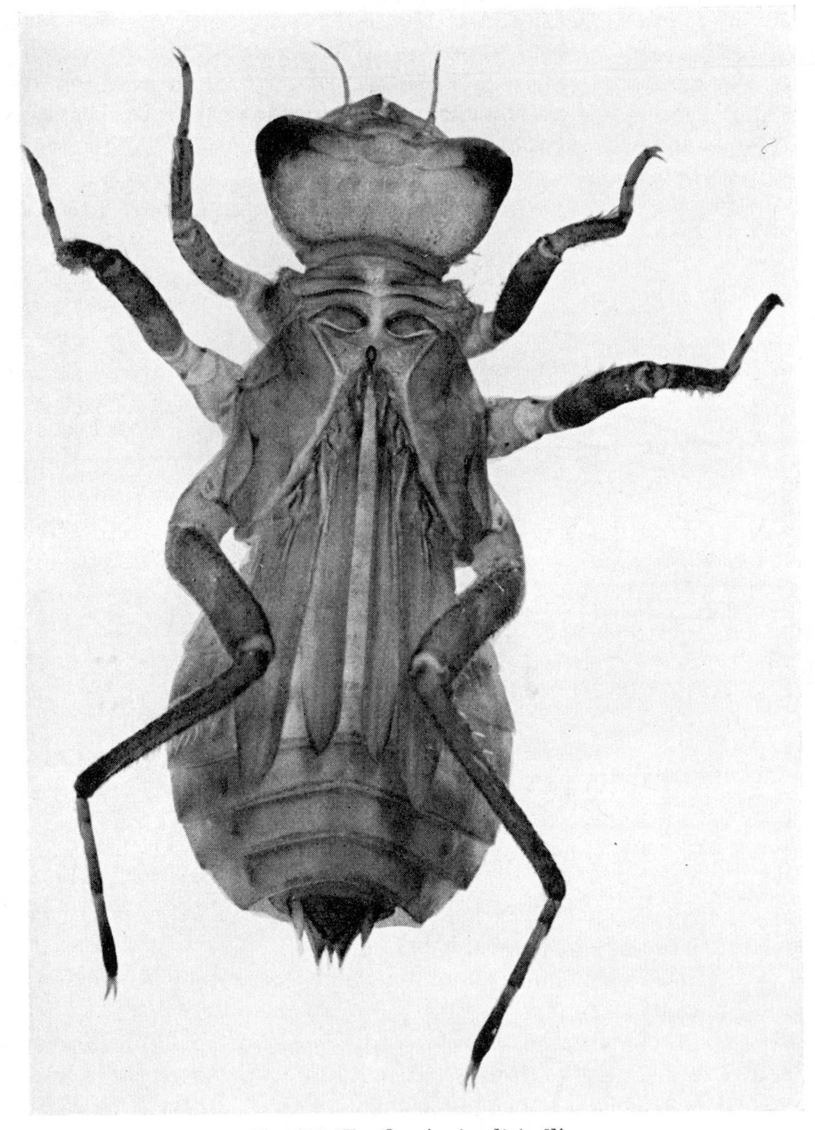

Fig. 313. *Erythemis simplicicollis.*

KEY TO THE SPECIES*

ADULTS

1—Body green, marked with brown, or (in old males) wholly pruinose blue; wings clear throughout 2
—Body black, brown, or red; wings generally with some brown or yellow in membrane at base 3

2—Face all green; caudal appendages yellow .... **simplicicollis**
—Face with black across frons; appendages blackish .... **collocata**

3—Front of thorax with wide middorsal yellowish or grayish band bordered on each side by black; wings with basal brown spot not extending outward beyond anal crossing (Ac) 4
—Front of thorax with no such band; wings of male with basal spot extending outward almost or quite to triangle .... **attala**

4—Hind wing generally with eleven antenodal crossveins; abdomen red **haematogastra**
—Hind wing generally with nine antenodal crossveins; abdomen of male all red .... **peruviana**
—Hind wing generally with seven antenodal crossveins; abdomen of male rusty brown or tawny .... **plebeja**

Erythemis attala Selys

Syn.: annulosa Selys, mithra Selys

1857. Selys, in Sagra, Hist. Cuba, Ins., p. 445 (in *Libellula*).
1861. Hagen, Syn. Neur. N. Amer., p. 172 (in *Mesothemis*).
1907. Calv., B. C. A., p. 330.
1923. Wmsn., Misc. Pub. Mus. Zool. Univ. Mich., 11:10.

Length 42–44 mm.; abdomen 26–30; hind wing 32–38.

An agile blackish species, with large blackish spots on base of hind wing. Head wholly blackish above and below. Thorax dark brown in front, marbled with paler, besprinkled with black hairs. Abdomen, at first paler on basal segments and on sides of 4 to 7, becomes deep shining black, with only caudal appendages partly yellow.

This is said to be a very agile and incessantly active species, hard to get within range of the collector's net.

Distribution and dates.—ANTILLES: Cuba, Haiti, Jamaica; also from Martinique and Mexico to Brazil and Paraguay.

June 4–April 25.

Erythemis collocata Hagen

1861. Hagen, Syn. Neur. N. Amer., p. 171 (in *Mesothemis*).
1895. Calv., Proc. Calif. Acad. Sci., (2)4:552 (figs.).
1905. Osburn, Ent. News, 16:195.
1913. Ris, Coll. Selys Libell., p. 600.
1923. Wmsn., Misc. Pub. Mus. Zool. Univ. Mich., 11:10.

* Also *E. mithroides* is reported from Tamaulipas by Calvert (see B. C. A., p. 335).

Length 40–42 mm.; abdomen 24–26; hind wing 31–32.

A Western form, generally considered a variety of *E. simplicicollis* (see p. 554); a little darker in color, a little more robust, more nearly parallel-sided in abdomen, and more depressed. Adult male has a smudge of black across frons hardly definite enough to be called a stripe. Caudal appendages darker, becoming more densely pruinose with age. On mid-dorsal carina of segment 10 is a dash of yellow, which, conjoined with yellow of intersegmental membrane, makes a yellow T mark on back.

Reported as inhabiting pools from warm springs.

Distribution and dates.—CANADA: B. C.; UNITED STATES: Ariz., Calif., Nev., N. Mex., Tex., Utah, Wash., Wyo.; MEXICO: Baja Calif., Chihuahua, Coahuila; also southward in Mexico.

May (Calif.) to September (Calif.).

Erythemis haematogastra Burmeister

1839. Burm., Handb., p. 857 (in *Libellula*).
1861. Hagen, Syn. Neur. N. Amer., p. 161 (in *Lepthemis*).
1909. Calv., B. C. A., p. 338.
1911. Ris, Coll. Selys Libell., p. 605.
1923. Wmsn., Misc. Pub. Mus. Zool. Univ. Mich., 11:9.

Length 45–50 mm.; abdomen 34–35; hind wing 35–37.

A rather stout, blackish species, with red abdomen. Labium pale, with a median black stripe. Face black, metallic, shining. Thorax dark, obscure, greenish, without definite color pattern. Legs black, femora red on inner side. Wings clear except for a black area across base of hind pair, and stigma in adult male red.

Distribution.—ANTILLES: Jamaica; also reported south to Paraguay. Hagen (1861) lists this species from "Georgia," but at least it has not been reported subsequently from the state of Georgia.

Erythemis peruviana Rambur

Syn.: bicolor Erichson, rubriventris Blanchard

1842. Rbr., Ins. Neur., p. 81 (in *Libellula*).
1907. Calv., B. C. A., p. 330.
1911. Ris, Coll. Selys Libell., p. 600.
1923. Wmsn., Misc. Pub. Mus. Zool. Univ. Mich., 11:10.

Length 36–43 mm.; abdomen 23–28; hind wing 28–34.

Face blackish in front on frons and labrum, intervening area paler; top of frons yellow; vertex and occiput brown.

Thorax hairy and rough on surface, its front largely covered by a band of yellow, bordered on each side by parallel black bands shading off insensibly on sides to a rusty brown. Legs black. Hind wings hyaline, with a dark brown spot covering base as far out as anal crossing; in fore wings, only a tinge of golden brown at extreme base. Hind wings with nine antenodal crossveins.

Abdomen all red, fiery red on dorsum. Segments 1 to 3 hairy and moderately swollen below. Remainder of abdomen smooth, parallel-sided, sharp-edged and prickly on all carinae. Beyond 3 it slowly tapers to 9. Caudal appendages red; genital parts of 2 black.

Adult males may sometimes be seen sitting in sun, with body flattened down against a floating leaf, flaming red abdomen erected and held steadily aloft, like a miniature obelisk.

Distribution and dates.—MEXICO: Tamaulipas; ANTILLES: Jamaica; also south to Argentina.

June and August (Tamaulipas).

Erythemis plebeja Burmeister

Syn.: verbenata Hagen

1839. Burm., Handb., p. 856 (in *Libellula*).
1861. Hagen, Syn. Neur. N. Amer., p. 162 (as *Lepthemis verbenata*).
1907. Calv., B. C. A., pp. 330, 336 (as *verbenata*).
1911. Ris, Coll. Selys Libell., p. 603.
1923. Wmsn., Misc. Pub. Mus. Zool. Univ. Mich., 11:9.
1927. Calv., Univ. Iowa Studies, 12:33 (nymph).
1929. N. & H., Handb., p. 247 (in *Mesothemis*).

Length 42–47 mm.; abdomen 31–37; hind wing 31–36.

A bold, aggressive, strong-flying blackish species. Face brown, varied with paler, yellowish on top of frons.

Thorax blackish, thinly clad with rusty brown hair. Front mainly occupied by a wide longitudinal yellow band, bordered on each side by a wide antehumeral black band. Latter fades out insensibly on otherwise unmarked sides. A black carina divides yellow band on front; edges of crest also narrowly black.

Legs pale brown, more blackish toward knees. Wings hyaline except for a blackish spot at base that in male extends from base outward far enough to cover anal crossing. Only a smaller wash of yellow in female. Generally seven (sometimes nine) antenodal crossveins in hind wing, and often no ankle cells between two rows of anal loop. First paranal cell of hind wing nearly twice as wide as second or third.

Abdomen very long and slender, and beautiful in its straightness and finish beyond its swollen base. Segments 1 to 3 expanded to potbellied form on lower side, and strongly compressed laterally; pale, with only carinae dark. Abdomen constricted at end of 3, and thereafter slowly widens to 8. Segments 4 to 7 black, with large lateral pale spots; 8 and 9 all black; middle of dorsum of 10 yellow; caudal appendages yellow. Old males become almost entirely black. Females paler, with much less black on abdomen and less brown in wing bases—sometimes merely a tint of gold. Scooplike, half-cone-shaped black subgenital plate of female projects almost directly downward, about as long as segment 9.

Distribution and dates.—UNITED STATES: Tex.; ANTILLES: Cuba, Dom. Rep., Haiti, Jamaica, P. R.; also south to Paraguay.

Apparently year-round.

Erythemis simplicicollis Say

Syn.: caerulans Rambur, gundlachii Scudder, maculiventris Rambur

Green Jacket

1839. Say, J. Acad. Phila., p. 28 (in *Libellula*).
1900. Wmsn., Odon. of Ind., p. 326.
1901. Ndm., Bull. N. Y. State Mus., 47:527 (nymph) (in *Mesothemis*).
1911. Ris, Coll. Selys Libell., p. 598.
1914. Whedon, Minn. State Ent. Rep., 13:102.
1923. Wmsn., Misc. Pub. Mus. Zool. Univ. Mich., 11:9, 10.
1929. N. & H., Handb., p. 246 (in *Mesothemis*).
1941. Bick, Ann. Ent. Soc. Amer., 34:215–230 (figs.).

Length 38–44 mm.; abdomen 24–30; hind wing 30–33.

A clear-winged species, with face and thorax of bright green, and abdomen half-ringed with black and green. These early colors may become wholly obscured with blue-gray pruinosity in old males. Greens fade to straw yellows in dried specimens. Face all green; labrum, vertex, and occiput olivaceous or darker.

Thorax green, with a black hairline edging on carina and crest, with touches of black in pits of first and third lateral sutures, and with some longitudinal blacker markings above bases of legs.

Abdomen green on moderately swollen basal segments. All carinae blackish; all longitudinal carinae sharp-edged and black. Narrowing on segment 3 is followed by gradual widening to 8, making it slightly spindle-shaped, as viewed from above, toward end. On 4 to 7, large quadrate yellow spots occupy about half of side; 8, 9, and 10 black; superior appendages yellow, inferior one black.

This species of wide distribution within our limits, sometimes very abundant. Frequents borders of clear ponds, where it flies intermittently, resting much of the time on low vegetation or on bare earth of some open path.

It is a vigilant species. It squats on bare ground or on floating logs or trash; or, if on a twig, it generally selects a low perch. Then it waits for suitable prey to come along, and darts out upon it. Many a damselfly is thus snapped up unawares. At night it hangs up among foliage of pond-side or roadside weeds. Female oviposits unattended, making descents to touch surface at points wide apart.

Williamson (1900) notes a peculiar aerial performance:

Two males ... flutter motionless, one a few inches in front of the other, when suddenly the rear one will rise and pass over the other, which at the same time moves in a curve downwards, backwards and then upwards so that the former position of the two is just reversed. These motions kept up with rapidity and regularity give the observer the impressions of two intersecting circles which roll along near the surface of the water.

Distribution and dates.—CANADA: Ont., Que.; UNITED STATES: Ala., Ariz., Calif., Conn., Fla., Ga., Ill., Ind., Iowa, Kans., Ky., La., Maine, Md., Mass., Mich., Minn., Miss., Mo., Mont., Nebr., Nev., N. H., N. J., N. Y., N. C., Ohio, Okla., Oreg., Pa., R. I., S. C., Tenn., Tex., Utah, Va., Wash., Wis.; MEXICO: Baja Calif., Tamaulipas; ANTILLES: Cuba, Haiti, Jamaica.

Year-round southward.

Genus LEPTHEMIS Hagen 1861

These are very elongate, clear-winged, bright green dragonflies, with the greatly swollen base of the abdomen very much compressed, then constricted, then parallel-sided and slender. The head is broad, with bulging frons. The thorax is green, almost without color pattern. The legs are long, blackish, and spiny.

This genus consists of one large, strong-flying American species. It is easily recognized by the three huge spines in the outer row on the femora of the middle and hind legs, and by the almost wasplike slenderness of the abdomen beyond the swollen basal segments.

The nymph is green in color and very like the nymph of *Erythemis*, very thick-set, with bulging eyes and downturned caudal appendages. It differs in having a minute lateral spine on abdominal segment 9, and several more lateral setae on the labium, as well as in its larger size.

Lepthemis vesiculosa Fabricius

Syn.: acuta Say

1775. Fabr., Syst. Ent., p. 421 (in *Libellula*).
1907. Calv., B. C. A., p. 749.
1911. Ris, Coll. Selys Libell., p. 607.
1929. N. & H., Handb., p. 248.
1932. Klots, Odon. of P. R., p. 56.

Length 56–59 mm.; abdomen 40–42; hind wing 39–40.

A wide-ranging Tropical American species that just enters our southern border. Face bright green, with only a wash of clay yellow on labrum. Top of frons, vertex, and occiput all green.

Thorax green, both front and sides, without stripes of dark color, and with a thin covering of short brown hair. Legs green toward base; tibiae, tarsi, some streaks on femora, and spines black.

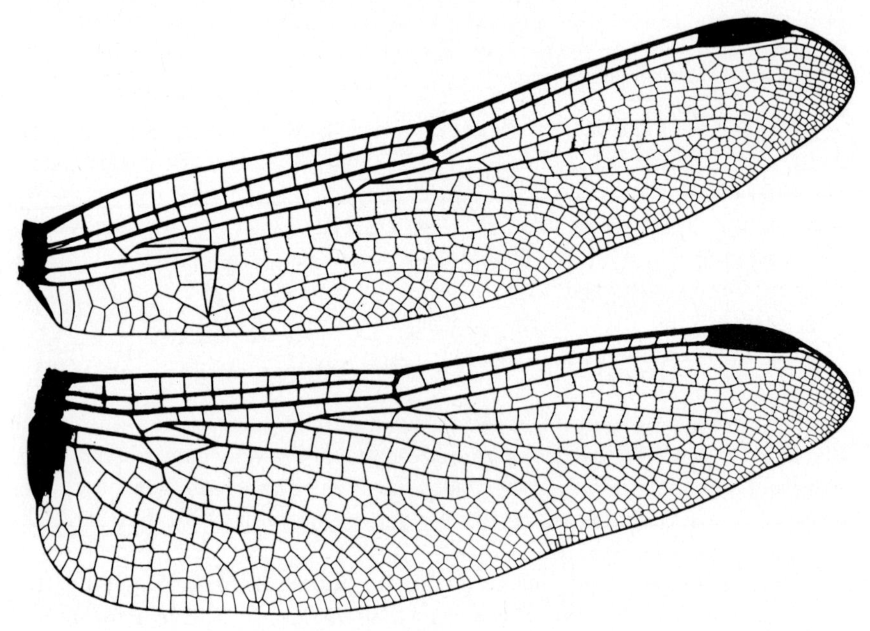

Fig. 314. *Lepthemis vesiculosa.*

Wings wholly hyaline. Under stigma are two crossveins, and proximal to them in same interspace are ten to twelve postnodal crossveins. Midrib of long anal loop at its base about four times as far from vein A1 as from vein A2.

Abdomen green on inflated basal segments, more or less blackish beyond on dorsum; remarkably contracted on segment 3, and thereafter slowly widened to end, with straight sides on long middle segments. A midlateral yellow band in black of each side interrupted on each segment by a wide black crossband, which gives abdomen a ringed appearance. Segments 8 and 9 all black above; 10 somewhat paler. Caudal appendages yellow.

A big, bold, piratical species, it attacks other insects viciously, taking butterflies and other dragonflies in flight, almost at mouth of collector's net. Flies threateningly near, but generally just beyond reach of net.

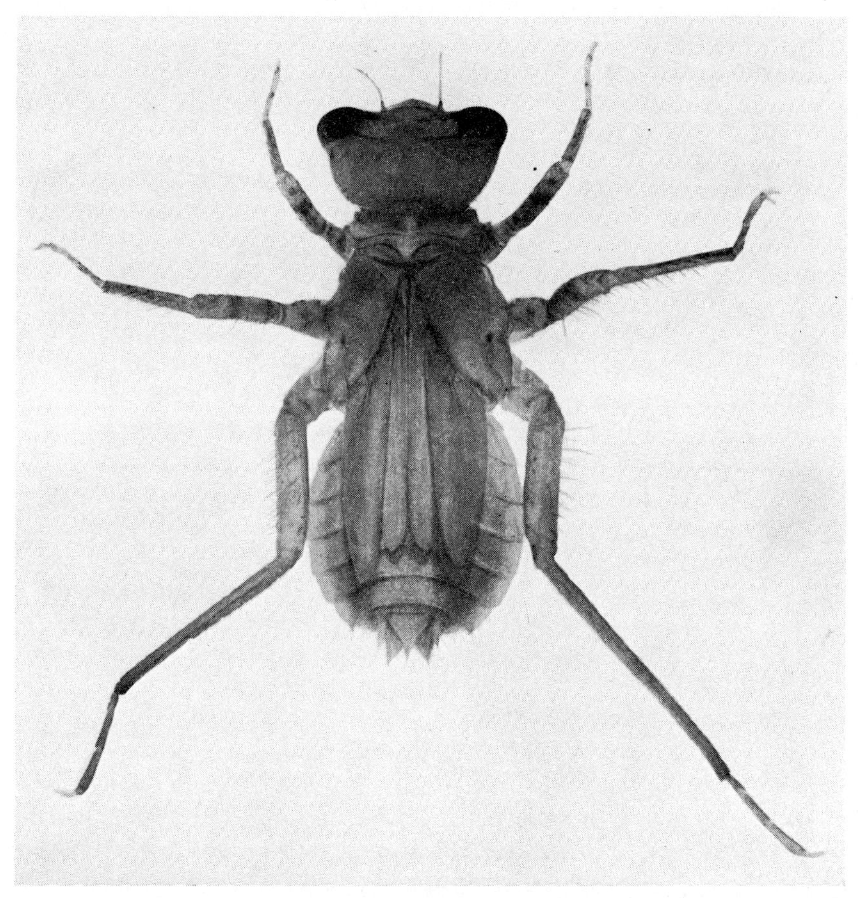

Fig. 315. *Lepthemis vesiculosa.*

It is both swift and agile, so that, while commonly seen near at hand, it is not easily taken.

Distribution and dates.—UNITED STATES: Fla., Okla., Tex.; MEXICO: Tamaulipas; ANTILLES: Cuba, Dom. Rep., Haiti, Jamaica, P. R.; also south to Paraguay.

Year-round (Fla.).

Genus BRACHYMESIA Kirby 1889

These are stocky dragonflies of moderate size, generally red or yellowish red. The prothorax is bilobed on its broadly elevated hind margin; the synthorax is patternless. The abdomen is very red.

The wings are hyaline; their venation is as shown in our table of genera (p. 426). The fore-wing triangle is nearly twice as long as wide,

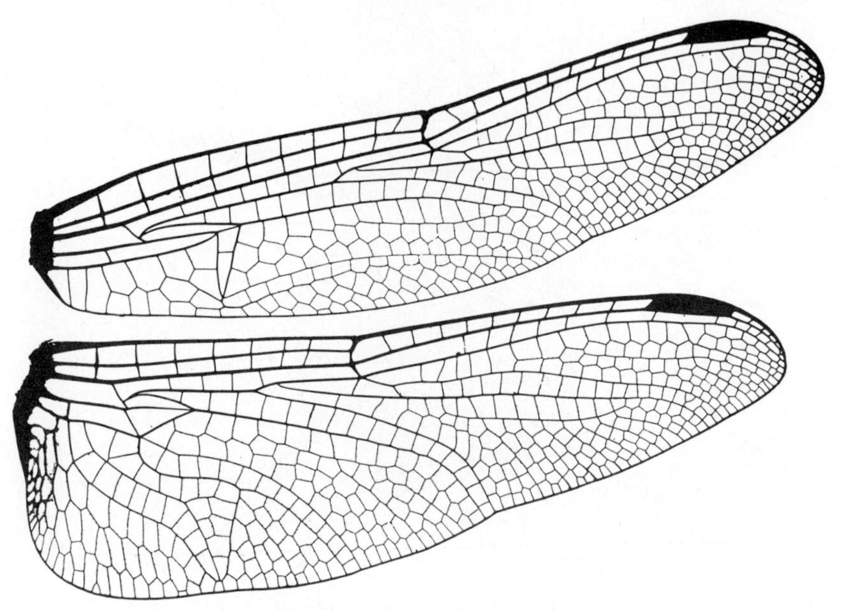

Fig. 316. *Brachymesia furcata.*

its hind angle strongly retracted, pointing inward. The apical planate subtends two rows of cells. The anal loop is rather short at the toe and wide at the heel, with one or two intercalary ankle cells on the distal side of the midrib.

The abdomen is rather short and tapers regularly to the end. It is black on the dorsum of segments 8 and 9. The caudal appendages are high-arched in the basal half, fusiform beyond, and slowly tapering to a long, upcurved point.

The nymph is a dweller in brackish pools. Its distinguishing features, in addition to those shown in our table (p. 432), are the broad, laterally flattened, subequal dorsal hooks on segments 6 to 9, and the strongly spinulose-serrate side margins of 9.

Fig. 317. *Brachymesia furcata.*

Brachymesia furcata Hagen

Syn.: australis Kirby, smithii Kirby

1861. Hagen, Syn. Neur. N. Amer., p. 169 (in *Erythemis*).
1889. Kirby, Ann. Nat. Hist., (6)14:266 (in *Cannacria*).
1895. Calv., Proc. Calif. Acad. Sci., (2)4:548 (figs.) (in *Cannacria*).
1899. Boright, The Nautilus, 3:30 (pl. 1, fig. 17).
1912. Ris, Coll. Selys Libell., p. 737.
1929. N. & H., Handb., p. 234 (figs.) (as *Sympetrum furcatum*).
1930. Ndm., Ent. News, 41:254.
1932. Klots, Odon. of P. R., p. 51 (figs.).
1938. Garcia, J. Agr. Univ. Puerto Rico, 22:58, 72, 74 (figs., nymph).

Length 41–46 mm.; abdomen 25–30; hind wing 32–36.

A single species of medium size and reddish coloration falls within our limits. Face red, with margins of labrum, sides of frons, tips of vertex, and occiput all yellow.

A dense fringe of longer hair clothes bilobed margin of hind lobe of prothorax. Thorax reddish brown, densely clothed with hair of same color. Front and sides without definite stripes.

Legs brown, paler basally and on under side of front femora. Wings hyaline, with only a touch of yellow in extreme base of front pair, and a wider partial basal crossband in hind pair extending out as far as anal crossing. Basal costal crossvein yellow; stigma tawny, within black bounding veins. Generally five crossveins postnodal in second series in fore wing, and six antenodals in hind wing. Three rows of cells beyond triangle out to level of nodus in fore wing. Radial and median planates (each subtends a single row); apical planate, two rows.

Abdomen strongly compressed on basal segments, suddenly contracted thereafter, and somewhat depressed, but tapers all the way to end. Very red on dorsum in old males; yellow in females. Black of dorsum mainly on segments 8 and 9, with hairlines of black on all carinae; segment 10, inferior appendage of male, and upper edge of superiors of male yellow; genitalia of 2 yellowish red. Inferior caudal appendage wide at tip and ending in a squarish notch. Superiors in side view strongly sigmoid, aslant upward in heavy terminal fourth, strikingly arched in basal half, and armed at end of arch with a row of half a dozen small denticles on under side.

Distribution and dates.—UNITED STATES: Fla., Mo., Tex.; MEXICO: Baja Calif., Tamaulipas; ANTILLES: Cuba, Haiti, Jamaica; also from Bahamas and south to Brazil.

Apparently year-round southward; dates include December 28 to April 22 (Jamaica), May 13 (Tex.), and September and October (Baja Calif.).

Genus CANNACRIA Kirby 1889

This is a genus of a few Neotropical species, two of which occur within our southern border. They are rather large, swiftly flying, graceful species. The colors are brown and yellow, and a band of black covers the greater part of the length of the abdomen. The eye-seam on top of the head is rather long. The frons is broadly rounded, with a deep longitudinal furrow.

The prothorax is broadly bilobed and densely hair-fringed on the elevated hind margin. The legs are slender. The wings are long and shining. The triangle of the fore wings is more than twice as long as wide, its hind angle strongly retracted and pointing inward. The apical planate subtends three rows of cells. The antenodal crossveins of the hind wings are generally seven, sometimes eight.

The slender abdomen is strongly compressed on its swollen basal segments and, viewed from above, is parallel-sided from end to end. It is black on the dorsum of segments 4 to 9. The superior caudal appendages of the male are slightly arched in their proximal half, cylindric and

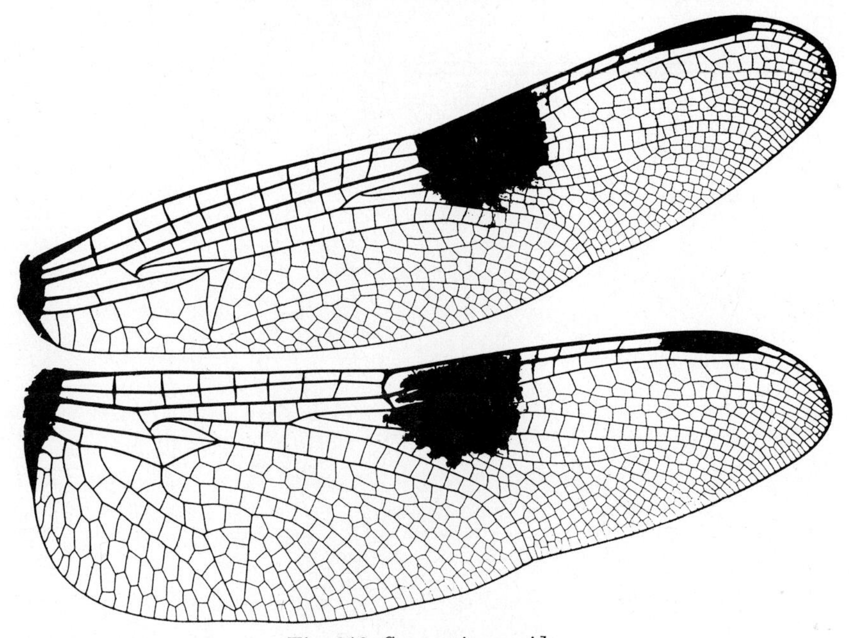

Fig. 318. *Cannacria gravida.*

straight in their outer half, and suddenly contracted to a short sharp point.

The nymphs are rather thin-skinned, clean, and smooth, with neatly patterned colors. The conspicuous spines of the abdomen all increase in length to rearward. The caudal appendages are as long as the last three abdominal appendages taken together.

KEY TO THE SPECIES

Adults

1—Apical planate subtends three rows of cells; trigonal interspace of fore wing with three rows of cells to level of middle fork (Mf), then four rows; face black and yellow . . . . **gravida**

—Apical planate subtends two rows of cells; trigonal interspace of fore wing with three rows of cells to level of nodus, then four rows; face yellow or red . . . . **herbida**

Fig. 319. *Cannacria gravida.*

NYMPHS

1—Lateral setae of labium 9, mental setae 9 + 3 . . . . **herbida**
—Lateral setae of labium 6, mental setae 6 + 3 . . . . **gravida**

Cannacria gravida Calvert

1890. Calv., Trans. Amer. Ent. Soc., 17:35 (in *Lepthemis*).
1912. Ris, Coll. Selys Libell., p. 735 (in *Brachymesia*).
1929. N. & H., Handb., p. 230.
1936. Byers, Ent. News, 47:35 (figs., nymph) (in *Brachymesia*).
1936. Ndm. & Fisher, Trans. Amer. Ent. Soc., 62:111 (figs., nymph).
1938. Garcia, J. Agr. Univ. Puerto Rico, 22:58, 74 (figs.).
1943. Wright, Carnegie Inst. Misc. Pub., 435:125–142.

Length 50–54 mm.; abdomen 32–40; hind wing 35–42.

A very handsome, slender, long-winged species of our southern coast. Face black and white; labrum and all above it black, save for two large spots on sides of postclypeus that are narrowly connected along fronto-clypeal suture, and two other round ones that cover prominences of frons. Top of vertex, at first pale, becomes black and shining in old males.

Front of thorax black; in tenerals and in females olive brown, with all carinae black. Black streaks above leg bases and in depths of first and third lateral sutures. Wings hyaline to nodus, where a conspicuous brown cloud of variable intensity begins. Cloud becomes black in old males, and in some females may be reduced to a postnodal yellow stain. Median planate subtends a double row of cells. Trigonal interspace of fore wing with three rows of cells about halfway to level of nodus, increasing to four rows thereafter.

Abdomen bare and shining, mostly black on dorsum, color deepening to rearward as far as segment 9; 10 yellow, with a small black basal middorsal triangle.

Haunts thin wave-worn outer rim of vegetation in Florida lakes.

Distribution and dates.—UNITED STATES: Ala., Fla., La., Md., Miss., Nebr., N. C., S. C., Tex.

March 29 (Fla.) to November 17 (Fla.).

Cannacria herbida Gundlach

Syn.: batesii Kirby, fumipennis Currie

1889. Gundlach, Contribucion a la Entomología Cubana, 2:261 (in *Libellula*).
1895. Calv., Proc. Calif. Acad. Sci., (2)4:548 (figs.).
1919. Calv., Trans. Amer. Ent. Soc., 45:365.
1930. Ndm., Ent. News, 41:254.
1932. Klots, Odon. of P. R., p. 51 (figs.).
1938. Garcia, J. Agr. Univ. Puerto Rico, 22:58, 74 (figs.).

Length 43–46 mm.; abdomen 32–34; hind wing 34–37.

A somewhat smoky-winged, slender brownish species, known only from Texas within the United States, but common in West Indies and Mexico. Basal half of frons above, vertex, and occiput brown.

Thorax brown, dull above, tawny beneath, and patternless. Wings subhyaline, more or less overspread with yellow in their membrane, this color deepest toward base and on a line from triangles to stigma. Forewing triangle nearly twice as long as wide.

Abdomen yellowish, with a wide black band covering dorsum of segments 4 to 9, widening to rearward to 8, absent on 10, which is yellowish. A line of yellow on upper side of superior caudal appendages of male.

Distribution and dates.—UNITED STATES: Tex.; ANTILLES: Cuba, Haiti, Jamaica, P. R.; also from Brazil.

Year-round.

Genus PACHYDIPLAX Brauer 1868

This genus consists of a single American species of medium size; it is olivaceous striped with brown, becoming pruinose blue with age. Whatever the outward change in appearance, the genus is easily recognized by a single venational character: the single crossvein under the distal end of the stigma has a vacant space, about as long as four normal cells, between it and the four postnodal crossveins of the second row. A combination of several additional venational characters when taken together might be equally distinctive: antenodal crossveins 6/5; intermedian crossveins 4/3; vein Cu1 springing from the outer side of the hind-wing triangle and not from its hind angle.

The body is rather stout, the eye-seam short; the hind lobe of the prothorax, wide, elevated, and notched in the middle; the thorax, striped with brown.

The abdomen is rather short, especially in the female, and depressed, continuing wide and parallel-sided in the female; moderately narrowed behind the little-enlarged basal segments in the male.

The nymph is smooth and depressed of body, with wide head. It has no dorsal hooks. The superior abdominal appendages are two-thirds as long as the inferiors. The nymphs clamber about among the trash and, when grown, transform within a few inches of the margin of the water if a suitable place is found so near; otherwise they may go a distance of several feet. They are generally dark in color, in a beautiful pattern with transverse banding of the femora.

Adults are swift of wing and somewhat difficult to capture with a net. The males hover near the surface of the water, darting hither and thither, meeting every newcomer, perching on a twig and immediately quitting it. When two males meet in mock combat, they have the curious habit of facing each other threateningly, then darting upward together into the air and flying skyward, often until lost from view.

The females are less in evidence. Except when foraging or ovipositing, they rest on trees back from the shore. When ovipositing over open water, they have an odd habit which we have not observed in other

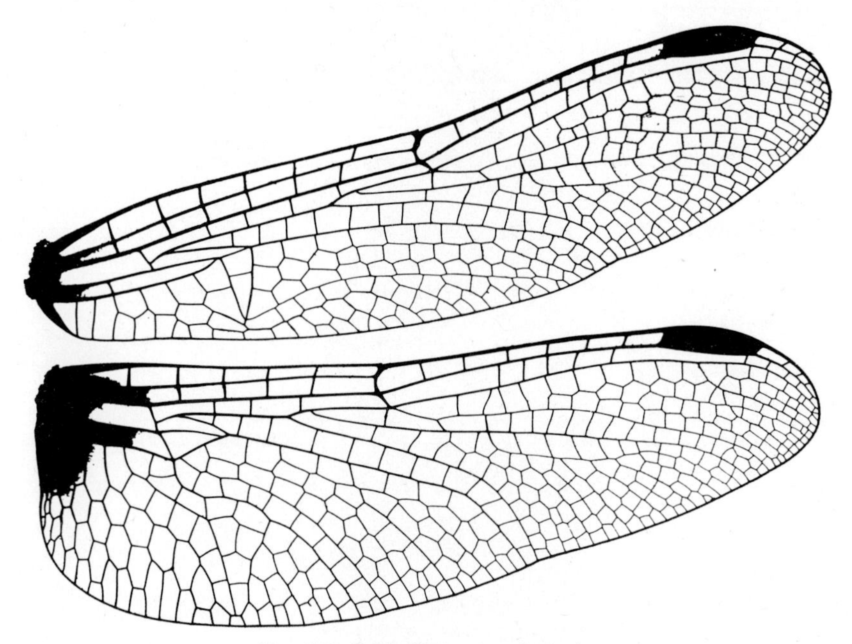

Fig. 320. *Pachydiplax longipennis.*

dragonflies: they do not rise and descend again between strokes of the abdomen against the surface of the water, but fly horizontally close to the surface and from time to time strike downward with the abdomen alone, presumably washing off eggs. In the midst of vegetation, however, they fly down and up again, as do other species.

Pachydiplax longipennis Burmeister

1839. Burm., Handb., p. 850 (in *Libellula*).
1901. Ndm., Bull. N. Y. State Mus., 47:527 (nymph).
1911. Ris, Coll. Selys Libell., p. 619.
1917. Kndy., Proc. U. S. Nat. Mus., 52:628.
1927. Garm., Odon. of Conn., p. 277.
1929. N. & H., Handb., p. 245.

Fig. 321. *Pachydiplax longipennis.*

Length 28–45 mm.; abdomen, male 28–35, female 23–25; hind wing 30–42.

A species of common occurrence and wide distribution in the United States. Face white, with labrum cream yellow. Top of frons and vertex both shining metallic blue at maturity in both sexes.

Prothorax brown above, its white-edged, elevated rear margin fringed with long pale hairs. Synthorax dark brown, with a pair of abbreviated pale stripes, divergent forward, laid upon it; another pair farther out (antehumeral), longer and narrower, connected by a pale line close under crest. Sides of thorax olivaceous or yellowish green, with three full-length brown stripes laid on three lateral sutures: first moderately wide and sinuous on its front border; other two straight-edged and narrower.

Legs black. Wings hyaline or sometimes a bit smoky, with at least a touch of flavescence in extreme base; more often, with a patch of yellow membrane there and two short parallel streaks of brown in the yellow, in subcostal and cubital bases, respectively. Often a large brownish cloud develops in wing membrane beyond nodus. Venation sparse, well adjusted, and relatively constant. It is as stated in our table of genera (p. 426). Length of wings not remarkable, though specific Latin name has reference to it. Shortness of abdomen in female gives impression of long wings.

Abdomen blackish, with parallel dashes (=) of yellow on dorsum of segments 3 to 8; 8 and 9 black above; 10 paler; caudal appendages black.

Distribution and dates.—CANADA: B. C., Man., Ont.; UNITED STATES: Ala., Ariz., Ark., Calif., Conn., Fla., Ga., Ill., Ind., Iowa, Kans., Ky., La., Maine, Md., Mass., Mich., Minn., Miss., Mo., Mont., Nebr., Nev., N. J., N. Y., N. C., Ohio, Okla., Pa., R. I., S. C., Tenn., Tex., Utah, Va., Wash., W. Va., Wis., Yellowstone; MEXICO: Baja Calif., Coahuila; also from Bermudas and Bahamas.

Collected every month in Florida.

Genus DYTHEMIS Hagen 1861

This Neotropical genus includes half a dozen species, three of which enter our southern border. They are shapely insects, larger than medium size, slender-legged, and with wings longer than the abdomen. Some of them are brightly colored. Often the wings are tinged with brown. The combination of venational characters by which the genus may be recognized is: first, an undulate vein M2; then a large anal loop with a strong curvature, whereby the midrib is much angulated at the ankle and the large heel is thrust to rearward well beyond the level of the hind-wing triangle. Behind the anal loop are two or three parallel rows of rather large hexagonal cells.

Our three species are strongly marked.

The nymphs are stoutly built stream dwellers, with the head flattened forward, and the abdomen rather high-ridged, with low, backward-curving dorsal hooks along the ridge. Short triangular lateral spines and caudal appendages converge at the blunt rear end. The nymphs are smooth and clean of body, ringed with brown on the legs, and neatly dappled with darker brown over the whole body in a pattern that matches the sand on the stream beds on which they sprawl. They are quite active. When the water is clear and the light good, they may be

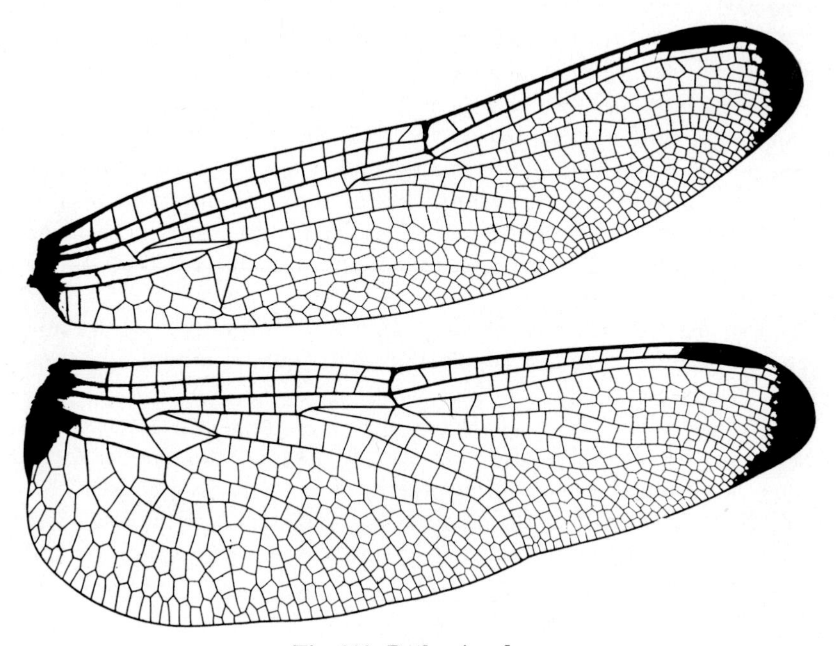

Fig. 322. *Dythemis velox.*

seen to run before the feet of a person wading in the stream. Their best distinguishing characteristics are shown in the table of genera (p. 432).

KEY TO THE SPECIES

Adults

1—Wings with red veins .... **rufinervis**
—Wings with blackish veins 2

2—Wings with wide basal crossband of brown out about to hind-wing triangle .. **fugax**
—Wings with little or no brown color in membrane at their base .... **velox**

Nymphs

1—Lateral spines of abdominal segment 9 much more than its middorsal length .. **fugax**
—Lateral spines of 9 much less than its middorsal length 2

2—Raptorial setae on lateral lobe of labium ten .... **velox**
—Raptorial setae on lateral lobe of labium seven .... **rufinervis**

Dythemis fugax Hagen

1861. Hagen, Syn. Neur. N. Amer., p. 163.
1904. Ndm., Proc. U. S. Nat. Mus., 27:700 (fig.).
1913. Ris, Coll. Selys Libell., p. 839.
1929. N. & H., Handb., p. 249.

Length 44–50 mm.; abdomen 31–34; hind wing 36–38.

A fine dark species, with pale half rings on abdomen, and brown bands on wing bases in which centers of the cells are partially clear. Face pale, becoming red on top in old males, with even upper third of compound eyes red.

Thorax nearly bare of hair except for a long erect hair fringe on collar. Front of thorax brown except for a half-length pale stripe each side. Sides of thorax pale, with four oblique brown stripes: front one narrow and sinuous, others exceedingly irregular in extent and cross connections. Rear end of thorax pale or even white beneath, as are adjacent spots on abdomen. Legs blackish, with pale bases. Wings hyaline except for basal brown crossband, and often a touch of brown on nodus; in female whole wing apex brown. Stigma black.

Slightly swollen basal segments of abdomen mostly yellow, ringed with brown on encircling carinae. Segments 4 to 9 brown, with lateral yellow spots: paired spots on 4 to 7, and single large and conspicuous ones on each side on 7; 8, 9, 10, and caudal appendages black.

Distribution and dates.—UNITED STATES: N. Mex., Okla., Tex.
May 13 (Tex.) to August 13 (Tex.).

Dythemis rufinervis Burmeister

Syn.: conjuncta Rambur, vinosa Scudder

1839. Burm., Handb., p. 850 (in *Libellula*).
1861. Hagen, Syn. Neur. N. Amer., p. 162.
1913. Ris, Coll. Selys Libell., p. 840.
1932. Klots, Odon. of P. R., p. 62 (figs., nymph).

Length 37–40 mm.; abdomen 25–30; hind wing 29–33.

A very pretty, slender species; wings veined with red and wing bases crossbanded with gold. Face pale at first, becoming yellow and then at maturity bright red, including top of frons and vertex.

Front of thorax brown, with a yellow line across it at crest, and another near it at top, with ends that run down at right angles to collar, making a quadrangular enclosure on middle of front. Enclosed brown area traversed by a hairline of yellow on middorsal carina. Sides of thorax more yellow than brown, with brown forming four oblique

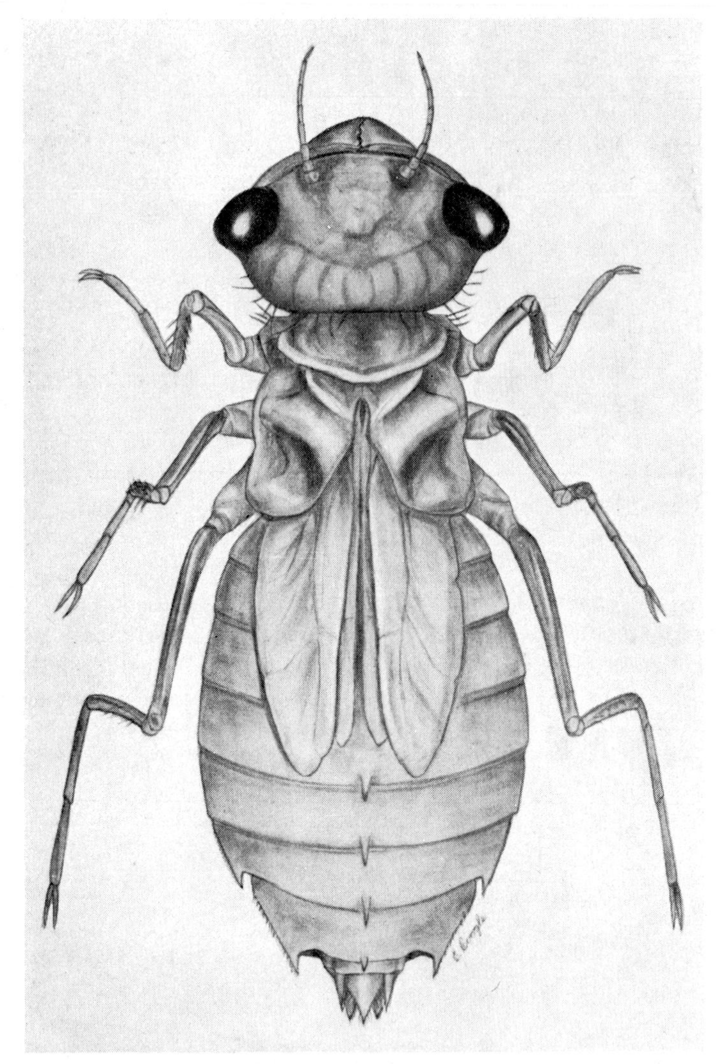

Fig. 323. *Dythemis rufinervis.* (Drawing by Esther Coogle.)

stripes: foremost a narrow antehumeral stripe; second a wide humeral stripe with sinuous front border; third and fourth narrower and very irregular, on a wide yellow field.

Legs blackish, with under side of front femur yellow. Wings broadly flavescent at base, with a short brown streak in both subcostal and cubital basal interspaces. Golden tint of base extends as far out as subtriangle in fore wing and triangle in hind wing.

Abdomen red its full length above in male, with streaks of black on middorsal and lateral carinae of segments 8 and 9; caudal appendages red. Female has a middorsal black band on 3 to 9; caudal appendages black.

Distribution and dates.—UNITED STATES: Fla.; ANTILLES: Cuba, Dom. Rep., Haiti, Jamaica, P. R.

Apparently year-round.

Dythemis velox Hagen*

1861. Hagen, Syn. Neur. N. Amer., p. 163.
1903. Ndm. & Cockerell, Psyche, 10:139 (nymph).
1916. Ris, Coll. Selys Libell., p. 1204.
1929. N. & H., Handb., p. 249.

Length 41–48 mm.; abdomen 26–30; hind wing 32–35.

A slender blackish species, with a striped thorax and spotted abdomen. Face yellowish, darkened across labrum, becoming shining metallic green in old males on top of frons and vertex.

Thorax brown in front, covered with close brown pubescence. Narrow hairline of yellow on middorsal carina, and two nearly parallel shorter and wider ones farther out. Sides of thorax yellow, with four irregular and inconstant brown stripes. The four often conjoined at their ends, forming rough block letters YIY. By further joining they may merely enclose unequal yellow spots.

Legs blackish, paler only at base and on under side of front femur. Wings hyaline, often with a touch of brown on base and tip on either or both pairs; sometimes smoky in female. Stigma blackish.

Abdomen moderately enlarged on pale basal segments, slender and darker beyond. Side of segments 1 and 2 with narrow vertical brown stripes; on 3 a broad stripe, narrowed above; on dorsum, paired yellow spots on 4 to 7, much the largest on 7; 8, 9, 10, and caudal appendages blackish.

This species sits on tall dry stems, perching, with the hinder half of the abdomen lifted high into the air. It deserves its name (*velox*, swift; by full right it might be *velocissima*). It flies off before you are near; but it invariably returns to the same place, even after several attempts have been made to catch it. It may take ten minutes for returning, coming into the collector's neighborhood frequently, but keeping always at a safe distance.—F. G. Schaupp, Texas (in litt.).

Distribution and dates.—UNITED STATES: Ala., Calif., Miss., N. Mex., Tex.; MEXICO: Baja Calif., Nuevo León, Tamaulipas; ANTILLES: Cuba.

April (Tex.) to September and October (Baja Calif.).

---

* Ris (1916) resurrected *D. multipunctata* Kirby, *nigrescens* Calvert, and *sterilis* Hagen, which in 1913 he had considered as synonyms of *velox*. Our southernmost records may apply to *multipunctata* or *sterilis*.

Genus MACROTHEMIS Hagen 1868

This is a Neotropical genus of about twenty-five species, only four of which occur within our limits. They are slender and delicate Libellulines. The head is rather small, its frons low and well rounded, with a wide frontal furrow. The thorax is rather narrow. The legs are slender. In the claws of most of the species, the tooth that is usually found on the under

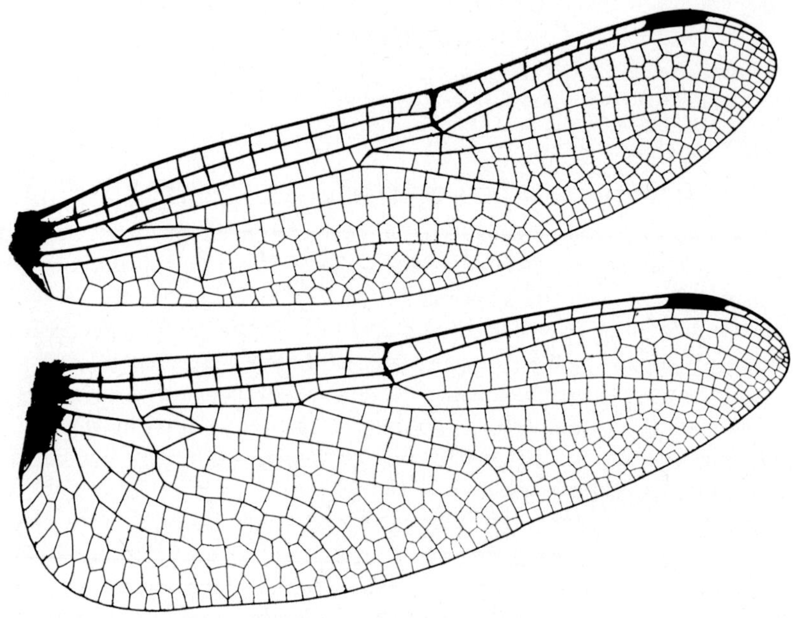

Fig. 324. *Macrothemis celeno.*

side of each claw is placed so far out that it is as long as the tip of the claw itself, sometimes slightly longer. The spines of the anterior row on middle and hind femora differ in male and female.

The principal characters of the venation of the wing are stated in our table of genera (p. 426). Additional vein characters may be noted here. There are two crossveins under the stigma. There are two long and very regular rows of cells in the trigonal interspace of the fore wing. The long anal loop has a short toe, the sole of the anal loop being hardly half as long as the gaff. The midrib of the loop is unusually close to vein A2 at its origin, and the paranal cells in the base of the loop are very large. There are two rows of cells between A2 and the irregular marginals of the hind wing.

The abdomen is somewhat swollen at the base, very slender on its middle segments, and variously shaped, according to species, on the end

Fig. 325. *Macrothemis celeno.*

segments. The hamules of the male are prominent on segment 2. The subgenital plate of the female is roundly notched in the middle of its low border.

The nymphal characters are given in our table of genera (p. 432).

The best account of adults of this genus is that in Dr. P. P. Calvert's *Macrothemis* paper of 1898. A more complete but much less detailed coverage of all the known species of the genus is that of Dr. F. Ris (1913) in his monograph on the Libellulinae.

KEY TO THE SPECIES

ADULTS

1—Tarsi with normal claws, each claw having its inferior tooth small and remote from tip .... **inequiunguis**
—Tarsi with inferior tooth enlarged and grown outward until its tip is on a level with tip of claw 2

2—Abdomen cylindric; middle and end segments of nearly equal width...... **celeno**
—Abdomen spatulate; end segments greatly widened and depressed 3

3—Superior abdominal appendage of male with short row of denticles under its enlargement .... **pseudimitans**
—This appendage with single large triangular tooth .... **leucozona**

Macrothemis celeno Selys

1857. Selys, in Sagra, Hist. Cuba, Ins., p. 454 (in *Libellula*).
1868. Hagen, Stettin. Ent. Ztg., 29:283.
1898. Calv., Proc. Boston Soc. Nat. Hist., 28:301 (figs.).
1913. Ris, Coll. Selys Libell., p. 879.
1932. Klots, Odon. of P. R., p. 59 (nymph).
1938. Garcia, J. Agr. Univ. Puerto Rico, 22:62.

Length 44–50 mm.; abdomen 30–32; hind wing 31–35.

A slender, delicate species. Face yellow across top of frons and down its sides. Below frons, face obscure; labrum blackish, with yellow spots. Eyes in life a deep translucent porcelain blue in male; blackish, with only a suggestion of blue in female.

Thorax brown in front, with a pair of erect and opposed **7**-shaped yellow marks, their inturned tops narrowly separated by the black middorsal carina. Edges of crest black. Sides of thorax brown, with two pairs of rather conspicuous, obliquely placed spots of yellow or white. In male, spines of front row on hind femur are reduced in number and so modified in form that they look almost like teeth: shortened, thickened, flattened, and obliquely truncated, their tips pointing inward toward body.

Abdomen black, spotted with white. Middle segments unusually long and slender, especially in male.

Common along nearly every little clear-flowing brook and rill in larger islands of West Indies, in hours of sunshine. It shifts in low fluttering flights from perch to perch and is easily taken in net.

A female seen ovipositing, unattended by the male, flew eight to twelve inches above a sandy riffle and made lightning-quick dashes to the water's surface and back again to the same level, hitting the water lightly with the tip of the abdomen each time, this descent and return made so quickly that the eyes could hardly follow. Although present in lower levels [altitudes], it is most common and easily seen along the water courses in the high levels. The nymphs in the rivers are found among the overhanging roots and stems of plants growing close to the edges.—Garcia (1938).

Distribution and dates.—ANTILLES: Cuba, Dom. Rep., Haiti, Jamaica, P. R.

Apparently year-round; recorded every month except August and September.

Macrothemis inequiunguis Calvert

Syn.: vulgipes Calvert

1895. Calv., Proc. Calif. Acad. Sci., (2)4:533 (figs.).
1906. Calv., B. C. A., p. 286.
1913. Ris, Coll. Selys Libell., pp. 689, 1212.

Length 33–36 mm.; abdomen 25–27; hind wing 24–29.

A smaller Mexican species, not yet reported from the United States. Frons and vertex metallic green in male, brown or olive in female. Dorsum of thorax dark brown, with pale middorsal carina and beside it a pair of green stripes convergent at their upper ends near crest. Sides of lighter brown, with four ill-defined subtriangular spots of pale green. Abdomen black above, with a pair of long, more or less interrupted lines of green.

Distribution and dates.—MEXICO: Baja Calif.; also south to Chile.

September to October.

Macrothemis leucozona Ris

1913. Ris, Coll. Selys Libell., p. 887 (as subsp. of *M. imitans* Karsch).
1916. Ris, Coll. Selys Libell., p. 1213.

Length 37 mm.; abdomen 26; hind wing 28.

A slender, clear-winged blackish species. Frons above and top of head are shining metallic green. Lower front and sides of frons yellow. Entire clypeus brownish and translucent; so is labrum, save for some irregular touches of bronzy green on its disc. Face clothed with short black hair, and thorax with longer tawny hair. Brown occiput thinly hairy and polished.

All pale spots on brown thorax creamy white, well defined on dark background; two on front subcuneate or trapezoidal, with long divergent antero-lateral corners. Midlateral stripe of thorax a single band, with its upper half lying athwart faint midlateral suture. On side of metathorax are three very unequal spots: two low down on side, and a smaller spot at a higher level.

Wing membrane clear except for touches of brown at extreme base of veins. In hind wing are nine antenodal crossveins. Truncated and proximally pointed spines of anterior row on hind femur fifteen in number, last two close-set.

Abdomen of male very slender, notably widened and depressed on segments 6 to 9, with very sharp middorsal and lateral carinae; very black, with usual midlateral and marginal streaks of pale translucent yellow widest but least defined on basal segments. Midlateral streak widened again on 6 and 7; a basal touch of yellow on 8; marginal streaks obscure on middle segments, but better defined on 7, 8, and 9; 10 and appendages black.

Distribution and dates.—UNITED STATES: Tex.

July 1, 1950 (Garner Park, Tex.); Paul N. Albright, collector.

Macrothemis pseudimitans Calvert

1895. Calv., Proc. Calif. Acad. Sci., (2)4:531 (figs.) (as *imitans*).
1898. Calv., Proc. Boston Soc. Nat. Hist., 28:329 (figs.).
1913. Ris, Coll. Selys Libell., p. 883.

Length 40–43 mm.; abdomen 26–29; hind wing 29–32.

A clear-winged Mexican species, male with a spatulate abdomen. Face blackish, with a yellow ring around top of metallic blue frons. Vertex also blue; occiput blackish.

Thorax blackish, pruinose in old males, with two faint pale stripes upon front and three oblique ones on each side. Legs black, with only under side of front femur greenish. Wings hyaline, shining.

Abdomen black, a little paler on basal segments, where thinly clothed with white hair; elsewhere bare and shining. Slightly swollen basal segments gradually tapered to rearward on segment 3, then widened again on 6, 7, and 8, and narrowed on 10. Long caudal appendages black. Superiors have a row of minute denticles on lower surface of middle third of their length, but no large tooth at end of row.

Distribution and dates.—MEXICO: Baja Calif., Tamaulipas; also south to Ecuador and Venezuela.

September and October (Baja Calif.).

Genus SCAPANEA Kirby 1889

This West Indian genus consists of but one species, for which the readiest recognition character is the broadly flattened posterior end of the abdomen. In the male, segments 7, 8, and 9 are twice as wide as the middle segments.

The venational characters by which the genus may be recognized are as shown in our table to genera of Libellulinae (p. 427). It may be further noted that the strongly slanted arculus is situated just before the second antenodal crossvein; under the stigma are three or four crossveins, with eight to ten postnodal crossveins in the space preceding them; veins M3 and M4 are elbowed at the outer end of the median planate; the radial

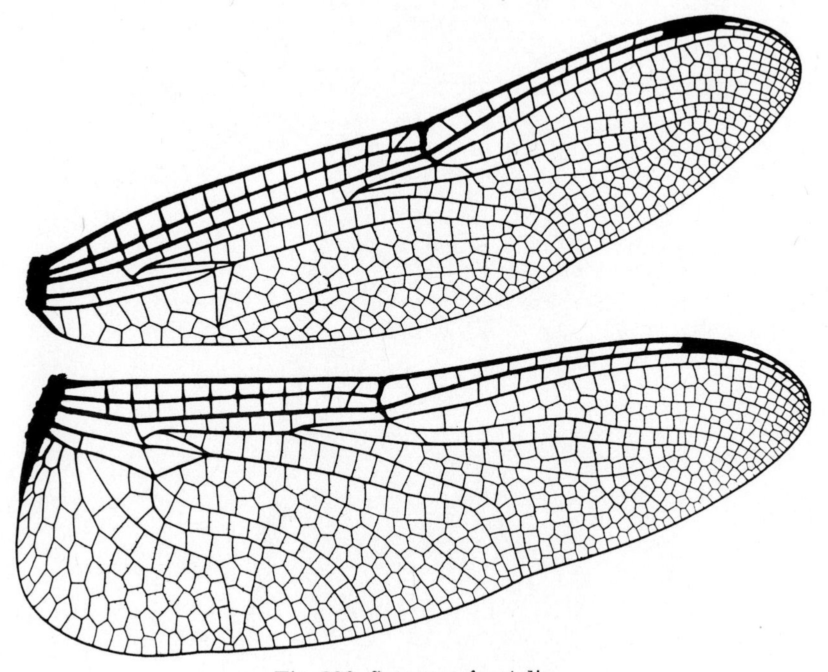

Fig. 326. *Scapanea frontalis.*

planate subtends two rows of cells, and an apical planate is generally wanting; there are four rows of cells between vein A2 and the hind angle of the wing. The inferior caudal appendage of the male is about as long as the superiors.

The nymph is stout, smoothly contoured, and compactly built, with the head sloping forward between the broadly rounded eyes. The antennae are very short and weak beyond a very stout basal segment. The very prominent supra-coxal processes of the prothorax fit closely against the prickly rear of the head; the epaulets on the disc are small, round, and rough-edged. The legs are short and bare; the tips of the tarsal segments are blackish. The caudal appendages are short, stout, and bare; the superior, viewed from above, is an equilateral triangle. Other characters are as shown in the table for nymphs (p. 433).

Fig. 327. *Scapanea frontalis.*

Scapanea frontalis Burmeister

1839. Burm., Handb., p. 857 (in *Libellula*).
1911. Wlsn., J. Inst. Jamaica, 2:50.
1913. Ris, Coll. Selys Libell., p. 848.
1932. Klots, Odon. of P. R., p. 66 (nymph).
1938. Garcia, J. Agr. Univ. Puerto Rico, 22:64.

Length 45–51 mm.; abdomen 31–35; hind wing 35–38.

A large blackish species that has its haunts in mountain streams. Face black and white: white on clypeus and sides of frons; black on labrum and remainder of frons; often with yellow spots or a cross stripe on labrum. Top of frons and vertex become shining metallic blue in old males.

Front of thorax dark brown, with black edges on middorsal carina and crest. A stripe of yellow at each side more or less interrupted above middle, with top end widened on inner side next to crest. Sides of thorax lighter brown, with three oblique stripes lying midway between lateral sutures. Middle stripe may not rise above level of spiracle. Legs black. Wings long and very clear in male; tinged with brown in their membrane in female. Vein M2 a little undulate.

In old males both top and sides of thorax become pruinose blue, but not abdomen. A cloud of pruinosity develops on all wings just before stigma. This white patch flashes reflected light conspicuously in flight.

Abdomen rather stout and, as noted above, widens at sides of segments 7 to 9 in male. Abdomen does not become pruinose, but remains black or blackish, save for some obscure streaks of dull yellow on basal half.

This is mainly a lotic species. . . . Nymphs in Puerto Rico have been collected, up to the present, only above the 500 foot level, becoming commoner higher up. . . . [The adults] are strong fliers, flitting close to the water up and down the course of the stream, loitering over pools. They are extremely shy, though approachable, but if one is missed by the net usually all go away. Most of them are caught on the wing. They copulate without coming to rest. . . . Their nymphs prefer rapid running water, clinging to the stones—some of them to the under surface.—Garcia (1938).

Distribution and dates.—ANTILLES: Cuba, Dom. Rep., Haiti, Jamaica, P. R.

Year-round.

Genus BRECHMORHOGA Kirby 1894

This is a genus of more than a dozen rather large Neotropical species, a single one of which enters our southwestern border. Its venational characters are as stated in our table of Libelluline genera (p. 426). Additional characters may be noted here. The anal loop is very long, with extra ankle cells on both sides of the midrib at the heel level, and with three cell rows behind it paralleling vein A2. Vein M2 is very slightly undulating. There is no development of an apical planate. Veins M3 and M4 have something of an elbow-like angulation at the outer end of the median planate, and there are but two rows of cells in the space beyond the triangle. The tooth on the under side of the tarsal claws is close to the tip, but does not reach its level.

The nymph is very short and thick of body, blunt at both ends, short-legged, and clean-surfaced, with bulging eyes and sharp spinelike dorsal hooks. It is found on sand-and-gravel beds of pools in torrential streams. Its distinctive characters are stated in our table for nymphs (p. 432).

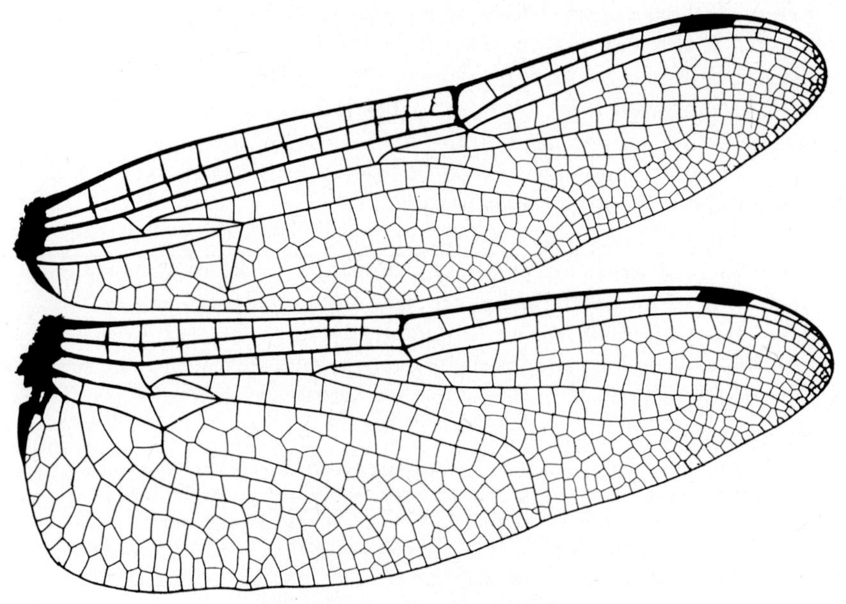

Fig. 328. *Brechmorhoga mendax.*

Brechmorhoga mendax Hagen

1861. Hagen, Syn. Neur. N. Amer., p. 164 (in *Dythemis*).
1898. Calv., Proc. Boston Soc. Nat. Hist., 28:313 (figs.).
1913. Ris, Coll. Selys Libell., p. 861.
1917. Kndy., Proc. U. S. Nat. Mus., 52:605, 627.
1929. N. & H., Handb., p. 250.

Length 53–62 mm.; abdomen 36–45; hind wing 34–43.

A stout grayish species with hoary appearance that seems to characterize many desert species. Face yellow, including vertex, clothed with short whitish pubescence.

Thorax clothed with a similar longer pubescence and with erect fringe of white hairs on collar. Front of thorax with two wide brown stripes beside pale carina, abbreviated above. Sides with a wide brown stripe upon humeral suture and two additional stripes on succeeding sutures farther back, confluent above, all on a yellow ground. Legs brown, paler

Fig. 329. *Brechmorhoga mendax.*

at base and on front femora to knees externally. Wings hyaline, with a short black stigma; tinge of brownish on extreme base, wider on hind wings.

Abdomen mostly pale on moderately swollen basal segments, blackish beyond; segments 2 and 3 narrowly annulate, with black encircling on carina; 3 and 7 with large, diffuse, paired dorsal spots, largest on 7; sides of 8 and 9 with a touch of yellow; 10 and appendages black.

These Brechmorhogas usually had short beats in the shade of the occasional large willow trees that grew on the gravel beaches. . . . The males were taken while flying on short beats over the stream. The female was captured while cutting S's and figure 8's through a swarm of small Diptera. She was indifferent to several passes I made at her before I succeeded in netting her. This species is the most graceful on the wing of any odonate with which I am familiar. Frequently they fly with a swinging mayfly-like motion. In the heat of the day they float around among the tree tops.—Kennedy (1917).

Distribution and dates.—UNITED STATES: Calif., Okla., Tex.; MEXICO: Baja Calif., Sonora; also from Tepic in Mexico.

June (Tex.) to September–October (Baja Calif.).

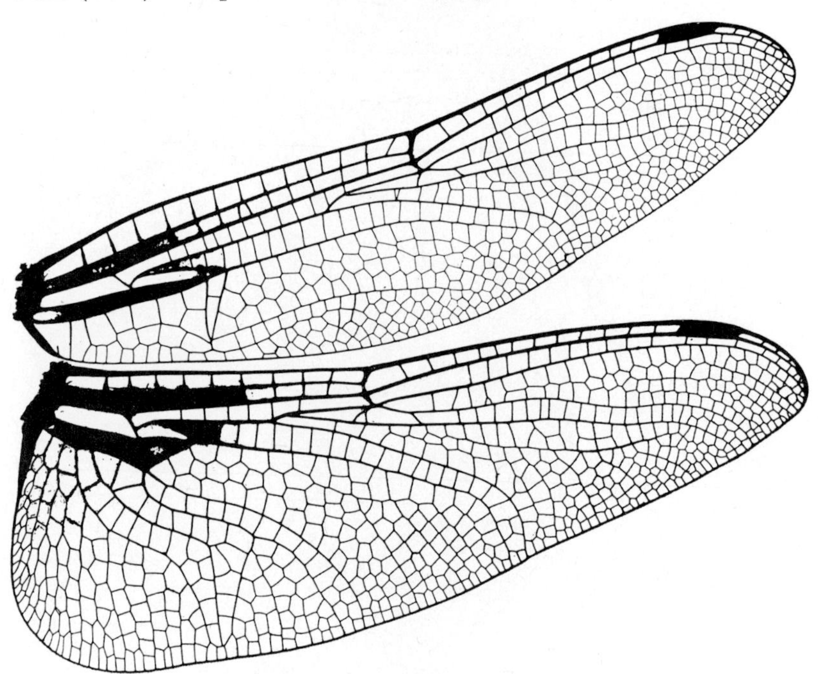

Fig. 330. *Paltothemis lineatipes.*

Genus PALTOTHEMIS Karsch 1890

This genus consists of the single species hereinafter described. The wings are very broad toward the base and tapered to the apex. The stigma is rather small, its outer end a little more oblique than the inner end. The triangle of the fore wing points inward, vein Cu being strongly bent. The toe of the anal loop is long and narrow. The basal cells of the anal loop are very large. The anal area behind the loop is not filled with rows of cells running parallel to A2, but with backwardly directed forks and columns of cells in double rows. There are generally ten antenodal crossveins in the hind wing.

The nymph is very smooth and very dark-colored. There are low dorsal hooks on abdominal segments 2 to 6, diminishing in size to rearward.

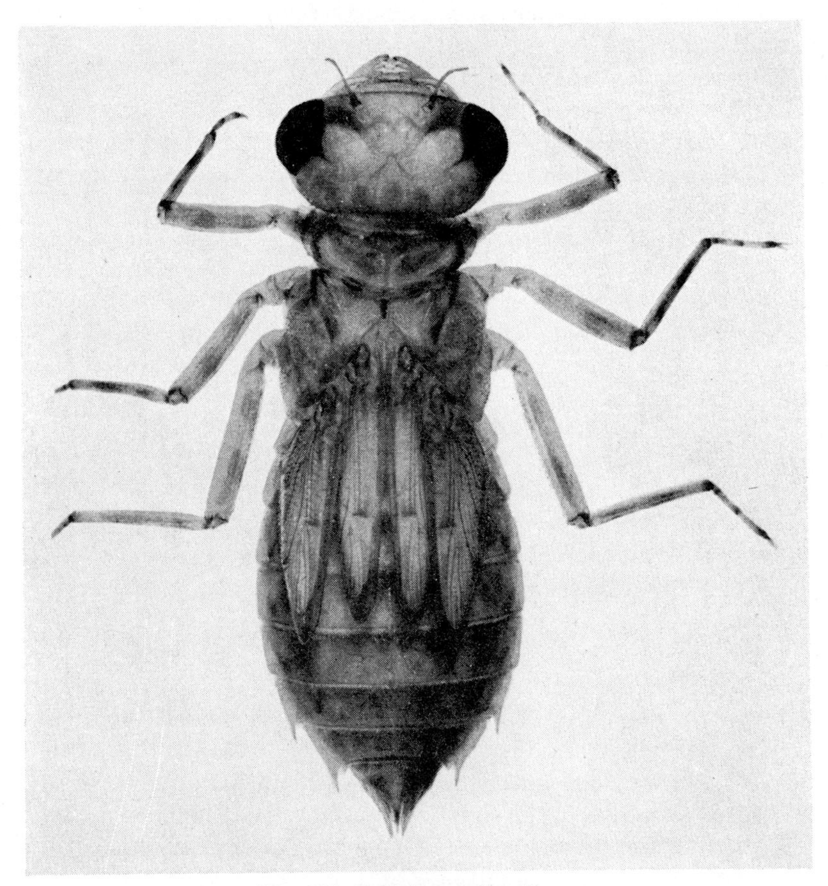

Fig. 331. *Paltothemis lineatipes.*

The lateral spines on 8 and 9 are short, sharp, and straight. The teeth on the front border of the lateral lobes of the labium are separated by deep notches.

Paltothemis lineatipes Karsch

Syn.: russata Calvert

1890. Karsch, Berlin Ent. Zeitschr., 33:362.
1899. Calv., Proc. Calif. Acad. Sci., (3)1:526 (figs.).
1904. Ndm., Proc. U. S. Nat. Mus., 27:699 (nymph).

1906. Calv., B. C. A., p. 292.
1913. Ris, Coll. Selys Libell., p. 846.
1929. N. & H., Handb., p. 251.

Length 47–54 mm.; abdomen 35–40; hind wing 43–46.

A fine large Southwestern species, rusty red in male and hoary gray in female. Face pale, becoming reddish with age, including vertex, especially in male. Frons bare, shining.

Thorax thinly clothed with short pale pubescence; collar with erect fringe of long whitish hairs. Front of thorax brown; sides with roundish spots of deeper brown in front of humeral suture and below. Rearward on olivaceous sides are three brown stripes of very irregular outline: two on lateral sutures, third (confluent below) on infero-lateral margin.

Slightly swollen basal segments of abdomen pale or rufous dorsally and blackish ventrally, with narrow crosslines of black on carinae; gradually widened segments beyond more extensively blackish, with basal and apical cross streaks of deeper black, faintly brown toward middle segments. Segment 10 narrower, partly pale. Appendages obscure yellowish brown to red.

Distribution and dates.—UNITED STATES: Ariz., Calif., N. Mex., Tex.; MEXICO: Baja Calif., Chihuahua, Sonora; also south to Brazil.

May (Calif.) to October (Baja Calif.).

Genus MIATHYRIA Kirby 1889

Syn.: Nothifixis Navas

This is a Neotropical genus of two species, both of which occur within our southern limits. The hind wings are broad at the base and pointed at the apex. The eye-seam is very long; the frons low, with a wide furrow. The legs are long and slender. The abdomen slowly tapers from the base of segment 3 to the end. The superior caudal appendages of the male, when viewed from the side, have a strongly sigmoid curvature.

The genus is easily recognizable by the small number of antenodal crossveins in the hind wing, by the very large size and unusual breadth of the paranal cells in the fore wing, and by the wide space and long cells subtended by the apical planate.

Nymphal characters, known for *marcella* only, are as shown in the table of genera (p. 432).

KEY TO THE SPECIES

ADULTS

1—Hind wing 29–33 mm.; radial planate in fore wing subtends five to seven cells; top of frons in mature male metallic violet.................... **marcella**

—Hind wing 23–26 mm.; radial planate subtends three to four cells; top of frons in mature male red.. **simplex**

Miathyria marcella Selys

1857. Selys, in Sagra, Hist. Cuba, Ins., p. 452 (in *Libellula*).
1861. Hagen, Syn. Neur. N. Amer., pp. 146, 316 (as *Tramea simplex*).
1906. Calv., B. C. A., p. 294.
1913. Ris, Coll. Selys Libell., p. 1009.

Length 37–40 mm.; abdomen 23–26; hind wing 29–33.

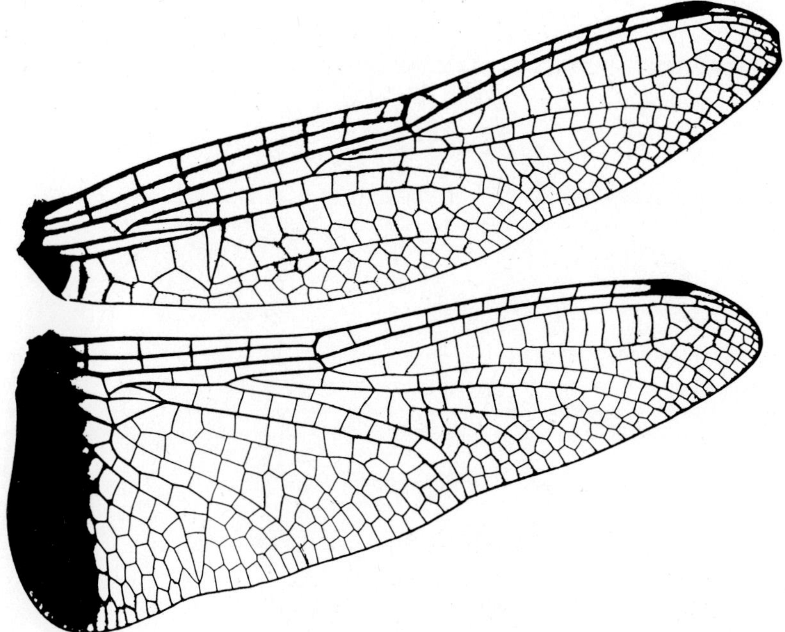

Fig. 332. *Miathyria marcella.*

A dainty little Southern species with greatly reduced and very highly specialized wing venation. Face pale; clypeus olive; front of frons and labrum yellow; top of frons and vertex metallic violet; occiput brown, fringed behind with white hairs.

Thorax brown in front, clad with tawny hairs. Sides olivaceous, with obscure clouds of brown on first and third lateral sutures, and a black ring around spiracle. Legs brown beyond their pale bases.

Unique venation as follows: antenodal crossveins 6–7/4; postnodals 5/5; two crossveins under stigma, with only three postnodals in interspace preceding them. Radial planate generally subtends and encloses six cells. Outer side of fore-wing triangle convex, its apex pointing inward. Front side half as long as inner side. Very large cells fill broad subtriangle.

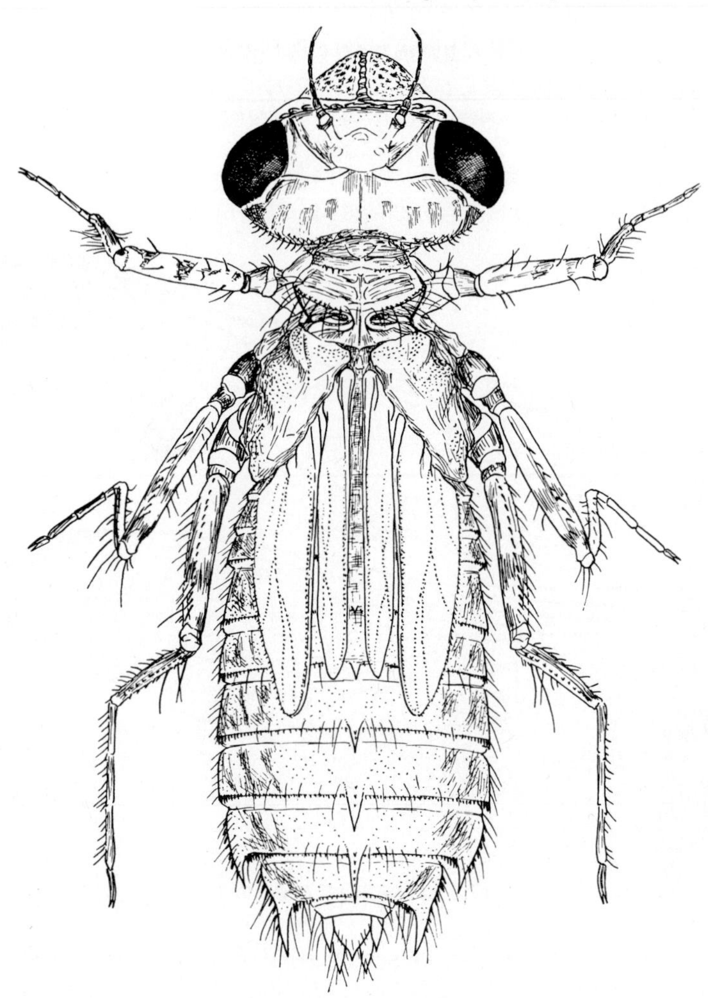

Fig. 333. *Miathyria marcella.* (Drawing by Esther Coogle; from a paper by the junior author, *Fla. Ent.*, 36:23, 1953).

Abdomen brown; its carinae and dorsum of segments 8 to 10 blackish.

Nymph of species obtained from Manatee pond of Botanic Garden, Georgetown, British Guiana. Adults were seen swarming over a pond-side hill slope near Havana, Cuba. Their effortless, swallow-like flight has been seen over coastal plains of Texas.

Distribution and dates.—UNITED STATES: Fla., La., Tex.; MEXICO: Tamaulipas; ANTILLES: Cuba, Jamaica, P. R.; also south to Argentina.

Apparently year-round.

Miathyria simplex Rambur

Syn.: pusilla Kirby

1842. Rbr., Ins. Neur., p. 121 (in *Libellula*).
1889. Kirby, Trans. Zool. Soc. London, 12:318 (fig.) (as *pusilla* n. sp.).
1907. Calv., B. C. A., p. 294.
1913. Ris, Coll. Selys Libell., p. 1010.

Length 28–32 mm.; abdomen 19–22; hind wing 23–26.

A delicate little Tropical species; a large brown spot on base of hind wing reaches outward to cover most of triangle.

Easily distinguished by characters stated in preceding key.

Distribution and dates.—MEXICO: Tamaulipas; ANTILLES: Cuba; also from Mexico south to Brazil.

November (Cuba); taken year-round in Mexico.

Genus TAURIPHILA Kirby 1889

This is a Neotropical genus of several species, one of which falls within our range. Ours is a large species, with a conspicuous band of brown across the base of the hind wing. The head is broad, the eye-seam long. The broad frons has a feeble carina around the front of its prominence, and a wide and shallow middle furrow. The legs are long and thin. The wings are long and pointed, with the base of the hind wings very broad. The main venational characters are as shown in our table of genera (p. 427). The superior caudal appendages of the male are not as long as segments 9 and 10 together.

This genus superficially resembles *Tramea,* but is distinct in a number of good structural characters. (1) The fore-wing triangle is two-celled, followed by three rows of cells; in *Tramea,* generally three-celled, followed by four rows of cells. (2) The subtriangle is well defined on its inner side; not broken or indistinctly developed on that side. (3) The stigma of the fore wing is about as long as that of the hind wing; not distinctly longer, as in *Tramea.* (4) The radial planate subtends a single row of cells; not two rows, as in *Tramea.* (5) In the anal loop of the hind wing, gaff and sole meet at the heel in an obtuse angle; in *Tramea,* in a rounded curve. (6) Vein A2 of the anal loop is a well-developed vein; not zigzagged and indistinct, as in *Tramea.* (7) The caudal appendages are of ordinary length; not extremely long and slender, as in *Tramea.*

The nymph is smooth or very thinly hairy, neatly patterned in green and brown. The head is strongly rounded behind the broad eyes, and thickly beset with spinules there externally. A zigzag band of darker brown extends across the head between the eyes. Two rows of diffuse brownish spots extend as a lateral band along the middle of the sides of the abdomen. The legs are faintly ringed with brown.

The abdomen is short and blunt, broadly depressed and sharp-edged, and smoothly contoured. The dorsal hooks form a high-arched ridge on the middle segments, where they are broad at base and laterally flattened; that line of decurvature is continued by the acute ridge of the superior caudal appendage.

Other characters are as shown in the table (p. 433).

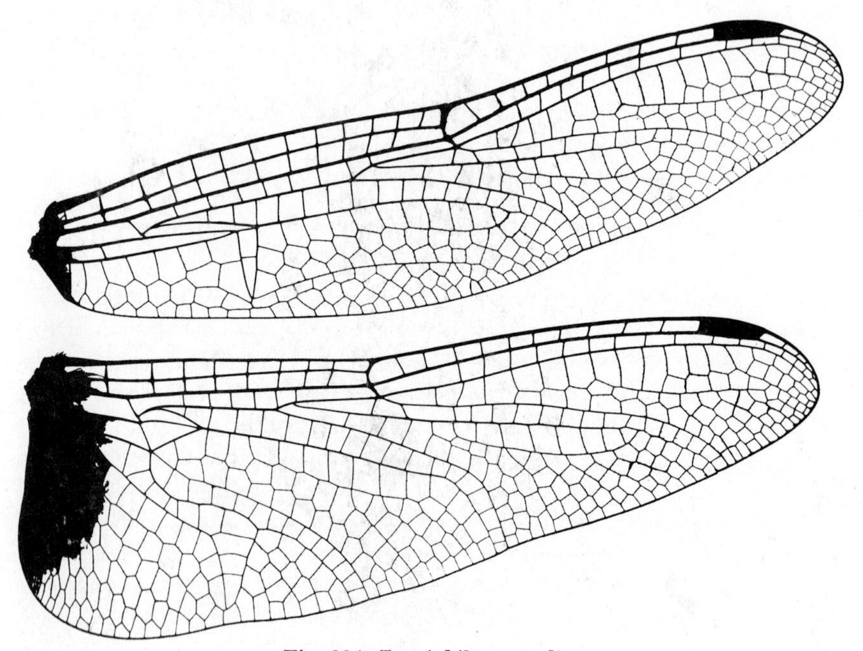

Fig. 334. *Tauriphila australis.*

Tauriphila australis Hagen

Syn.: iphigenia Hagen

1867. Hagen, Stettin. Ent. Ztg., 28:229 (in *Tramea*).
1906. Calv., B. C. A., pp. 296, 297.
1913. Ris, Coll. Selys Libell., p. 1001.

Length 42–47 mm.; abdomen 28–32; hind wing 36–38.

A strong-flying brownish species. Face brown, with two paler cross streaks: one on anteclypeus, other on fronto-clypeal suture. Top of head dark metallic violet in mature male; reddish yellow in female.

Hind lobe of prothorax narrow from side to side, low, and bare. Front of synthorax shining black at maturity, including crest, and covered

Fig. 335. *Tauriphila australis.*

with tawny hair. Sides of thorax blackish behind humeral suture, becoming gradually paler to rearward, without definite stripes or spots.

Legs black from middle of femora outward. Wings hyaline, or often in female strongly tinged with yellow throughout membrane; more deeply at base of hind wing. When clear, always a tinge of brown at extreme base of fore wing. Always a blackish band across base of hind wing spread out to level of first antenodal crossvein, and to tip of fifth paranal cell; then rounded backward to hind margin at anal angle of wing. No marginal clear spot next to black membranule.

Abdomen in male reddish brown, darker on joinings of segments and on all carinae; more broadly black on 8 and 9; 10 and caudal appendages paler. Female much paler throughout than male, mostly pale brown, with black carinae, and dimly showing pale areas on sides of middle segments. These become more reddish on 8 and 9. Segment 10 and caudal appendages blackish.

Distribution and dates.—UNITED STATES: Fla.; ANTILLES: Cuba, Dom. Rep., Haiti; also from Mexico south to Brazil.

May 15 (Cuba) to September 7 (Cuba).

Genus THOLYMIS Hagen 1867

This is a Tropical genus of a few species, none of which is as yet known in the United States; one occurs in Cuba and one in Mexico. The eyes are large. The eye-seam on the top of the head is long, a little longer than the middorsal length of the occiput.

The hind lobe of the prothorax is low and bare. The legs are long and thin. The wings are hyaline save for a thin cloud of amber brown at the nodus, very much smaller in the fore wing than in the hind. The stigma is trapezoidal, being longer on the front side than on the rear. The front-wing triangle is very long and narrow, its apex strongly retracted so that it points notably inward. The anal loop is long and sinuous, wide and broadly rounded at the heel; its toe is open, veins A1 and A2 ending at the hind margin of the wing. A deep sag in the middle of the radial and median planates, where a cell row is added, is also quite characteristic of this genus. The first radial interspace has no long vacancy between the postnodal crossveins of the second series and the crossveins behind the stigma.

The caudal appendages of the male are of the usual form, but long and slender and with a very long line of denticles on the under side of the superiors. The hamules are simple, lacking an outer branch. The subgenital plate of the female extends to rearward about half the length of the sternum of segment 9, and is roundly notched at its tip. Behind the plate, in place of the usual pair of little nipple-shaped palps, a channeled egg slide extends like an inverted trough to rearward to the level of the tip of the abdomen. The raised edges of this slide are beset with a line of stiff hairs.

The nymph of our species apparently has not yet been reared. Fraser described and figured the nymph of an Oriental species, *Tholymis tillarga* (Records Mus. Indian, 16: 460, 1919).

Tholymis citrina Hagen

1867. Hagen, Stettin. Ent. Ztg., 28:218.
1906. Calv., B. C. A., p. 220 (figs.).
1913. Ris, Coll. Selys Libell., p. 915 (in *Sympetrum*).

Length 48–53 mm.; abdomen 32–40; hind wing 36–39.

A large, strong-flying insect of very graceful form. Much like *Cannacria gravida* in superficial appearance, but with a smaller cloud of

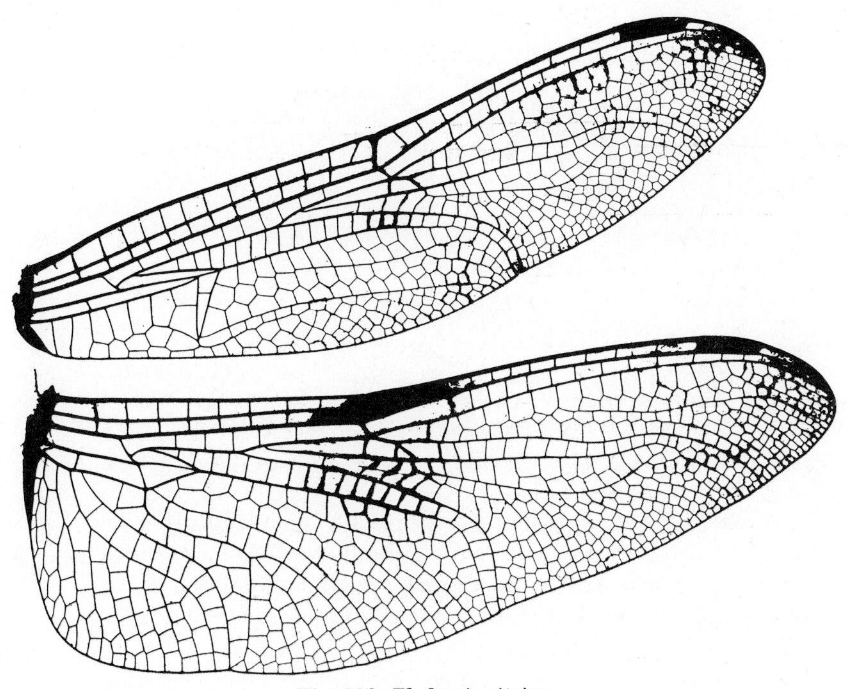

Fig. 336. *Tholymis citrina.*

amber brown at nodus of hind wing. Also, three paranal cells on proximal side of anal loop where *Cannacria* has but two.

Face and frons above yellowish, latter and vertex becoming brown and, at full maturity in male, shining blue-black. Thorax nearly uniform olive brown, with all carinae blackish; thinly clad with short brown hair. Sides become brown with age, darker in front, where, at maturity in males, all is black with bluish reflections.

Legs pale yellowish, at first with blackish spines, later becoming brown. Wings hyaline, with brown veins and tawny stigma. At nodus in hind wings, a thin cloud of brown covers half of width of wing; in fore wing it is often a mere diffuse spot. Venation as stated in table

(p. 427). Open end of long anal loop readily distinguishes this species from all others in our fauna.

Abdomen pale brown phalerate with black carinae and a diffuse dorsal band of brown that deepens rearward to segment 9; 10 and appendages at first yellow; at maturity all becomes brown in female, black or blackish in male.

Distribution and dates.—ANTILLES: Cuba, Jamaica; also from Mexico to Brazil.

January (Cuba) to July (Cuba).

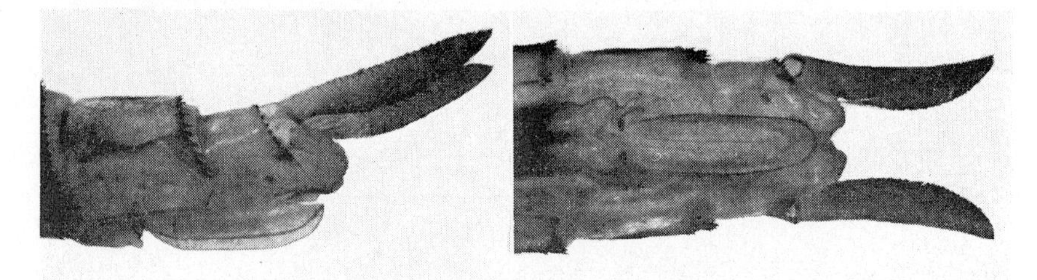

Fig. 337. *Tholymis citrina,* female.

Genus TRAMEA Hagen 1861

Syn.: Trapezostigma Hagen

Saddle Bags

These are large, wide-ranging dragonflies, conspicuously marked with bands of brown across the base of the hind wing. The head is large; the eye-seam long; the frons broad and prominent, with a faint rimlike carina around its front and a broad median furrow. The hind lobe of the prothorax is low and narrow.

The legs are slender and spiny. The wings are long, strong, and pointed, the hind ones very wide across the base. The stigma of the front wing is distinctly longer than that of the hind wing. The forewing triangle is generally three-celled, with four rows of cells beyond it. The subtriangle is large and very irregular on its inner side, the course of vein Cu2 to the hind angle of the triangle devious. The apical planate subtends first one row and then two rows of very elongate cells. The radial planate subtends two rows and turns forward and joins vein Rs, causing that vein to become zigzagged at the junction. In the long anal loop of the hind wing the gaff joins the sole in a rounded curve at the heel. The anal area is very broadly expanded, and the anal lobe is traversed by many radiating branches forking repeatedly as they run out to the hind margin. Other characters are as shown in our table of genera (p. 427).

The species of this splendid genus range over the Temperate and Torrid zones of the whole earth. Six of them occur within our limits.

The nymphs clamber about actively among the stems of green water-weeds and half-floating masses of algae. They are green, with delicately patterned markings of brown. The head is large; the eyes are very prominent, with their outer margins sloping to rearward. The labium is very

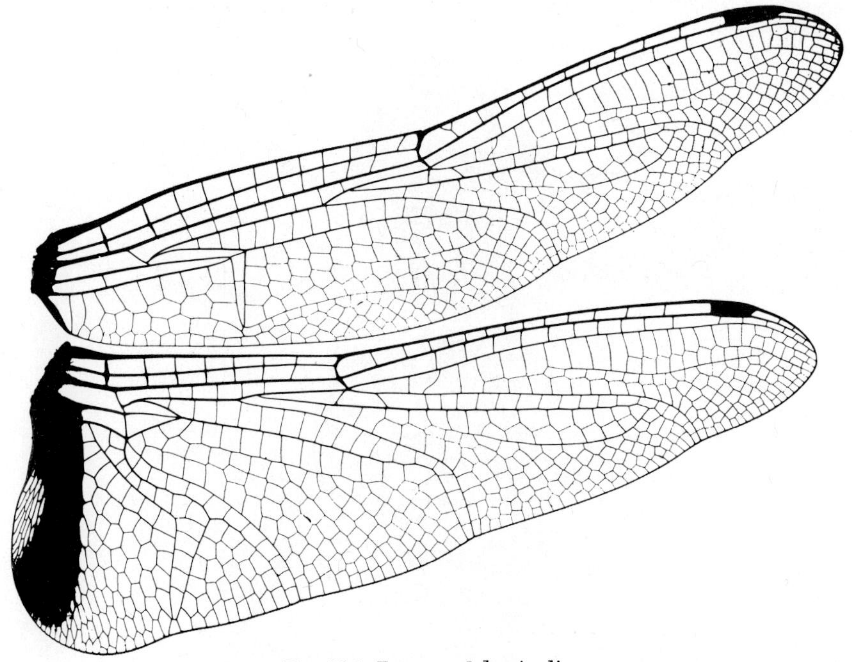

Fig. 338. *Tramea abdominalis.*

wide, armed with numerous raptorial setae and a very long and unusually straight movable hook.

The abdomen is without dorsal hooks and is wide all the way back to the big lateral spines on segments 8 and 9. These spines are flattened and incurving, are beset on their outer margin with large and numerous spinules. The caudal appendages are acuminately pointed, the superior distinctly shorter than the inferiors; the laterals are bare, the others beset on the outer edges with conspicuous spinules. Other characters are as shown in our table of genera (p. 433).

The best recent work is, as usual, that of Ris (*Coll. Selys Libell.*). Of historic interest is the treatment of the species by Hagen (*Stettin. Ent. Ztg.*, 28:222, 1867).

KEY TO THE SPECIES

Adults

1—Base of hind wing with narrow crossband of dark color extending outward to about level of anal crossing; outer edge of middle third nearly straight.. 2

—This band wide, extending to distal angle of triangle; outer edge of middle third jagged 4

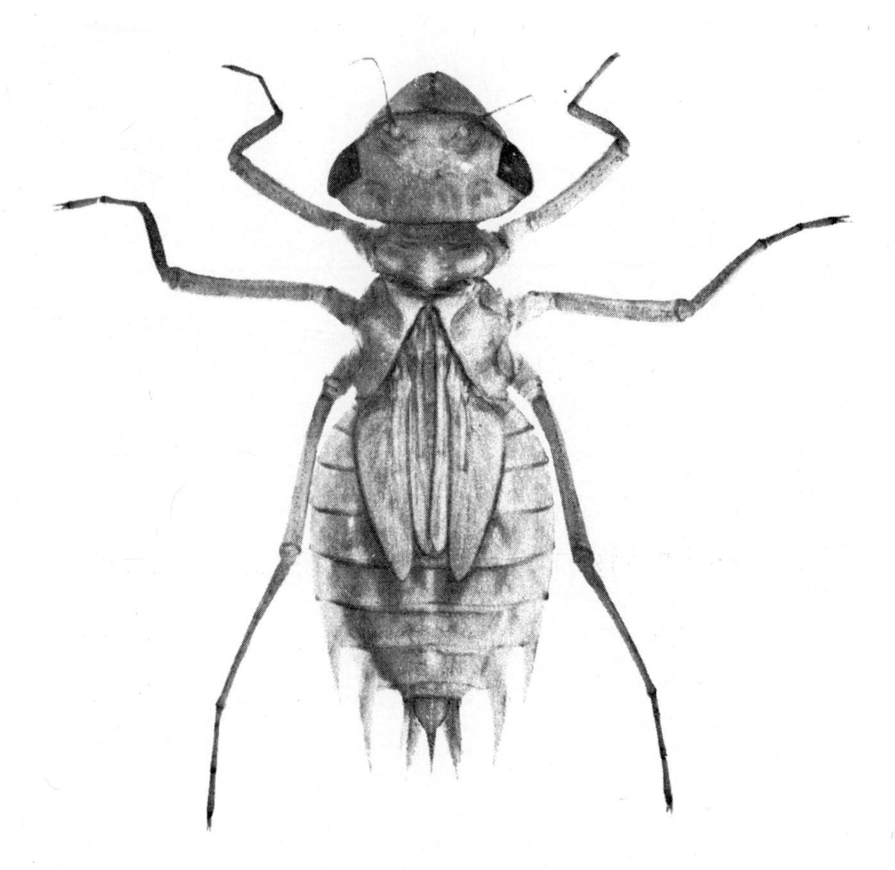

Fig. 339. *Tramea carolina.*

2—Sides of synthorax with two wide yellow stripes on dark background...**cophysa**

—Sides brownish, without stripes of yellow.................................. 3

3—Top of frons metallic violet.....................................**binotata**

—Top of frons red...**abdominalis**

4—Fore-wing triangle points straight to rearward; vein A2 in hind wing is zigzagged, but traceable to toe in anal loop; anal loop beyond midrib with single row of cells (sometimes single extras).......................... 5

—Fore-wing triangle inclines inward; vein A2 in hind wing not traceable to toe of anal loop, being lost midway among cells of irregular arrangement; anal loop with short extra row of cells on each side of midrib at level of heel; two rows of cells between veins Cu1 and Cu2 for a distance along deep bend in gaff . . . . **carolina**

5—Basal wing band reddish; top of head red; hamule of male projects well beyond level of tip of genital lobe . . . . **onusta**

—Basal wing band blackish; top of head black; hamule of male does not project beyond level of tip of genital lobe . . . . **lacerata**

NYMPHS

1—Lateral spines of segment 8 directed straight to rearward; two rows of spinules on upper surface of superior caudal appendage . . . . **carolina**

—Lateral spines of 8 incurved; a few scattered spinules on upper surface of superior caudal appendage 2

2—Lateral caudal appendages as long as superior . . . . **abdominalis**

—Laterals shorter than superior 3

3—Laterals about nine-tenths as long as superior . . . . **onusta**

—Laterals about four-fifths as long as superior . . . . **lacerata**

Nymphs unknown: **binotata** and **cophysa.**

Tramea abdominalis Rambur

Syn.: basalis Burmeister

1842. Rbr., Ins. Neur., p. 37 (in *Libellula*).
1890. Cabot, Mem. M. C. Z., 3:45 (figs., nymph).
1913. Ris, Coll. Selys Libell., p. 994.
1927. Garm., Odon. of Conn., p. 290.
1929. N. & H., Handb., p. 256 (fig.).
1932. Klots, Odon. of P. R., p. 69.
1938. Garcia, J. Agr. Univ. of Puerto Rico, 22:65.

Length 44–50 mm.; abdomen 28–33; hind wing 38–42.

A big reddish species that ranges our southern borders. Top of frons red in male, with narrow black brow band at its base. Broad vertex blackish.

Front of thorax olivaceous under a heavy coat of pallid hair. Sides of thorax paler, less hairy; marked with black on subalar carina, on spiracle, on pit in humeral suture, and above leg bases.

Legs blackish beyond their pale bases. Wings subhyaline, with reddish veins and stigma, and with a blackish band across base of hind wings. This nearly continuous band extends outward about as far as anal crossing, thence directly backward to hind angle of wing, with only a narrowly crescentic clear strip along margin and below membranule.

Abdomen red—pale red in female—with much black on dorsum of segments 8 and 9 and with a little basal ring of black on 10. Caudal appendages nearly or quite as long as 9 plus 10.

Distribution and dates.—UNITED STATES: Fla., Mass. (?); ANTILLES: Cuba, Dom. Rep., Haiti, Jamaica, P. R.; also from Bermuda and south to Brazil.

Year-round.

Tramea binotata Rambur

Syn.: brasiliana Brauer, insularis Hagen, longicauda Brauer, paulina Förster, subbinotata Brauer

1842. Rbr., Ins. Neur., p. 36 (in *Libellula*).
1913. Ris, Coll. Selys Libell., p. 991.
1930. Byers, Odon. of Fla., p. 146.
1932. Klots, Odon. of P. R., p. 70.

Length 45–51 mm.; abdomen 31–34; hind wing 38–43.

A handsome species, with face pale at first, becoming metallic violet on top of frons in mature males. On face is a narrow yellow line across frons near its top.

Thorax brown in front and moderately hairy; crest above it black. Sides pale brown and less hairy; indistinctly blackish in depths of first and third lateral sutures and above leg bases.

Abdomen blackish in male, with a tinge of red on middle segments and with 8, 9, and 10 wholly black. Caudal appendages black, with reddish bases. Female with pale colors generally more extended, wing membrane smoky, and dorsum of middle abdominal segments largely tawny. Caudal appendages of male (4–5 mm. long) almost as long as 8 plus 9 plus 10.

Distribution and dates.—UNITED STATES: Fla.; ANTILLES: Cuba, Dom. Rep., Haiti, Jamaica, P. R.; also south to Argentina.

Year-round.

Tramea carolina Linnaeus

1763. Linn., Amoenit. Acad., 6:411 (in *Libellula*).
1890. Cabot, Mem. M. C. Z., 3:46 (nymph).
1913. Ris, Coll. Selys Libell., p. 997.
1920. Dozier, Ann. Ent. Soc. Amer., 13:354.
1927. Garm., Odon. of Conn., p. 290.
1929. N. & H., Handb., p. 255 (fig.).
1930. Byers, Odon. of Fla., p. 147.

Length 48–53 mm.; abdomen 33–35; hind wing 44–45.

A fine big red species. Top of frons in male metallic violet; in female dark blue at maturity. General color of body reddish brown. Front of thorax reddish brown, without stripes, and thickly clothed with tawny hair; crest above, also, reddish brown. Sides, including subalar carina, a little paler reddish brown; and there is brown in pits of first and third lateral sutures. Legs brown to middle of femora and black beyond.

Wings red. In addition to venational characters stated above for genus: between veins Cu1 and Cu2 are two rows of cells for a distance along a notable concavity in gaff; also, anal loop has more extra cells than other species, variously disposed between its two regular marginal rows.

Abdomen reddish brown dorsally, olivaceous on sides of enlarged basal segments, black on lateral margins and beneath, and deeper black across dorsum of segments 8 and 9; 10 and caudal appendages paler, with black tips.

Female differs in having front of frons and tip of vertex orange, with a broad greenish-black brow band between the two. Rear of vertex and occiput olivaceous.

Distribution and dates.—CANADA: Ont.; UNITED STATES: Ala., Conn., Fla., Ga., Ind., Iowa, Kans., Ky., La., Mass., Mich., Miss., Mo., N. J., N. Y., N. C., Ohio, Okla.(?), Pa., R. I., S. C., Tenn., Tex., Va., Wis.

Year-round in Florida.

Tramea cophysa Hagen

Syn.: calverti Muttkowski, darwini Kirby

1867. Hagen, Stettin. Ent. Ztg., 28:226.
1906. Calv., B. C. A., p. 300.
1913. Ris, Coll. Selys Libell., p. 988 (in *Sympetrum*).
1943. Whts., Bull. Inst. Jamaica, Sci. Ser., 3:43.

Length 49 mm.; abdomen 30; hind wing 41.

Another Neotropical red-veined species. Face pale, with a black-edged labrum. Top of frons in mature male metallic violet; in female red. Occiput brown, beset with bristling black hair.

Thorax reddish brown in front, including crest, with black carinae. Sides yellow, each crossed with two oblique yellow bands that distinguish this species from others herein described.

Legs blackish. Wings pale, with reddish veins; hind wings have a basal band of brown like that of *abdominalis* in form and place, but slightly smaller and more convex on its outer margin.

Abdomen reddish or yellowish above, with black carinae, and black along lateral carina spreading upward to rearward and covering entire segments 8 and 9. Caudal appendages reddish or yellowish.

Whitehouse (1943) says that this species is distinguishable from *abdominalis* in flight by its red venation and black subterminal abdominal segments.

Distribution and dates.—UNITED STATES: Tenn. (?), Tex.; MEXICO: Baja Calif.; ANTILLES: Cuba, Haiti, Jamaica; also south to Brazil.

Apparently year-round; dates include January 7 (Jamaica), May 17 (Tex.), and September 27 (Tenn.).

Tramea lacerata Hagen

1861. Hagen, Syn. Neur. N. Amer., p. 145.
1890. Cabot, Mem. M. C. Z., 3:46 (nymph).
1913. Ris, Coll. Selys Libell., p. 998.
1923. Ndm., J. Ent. Zool. Claremont, Calif., 16:130.
1927. Garm., Odon. of Conn., p. 291.
1927. Seemann, J. Ent. Zool. Claremont, Calif., 19:31.
1929. N. & H., Handb., p. 254 (fig.).
1930. Byers, Odon. of Fla., p. 148.

Length 51–55 mm.; abdomen 35–36; hind wing 45–47.

A fine blackish species of wide distribution. Face yellowish, with labrum black. Top of frons very dark metallic violet; vertex and occiput blackish, latter beset with thin brown hair.

Front of thorax dull black under a thin coat of pale hair. Crest also dull black. Sides paler, with less hair, and with diffuse spots of deeper black overspreading pits of first and third lateral sutures. Black on spiracle and above leg bases.

Legs black. Wings hyaline, with broad basal bands of black. On fore wing, color very scanty—a wash of brown in subcostal and cubital interspaces. On hind wings, band broad and black, covering entire breadth of wing out at least as far as outer angle of triangle, thence to ankle of anal loop, thence rounded to hind angle of wing. A large roundish clear spot occupies inner margin of wing next to membranule.

Abdomen black at maturity, with broad dorsal yellowish spots, somewhat obscure on middle segments, becoming broad crossbands of brighter yellow on segments 7 and 8; 9 black; 10 paler. Caudal appendages blackish in both sexes; superiors of male at least 5 mm. long.

Distribution and dates—CANADA: Ont.; UNITED STATES: Ala., Ariz., Calif., Conn., Fla., Ga., Ill., Ind., Iowa, Kans., Ky., La., Md., Mass., Mich., Miss., Mo., Nebr., Nev., N. J., N. Y., N. C., Ohio, Okla., Pa., S. C., Tenn., Tex., Utah, Va., Wis.; also from Mexico and Hawaiian Islands.

March 1 (Miss.) to January 1 (Fla.).

Tramea onusta Hagen

1861. Hagen, Syn. Neur. N. Amer., p. 144.
1913. Ris, Coll. Selys Libell., p. 996.
1927. Byers, J. N. Y. Ent. Soc., 35:72 (nymph).

1927. Seemann, J. Ent. Zool. Claremont, Calif., 19:31.
1929. N. & H., Handb., p. 254 (fig.).
1930. Byers, Odon. of Fla., p. 149.

Length 41–49 mm.; abdomen 29–34; hind wing 38–43.

Another red species, of a lighter tint of red than in *carolina*. Face red, including labrum, and clad with short bristling black hairs. Frons and occiput greenish, rear margin of latter with a fringe of white hairs.

Front of thorax reddish or tawny, with crest above it of same color; rather thinly hairy. Sides of thorax olivaceous, without black markings.

Legs pale brown almost to knees, and blackish thereafter. Wings reddish-veined; stigma tawny. Colored basal crossband on hind wing divided by a clear strip through midbasal space and beyond. Thereafter brown, with red veins. It extends outward about as far as apex of triangle, thence rearward along midrib in toe of anal loop, and then rounded, just short of hind margin, to hind angle of wing. A large clear spot on inner expanded margin next to white membranule. A few extra cells within anal loop at level of heel; much fewer than in *carolina*.

Abdomen olivaceous on enlarged basal segments, with encircling red carinae; reddish on dorsum of middle segments, black on dorsal half of 8 and 9, and narrowly black on 10. Caudal appendages yellowish red.

Distribution and dates.—CANADA: Ont.; UNITED STATES: Ala., Calif., Fla., Ga., Ill., Ind., Kans., Ky., La., Miss., Mo., Nebr., Nev., N. Mex., N. C., Ohio, Okla., S. C., Tenn., Tex., Utah; MEXICO: Baja Calf., Tamaulipas; ANTILLES: Cuba, P. R.; also south to Panama.

March 13 (P. R.) to November 25 (Tex.).

Genus PANTALA Hagen 1861

These are large, strong-flying brownish dragonflies, with an overcast of red on the faces of the males at full maturity. The face is pale. The eye-seam is long; the vertex very broad.

The thorax is strongly braced internally, and bulging in front and at the sides. The hind legs are long and strong. The wings are very broad at base and pointed toward the tip. The stigma is rather small, its outer end more oblique than the inner. Vein M2 is strongly undulating. Veins M3 and M4 are elbowed at the outer end of the median planate. The anal loop is large and strongly sigmoid in shape, with a low heel and a long toe. Vein A2 is angulated at the knee beside a very long patella. A second cubito-anal crossvein is aslant toward the triangle, and generally joins the inner side of the triangle below its upper end. The fore wing has an extra row of cells between the paranal and marginal rows.

The abdomen is stout, with segments 1 to 3 short-spindle-shaped, and 4 to 10 regularly tapering to rearward. There are encircling carinae on three segments: two each on 3 and 4, and one on 5. The caudal appendages are longer than 9 and 10 taken together, in both sexes.

The nymphs are clean, smooth, depressed in form, and greenish, varied and beautifully patterned with brown. There are apical crossrows of from

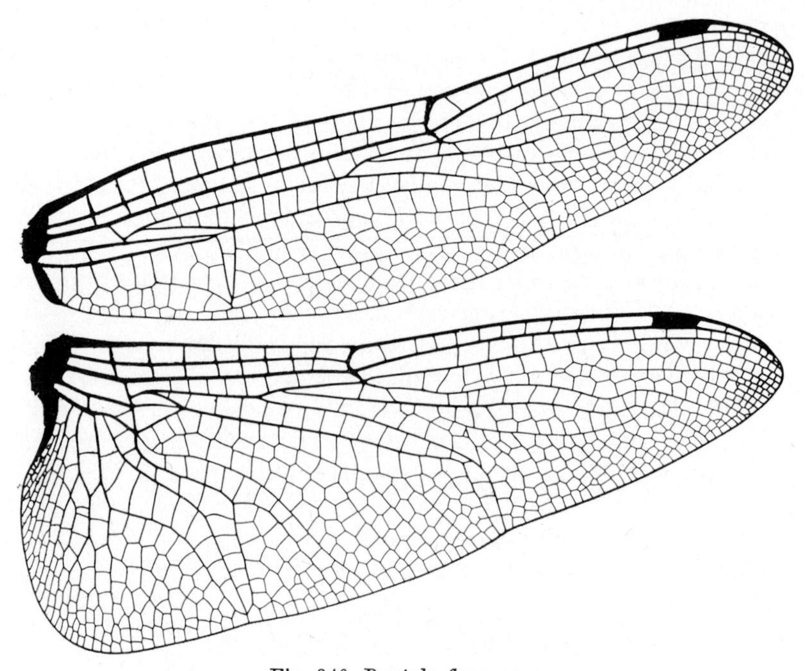

Fig. 340. *Pantala flavescens.*

four to six black dots on segments 5 to 8, and paired blotches at the middle of the sides of 7, at the lateral margins of 8, and near the middle line of 9; 10 is entirely suffused with black. The raptorial setae of the labium are numerous and crowded in line: twelve to sixteen lateral setae, fifteen to seventeen mental setae. There are about ten teeth on the opposed edges of each of the lateral lobes. These teeth are obliquely oval, aslant inward, deeply cleft apart, the clefts diminishing in depth proximally.

There are no dorsal hooks. The lateral spines are very long and incurvate, those of segment 9 reaching the level of the tips of the appendages.

These nymphs are very similar superficially to those of *Tramea.* They may be distinguished by the great length of the superior caudal appendage; by the greater depth of the incisions between the teeth on the lateral lobes; and by the extreme shortness of the movable hook, which in *Tramea* is nearly as long as the setae.

There are but two species in this genus. Both occur in North America,

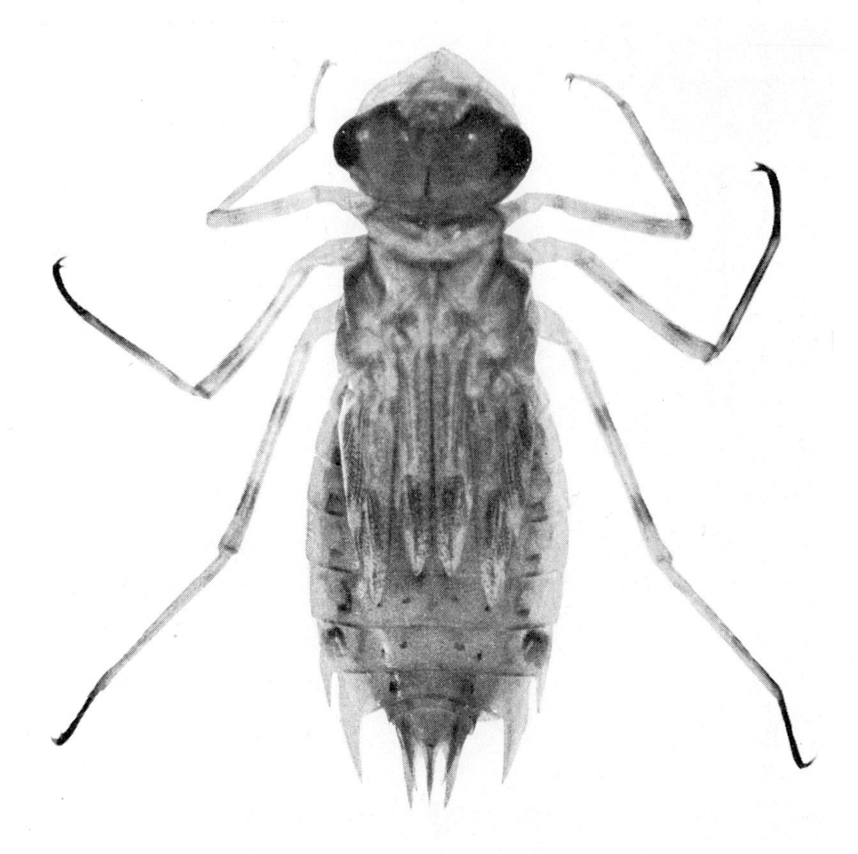

Fig. 341. *Pantala flavescens.*

one of them cosmopolitan and well known the world around for its sudden appearance in vast numbers in flight. Because of its superb powers of flight and wide-ranging habits, this is one of the first species to take possession of newly exposed waters in temporary pools, in tanks and water troughs where there is rich growth of green algae. It gets in before the larger aquatic plants can gain a foothold. In such places the growth of this nymph is very rapid.

KEY TO THE SPECIES

ADULTS

1—Hind wings with large brown spot near broadly rounded anal angle....**hymenea**
—Hind wings with no spot near anal angle............................**flavescens**

NYMPHS

1—Body pattern of brown conspicuous; superior caudal appendage as long as inferiors ....**hymenea**
—Body pattern pale; superior caudal appendage longer than inferiors...**flavescens**

Pantala flavescens Fabricius

Syn.: analis Burmeister, sparshallii Curtis, terminalis Burmeister, viridula Beauvais

Globe Trotter

1798. Fabr., Ent. Syst., Suppl., p. 285 (in *Libellula*).
1890. Cabot, Mem. M. C. Z., 17:44 (fig., nymph).
1913. Ris, Coll. Selys Libell., p. 917.
1920. Howe, Odon. of N. Eng., p. 89.
1927. Garm., Odon. of Conn., p. 293.
1929. N. & H., Handb., p. 252.
1940. Byers, Proc. Fla. Acad. Sci., 5:14–25.

Length 47–50 mm.; abdomen 27–34; hind wing 36–42.

The one cosmopolitan dragonfly. It is found in Temperate and Tropical regions, and is often taken on ocean-going vessels far out at sea. Face yellow, becoming red with age in old males; so also, top of frons and vertex. Occiput olivaceous, and bare of hair.

Stout thorax tawny in front, with black carinae, and ample growth of hair. Sides paler, with black streaks in pits of both first and third lateral sutures and a black ring around spiracle; also, three black crescents above leg bases.

Legs blackish beyond basal segments. Wings hyaline, sometimes with faintly brown tips. Stigma tawny.

Abdomen rather short and stout. Basal segments streaked with black below bulge. Some diffuse black middorsal markings that grow darker on 8, 9, and 10. Caudal appendages yellowish, becoming blackish beyond mid-length in male.

Distribution and dates.—CANADA: Man., Ont., Que.; UNITED STATES: Ala., Ariz., Ark., Calif., Fla., Ga., Ill., Ind., Iowa, Kans., Ky., La., Maine, Md., Mass., Miss., Mo., Nebr., Nev., N. H., N. J., N. Y., N. C., Ohio, Okla., Pa., R. I., S. C., Tenn., Tex., Va., Wis.; MEXICO: Baja Calif., Chihuahua; ANTILLES: Cuba, Dom. Rep., Haiti, Jamaica, P. R.; found on all continents except Europe (Mtk.)

Found every month southward.

Pantala hymenea Say

1839. Say, J. Acad. Phila., p. 19 (in *Libellula*).
1912. Wlsn., Proc. U. S. Nat. Mus., 43:194.
1913. Ris, Coll. Selys Libell., p. 921.
1923. Kndy., Can. Ent., 54:36 (figs., nymph).
1927. Garm., Odon. of Conn., p. 294.
1929. N. & H., Handb., p. 252.

Length 45–50 mm.; abdomen 29–34; hind wing 40–45.

Much more limited in distribution than *flavescens,* which it much resembles, but from which it is at once distinguishable by spot on hind wings; also somewhat lighter in general coloration.

Little pattern on thorax, but sometimes on sides there are diffuse broad bands of brown on first and third lateral sutures, and these may be connected at their lower ends.

Distribution and dates.—CANADA: Man., N. B., Ont.; UNITED STATES: Ala., Ariz., Calif., Fla., Ga., Ill., Ind., Kans., Ky., La., Maine, Md., Miss., Mo., Nebr., Nev., N. J., N. Mex., N. C., Ohio, Okla., Pa., S. C., S. Dak., Tenn., Tex., Wis.; MEXICO: Baja Calif., Coahuila; ANTILLES: Cuba; also south to Chile.

April 3 (Ohio) to October 15 (Ariz.); Whitehouse collected it on St. John, Virgin Island, January 5.

GLOSSARY

The more or less technical terms currently used to describe dragonflies are assembled here for ready reference.

acuminate ending in a long, tapering point.
acute acute-angled (less than 30°) and sharp-pointed.
annular ringlike.
anterior front.
arcuate curved, as a bow.
bidentate two-toothed.
bifid two-cleft, forked.
bilineate two-lined.
carinate keeled or sharp-ridged.
caudal pertaining to the tail, to the rear end.
ciliated fringed with hairs.
coalesce fuse.
compressed flattened laterally (on the side).
confluent flowing together.
conjoined connected, joined together.
constricted compressed or drawn together, narrowed.
contiguous touching or in contact.
converge to approach; opposed to *diverge.*
cordate heart-shaped.
crenate scalloped or toothed by even, rounded notches.
crenulate minutely crenate.
cultriform sickle-shaped, or like a parrot's beak.
dentate toothed.
denticulate minutely dentate.
depressed flattened dorsally (on the back.)
differentiate grow different.
dilated widened, expanded.
distal away from the base, outward; opposed to *proximal.*
divaricate widely diverging, branching off at a large angle and spreading apart.
diverge to spread apart; opposed to *converge.*
dorsal pertaining to the back or upper side; opposed to *ventral.*
elliptical oblong with rounded ends.
emarginate notched.
excised cut out.
falcate, falciform curved like a hawk's beak.
ferruginous rust-colored.
filiform threadlike.
flavescent somewhat yellow.
forcipate forceps-like.
fulvous tawny, color of common deer.
furcate forked.
fuscous dark brown, approaching black.
geminate occurring in pairs, two side by side, twinned.
glaucous of a sea-green color.
hirsute covered with short soft hairs.
humeral pertaining to the humerus or shoulder.
hyaline transparent, glasslike.
incised cut in a slit.
interalar between the wings.
lamellate, lamelliform shaped like a plate.
lateral at the side, pertaining to the sides.
lenitic living in still water; opposed to *lotic.*
lobed having rounded divisions that extend not more than halfway to the center.
lotic living in moving water, streams, and wave-washed shores.
luteous egg yellow.
median in the middle.
Neotropical, Tropical, Subtropical areas of the Americas.
obovate ovate but widened distally; opposed to *ovate,* widened proximally.
olivaceous color of a green olive.
oval, ovate egg-shaped, widened proximally; opposed to *obovate.*
patellar pertaining to the patella, a paranal cell of hind wing.
petiolate narrowed into a handle-like base.

phalerate harness-striped.
pilose covered with fine soft hair.
plantar pertaining to the planta or sole of the anal loop.
plicated folded.
process an upstanding prominence.
produced drawn out or prolonged.
proximal toward the base, inward; opposed to *distal.*
pruinose covered with a bluish-white bloom.
pubescent clothed with short soft fine hair or down.
quadrate squarish.
raptorial adapted for seizing prey.
recurved curved backward.
reniform kidney-shaped.
rufescent somewhat reddish.
rufous reddish.
rugose wrinkled.
sanguineous blood red.
serrate toothed or notched along the edge, like a saw.
serrulate minutely serrate.
setaceous, setiform bristle-like.
sinuate, sinuous wavy, sigmoid.
spinulose having small spines.
spinulose-serrate short spinules in serrate order on margin.
striate grooved, having fine linear markings.
style, stylet a stiff process.
submarginal just under the margin.
subtend to stretch underneath.
subulate shaped like an awl.
sulcate grooved, having long narrow channels or furrows.
surpassing extending beyond.
teneral newly emerged and not fully colored.
tomentose covered with tomentum, or matted woolly hairs.
transverse lying or being across.
trifid three-cleft.
trigonal pertaining to the triangle of the wing.
triquetral three-ridged and triangular in cross section.
triradiate having three rays or radiate branches.
truncate cut off squarely.
tumid swollen.
unilateral one-sided.
ventral pertaining to the belly or underside; opposed to *dorsal.*
vermiculate wormlike in shape.
villous clothed with long soft hair.
violaceous violet-colored.

SYNONYMS

INDEX